AF598483

Later Proterozoic Stratigraphy of the Northern Atlantic Regions

LATER PROTEROZOIC STRATIGRAPHY OF THE NORTHERN ATLANTIC REGIONS

edited by
J. A. WINCHESTER
Lecturer in Geology
University of Keele

Blackie
Glasgow and London

Published in the USA by
Chapman and Hall
New York

Blackie and Son Ltd
Bishopbriggs, Glasgow G64 2NZ
7 Leicester Place, London WC2H 7BP

Published in the USA by
Chapman and Hall
in association with Methuen, Inc.
29 West 35th Street, New York, NY 10001–2291

First published 1988

British Library Cataloguing in Publication Data

Later Proterozoic stratigraphy of the Northern
Atlantic regions.
1. Geology, Stratigraphic—Pre-Cambrian
2. North Atlantic Region
I. Winchester, J. A.
551.7′15′091821 QE653

ISBN 0–216–92263–1

Library of Congress Cataloging-in-Publication Data

Later Proterozoic stratigraphy of the northern Atlantic
regions.

Bibliography: p.
Includes index.
1. Geology, Stratigraphic—Pre-Cambrian.
2. Geology—Great Britain. 3. Geology—North Atlantic
Ocean Region. I. Winchester, J. A.
QE653.L275 1987 551.7′12′091821 87–6333
ISBN 0–412–01591–9 (Chapman & Hall)

Photosetting by Thomson Press (India) Limited, New Delhi
Printed in Great Britain by Bell & Bain (Glasgow) Ltd

Preface

Later Proterozoic Stratigraphy of the Northern Atlantic Regions aims to produce a concise and up-to-date synthesis of the later Proterozoic geology of those lands bordering the North Atlantic that were once situated north of the Iapetus Suture and the Tornquist Line. Proterozoic rocks deposited between 1150 and 650 Ma (the latter date marked by the Varanger glaciation) are the main subject of the book, although reference is also made to deposits laid down at the end of the Proterozoic in Scandinavia, Newfoundland and Greenland. The need for such a comprehensive review has become increasingly apparent in recent years, because the introduction of many new methods of resolving problems in complex metamorphic terrains has unlocked a vast store of new information.

This book is not the result of a specially-convened conference, but is a collection of specially commissioned articles drawing upon the expertise of 23 scholars who are all actively investigating the rocks described. The co-operation of these contributors has been immensely stimulating and their prompt submission of text and illustrative material has enabled rapid production of the book. Funding for this research has come from many sources, including the Natural Environment Research Council, the British Geological Survey, the Geological Surveys of Greenland and Newfoundland and many universities.

Many of the chapters use differing and interesting methods of approach, including structural analysis, sedimentology, whole-rock trace element geochemistry, geophysics, and isotopic age dating. The scope of the original research was extended to include formerly adjacent areas and, as a result, a number of useful correlations between these regions can be made. The book concludes with a somewhat speculative model for the stratigraphic evolution of the entire region in the later Proterozoic, which identifies many outstanding problems. Consequently, it is hoped and expected that this book will provide a valuable synthesis of past knowledge and a basis for future research.

JAW

Contents

Contributors

D. Barr
Britoil PLC
150 St Vincent Street
Glasgow G2 5LJ
UK

M. Bentley
Shell Exploration and Production
Shell-Mex House
The Strand
London WC2R 0DX
UK

J. Bertrand-Sarfati
Centre National de la Recherche Scientifique
Centre Géologique et Géophysique
Place Eugene Bataillon
34060 Montpellier
France

R. Caby
Centre National de la Recherche Scientifique
Centre Géologique et Géophysique
Place Eugene Bataillon
34060 Montpellier
France

D. Flinn
Department of Geology
University of Liverpool
Brownlow Street
PO Box 147
Liverpool L69 3BX
UK

N. R. W. Glendinning
Department of Geology
Royal Holloway and Bedford New College
(University of London)
Egham, Surrey TW20 OEX
UK

B. W. Glover
Department of Geology
University of Keele
Keele, Staffordshire ST5 5BG
UK

C. F. Gower
Newfoundland Department of Mines and Energy
PO Box 4750
St John's
Newfoundland, Canada A1C 5T7

J. Hibbard
Department of Geological Sciences
Snee Hall
Cornell University
Ithaca, NY 14853
USA

A. K. Higgins
Grønlands Geologiske Undersøgelse
Øster Voldgrade 10
DK-1350 Copenhagen K
Denmark

C. B. Long
Geological Survey of Ireland
Beggars Bush
Haddington Road
Ballsbridge
Dublin 4
Eire

M. D. Max
Acoustics Division
Naval Research Laboratory
Code 5110
Washington DC 20375–5000
USA

F. May
British Geological Survey
Murchison House
West Mains Road
Edinburgh EH9 3LA
UK

S. J. Moorhouse
Geological Sciences
School of Natural Sciences
The Hatfield Polytechnic
Hatfield, Hertfordshire AL10 9AB
UK

Valerie E. Moorhouse
Geological Sciences
School of Natural Sciences
The Hatfield Polytechnic
Hatfield, Hertfordshire AL10 9AB
UK

J. P. Nystuen
Saga Petroleum a.s.
Maries vei 20
Postboks 9
N-1322 Høvik
Norway

M. A. J. Piasecki
Department of Geology
University of Hull
Hull HU6 7RX
UK

Anna Siedlecka
Saga Petroleum a.s.
Maries vei 20
Postboks 9
N-1322 Høvik
Norway

A. D. Stewart
Department of Geology
University of Reading
Whiteknights
Reading RG6 2AB
UK

R. A. Strachan
Department of Geology and Physical Sciences
Oxford Polytechnic
Oxford OX3 0BP
UK

S. Temperley
Department of Geology
University of Hull
Hull HU6 7RX
UK

A. E. Wright
Department of Geological Sciences
University of Birmingham
Birmingham B15 2TT
UK

1
Introduction

J. A. WINCHESTER

1.1 Aim of this volume

The contributions to this book are designed to bring together current knowledge about the later Proterozoic stratigraphy in the northern Atlantic borderlands. The intention is to facilitate correlation and thus increase understanding of the geology of this now sundered area.

Many detailed geological studies of the later Proterozoic rocks cropping out in this region have been confined to relatively small areas. This has been true particularly in the British Isles, where numerous geologists have been intensively studying a small portion of the northern Atlantic region for almost 200 years. In Phanerozoic sequences, correlation between scattered areas is greatly facilitated by the presence of fossils: the scarcity or absence of fossils in Precambrian rocks has conversely retarded correlation of older sequences. Hence the aim of this volume is to clarify the regional stratigraphy of the lands bordering the northern Atlantic Ocean, by bringing together numerous detailed descriptions of rock series of middle and late Proterozoic age. In the British Isles this considerable span of geological time, ranging from approximately 1200 to 570 Ma, encompasses a stratigraphy which has remained poorly understood because most of the rocks formed during this interval have been intensely deformed and metamorphosed during the Lower Palaeozoic Caledonian orogeny, and laterally displaced by major Caledonide wrench faulting, and which is only now being interpreted. In eastern Canada relatively few rocks were formed during the same period and a very incomplete picture of the later Proterozoic geological history emerges, and to a lesser extent this is true of Scandinavia. Between these two areas, representing stable cratonic blocks, already established in their present form before the late Proterozoic, the record of the later Proterozoic stratigraphy is much more complete, and hence a large proportion of the work described in this book relates to the NW portion of the British Isles where most fragments of this record are preserved. For this reason an understanding of the later Proterozoic geology of the NW British Isles is vital, not only to British geologists, but also to students of the later Proterozoic geology of eastern Canada, Greenland and Scandinavia.

As our comprehension of the effects of plate-tectonic movements in the Phanerozoic has grown, it has become increasingly clear that the later Proterozoic stratigraphy of all the northern Atlantic borderlands is understood better by studying them in a reconstruction of their former relative positions, prior to both the opening of the Atlantic Ocean and the movements associated with the Caledonides.

1.2 Geographical limits

The Atlantic Ocean is geologically young. A product of the present cycle of sea-floor spreading, it continues to widen and separate portions of continental crust which were formerly contiguous. Many lands now scattered around the northern Atlantic Ocean contain exposures of rocks formed during the later Proterozoic (Fig. 1.1). These exposures have generally been studied in isolation because correlation with other Proterozoic rocks across considerable distances has usually been highly speculative. Lithological similarity has rarely proved a reliable means of correlation where attempts to link rock groups have been attempted for these ancient rocks. Consequently, although many of these later Proterozoic rocks may have been formed originally on the same continental mass at approximately the same time, they have not been classified as related formations and thus the later Proterozoic history of such areas remains poorly understood. Further complications have resulted from the opening and subsequent closure during the early Palaeozoic of an earlier ocean (now usually termed the Iapetus Ocean, after Harland and Gayer, 1972), which brought together continental masses that were on opposite sides of this ocean, in a different configuration to any which existed before the ocean was formed. Following closure, major wrench faulting within the Caledonide belt further altered the relative positions of areas where later Proterozoic rocks are now exposed. Consequently, a reconstruction of the relative positions of the continental masses prior to the opening of the Atlantic Ocean (Fig. 1.2) can only give a very approximate idea of the relative depositional positions of later Proterozoic rock groups. The sutures marking the former location of the Iapetus Ocean tend to occur within the central portion of the Caledonide mobile belt, but within the northern Atlantic area their positions are obscured, either because the sea has drowned large tracts of the continental margins (for example the North Sea, the Norwegian Sea, the Grand Banks and the Gulf of St Lawrence), or because of the existence of numerous basins filled with thick sequences of younger rock, or as a result of major wrench faulting, separating 'exotic terranes' of variable sizes. Because the original relative positions of areas of later Proterozoic rocks around the northern Atlantic are uncertain, this volume is restricted to a description of the later Proterozoic rocks in those areas which are believed to have been formerly situated on the 'American' side of the Iapetus Ocean, or to the north of 'Tornquist's Sea'. However, since the Atlantic Ocean has not opened along the same line as the Iapetus suture (Fig. 1.1), many of the areas described in this volume are now

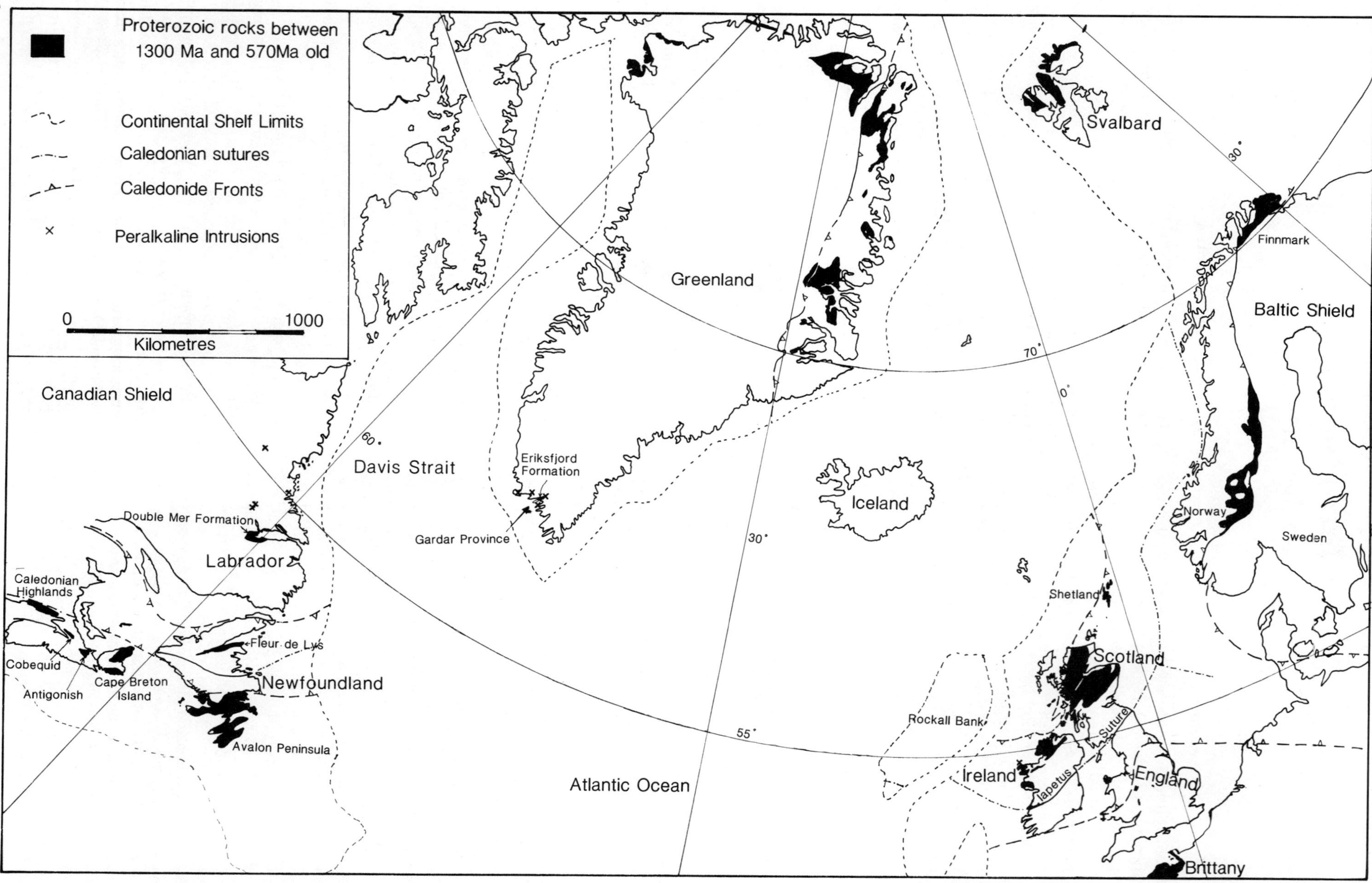

Figure 1.1 Present-day distribution of exposed later Proterozoic rocks around the northern Atlantic Ocean.

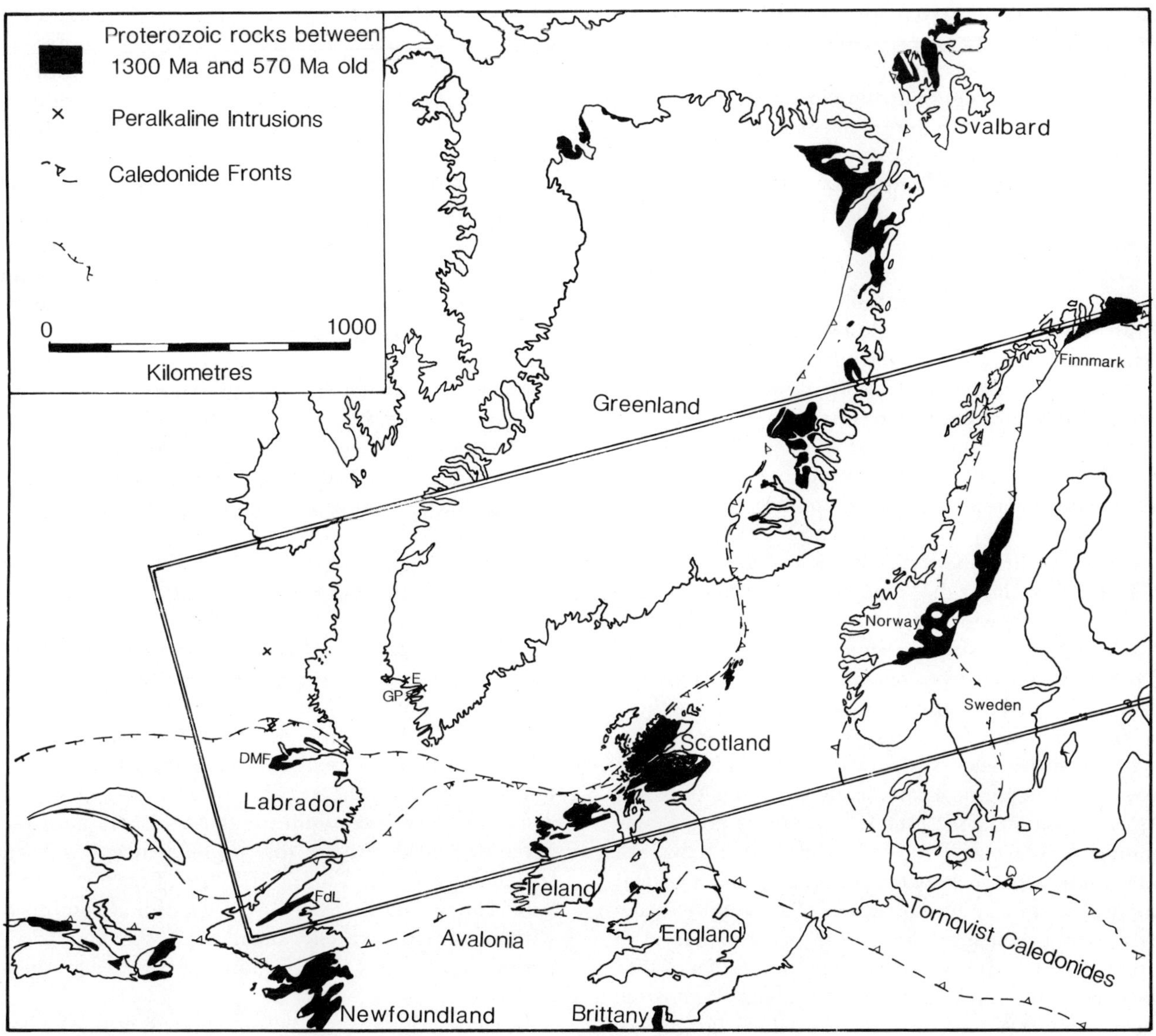

Figure 1.2 A reconstruction of the configuration of the northern Atlantic lands before the opening of the Atlantic Ocean. Crosses indicate the locations of mid-Proterozoic peralkaline granite complexes. *DMF*—Double Mer Formation, *FdL*—Fleur de Lys Supergroup, *GP*—Gardar Province. The area outlined is described in this book.

situated on the European side of the Atlantic Ocean. Similarly, some of the later Proterozoic rocks cropping out along the Atlantic seaboard of North America, notably those forming part of the Avalon block, are not described in this volume because they were apparently formed on the southeastern (non-American) side of the Iapetus Ocean and hence form part of basement massifs with an entirely different geological history.

The later Proterozoic rock groupings described in this volume are in a sequence of generally diminishing age, but because of the uncertainties in dating those groups affected by later metamorphic overprinting, it is recognized that some of the groups may have been described out of order. Many of the groups described occupy very small areas, but are included because they add information vital to the construction of any model of the palaeogeography of the North Atlantic region during the later Proterozoic.

The following areas are described and compared. In easternmost North America, later Proterozoic rocks occur as a sedimentary sequence preserved in outliers and grabens in Labrador (Gower *et al.*, 1986), and as highly deformed metamorphic rocks in NW Newfoundland, where they are collectively described as the Fleur de Lys Supergroup (Kennedy, 1971, 1975; Williams, 1977; de Wit, 1980), and occur NW of the Baie Verte–Brompton lineament. The later Proterozoic rocks in the Avalonian terrane of SE Newfoundland (Rast *et al.*, 1976; Williams, 1976, 1978; O'Brien *et al.*, 1983; Keppie, 1985), the Cape Breton Highlands, the Antigonish and Cobequid terranes of Nova Scotia (Keppie, 1985) and the Caledonian Highlands of southern New Brunswick (Keppie, 1985) (Fig. 1.1) occur SE of the Baie Verte–Brompton lineament. Further SW, in the NE Appalachians of the USA, other areas of later Proterozoic rocks occur ESE of the main Appalachian fold belt, for example the Boston terrane of Massachusetts (O'Hara and Gromet, 1985; Skehan and Murray, 1980; Skehan and Rast, 1980) and these are also thought to relate to the Avalonian terrane. Because this terrane is situated SE of the Iapetus suture, which is probably continued in North America as the Baie Verte–Brompton lineament, it formed part of a separate continental block during the later Proterozoic, and hence cannot at present be spatially related to contemporaneous rocks deposited on a North American basement.

In Greenland, now separated from North America by the minor oceanic re-entrant formed by the Davis Strait, correlation of older Proterozoic and Archaean rocks with Labrador is well established (Collerson, 1982). However, most mid- to late Proterozoic rocks in Greenland occur on the eastern margin and appear to bear little relationship to rocks in either Labrador or NW Newfoundland, partly because, even with the continental masses reassembled in their pre-Atlantic configuration, they were formed over 1500 km apart.

However, the same reconstruction of the continents (Fig. 1.2) also shows that the later Proterozoic rocks in East Greenland, which crop out over an area exceeding 20 000 km^2, were situated much closer to the later Proterozoic rocks deposited in the NW part of the British Isles, and the latter rocks were also closer to those in Newfoundland. The reconstruction shows, therefore, that the British Isles occupy a position linking all the main areas along the former 'eastern' margin of North America, and straddling the Iapetus suture in a similar manner to Newfoundland. This position shows that the later Proterozoic rocks of the British Isles are of critical importance in any interpretation of later Proterozoic palaeogeographical evolution of the eastern margin of North America. It is therefore perhaps significant that the variety and thickness of preserved later Proterozoic rocks in the NW part of the British Isles is far greater than that exposed elsewhere in the northern Atlantic region. In this part of the British Isles, major Caledonian wrench faulting has juxtaposed Proterozoic rock groups which were originally deposited in more widely scattered localities, and consequently the importance of the British Isles in understanding the nature of later Proterozoic environmental evolution in the northern Atlantic region is far greater than their relatively small area might suggest. Hence, much of this book is devoted to detailed descriptions of and correlations between the varied later Proterozoic rocks of the NW British Isles.

In the British Isles, as in parts of East Greenland and NW Newfoundland, most late Proterozoic rocks have been affected by deformation and metamorphism associated with the closure of the Iapetus Ocean and the formation of the Caledonides. Within the British Isles there are at least eleven distinct groupings of later Proterozoic rocks which are isolated from each other, either geographically, tectonically or stratigraphically. On the northwestern seaboard of Scotland there occur the only later Proterozoic sediments which have not been significantly modified by subsequent metamorphism: the Stoer, Torridon and Colonsay Groups. To the SE, within the Caledonian orogenic belt, there exist major rock series which were also affected by the mid- to late Proterozoic metamorphism contemporary with the Grenville event in eastern North America, in addition to a Caledonian overprinting. The tectonically isolated divisions of the Moine Assemblage in Scotland, and the Erris Complex of NW Ireland, have all yielded isotopic ages interpreted as evidence that they were originally formed during mid-Proterozoic times. In apparent tectonic contact with these rocks is the Dalradian Supergroup, now considered by some to include both the Grampian Group of Scotland (Harris *et al.*, 1978) and the Erris Group of NW Ireland. The tectonic contact at the base of the Grampian Group, termed the Grampian Slide by Piasecki (1980), may mask an unconformity (Piasecki, 1980), and there is no evidence that any part of the Dalradian Supergroup was affected by the later Proterozoic metamorphic event which affected the Moine Assemblage. Comparisons with East Greenland suggest that the Dalradian Supergroup is broadly contemporaneous with the Eleonore Bay succession there.

Not all the Dalradian Supergroup is described in this volume. The uppermost group (Southern Highland Group) is Cambrian in age, while parts of the underlying Argyll Group may also be Cambrian. Instead, because of its widespread use as a marker horizon, the tillite at the base of the Argyll Group has been chosen as a suitable upper age limit for the British Proterozoic rocks described. Above this tillite, which has been correlated with the Varanger Tillite of northern Norway (Harris and Pitcher, 1975), deposition continued without a stratigraphic break into the Cambrian period, and many descriptions of this very thick succession exist.

Brief descriptions are also included of somewhat enigmatic rock groups in Ireland, which are currently interpreted as 'pre-Caledonian' or 'pre-Dalradian'. Varied rocks from the Slishwood Division of Co. Sligo and metasediments from the Inishkea Division of the Erris Complex appear, on the evidence of poorly constrained isotopic ages, to be intermediate in age between the Moine and Dalradian rocks.

By contrast, descriptions of the late Proterozoic rocks exposed in small and isolated inliers in parts of Wales and the English Midlands are omitted, as these rocks lie SE of the Iapetus suture and are believed to have affinities with the Avalon composite terrane of SE Newfoundland, and thus record an entirely different geological history from the Proterozoic rocks in the NW British Isles. For the same reason, descriptions of the Proterozoic rocks of Brittany, the Channel Isles and the Rosslare Complex of SE Ireland are also omitted.

Recent work has suggested that the Caledonide orogenic belt bifurcates east of Scotland, with branches passing both to the NW of the Baltic Shield and to the SE, the latter belt passing through Poland and Romania (Ziegler, 1984, 1986). Palaeomagnetic evidence suggests that in the late Proterozoic the Baltic Shield formed part of the same supercontinent as North

America, albeit in a somewhat different relative position to the one established after the formation of the Caledonides (Patchett and Bylund, 1977), and for this reason a discussion of later Proterozoic rocks from Scandinavia is included. Further correlation with areas further to the north and east may be possible, and, in particular, rocks present in NE Greenland and Svalbard may be correlated with those described in this volume. However, because inclusion of these rocks would greatly increase the area covered in the book, and because no clear late Proterozoic palaeogeographical limits occur in that direction, a limit must be drawn and these distant rocks are not described here.

1.3 Proterozoic time-span covered

1.3.1 *Terminology*

In order to understand the interval of time over which the rocks described in this volume were formed, the chronological relationship between the various terms used in the different areas must be clarified. As a result of independent work on Precambrian complexes in different areas, a confusing plethora of names has emerged. In the North Atlantic area different names have been used in Scandinavia and North America for the same time intervals, but in the British Isles, reliable dating of most Proterozoic rocks has proved so difficult, on account of extensive metamorphic overprinting during the Caledonide event, that, in general, these rocks have not been firmly assigned to specific time intervals within the Proterozoic.

The Proterozoic aeon covers the immense time-span between the end of the Archaean at 2500 Ma to the beginning of the Cambrian period, dated at approximately 570 Ma (Harland *et al.*, 1982). As the oldest rocks described in this volume were probably formed about 1200 Ma ago, all were products of the later half of the Proterozoic aeon. However, on the recommendation of the Subcommission on Precambrian Stratigraphy of the IUGS, the Proterozoic aeon is now divided into three eras, and as a result, while most of the rocks described here were formed during the late Proterozoic era, ranging from 900 Ma to 570 Ma, some were formed during the mid-Proterozoic era, which spans the time interval from 1600 Ma to 900 Ma. For this reason, all the rocks described here are collectively described as 'later' proterozoic, rather than being split and assigned more specifically to either the mid- or late Proterozoic eras.

In Canada, the late Proterozoic era broadly corresponds with the Hadrynian tectonic division, with its older time limit set variously at 880 Ma (Stockwell, 1964) and 1000 Ma (Douglas, 1980). Likewise, the latter part of the mid-Proterozoic era corresponds in time with the formation of the Neo-Helikian tectonic division of the Canadian Shield. Clearly, if these division are used, rocks from the Stoer Group and Moine Assemblage of Scotland, the Erris Complex of NW Ireland, and perhaps the Krummedal supracrustal sequence of East Greenland were formed during the mid-Proterozoic era. Wherever exposure reveals the contact, in all these areas these rocks are seen to have been deposited upon much older basement, and hence it is logical to include them within this study, rather than excluding them arbitrarily as 'basement' to late Proterozoic rocks. Indeed in Scotland, because the Moine Assemblage and the Dalradian Supergroup have so many shared tectonic and metamorphic characteristics, that they were for many decades described together and were only distinguished on a lithological basis.

The rock groups described in this volume fall into three broad categories, based upon shared tectonic or metamorphic histories:

(i) Rocks affected by metamorphism and deformation at the end of the mid-Proterozoic era (Grenville or Sveco-Norwegian event)
(ii) Rocks deposited upon an Archaean or early Proterozoic metamorphic basement, which have not since been metamorphosed
(iii) Rocks deposited during the late Proterozoic era which have only been affected by very late Precambrian or Palaeozoic metamorphism.

Accordingly, the order of description of each rock group is based on shared post-depositional history, rather than present geographical distribution, and the book is divided into three sections corresponding with the categories defined above.

Within the first section are descriptions of the thickest mid-Proterozoic sedimentary accumulations in the area—now preserved in northern Scotland and perhaps East Greenland—where a thickness of over 10 km of clastic sediment accumulated. Each tectonic division is described separately, because correlation between them still remains somewhat unclear, although likely relationships are discussed.

Within the second section are descriptions of sedimentary rock groups of very disparate ages, deposited directly on a much older metamorphic basement. The Stoer Group is discussed separately, as it is very much older than the Torridon, Sleat and Iona Groups of NW Scotland. Correlation between the latter and the Colonsay Group is obstructed by the large displacement on the Great Glen Fault, and hence the latter is described separately. The Double Mer Formation of Labrador is also included; this shows some lithological resemblance to the Torridon Group, although it may be substantially younger (Gower *et al.*, 1986).

The final section is devoted to rocks affected by the Caledonide event, which generally were deposited after the Grenville tectonic event. The older basement of NW Ireland is discussed in this section, although older rocks occur within it. Reference is made to the pre-Grenville Annagh Division, and its similarities to some aspects of the mid-Proterozoic geology of Labrador, allowing for modifications to the Irish rocks stemming from the effects of both the Grenville and Caledonian events. However, the sections on the Grampian Group, the Erris Group and the 'Moine' of Shetland all describe dominantly psammitic sequences—which are probably lateral equivalents forming a distinct lithological grouping at the base of the Dalradian Supergroup. Whether or not they should be included as part of the Dalradian Supergroup is still under discussion. The lower (Appin) group within the Dalradian is also described, together with accounts of the Fleur de Lys Group of Newfoundland and the Eleonore Bay Group of East Greenland, which may be interpreted as equivalents of parts of the Dalradian Supergroup. The final descriptive section on the later Proterozoic rocks of Scandinavia is included as a study of rocks thought to be broadly contemporaneous with the Dalradian Supergroup, formed in a somewhat different environment—either in a different basin, or occurring within the same basin as the Dalradian—but consisting of material derived from the Baltic Shield.

1.4 Tectonometamorphic events

Correlation within the later Proterozoic rock groups in the northern Atlantic region has not only been rendered difficult by their present geographical dispersion, but also by widespread metamorphism and deformation, which has made comparisons of original sedimentary facies almost impossible. In addition, the recent application of the terrane concept to the Caledonides has revealed that considerable strike-slip movement along major tectonic lineaments almost certainly accompanied closure of the Iapetus Ocean (Bluck, 1984; Curry *et al.*, 1984). Equally, the concept of transtensional and transpressional movements that accompany ocean creation and destruction has revealed the difficulty of reconstructing geometrical fits between continental masses with margins which suffered major distortion during the ocean closure.

It is important, therefore, to assess the effect which each tectonometamorphic event may have had upon each of these rock groups and their relative geographical positions. Both the Grenville Front and the Caledonide Fronts traverse the area covered by these rocks, so that most groups were affected by the Caledonide metamorphism, and some were also metamorphosed during the Grenville event. In order to clarify the likely effects of these events on the later Proterozoic in the northern Atlantic area, a brief summary of these events and their effects is needed. Two principal far-reaching and multiple events are well documented: the Grenville or Sveco-Norwegian event and the Caledonide (*sensu lato*) event. In addition, sporadic hints have emerged from isotopic data, that parts of Scotland, East Greenland, and the Precambrian rocks beneath the Mesozoic cover of Denmark may also have been affected by a later Proterozoic event, although the nature and even the very existence of this event are still disputed.

1.4.1 *The Grenville–Sveco-Norwegian event*

Within the Grenville Province of Canada, two metamorphic events have been recognized: an earlier, more intense event, dated at approximately 1100–1150 Ma, which was of limited extent, and a more widespread but less intense later event dated at about 950 Ma (Baer 1981*a*, 1981*b*). The Grenville event, in the broad sense, was therefore probably a multiple sequence of events ranging over a 200 Ma interval of time.

Within the Caledonide belt of East Greenland, metamorphic complexes have yielded some pre-Caledonian isotopic ages. Within the Krummedal supracrustal sequence of the Scoresby Sund area (Higgins, 1974; Escher and Watt, 1976; Higgins *et al.*, 1978, 1981) a major metamorphic event was dated at 1162 ± 85 Ma (Hansen *et al.*, 1978), while Higgins (1974) recorded dates within the 950–1150 Ma range. Hence the range of Proterozoic dates obtained from East Greenland is broadly comparable with those from the Canadian Grenville Province, although the existence of clearly distinguishable metamorphic events between 1162 and 950 Ma is less clearly defined.

In the British Isles, a scatter of comparable dates has emerged in recent years, but all are from areas strongly affected by Caledonide metamorphism. The Archaean and early Proterozoic rocks of the Lewisian 'foreland' NW of the Caledonian Front were unaffected by 'Grenville' metamorphism, and all the evidence that parts of the British Isles were affected by 'Grenville' metamorphism, has been obtained from rocks subsequently subjected to Caledonide metamorphism during the early Palaeozoic era. Consequently, the early dates obtained from both Scotland (Brook *et al.*, 1976, 1977; Brewer *et al.*, 1979; Piasecki and van Breemen, 1979) and Ireland (van Breemen *et al.*, 1978; Max and Sonet, 1979; Aftalion and Max, 1987) are at present too few and scattered to identify clearly any discrete metamorphic peaks during the 'Grenville' event in the British Isles.

In Scandinavia, the Sveco-Norwegian Province of SW Sweden and southern Norway has long been considered a geochronological equivalent of the Canadian Grenville Province (Kratz *et al.*, 1968). As in Canada, two main tectonic and geochronological events are indicated: an earlier event, dated around 1100 Ma in Rogaland, and a more widespread event dated at 900–1000 Ma (Pedersen *et al.*, 1978; Baer, 1981*b*).

With the exception of the Double Mer Formation and the various rock groups from the NW foreland of Scotland, all the rocks described in this volume were obtained from areas apparently affected by the Grenville–Sveco-Norwegian event. However, details of any major tectonic dislocations active during the Grenville event, and indeed the very nature of the 'event' itself, are obscure. Hence any calculation of the relative depositional positions of pre-Grenville metasedimentary rocks is bound to be somewhat speculative. Indeed, differences between Phanerozoic and Proterozoic plate-tectonic models (Baer, 1977, 1981*a*) raise the possibility that application of the terrane concept to Proterozoic mobile belts may prove invalid. Equally, however, if Proterozoic mobile belts are the product of localized intracratonic deformation, reconstruction of pre-deformational geometry may prove simpler than in models involving large-scale subduction. Hence, reconstruction of, for example, the original basin in which rocks of the Moine Assemblage were deposited, may not prove impossibly complex.

1.4.2 *Late Precambrian events*

The occurrence of pegmatites within the Morar Division of the Moine Assemblage of Scotland, yielding ages in the range 780–700 Ma, led to the conclusion that a metamorphic event had affected the NW Highlands of Scotland at this time. This event, originally named the 'Knoydartian orogeny' (Bowes, 1968), is more commonly referred to as the 'Morarian orogeny' (Lambert, 1969; van Breemen *et al.*, 1974; Johnstone, 1975; van Breemen *et al.*, 1979). Further pegmatites have yielded similar ages in other parts of the Scottish Highlands, notably from within the Glenfinnan Division and from a single deformed pegmatite at the base of the Grampian Group (Piasecki and van Breemen, 1979; Piasecki, 1980; Piasecki *et al.*, 1981), but interpretation of these ages as a record of a separate event has been questioned since the evidence of earlier metamorphism of the Moine Assemblage emerged. It is not currently agreed whether these late Proterozoic dates indicate a separate event, or merely record cooling of a system which had been at elevated temperatures for a considerable time. In recent years, similar dates have been obtained from basement rocks beneath southern Denmark (Ziegler, 1986), and from East Greenland (Rex and Gledhill, 1981), and in the latter area some of these dates seem to record an episode of igneous intrusion. This additional information shows that a late Proterozoic event may have occurred across the region, but any associated metamorphism must have been limited in extent.

Distinct from the Morarian event is the evidence for a very late Proterozoic deformational event, which affected the Colonsay Group (Bentley, this volume). At present it is only recorded in Colonsay, but it raises the possibility that some of the 'early' Caledonian deformation recorded in the Dalradian Supergroup may be earlier than currently supposed.

1.4.3 *Palaeozoic events*

Almost all the later Proterozoic rocks in the northern Atlantic region were affected by Caledonian (*sensu lato*) metamorphism. Extensive recent work has shown that the 'Caledonian' event was multiple, and is divisible into numerous constituent episodes, their ages varying from place to place within the orogen. Each may be related to different aspects of the closure of the Iapetus Ocean, which occurred at different times in different areas. Furthermore, while the existence of some major strike-slip faults, such as the Great Glen Fault (Kennedy, 1946), has been known as a major feature of the Caledonides for many years, it is only recently that the scale and significance of the movement on some of these fractures has been appreciated and used in the application of the terrane concept to the Caledonides (Dewey and Shackleton, 1984; Soper and Hutton, 1984). Mounting evidence suggests that other major fractures, such as the Highland Boundary Fault and Southern Uplands Fault in Scotland, also underwent major lateral movement (Bluck, 1983, 1984), which have juxtaposed regions with unrelated geological histories. These areas, each possessing a unique geological record, constitute the terranes within the Caledonides. Without precise knowledge of the displacements on the principal transcurrent fractures, it is difficult to match the Proterozoic history of each terrane, although recent work in the Caledonides has suggested that during the early Palaeozoic era the major transcurrent movements were sinistral (Kennedy, 1946; Winchester, 1973; Storetvedt, 1974; Dewey and Shackleton, 1984). Hence the rocks of the Midland Valley composite terrane and the Southern Uplands terrane in Scotland may overlie slivers of Proterozoic basement related to that of NW Newfoundland, which have been translated far to the NE of their original position during the Caledonian orogeny.

In the British Isles, terranes south of the Iapetus suture (Phillips *et al.*, 1976) cannot be related to rocks on the Baltic Shield, from which they are now separated by the Caledonian belt extending SE into Eastern Europe, referred to by Ziegler (1986) as the Mid-European Caledonides. These terranes are now generally interpreted as parts of separate microcontinents derived originally from the northern margin of Gondwana (Cocks and Fortey, 1982; Ziegler, 1986), and hence their Proterozoic record is totally unrelated to that of the remaining northern Atlantic lands. For this reason they are not included in this volume, but their present locations, close to other Proterozoic terranes, graphically illustrate the problems in correlation arising from the widespread effects of the Caledonide event.

1.5 Stratigraphic methods

Owing to the lack of widespread fossils, stratigraphic correlation in the Proterozoic is more difficult than in Phanerozoic rocks. Additional difficulties, arising from the loss of sedimentary structures and other original features of sedimentary rocks during metamorphism, compound the problem. Hence stratigraphic correlation of Proterozoic rocks is frequently less certain than in younger rocks, and a wider range of correlation methods must be employed.

1.5.1 *Lithological correlation*

Correlation on the basis of lithological similarity in sequences has been used as a first method. Where marked lithological changes occur regularly in the stratigraphic column and the lithological formations are laterally persistent, as in much of the Dalradian Supergroup, this method can be successfully applied. Elsewhere, however, where enormous thicknesses of rock showing little overt lithological variation occur, as in the Grampian Group or Moine Assemblage of Scotland, this method is notoriously unreliable.

1.5.2 *Chemostratigraphic correlation*

In recent years, many changes in the interpretation of the stratigraphic relationships in the Scottish Highlands have resulted from the use of other correlation methods. With the recent widespread availability of rapid chemical analysis techniques, sufficient numbers of rock analyses can now be made, which permit the use of chemistry in correlation. Studies undertaken in the Scottish Highlands have shown that the concentrations and proportions of various elements, particularly many trace elements, tend to be characteristic of individual formations or groups, and the resulting 'chemical signatures' may persist for considerable distances along strike within a basin (Lambert *et al.*, 1981, 1982; Winchester *et al.*, 1981, 1983; Hickman and Wright, 1983). It is these 'chemical signatures' that may be used in chemostratigraphic correlation.

The reliability of this method depends upon a combination of three factors. Where the formations to be compared have been subjected to later metamorphism, the elements used should have concentrations or proportions that have been relatively unaffected by diagenetic or metamorphic processes. It also must be assumed that the provenance of the sediment supplied to each formation is unique, as, with time and erosion in the source area, the rocks being exposed and eroded will change progressively. A final chemical influence upon the unique composition may also be a product of the sedimentary environment. While sedimentary environments may be repeated, the source of the clastic material supplied is unlikely to be precisely the same. However, this method can only be applied readily in areas where formations are laterally persistent. Where rapid lateral facies changes occur, chemostratigraphic correlation becomes very difficult.

1.5.3 *Isotopic age-dating methods*

The improvement in isotopic age-dating precision and methods in recent years has enabled further checks to be made. Indeed, the importance of isotopic dating in resolving the geological problems in the northern Atlantic region cannot be overestimated. In several of the later Proterozoic rock groups described in this volume, isotopic dating has been crucial in revealing the approximate age of formation. Where the rocks are unmetamorphosed, as in the Stoer and Torridon Groups of the NW Scottish foreland, isotopic dating of fine-grained clastic sediment revealed that the Stoer Group yielded an isochron age of 968 Ma, while the Torridon Group Applecross Formation gave an isochron age of 777 Ma. In this example, isotopic dating techniques could establish both the considerable difference in age between the two groups, and also the time interval represented by the unconformity below the Cambrian quartzites, which had hitherto been the only geological clue to the age of the Torridon Group (Moorbath *et al.*, 1967; Moorbath, 1969; Steiger and Jäger, 1979). Interpreting these dates may be less simple. Although they were originally interpreted as deposition dates, they may relate to mild events post-dating deposition, such as diagenesis or gentle heating, unaccompanied by metamorphism.

Interpreting isotopic dates in areas which have been metamorphosed is much more difficult, and yet, it is in these areas that isotopic dating has had the most impact recently. In the Scottish Highlands, concepts of geological relationships have been revolutionized by the isotopic dates obtained recently, and much of the nomenclature of the rocks has been consequently modified. In particular, isotopic dating first revealed the Precambrian age of Moine Assemblage rocks within the Caledonides of the Northern Highlands of Scotland (Long and Lambert, 1963). Since then, it has played a fundamental role in unravelling the complex sequence of metamorphic events in that area. It was only by isotopic dating that the presence of a metamorphic event contemporary with the Grenville event in Canada was found to have affected the rocks of the Moine Assemblage (Brook *et al.*, 1976, 1977; Brewer *et al.*, 1979) in the Northern Highlands. In addition, the discovery of rocks yielding similar ages from other divisions of the Moine Assemblage in the Northern Highlands (Brewer *et al.*, 1979; Piasecki and van Breemen, 1979) has since been used in conjunction with detailed structural and stratigraphic studies (Roberts and Harris, 1983; Barr *et al.*, 1985) to establish that all three divisions of the Moine Assemblage in the Northern Highlands were affected by metamorphism around 1000 Ma ago, and were thus deposited during the mid-Proterozoic era.

In the Central Highlands of Scotland, east of the Great Glen Fault, isotopic dating has emphasized the existence of an apparent structural discontinuity (Piasecki and van Breeman, 1979; Piasecki, 1980), and revealed that the Central Highland Division (Piasecki and van Breeman, 1979), which had previously been mapped as part of an undivided 'Central Highland Granulites' group correlated with the Moine Assemblage, yielded an isochron age in excess of 1000 Ma (Piasecki, 1980). By contrast, the bulk of the area formerly mapped as 'Central Highland Granulites' overlies this structural discontinuity, and has yielded no evidence of such antiquity. The earliest age obtained, at 720 Ma, provides no evidence that these rocks were subjected to Grenville metamorphism, and suggests that their former correlation with the Moine Assemblage is invalid. Ironically, chemostratigraphic work (Lambert *et al.*, 1982) emphasized their similarity to Moine rocks, and the resulting stratigraphic revision led to their description variously as 'Younger Moine', 'Grampian Division' (Piasecki, 1980; Piasecki *et al.*, 1981; Haselock *et al.*, 1982; Lambert *et al.*, 1982), or 'Grampian Group' (Harris *et al.*, 1978). The latter name was proposed by those who favoured the inclusion of the Grampian Group within the Dalradian Supergroup, with which it appears to be structurally continuous, despite its lithological and geochemical differences. However, the stratigraphic revision of this major section of the Scottish Proterozoic succession is a direct result of the impact of isotopic dating as an instrument of correlation.

In NW Ireland, where the extent of Caledonian overprinting and dislocation has been as marked as in Scotland, the impact of isotopic dating has been almost as dramatic. In NW Mayo the Erris Complex is now recognized as an inlier of pre-Caledonian basement (Max and Long, 1985; Sutton and Max, 1969), although it was long thought to consist of Caledonian migmatites (Kilroe, 1907; Trendall and Elwell, 1963; Phillips *et al.*, 1969); isotopic dating has provided the proof. Furthermore, while the Annagh Division of the Erris Complex (Winchester and Max, 1984) was originally thought to be related to the Lewisian rocks of NW Scotland (Crow *et al.*, 1971)—isotopic dating has revealed that the principal metamorphism was Grenville (Max and Sonet, 1979) and that the original intrusive rocks were probably formed during the mid-Proterozoic era. By contrast, the rocks constituting the islet of Inishtrahull, immediately N of Ireland, have yielded ages averaging 1710 Ma, with no trace of a later Grenville metamorphic overprint (Roddick and Max, 1983), suggesting that the trace of a 'Grenville Front' beneath the British Isles should pass between the Erris Complex and Inishtrahull.

In East Greenland, where the Proterozoic rocks have also been metamorphosed during major Caledonian diastrophism, isotopic work has also played a fundamental role in providing evidence for the extent of areas also affected by an earlier 'Grenville' metamorphism (Hansen *et al.*, 1978; Rex and Gledhill, 1981), although the precise affinites of the rocks directly affected is currently in dispute (Peucat *et al.*, 1985).

1.5.4 *Geophysics in correlation*

Geophysical techniques have played an important role in recent years, in resolving the relationships between the later Proterozoic rocks in the lands around the northern Atlantic Ocean. In particular, two applications of geophysics have been significant: the use of gravity and magnetic surveys to trace the courses of concealed major lineaments, and the use of palaeomagnetism in assessing the relative motion of the major terranes.

1.5.4.1 *Tracing basement features.* Geophysical techniques have achieved the most dramatic results when applied across the broad continental shelves around Newfoundland, the British Isles and the Rockall Microcontinent (Fig. 1.3). Using the geophysical information obtained, the likely positions of both the Grenville Front and the Iapetus suture between Canada and the British Isles have been plotted, and the scanty geological information so far obtained from basement out-

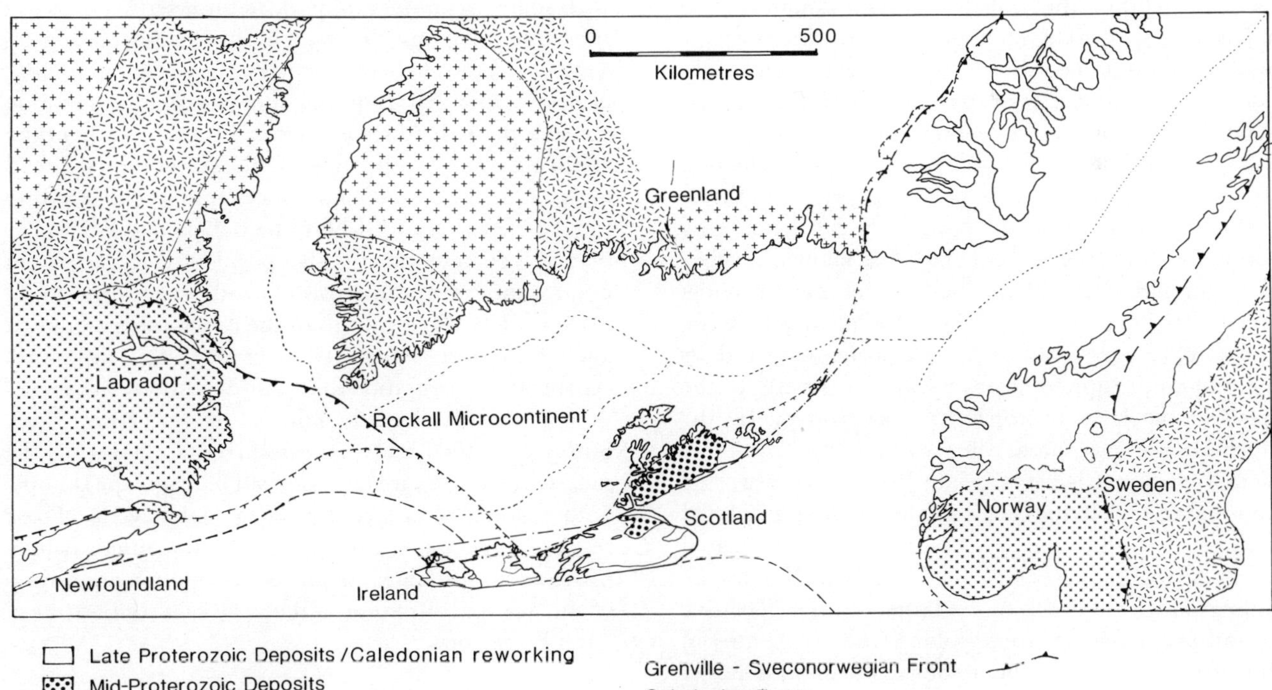

Figure 1.3 A reconstruction of the pre-Devonian distribution of the later Proterozoic deposits and their local basement provinces within the main area discussed in this volume. The map portrays the relative positions of major terranes after Caledonian nappe transport, but before the principal strike-slip movements on the Great Glen Fault and associated shears.

crops on these shelves seems to back up the geophysical results. Hence, the position of the Grenville Front on the Rockall Microcontinent (Fig. 1.3) is based on the position of an east–west gravity lineation (Roberts, 1970) which appears to separate gneisses yielding Lewisian and Grenville isotopic ages. However, tracing the Grenville Front on to the continental shelf north of Ireland has proved more difficult, as the presence of narrow Mesozoic basins—their limits apparently controlled by basement dislocations—complicates the pattern. The trace of the Iapetus suture north of Newfoundland is based on gravity and magnetic data (Haworth *et al.*, 1976; Haworth and Jacobi, 1983). However, proving the continuity of this trace with the postulated course of the Iapetus suture west of Ireland (Phillips *et al.*, 1976) is difficult. Lefort (1984) suggested that it may be marked by a magnetic feature on the SW end of the Rockall Microcontinent (Vogt and Avery, 1974), and, if so, a considerable bend in the trace of the Caledonide Belt between Newfoundland and western Ireland is indicated. Perhaps because it appears to be unaccompanied by major intrusions and is generally marked by low-angle thrusting, the position of the NW Caledonide Front is not clearly defined by geophysics in the same area.

1.5.4.2 *Palaeomagnetism.* Palaeomagnetic information has also contributed greatly to the understanding of the relationships in both space and time of the later Proterozoic rocks from the North Atlantic lands. Where later metamorphism has reset the magnetism in the rocks, as within the Caledonide belt, the value of palaeomagnetic data may be limited: certainly the information obtained should be interpreted with care, but enough evidence may survive to permit an assessment of the relative motion of terranes. However, there is a secondary role for palaeomagnetism, which is equally important.

Where the pattern of apparent polar wandering curves for the major cratonic blocks is known, a check on the apparent ages of some formations may be made. Sometimes, there appears to be a discrepancy between the isotopic age obtained from a formation, and its position on the apparent polar wandering path for the cratonic block with which it is associated. In the northern Atlantic area, the two principal cratonic blocks which existed in the late Proterozoic were the Laurentian and Baltic Shields. The small and isolated cratonic block forming NW Scotland has also yielded an apparent polar wandering path which may be compared with those from the larger cratons. Relative deviations of the apparent polar wandering paths in different cratons may prove to be significant evidence of the development of major intracratonic basins at different times within the area, as they may record periods of crustal tension in the weakened zones between the main rigid blocks. Piper (1982) has claimed that Proterozoic apparent polar wandering loops from several continental shields formed similar sequences of palaeopoles, and could be matched. From this he deduced that a single 'supercontinent' was in existence during almost the entire Proterozoic aeon, and that within this composite supercontinent the Baltic Shield was situated adjacent to the Laurentian Shield (Piper, 1982; Smith *et al.*, 1983). From a study of palaeomagnetic poles obtained from mid-Proterozoic dolerites collected in Canada, SW Greenland, Sweden and Finland, it was deduced that in the period 1290–1155 Ma, the Baltic Shield adjoined the Laurentian Shield but in a different orientation to that established for the late Proterozoic era. Comparison with palaeopoles from later Proterozoic rocks formed in the period 1000–850 Ma has argued for a rotation of the Baltic Shield through 90° in the intervening period (Patchett *et al.*, 1978). Subsequently, Piper (1980) revised this calculation and suggested that the optimum reconstruction allowed for an anticlockwise rotation of the Baltic Shield relative to the Laurentian Shield during this period. Whatever the precise amount of rotation, palaeomagnetism seems to be providing evidence of an early dislocation between the Baltic and Laurentian Shields, which might have formed the locus for Proterozoic intracratonic basin development.

Within the microcraton of NW Scotland, only the Stoer and Torridon Groups were formed in the mid- to later Proterozoic and subsequently escaped metamorphism. Their ages were thought to have been carefully calculated and therefore they were included on the apparent polar wandering path for the Laurentian Shield, with corrections applied for their present position. Usually, where the age of a rock group is known with some precision, and its structural setting is free of subsequent complications, a good correlation is achieved. However, the primary palaeomagnetic pole positions obtained from the Stoer Group were found to correspond with an age exceeding 1100 Ma, rather than confirming an isotopic age of 968 Ma obtained previously. A similar discrepancy was found to occur within the Torridon Group, where the palaeomagnetic polar positions corresponded with a position on the apparent polar wandering path consistent with an age of approximately 1040 Ma, rather than the previously calculated age of 777 Ma. Attempts to explain this difference centre on the suggestion that post-depositional diagenetic reactions between the mineral phases (Piper, 1982; Smith *et al.*, 1983) may have occurred, so that while the magnetization records the depositional age, the isotopic date records a much later diagenetic event. These results, therefore, suggest that palaeomagnetism may prove an additional valuable tool in resolving questions of the age and situation of these rocks in the northern Atlantic region.

1.6 External links

The geographical limits placed upon the rock groups described in this volume are clearly artificial. It is almost certainly possible to link some of the rocks with contemporary sequences outside the area covered. For instance, the later Proterozoic sequences of NE Greenland and Svalbard may be linked with those further south in East Greenland. Likewise, a plausible westward correlation with the Keeweenawan Group of North America is indicated by the palaeopole dates obtained from the Stoer Group of NW Scotland, hinting that both rock groups were formed under similar circumstances in response to a continent-wide stress field. At present, such correlations are speculative, but it is to be hoped that this volume, with its detailed accounts of several later Proterozoic rock groups in the northern Atlantic lands, written by contributors who are currently investigating them, will facilitate correlation and clarify the later Proterozoic history of this complex and fragmented area.

References

Aftalion, M. and Max, M. D. (1987) U–Pb zircon geochronology from the Precambrian Annagh Division gneisses and the Termon Granite, NW Co. Mayo, Ireland. *J. geol. Soc. London* **144**, 401–406.

Baer, A. J. (1977) The Grenvillian Province as a shear zone. *Nature* **290**, 129–131.

Baer, A. J. (1981*a*) A Grenvillean model of Proterozoic plate tectonics. In Kroner A. (ed.), *Precambrian Plate Tectonics*. Elsevier, Amsterdam.

Baer, A. J. (1981*b*) Two orogenies in the Grenville Belt. *Nature* **290**, 129–131.

Barr, D., Roberts, A. M., Higton, A. J., Parson, L. M. and Harris, A. L. (1985) Structural setting and geochronological significance of the West Highland Granitic Gneiss, a deformed early granite within Proterozoic Moine rocks of northwest Scotland. *J. geol. Soc. London* **142**, 663–675

Bluck, B. J. (1983) Role of the Midland Valley of Scotland in the Caledonian orogeny. *Trans. R. Soc. Edinburgh: Earth Sci.* **74**, 119–136.

Bluck, B. J. (1984) Pre-Carboniferous history of the Midland Valley of Scotland. *Trans. R. Soc. Edinburgh: Earth Sci.* **75**, 275–295.

Bowes, D. R. (1968) The absolute time-scale and the subdivision of Precambrian rocks in Scotland. *Geol. Fören. Stockholm Förh.* **90**, 175–188.

Brewer, M. S., Brook, M. and Powell, D. (1979) Dating of the tectonometamorphic history of the southwestern Moine, Scotland. In Harris, A. L., Holland, C. H. and Leake, B. E. (eds.), The Caledonides of the British Isles–Reviewed. *Spec. Publ. geol. Soc. London* **8**, 129–137.

Brook, M., Powell, D. and Brewer, M. S. (1976) Grenville age for rocks in the Moine of northwestern Scotland. *Nature* **260**, 515–517.

Brook, M., Powell, D. and Brewer, M. S. (1977) Grenville events in the Moine rocks of the Northern Highlands, Scotland. *J. geol. Soc. London* **133**, 489–496.

Cocks, L. R. M. and Fortey, R. A. (1982) Faunal evidence for oceanic separation in the Palaeozoic of Britain. *J. geol. Soc. London* **139**, 465–478.

Collerson, K. D. (1982) Geochemistry and Rb–Sr geochronology of associated Proterozoic peralkaline and subalkaline anorogenic granites from Labrador. *Contrib. Miner. Pet.* **81**, 126–147.

Crow, M. J., Max, M. D. and Sutton, J. S. (1971) Structure and stratigraphy of the metamorphic rocks in part of NW Co. Mayo, Ireland. *J. geol. Soc. London* **127**, 579–584.

Curry, G. B., Bluck, B. J., Burton, C. J., Ingham, J. K., Siveter, D. J. and Williams, A. (1984) Age, evolution and tectonic history of the Highland Border Complex, Scotland. *Trans. R. Soc. Edinburgh: Earth Sci.* **75**, 113–133.

Dewey, J. F. and Shackleton, R. M. (1984) A model for the evolution of the Grampian tract in the early Caledonides and Appalachians. *Nature* **312**, 115–121.

de Wit, M. J. (1980) Structural and metamorphic relationships of pre-Fleur de Lys and Fleur de Lys rocks of the Baie Verte Peninsula, Newfoundland, *Can. J. Earth Sci.* **17**, 1559–1575.

Douglas, R. J. W. (1980) Proposals for time classification and correlation of Precambrian rocks and events in Canada and adjacent areas of the Canadian Shield. *Geol. Surv. Can. Pap.* **80–24**, 19 pp.

Escher, A. and Watt, W. S. (1976) *Geology of Greenland*. Andelsbogtrykkeriet, Odense.

Gower, C. F., Erdmer, P. and Wardle, R. J. (1986) The Double Mer Formation and the Lake Melville rift system, eastern Labrador. *Can. J. Earth Sci.* **23**, 359–368.

Hansen, B. T., Higgins, A. K. and Bar, M.-T. (1978) Rb–Sr and U–Pb age patterns in polymetamorphic sediments from the southern part of the East Greenland Caledonides. *Bull. geol. Soc. Denmark* **27**, 55–62.

Harland, W. B. and Gayer, R. A. (1972) The Arctic Caledonides and earlier oceans. *Geol. Mag.* **109**, 289–314.

Harland, W. B., Cox, A. V., Llewellyn, P. G., Pickton, C. A. G., Smith, A. G. and Walters, R. (1982) *A Geologic Time Scale*. Cambridge University Press.

Harris, A. L. and Pitcher, W. S. (1975) The Dalradian Supergroup. In Harris, A. L., Shackleton, R. M., Watson, J., Downie, C., Harland, W. B. and Moorbath, S. (eds.), A Correlation of Precambrian Rocks in the British Isles. *Geol. Soc. Spec. Rep.* **6**, 52–75.

Harris, A. L., Baldwin, C. T., Bradbury, H. J., Johnson, H. D. and Smith, R. A. (1978) Ensialic basin sedimentation: the Dalradian Supergroup. In Bowes, D. R. and Leake, B. E. (eds.), Crustal Evolution in Northwestern Britain and Adjacent Regions. *Geol. J. Spec. Issue* **10**, 115–138.

Haselock, P. J., Winchester, J. A. and Whittles, K. H. (1982) The stratigraphy and structure of the southern Monadhliath Mountains between Loch Killin and upper Glen Roy. *Scott. J. Geol.* **18**, 275–290.

Haworth, R. T. and Jacobi, R. D. (1983) Geophysical correlations between the geological zonation of Newfoundland and the British Isles. In Hatcher, R. D., Zietz, I. and Williams, H. (eds.), *Tectonics and Geophysics of Mountain Chains*.

Haworth, R. T., Grant, A. C. and Folinsbee, R. A. (1976) Geology of the continental shelf off southeastern Labrador. *Pap. Geol. Surv. Can.* **76–1**, 61–70.

Hickman, A. H. and Wright, A. E. (1983) Geochemistry and chemostratigraphical correlation of slates, marbles and quartzites of the Appin Group, Argyll, Scotland. *Trans. R. Soc. Edinburgh: Earth Sci.* **73**, 251–278.

Higgins, A. K. (1974) The Krummedal supracrustal sequence around inner Nordvestfjord, Scoresby Sund, East Greenland. *Grønlands Geol. Unders. Rapport* **68**, 34 pp.

Higgins, A. K., Friderichsen, J. D., Rex, D. C. and Gledhill, D. R. (1978) Early Proterozoic isotopic ages in the East Greenland Caledonian fold belt. *Contrib. Miner. Pet.* **67**, 87–94.

Higgins, A. K., Friderichsen, J. D. and Thysted, T. (1981) Precambrian metamorphic complexes in the East Greenland Caledonides (72–74°N)—their relationships to the Eleanore Bay Groups, and Caledonian Orogenesis. *Grønlands Geol. Unders. Rapport* **104**, 4–46.

Johnstone, G. S. (1975) The Moine succession. In Harris, A. L., Shackleton, R. M., Watson, J., Downie, C., Harland, W. B. and Moorbath, S. (eds.), A Correlation of the Precambrian Rocks in the British Isles. *Geol. Soc. Lond. Spec. Rep.* **6**, 30–42.

Kennedy, M. J. (1971) Structure and stratigraphy of the Fleur de Lys Supergroup in the Fleur de Lys area, Burlington Peninsula, Newfoundland. *Proc. Geol. Ass. Can.* **24**, 59–71.

Kennedy, M. J. (1975) The Fleur de Lys Supergroup: stratigraphic comparison of Moine and Dalradian equivalents in Newfoundland with the British Caledonides. *J. geol. Soc. London* **131**, 305–310.

Kennedy, W. Q. (1946) The Great Glen Fault. *J. geol. Soc. London* **102**, 41–76.

Keppie, J. D. (1985) The Appalachian collage. In Gee, D. G. and Sturt, B. (eds.), *The Caledonide Orogen, Scandinavia and Related Areas*. Wiley, New York.

Kilroe, J. R. (1907) The Silurian and metamorphic rocks of Mayo and North Galway. *Proc. R. Ir. Acad.* **108**, 129–160.

Kratz, K. O., Gerling, E. K. and Lobach-Zhuchenko, S. B. (1968) The isotope geology of the Precambrian of the Baltic Shield. *Can. J. Earth. Sci.* **5**, 657–660.

Lambert, R. StJ. (1969) Isotope studies relating to the Precambrian history of the Moinian of Scotland. *Proc. geol. Soc. London* **1652**, 243–245.

Lambert, R. StJ., Winchester, J. A. and Holland, J. G. (1981) Comparative geochemistry of pelites from the Moinian and Appin Group (Dalradian) of Scotland. *Geol. Mag.* **118**, 477–490.

Lambert, R. StJ., Holland, J. G. and Winchester, J. A. (1982) A geochemical comparison of the Dalradian Leven Schists and the Grampian Division Monadhliath Schists of Scotland. *J. geol. Soc. London* **139**, 71–84.

Lefort, J.-P. (1984) The main basement features recognized in the northern part of the North Atlantic area. *Init. Rep. DSDP* **80**, 1103–1114.

Long, L. E. and Lambert, R. StJ. (1963) Rb–Sr isotopic ages from the Moine Series. In Johnson, M. R. W. and Stewart, F. H. (eds.), *The British Caledonides*. Oliver and Boyd, Edinburgh, 217–247.

Max, M. D. and Sonet, J. (1979) A Grenville age for pre-Caledonian rocks in NW Co. Mayo, Ireland. *J. geol. Soc. London* **136**, 379–382.

Moorbath, S. (1969 Evidence for the age of deposition of the Torridonian sediments of northwest Scotland. *Scott. J. Geol.* **5**, 154–170.

Moorbath, S., Stewart, A. D., Lawson, D. E. and Williams, G. E. (1967) Geochronological studies on the Torridonian sediments of northwest Scotland. *Scott. J. Geol.* **3**, 389–412.

O'Brien, S. J., Wardle, R. J. and King, A. F. (1983) The Avalon Zone: a Pan-African terrane in the Appalachian orogen of Canada. *Geol. J.* **18**, 195–222

O'Hara, K. and Gromet, L. P. (1985) Two distinct late Precambrian (Avalonian) terranes in southeastern New England and their late Palaeozoic juxtaposition. *Am. J. Sci.* **285**, 673–709.

Patchett, P. J., Bylund, G. and Upton, B. G. J. (1978) Palaeomagnetism and the Grenville Orogeny: new Rb–Sr ages from dolerites in Canada and Greenland. *Earth Planet. Sci. Letters* **40**, 349–364.

Pedersen, S., *et al.* (1978) Rb–Sr dating of the plutonic and tectonic evolution of the Sveconorwegian Province, southern Norway. In Zartman, R. E. (ed.) Short Papers of the Fourth International Conference on Geochronology, Cosmochronology and Isotope Geology. *US Geol. Surv. Open-File Rep.* 78–701.

Peucat, J. J., Tisserant, D., Caby, R. and Clauer, N. (1985) Resistance of zircons to U–Pb resetting in a prograde metamorphic sequence of Caledonian age in East Greenland. *Can. J. Earth Sci.* **22**, 330–338.

Phillips, W. E. A., Kennedy, M. J. and Dunlop, G. M. (1969) Geologic comparison of western Ireland and northeastern Newfoundland. In Kay, M. (ed.), North Atlantic—Geology and Continental drift, a Symposium. *Mem. Am. Ass. Pet. Geol.* **12**, 194–211.

Piasecki, M. A. J. (1980) New light on the Moine rocks of the central Highlands of Scotland. *J. geol. Soc. London* **137**, 41–59.

Piasecki, M. A. J. and van Breemen, O. (1979) The 'Central Highland Granulites': cover–basement tectonics in the Moine. In Harris, A. L., Holland, C. H. and Leake, B. E. (eds.), The Caledonides of the British Isles—Reviewed. *Spec. Publ. geol. Soc. London* **8**, 139–144.

Piasecki, M. A. J., van Breemen, O. and Wright, A. E. (1981) Late Precambrian geology of Scotland, England and Wales. In Kerr, J. W. and Fergusson, A. J. (eds.) Geology of the North Atlantic Borderlands, *Mem. Can. Soc. Pet. Geol.* **7**, 57–94.

Piper, J. D. A. (1980) Palaeomagnetic study of the Swedish Rapakivi Suite: Proterozoic tectonics of the Baltic Shield. *Earth Planet. Sci. Letters* **46**, 443–461.

Piper, J. D. A. (1982) The Precambrian palaeomagnetic record: the case for the Proterozoic Supercontinent. *Earth Planet. Sci. Letters* **59**, 61–89.

Plant, J. A., Watson, J. V. and Green, P. M. (1984) Moine–Dalradian relationships and their palaeotectonic significance. *Proc. R. Soc. London* **A395**, 185–202.

Rast, N., O'Brien, B. H. and Wardle, R. J. (1976) Relationships between Precambrian and Lower Palaeozoic rocks of the 'Avalon Platform' in New Brunswick, the northeast Appalachians and the British Isles. *Tectonophysics* **30**, 315–338.

Rex, D. C. and Gledhill, A. R. (1981) Isotopic studies in the East Greenland Caledonides (72–74°N)—Precambrian and Caledonian ages. *Grønlands Geol. Unders. Rapport* **104**, 47–72.

Roberts, D. G. (1970) Recent geophysical studies on the Rockall Plateau and adjacent areas. *Proc. geol. Soc. London* **1662**, 87–92.

Roberts, A. M. and Harris, A. L. (1983) The Loch Quioch Line—a limit of early Palaeozoic crustal reworking in the Moine of the Northern Highlands of Scotland. *J. geol. Soc. London* **140**, 883–892.

Roddick, C. and Max, M. D. (1983) A Laxfordian age from the Inishtrahull Platform, Co. Donegal, Ireland. *Scott. J. Geol.* **19**, 97–102.

Skehan, J. W. S. J. and Murray, D. P. (1980) Geologic profile across southeastern New England. *Tectonophysics* **69**, 285–319.

Skehan, J. W. S. J. and Rast, N. (1982) Relationship between Precambrian and Lower Palaeozoic rocks of southeastern New England and other North Atlantic Avalonian terrains. In Schenk, P. E. (ed.), *Regional Trends in the Geology of the Hercynide–Mauritanide Orogen*. D. Reidel, Dordrecht, 131–162.

Smith, R. L., Stearn, J. E. F. and Piper, J. D. A. (1983) Palaeomagnetic studies of the Torridonian sediments, NW Scotland. *Scott. J. Geol.* 29–45.

Soper, N. J. and Hutton, D. H. W. (1984) Late Caledonian sinistral displacements in Britain: implications for a 3-plate collision model. *Tectonics* **3**, 781–794.

Steiger, R. H. and Jäger, E. (1977) Subcommission on Geochronology: Convention on the use of decay constants in geo- and cosmochronology. *Earth Planet. Sci. Letters* **36**, 359–362.

Stockwell, C. H. (1964) Fourth report on structural provinces, orogenies and time classification of rocks of the Canadian Precambrian shield. *Geol. Surv. Can. Spec. Paper* 64–77.

Storetvedt, K. M. (1974) A possible large-scale sinistral displacement along the Great Glen Fault in Scotland. *Geol. Mag.* **111**, 23–30.

Trendall, A. F. and Elwell, R. W. D. (1963) The metamorphic rocks of northwest Mayo. *Proc. R. Ir. Acad.* **62B**, 217–247.

van Breemen, O., Pidgeon, R. T. and Johnson, M. R. W. (1974). Precambrian and Palaeozoic pegmatites in the Moines of northern Scotland. *J. geol. Soc. London* **130**, 493–507.

van Breemen, O., Halliday, A. N., Johnson, M. R. W. and

Bowes, D. R. (1978) Crustal additions in late Precambrian times. In Bowes, D. R. and Leake, B. E. (eds.), Crustal Evolution in Northwestern Britain and Adjacent Regions. *Geol. J. Spec. Issue* **10**, 81–106.

van Breemen, O., Aftalion, M., Pankhurst, R. J. and Richardson, S. W. (1979) Age of the Glen Dessary Syenite, Inverness-shire: diachronous Palaeozoic metamorphism across the Great Glen. *Scott. J. Geol.* **15**, 49–62.

Vogt, P. R. and Avery, O. E. (1974) Detailed magnetic surveys in the NE Atlantic and Labrador Sea. *J. Geophys. Res.* **79**, 363–389.

Williams, H. (1976) Tectonostratigraphic subdivision of the Appalachian orogen. *Geol. Soc. Am. Abstr.* **8**, 300.

Williams, H. (1977) Ophiolitic melange and its significance in the Fleur de Lys Supergroup, northern Appalachians. *Can. J. Earth Sci.* **14**, 987–1003.

Williams, H. (1978) Tectonic lithofacies map of the Appalachians. Memorial Univ. Map 1. Geol. Dept., St John's, Newfoundland.

Winchester, J. A. (1973) Pattern of regional metamorphism suggests a sinistral displacement of 160 km along the Great Glen Fault. *Nature* **246**, 81–84.

Winchester, J. A. and Max, M. D. (1984) Geochemistry and origins of the Annagh Division of the Precambrian Erris Complex, NW Co. Mayo, Ireland. *Precamb. Res.* **25**, 397–414.

Winchester, J. A., Lambert, R. StJ. and Holland, J. G. (1981) Geochemistry of the western part of the Moinian assemblage. *Scott. J. Geol.* **17**, 281–294.

Winchester, J. A., Lambert, R. StJ. and Holland, J. G. (1983) Chemical correlation of Moinian rocks in the northern Highlands of Scotland. *Geol. Mag.* **120**, 165–174.

Ziegler, P. A. (1984) Caledonian and Hercynian crustal consolidation of Western and Central Europe—a working hypothesis. *Geol. Mijnbouw* **63**, 93–108.

Ziegler, P. A. (1986) Geodynamic model for the Palaeozoic crustal consolidation of Western and Central Europe. *Tectonophysics* **126**, 303–328.

2

Sedimentary structures and sequences within a late Proterozoic tidal shelf deposit: the Upper Morar Psammite Formation of northwestern Scotland

N. R. W. GLENDINNING

2.1 Introduction

The interpretation of thick, sandy, dominantly cross-bedded sequences of Precambrian age is problematical, as they may represent either fluvial or shallow-marine deposits (Long, 1978). Prior to the appearance of stabilizing vegetation during the Palaeozoic, sandy fluvial systems could have adopted an unconfined sheet-braided style, depositing laterally extensive cross-bedded sheet sands (Schumm, 1968; Cotter, 1978). Geometrically similar sand beds could have formed in a shallow-marine setting under the influence of tide and/or storm generated currents. Furthermore, given a strongly asymmetrical tidal flow, these may lack the variable or reversing palaeocurrents associated with tidal deposits, and, like the fluvial braid-sheets, record only unidirectional flows.

This chapter is concerned mainly with such a sequence, the Upper Psammite Formation within the Morar Division of the late Proterozoic Moine Assemblage. A facies scheme for the formation is outlined and an attempt is made to justify a shallow-marine origin.

The Upper Morar Psammite Formation crops out along both limbs of a major N–S-trending Caledonian fold—the Morar Antiform. Along the western limb the metasediments are relatively undeformed and consist mainly of cross-bedded sheet sands with subordinate silty beds. Within the sand units, bimodal, bipolar palaeocurrents, complex sand waves, winnowed gravel lags and thin waning flow sequences (tempestites?) indicate a tide-dominated shallow-marine palaeoenvironment, possibly within, or linked to, a large gulf. Parts of the underlying Morar Striped Schist Formation may be intertidal or shallow subtidal. Within the Upper Psammite, transitions along the dominant S–N transport path (reduction in grain size of sands, increase in the proportion of silt, disappearance of large sand waves and a change in the dominant style of cross-bedding from strongly three-dimensional to more open forms) are consistent with changes down a tidal, net sand transport–deposition path. In addition, proximal facies progressively interfinger with—and overlie—distal facies, suggesting that the sequence is, overall, regressive. Comparisons with the more highly deformed structurally complex rocks along the eastern limb of the Morar Antiform are conjectural. However, it appears that the S to N facies and grain-size changes observed along the western limb also run from W to E and are accompanied by a decrease in thickness from 5 km to 1.5 km. This may reflect an original stratigraphic thinning. Therefore the sediment source and basin depocentre lay to the W. In addition, the concentration of soft-sediment deformation structures along the western limb can be interpreted as a response to earthquakes generated by movements along basin margin fault(s). Hence the Upper Psammite was probably deposited in a N–S-trending half-graben with a major growth fault along its western margin.

2.2 Geological setting

The Moine Assemblage occupies a broad belt running NNE to SSW across much of the Scottish Highlands. In the southern part of this belt, the metasediments have been subdivided into three tectonostratigraphic units: from W to E the Morar, Glenfinnan and Loch Eil Divisions (Johnstone *et al.*, 1969). This chapter deals with the Morar Division, which is separated from the Glenfinnan and Loch Eil Division (or Sgurr Beag Nappe of Roberts *et al.*, 1987) by a major Caledonian ductile shear zone, the Sgurr Beag Thrust (Tanner, 1971). It is the least deformed part of the Moine Assemblage, sedimentary structures are commonly preserved, and on their basis a fourfold stratigraphy has been long established (Richey and Kennedy, 1939; Ramsay and Spring, 1962; Lambert and Poole, 1964) (Fig. 2.1). For the sake of brevity, the four formations will be referred to as the Basal Pelite, Lower Psammite, Striped Schist and Upper Psammite.

Many detailed accounts of the structural geology of this area exist (Powell, 1974; O'Brien, 1985; Holdsworth *et al.*, 1987, in press); aspects which particularly constrain sedimentological studies are outlined here. The outcrop pattern is a product of Caledonian deformation; the above formations are repeated along the limbs of major N–S folds, the Morar Antiform in Morar and Ardnamurchan and the Assapol Synform on the Ross of Mull (Fig. 2.1). This structural pattern might appear to provide both a N–S (along limbs) and W–E (across limbs) stratigraphic control but correlations/transitions based on this premise must be treated with caution as other structural factors complicate this simple pattern. Principally these are:

(i) On the western limb of the Morar Antiform the contact between the Upper Psammite and Striped Schist is stratigraphic at Morar but tectonic at

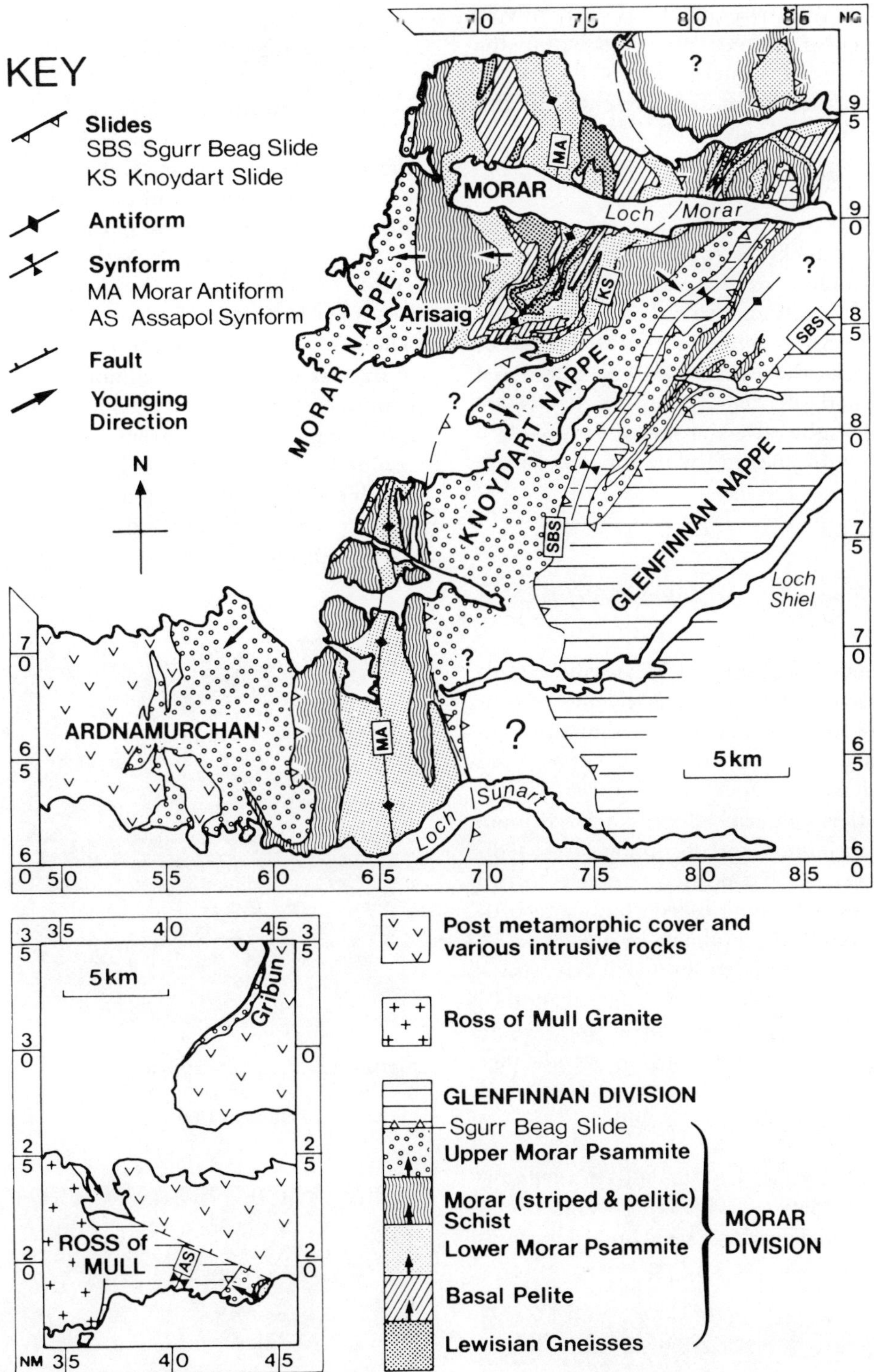

Figure 2.1 Geological map showing outcrop of Morar Division rocks and the major structural features from Morar southwards to the Ross of Mull.

Ardnamurchan (the 'Gortenfern Disturbance' of Bailey *et al.*, 1922) where the lower part of the Upper Psammite may be missing.

(ii) A Caledonian shear zone, the Knoydart Slide, can be traced down the eastern limb of the Morar Antiform (Fig. 2.1). Prior to shortening across this slide, the geographical separation between the Upper Psammite sediments along the two limbs may have been much greater. (As a comparison Kelley and Powell (1985) postulated a minimum displacement of 50 km across the Sgurr Beag Thrust in the Fannich–Ullapool area.) Alternatively, these formations possibly maintain their initial relationship, as the Knoydart Slide may have been refolded around the Morar Antiform to occupy a similar stratigraphic level on the western limb (Poole and Spring, 1974).

(iii) The degree of strain and resultant flattening of sedimentary structures is variable, showing a general decrease from E to W. Consequently, thick relatively undeformed sequences are only preserved within the Upper Psammite on the western limb of the Morar Antiform. Most of the information presented in this chapter is drawn from this area.

2.3 Petrography

The Morar division rocks consist of psammites, semipelites and pelites. These are the metamorphic derivatives of original arkosic and subarkosic sands, and finer-grained sediments ranging from silts to clays. For the purposes of this discussion they are considered simply as sands and silts. Heavy-mineral bands are common: the chief mineral being sphene (probably a breakdown product of ilmenite) with minor zircon.

2.4 Sedimentology

2.4.1 *The Upper Psammite on the western limb of the Morar Antiform*

The Upper Psammite consists of thick, laterally extensive sand sheets (see Fig. 2.2), separated by massive silts or finely interbedded sand and silt. Within the sand sheets, sedimentary structures are well preserved, and on their basis five facies (1 to 5) have been identified. The intervening finer-grained sediments proved more susceptible to deformation and metamorphism, sedimentary structures are only locally preserved, so two broad 'bucket facies' are distinguished, thinly interbedded sand and silt (facies 6), and massive silt (facies 7).

Certain facies are restricted to particular stratigraphic levels. Others recur throughout the sequence. In the following section they are described and, where appropriate, their palaeoenvironmental significance discussed. Brief descriptions of the Striped Schist and eastern limb Upper Psammite are included.

2.4.1.1 *Facies* 1. This facies has small- to medium-scale cross-bedding. Sets vary between 3 and 45 cm, averaging 10 cm in thickness, and are arranged in cosets generally about 1 m thick. Three subfacies, 1A, 1B and 1C, are distinguished, mainly on the criterion of cross-bedding geometry (i.e. the relative proportions of set height, length and breadth).

1A: tabular sets only form a significant proportion of the sequence at Gribun, Ross of Mull. The subfacies consists of laterally continuous sets which frequently climb upcurrent, thereby passing into subfacies 2B.

1B: trough-parallel sets (Michelson and Dott, 1973) (PR* 32–170, mean 70); this is the dominant cross-bedding type in the lower parts of the Upper Psammite. The subfacies generally occurs in fine- and medium-grained sands. Sets are of uniform thickness and appear continuous, but when traced laterally, trough margins and contacts are apparent. Foresets are low angle (around 12°) and strongly concave, with well-developed bottomsets. Sets do not climb markedly up or down current, but are subparallel to set boundaries. Fig. 2.3 provides a typical view of this subfacies.

Figure 2.2 General view of laterally extensive sand sheet facies (Gribun, Ross of Mull).

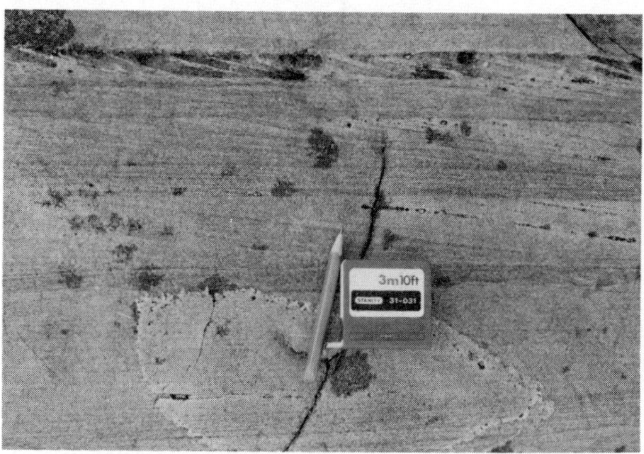

Figure 2.3 Sets of subfacies 1B, note well-developed bottomsets and unit of plane bedding (facies 4) at base of photograph. Current from left to right. (Fig. 2.22*B*, Loc. B.)

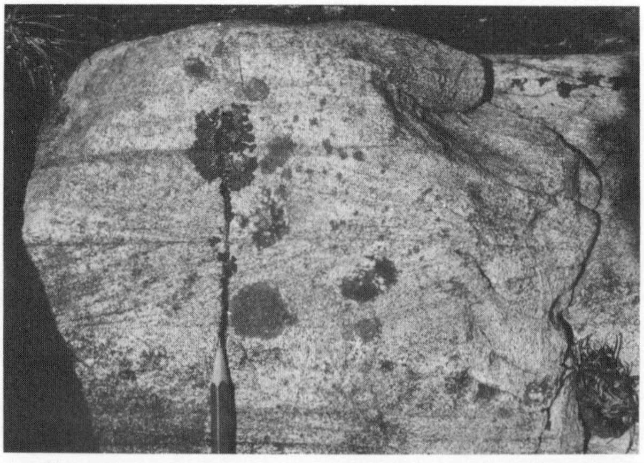

Figure 2.4 Isolate reversed cross-set (immediately above pencil point) within subfacies 1B; lateral persistence over several metres confirmed that this is a true palaeocurrent reversal.

*PR Persistence ratio = set length/set height (Anderton, 1976)

Within the Upper Psammite, palaeocurrents are largely unidirectional; subfacies 1B is notable in that it records occasional reversals. These occur as isolate reversed sets within cosets (Fig. 2.4) or, more commonly, as an entire reversed coset.

1C: trough-festoon sets (Michelson and Dott, 1973) (PR* 7–53, mean 18); tend to occur in coarse, often gravelly sands towards the top of the Upper Psammite. Compared to 1B, this subfacies comprises relatively short, narrow, deep cross-sets (Fig. 2.5), trough contacts are numerous and strongly erosive; this is reflected in a greater variability of set thickness. Sets often climb up and down current. Foresets are concave and tangential in section, strongly curved in plan (Fig. 2.6) and dip about 20°.

Clearly, this facies was produced by the migration of dunes which became increasingly three-dimensional in response to increasing flow power (Allen, 1970; Dalrymple *et al.*, 1978). The dominance of 1B and 1C in the lower and upper portions of the sequence respectively probably reflects a hydrodynamic change with time, and may also be grain-size controlled. The latter is clearly reflected in the lower foreset angles of subfacies 1B; higher proportions of fine-grained sands are carried in suspension; this favours deposition on toesets and bottomsets via the separation/back-flow zone, as opposed to simple avalanching down steeper, angle-of-rest foresets (Jopling 1963, 1965).

Figure 2.5 Erosive based trough-sets of subfacies 1C; note scattered granules within sandstone. Current approximately right to left. (Fig. 2.22*B*, Loc H.)

Figure 2.6 Strongly curved foresets of subfacies 1C exposed on a bedding plane. Current from right to left. (Fig. 2.22*B*, Loc. E.)

The main palaeoenvironmental evidence within this facies is the bimodal, bipolar palaeocurrent pattern of subfacies 1B. This supports a tidal origin, whereas unidirectional cosets of small- to medium-scale cross-bedding could equally be deposited in a fluvial or shallow-marine setting.

2.4.1.2 *Facies 2.* This facies is volumetrically subordinate but of considerable palaeoenvironmental importance. It is developed in medium to coarse, gravelly sands and comprises tabular cross-sets between 20 cm and 2 m in thickness. These are usually isolate and laterally persistent, extending over the length of the outcrop. Two subfacies have been distinguished.

2A: Simple tabular sets: the commonest subfacies, characterized by simple avalanche foresets.

2B: Compound tabular sets consist of sets of subfacies 1A which dip downcurrent at low angles, on average at 12°.

These subfacies are not exclusive; sets of subfacies 2B can be traced laterally into 2A and vice versa (for example as sketched in Fig. 2.7*A*; note the steepening which accompanies the transition from compound to avalanche foresets). Subfacies 2B and 1A are similarly related.

This facies is interpreted as the product of straight-crested sand waves (Southard, 1975, page 24). In subfacies 2A, these advanced by simple lee-slope avalanching, in 2B by the migration of superimposed small-scale dunes. Allen and Collinson (1974) view the latter behaviour as a lag effect in response to variable unidirectional flows; a large sand wave in equilibrium at high current speeds will not rapidly degenerate and re-equilibrate in accordance with decreased velocity, instead, smaller-scale bedforms will form and migrate across its upper surface, with down-current dipping cross-stratification preserved on the lee slope. However Dalrymple *et al.* (1978) report considerable overlap in sand wave/dune stability fields, supporting Banks' (1973*a*) contention that this structure results from the relatively rapid migration of dunes across the surface of a slower moving sand wave.

The simple structures thus described occur in both fluviatile (e.g. Collinson, 1970; Banks, 1973*a*) and shallow-marine settings (Levell 1980*b*; Reineck and Singh, 1980). The latter environment is favoured by the presence of fine sediment drapes and reactivation surfaces: features attributed to tidal activity (Allen and Narayan, 1964; Barnes and Klein, 1975; Thompson, 1975; Levell 1980*b*; Visser, 1980).

Examples of drapes are shown in sketch form in Figs 2.7*A* and *B*; they separate foresets and downcurrent dipping sets, are generally massive and less than 5 cm in thickness, thicker drapes are of finely interbedded sand and silt. In one set, 23 drapes were recorded over a 30 m distance. Reactivation surfaces are less common,

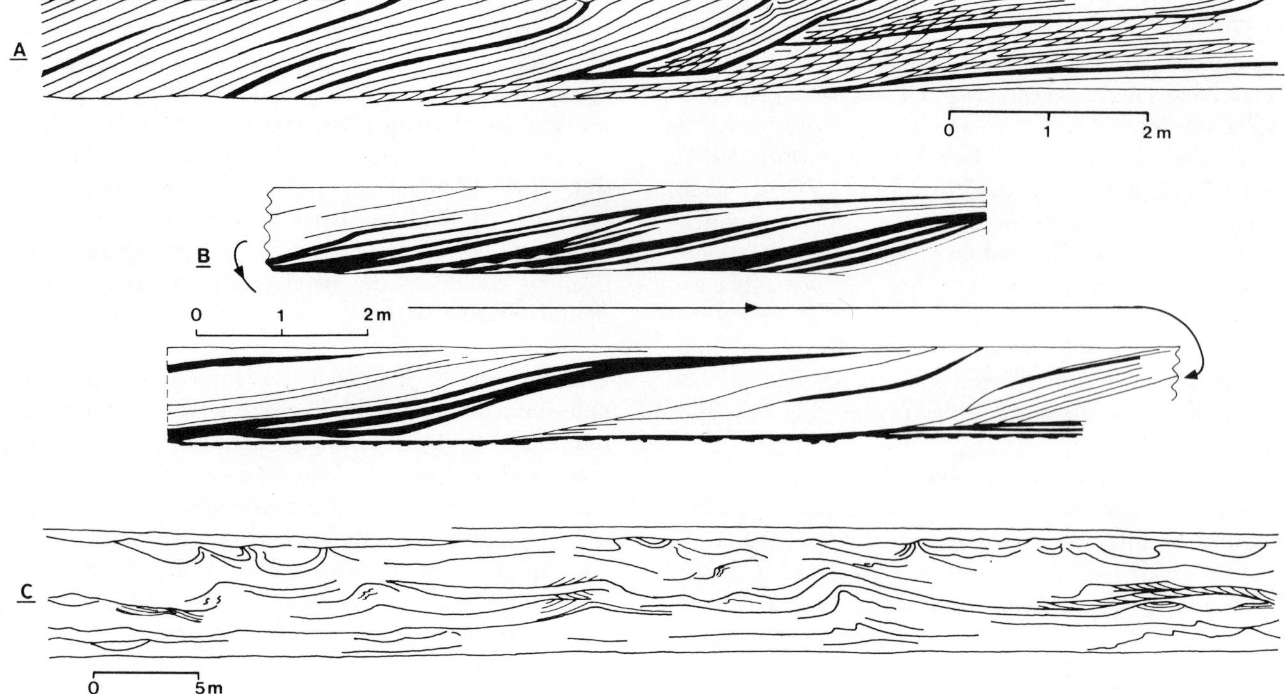

Figure 2.7 (*A*) Facies 2: compound tabular set (2B) passing downcurrent into a simple tabular set (2A). Note thick mud drapes developed on avalanche foresets. (Fig. 2.22*B*, Loc. H.) (*B*) Facies 2: complex sand wave with mud drapes, rippled toesets and reactivation surfaces (Gribun, Ross of Mull). (*C*) Large-scale convoluted bedding. (Fig. 2.22*B*, Loc. A.)

Fig. 2.7*B* shows drapes and foreset/set boundaries erosively truncated downcurrent.

Drapes develop in asymmetrical tidal regimes (Allen, 1980); the subordinate reversed flow is too weak to erode the fine-grained sediments which are deposited from suspension during the slack water periods between successive tides. Allen (1981*a*, *b*; 1982) and Visser (1980) have related these features to neap–spring tidal cycles; at neaps, reduced tidal ranges and current speeds enhance suspension over traction deposition, producing thinner tidal bundles and thicker drapes. Reactivation surfaces reflect a stronger reversed flow which erosively modifies the sand wave lee slope.

2.4.1.3 *Facies 3: Large-scale trough cross-bedding.* This facies developed in medium to coarse, gravelly sand, and comprises thick (> 50 cm) erosively based trough sets. These are isolate or amalgamated into thick cosets. The latter are interpreted as the products of trains of large, strongly three-dimensional dunes; the former may represent channel-fills.

2.4.1.4 *Facies 4: Upper phase plane bed.* This facies consists of well-sorted fine and medium-grained sands (i.e. grain size less than 0.5 mm). Units tend to be thin, around 24 cm thick, but at the base of the Upper Psammite they are much thicker, ranging up to 3.2 m thick. Commonly, this facies is associated with cross-bedded and cross-laminated sands (facies 1 and 5), suggesting that fluctuations in current velocity caused overlaps into the ripple and dune stability fields.

Flume studies and the resultant bedform plots show that plane beds in this grade of sediment were deposited under upper flow regime conditions, at Froude numbers (F) greater than 1 (e.g. Simons and Richardson, 1961; Southard and Boguchwal, 1973). Therefore sedimentation occurred in very shallow water; for example to obtain $F = 1$ in 0.5 m of water, a current velocity of 2.2 m/s is required, in 10 m of water 9.9 m/s (although in natural environments upper flow regime conditions can be achieved at $F = 0.6$–0.7). This has considerable palaeoenvironmental implications; on modern shelves, in water depths around 10 m, current speeds are usually less than 2 m/s and rarely as high as 3 m/s, so upper phase plane bed is unlikely. It has been reported from intertidal zones (Knight and Dalrymple, 1975) and the subtidal parts of inlets/estuaries (Terwindt, 1971, 1975), but is more typical of high-energy shorelines (Clifton, 1969; Clifton *et al.*, 1971), and fluvial deposits, particularly shallow, ephemeral streams (McKee *et al.*, 1967; Tunbridge, 1984).

2.4.1.5 *Facies 5: Cross-lamination.* Cross-lamination (set thickness < 3 cm) is restricted to fine-grained sands. Cosets are thin, on average 20 cm. Two subfacies are distinguished.

5A: Planar cross-lamination was deposited by migrating straight-crested ripples, and occurs as isolate cosets or interbedded with facies 1 and 4. Sets occasionally climb downcurrent.

5B: Trough cross-lamination: this resembles the 'micro-cross-lamination' of Hamblin (1961). It exists as dis-

Figure 2.8 Subfacies 5B: trough cross-lamination, current from right to left. (Gribun, Ross of Mull.)

crete cosets deposited by trains of three-dimensional ripples. Internally, numerous trough contacts are preserved (Fig. 2.8).

2.4.1.6 *Facies 6: interbedded sands and silts.* The facies consists of thin layers and lenses of sand (< 15 cm thick) enclosed by silts. Sedimentary structures are preserved within low-strain zones, the sand layers contain either cross-bedding, plane-bedding or cross-lamination (facies 1, 4 and 5). Rarely layers are composite; Fig. 2.9 shows a portion of a log through this facies at Morar; several sand beds show evidence for deposition from waning flows with locally erosive bases and massive basal units, passing upwards into upper phase plane bed

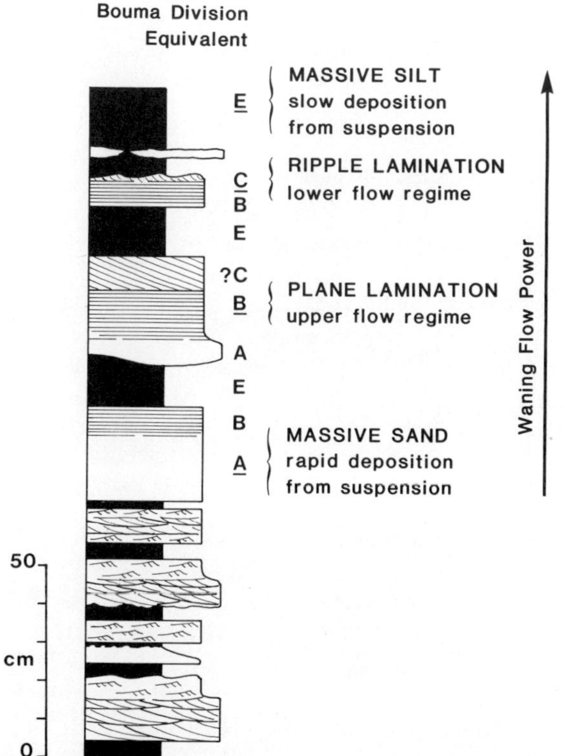

Figure 2.9 Thin, waning-flow sequences in facies 6 which mimic the Bouma sequence. (Fig. 2.22*B*, Loc. C.)

Figure 2.10 Facies 6: ?wave-rippled upper surface of thin sandstone bed. The ripple to the right, although only partially exposed, may preserve a form-discordant internal structure. (Fig. 2.22*B*, Loc. C.)

and rippled tops. The latter are normally asymmetrical current-ripples, but symmetrical, possibly wave-generated forms with form-discordant internal structures are also found (Fig. 2.10).

Allen (1984) outlines how thin waning flow sequences can be produced in a marine environment during storms. As a storm moves onshore, the sea-level is pushed up against the coast; this generates a compensatory bottom return flow, which transports material offshore to be deposited as a thin graded bed (storm layer or tempestite). This may resemble a thin Bouma sequence (cf. Fig. 2.9). Similar layers attributed to storms have been described from both modern (Kumar and Sanders, 1976; Nelson, 1982; Aigner, 1985) and ancient (Goldring and Bridges, 1973; Brenchley *et al.*, 1979; Tucker, 1982; Aigner 1982, 1985) shallow-marine sediments.

However, comparable waning-flow sequences interbedded with silts have also been reported from fluviatile settings (Stanley, 1968; Steel and Aasheim, 1978) where they are likely to represent crevasse-splay or sheet-flood deposits.

2.4.1.7 *Facies 7: massive silts.* Massive silt units tend to be thin (< 15 cm) and overlie cosets of facies 1 and 2. These may be interpreted as fine sediment drapes. Thicker units, up to a maximum of 1.7 m, which represent the prolonged fall-out of fine suspended sediment, are rare.

2.5 Other features

2.5.1 *Palaeocurrents*

Figure 2.11 summarizes the palaeocurrent measurements collected from the cross-bedded facies within the Upper Psammite. These are expressed as vector means, and additionally—where bi- or polymodality is developed—as modes. The dominant transport direction was roughly SSW to NNE, subparallel to the

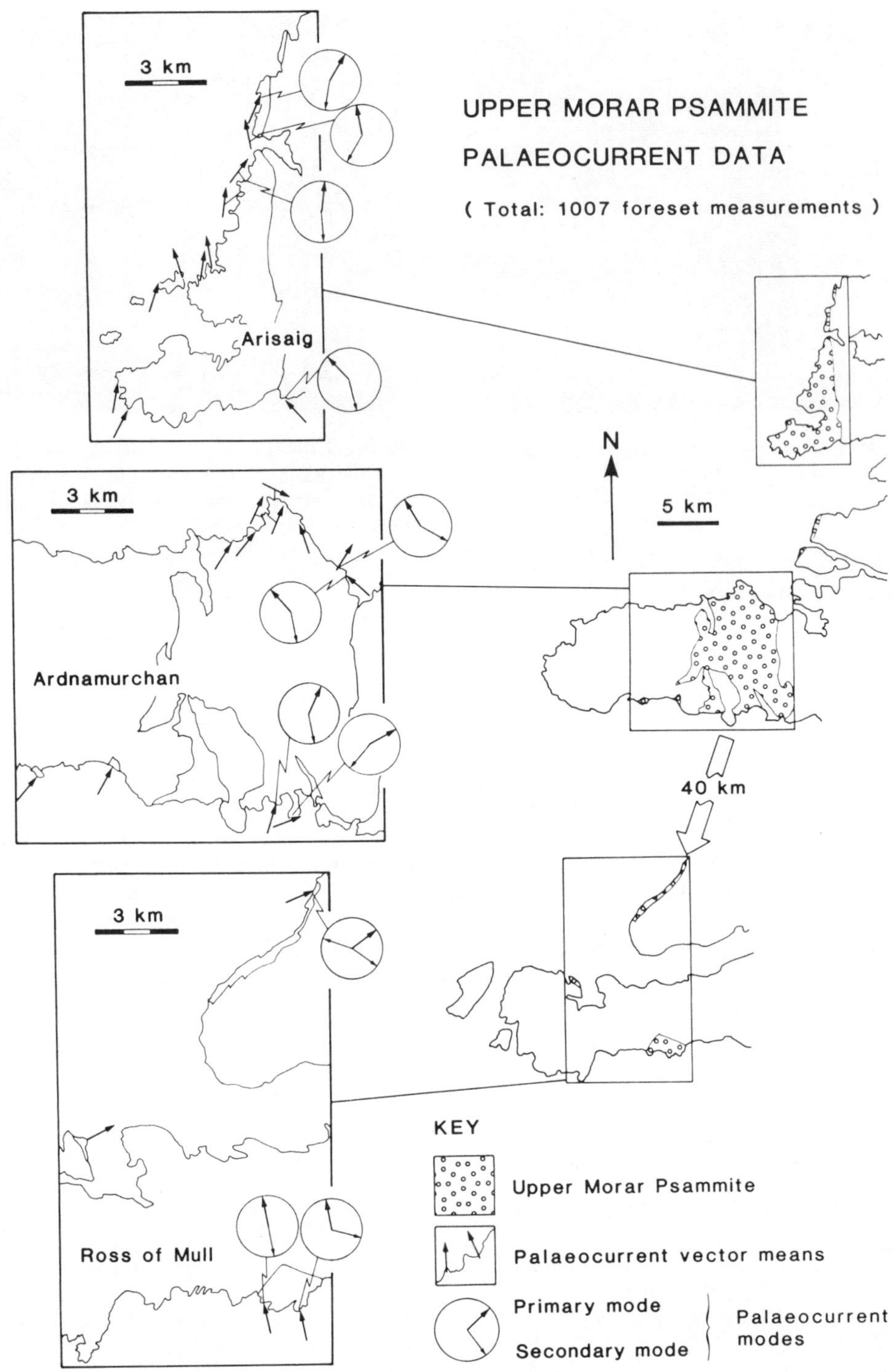

Figure 2.11 Upper Psammite palaeocurrent data.

regional strike. This agrees with observations recorded throughout the Moine Assemblage (e.g. Wilson *et al.*, 1953; Soper, 1960; Strachan, 1986). The presumed tidally generated, bimodal bipolar palaeocurrents previously mentioned are almost exclusive to subfacies 1B (trough-parallel sets). Consequently, within the main Upper Psammite outcrops of Morar and Ardnamurchan they share the same distribution and are present in the lower parts of the formation, but disappear upwards.

2.5.2 *Soft-sediment deformation*

This is very common within the Upper Morar Psammite (e.g. Figs 2.12, 2.15, 2.22*A*). A variety of structures are preserved. Small-scale structures include overturned and recumbently folded cross-bedding, load structures (particularly associated with heavy-mineral bands) and sandstone dykes and sills (Fig. 2.17). These are, however, relatively uncommon; the most widespread and spectacular examples of soft-

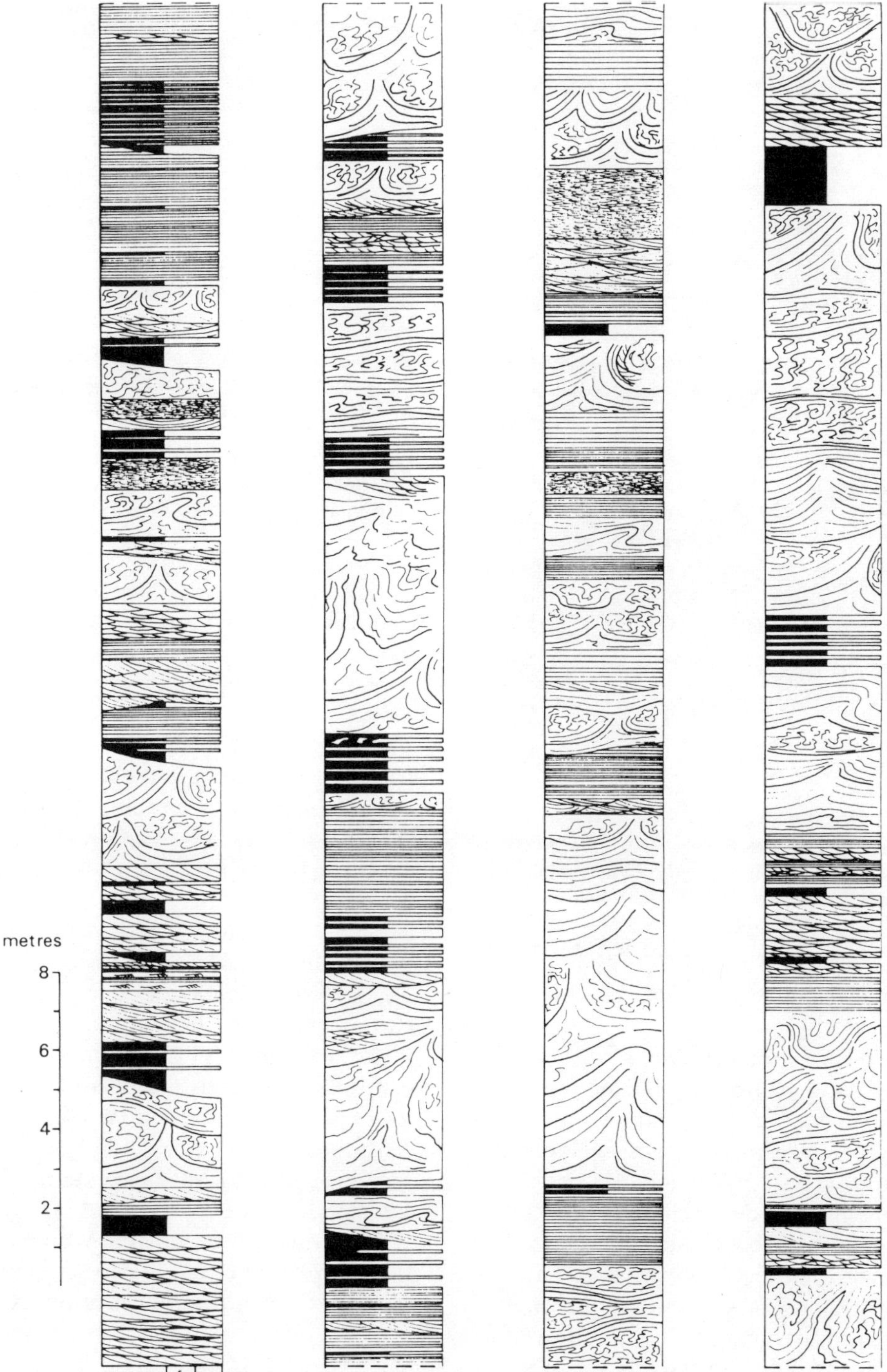

Figure 2.12 Graphic log from base of Upper Psammite at Morar. (Fig. 2.22*B*, Loc. A.)

sediment structures are thick, laterally persistent sheets of highly contorted sand. At certain localities (Fig. 2.12) these constitute over half of the sequence, and attain a maximum thickness exceeding10 m. Internally they have the following characteristics: major bedding planes are well preserved, the intervening sands are mostly massive and contorted, while in places relict structures demonstrate that the sheets were originally trough cross-bedded. The major bedding planes are curved and consist of broad synclines separated by sharp anticlines (Fig. 2.2*C*). This structure is interpreted as large-scale convolute bedding; escaping porewater migrated away from synclinal areas into the developing anticlines, which served as water-escape paths (Fig. 2.13). These are erosively truncated by later beds within the same sand unit, therefore the convolute bedding is syndepositional. Note that in Fig. 2.12, adjacent units of upper phase plane bed (facies 4) are undisturbed. This reflects their low depositional porosity; in comparison the cross-bedded sands were more

Figure 2.13 Water-escape structure. The bedding (near vertical) erosively truncates the structure. Rucksack (to left of photograph) is approx. 50 cm across. (Fig. 2.22*B*, Loc. G.)

Figure 2.14 Winnowed gravel layer on upper surface of sandstone bed. This is overlain by a thin silt bed, now cleaved semi-pelite. (Fig. 2.22*B*, Loc. H.)

loosely packed and readily compacted and dewatered (Allen, 1972; Lowe, 1975).

Soft-sediment deformation structures can be triggered in a number of ways. Leeder (1987, in press) has related certain classes of structure (catastrophic dewatering pipes, recumbently folded cross-bedding and sandstone ball and pillows) to earthquakes generated along nearby faults and—as discussed later—within the Upper Psammite, their regional distribution may support this contention. However, on the outcrop scale simple, rapid deposition of saturated sands provides an adequate mechanism, particularly as the intervening finer sediments (facies 6 and 7) are largely undisturbed. Thinly interbedded sands and silts are susceptible to both seismic activity (e.g. Sims, 1975; Mayall, 1983) and slumping (Adams and Cossey, 1978; E. W. Johnson, 1981).

2.5.3 *Erosion surfaces*

Obvious contemporaneous erosion surfaces within the Upper Psammite are rare. Two forms have been recognized.

2.5.3.1 *Basal scour surfaces.* Occasionally, sand sheets, mainly facies 1C, 3 and sand units within Facies 6 (Fig. 2.9) have undulatory, erosive bases and may incorporate rip-up clasts of the underlying silts.

2.5.3.2 *Winnowed gravel lags.* These are one to a few granules thick, and are developed both within and on the upper surfaces of sand sheets (facies 1C and 2). This suggests that they were produced by preferential removal, or winnowing, of sand to leave a residual lag. They persist over the width of the outcrop (up to a few tens of metres) and are usually planar, rarely channelled. Top-surface lags are draped with thin silts (Fig. 2.14). The common heavy-mineral bands may also have originated through winnowing of the lighter quartz and feldspar grains.

Gravel lags are most abundant in the upper parts of the sequence on the N coast of Ardnamurchan, Fig. 2.15 shows a 45 m section which contains twenty-two gravel layers, but only four examples have been found at Morar. This distribution may be grain-size controlled. Gravelly sediment was absent from the lower parts of the formation, and therefore winnowing could not produce a residual lag.

Similar features have been described from, and may be characteristic of, ancient shallow-marine sands (Levell, 1980*a*, *b*; Wright and Walker, 1981). They have been attributed to tidal scour (Anderton, 1976), storm erosion (Cotter, 1985) or a combination of the two.

2.5.4 *Shrinkage cracks?*

Figure 2.16 shows a plan view of a series of persistent, branching sand-filled cracks within a thin silt horizon. These have been found on several bedding planes at only one locality (point *D* on Fig. 2.22*B*). Richey and Kennedy (1939) attributed such structures to shrinkage and this interpretation has been quoted in more recent reviews (M. R. W. Johnson, 1983, p. 56). If this is the case, an origin by desiccation through exposure appears likely; subaqueous shrinkage or synaeresis cracks tend to follow more irregular, radiating patterns with individual cracks lenticular and discontinuous in plan (Collinson and Thomson, 1982), although according to Allen (1984), there is no mechanical reason for the two forms to differ.

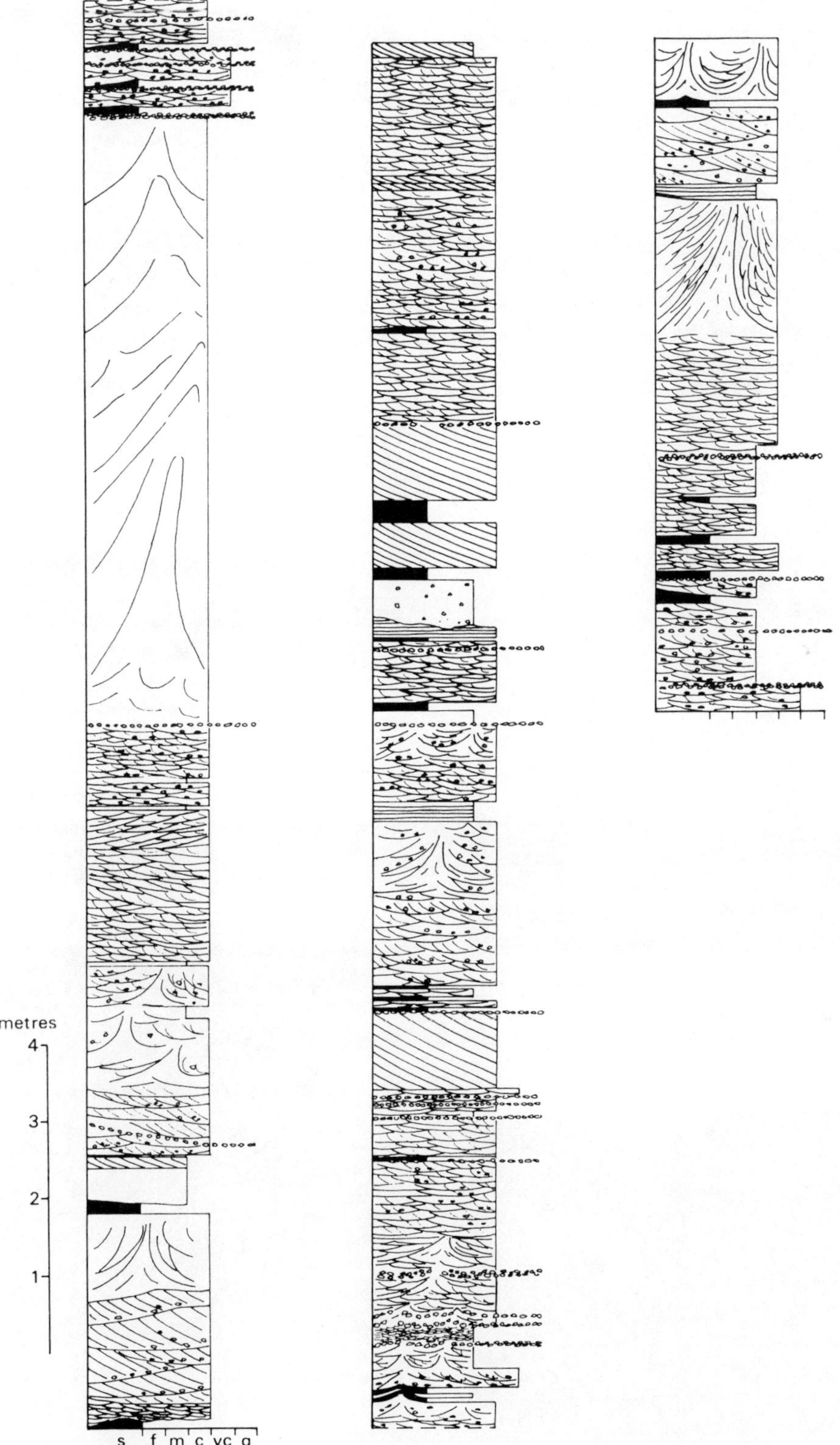

Figure 2.15 Graphic log of the Upper Psammite (upper part) from the north coast of Ardnamurchan. (Fig. 2.22*B*, Loc. H.)

However, in section no examples of tapering-downwards sand-filled cracks, or of crack-fills originating from an overlying sand bed, could be found. Only vertical and inclined sand dykes which were clearly intruded from below were observed (e.g. Fig. 2.17). Therefore these branching sand-filled cracks are probably the equivalent of structures described by Leeder (1987, in press Fig. 3); he noted that dykes produced by the expulsion of overpressured, liquified sand will intersect bedding planes to produce a crudely polygonal pattern.

2.5.5 *Calc-silicates*

These are developed in all three divisions of the SW Moine (Johnstone *et al.*, 1969), and have long been

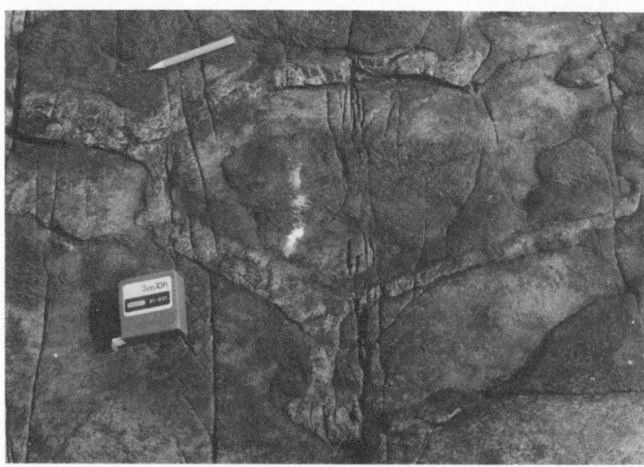

Figure 2.16 Polygonal pattern of sand-filled cracks in a thin silt horizon. These are either desiccation cracks or sedimentary dykes. (Fig. 2.22*B*, Loc. D.)

Figure 2.17 Sandstone dyke intruding upwards into silts. The folding of the dyke was probably produced by differential compaction between the sand and silts. (Fig. 2.22*B*, Loc. D.)

Figure 2.18 Pale calc-silicate lenses preferentially developed along sandier beds within the Striped Schist.

interpreted as the metamorphosed equivalents of former calcareous-rich bands (Richey and Kennedy, 1939). Details of their present metamorphic mineralogy are described by Tanner (1976). Within the Morar Division their distribution is patchy. They are most abundant in the upper portion of the Striped Schist, but absent from its lower parts and relatively uncommon in the Upper Psammite. They are thin (< 10 cm) and are preferentially developed along cross-bedded sand units (Fig. 2.18), which presumably possessed higher porosity and permeability than the surrounding silts and were exploited by migrating pore-fluids with resultant carbonate cementation.

Figure 2.19 Striped Schist: interbedded sands and silts. Original cross-bedding (facies 1) within the sands has been flattened and now resembles plane-bedding (facies 4).

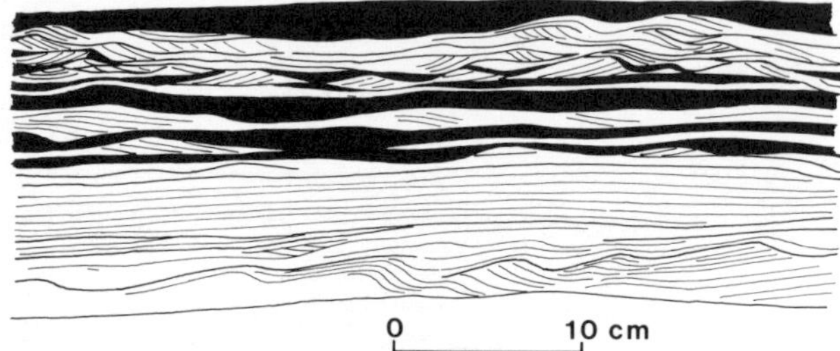

Figure 2.20 Flaser bedding from a low-strain zone within the Striped Schist. Dominant flow from left to right, with subordinate reversed flow. (Fig. 2.22*B*, Loc. A.)

Figure 2.21 Striped Schist: thin layers and lenses of fine sand interbedded with silt. (Fig. 2.22*B*, Loc. A.)

2.5.6 *Morar Striped and Pelitic Schist*

Figure 2.19 illustrates and 'striped' nature of this formation; again these mixed lithologies have been strongly deformed, and unmodified sedimentary structures are rare. The following description is drawn mainly from a low-strain zone immediately below the Upper Psammite (near point *A* in Fig. 2.22*B*) and from the coastal section to the south of Arisaig (Fig. 2.1). In terms of the scheme devised for the Upper Psammite, the following facies are developed.

2.5.6.1 *Facies 1: cross-bedded sand sheets.* These contain small- and medium-scale cross-bedding (mainly subfacies 1B) with subordinate plane bed (4). The sheets are up to 3 m in thickness but generally much thinner, 5 to 50 cm. Palaeocurrents were only obtained at one locality and proved to be bimodal with a divergence of 145° between the two modes (Fig. 2.11).

2.5.6.2 *Facies 6: interbedded sands and silts.* Thin layers and lenses (mm to cm scale) are separated by silts. Cross-lamination and climbing-ripple lamination (facies 5) are occasionally well preserved. In appearance these layers resemble the flaser and lenticular bedding (see Figs 2.20 and 2.21 respectively) described from intertidal and subtidal sequences (Terwindt, 1981). In this context the palaeocurrent variability shown in Fig. 2.20 may be of tidal origin.

2.5.6.3 *Facies 7: massive silts.* The greatest development of this facies is in the middle of the Striped Schist as an approximately 100 m thick silt band with very subordinate sand (on BGS Sheet 61 this is shown as a separate pelitic unit). Generally this facies is restricted to beds less than 2 m thick.

2.5.6.4 *Interpretation.* Palaeocurrent variability again supports a tidal origin, and the presence of lenticular and flaser bedding indicates that parts of the Striped Schist may be intertidal or shallow subtidal. An obvious drawback to the latter interpretation is the lack of evidence for the channelling common in modern intertidal/shallow subtidal settings (Reineck, 1975). Comparison with Thompson's (1968, 1975) work on the Colorado Delta may explain this. He suggested that under conditions of low wave energy (cf. the paucity of wave ripples in the Upper Psammite), barrier island development is insignificant, tides are unconstricted and channel systems do not develop. Instead the tide inundates the flats as a broad, uniform flow.

An intertidal/shallow subtidal origin may also explain the preferential development of calc-silicates in the Striped Schist. In shallower water, relatively higher evaporation rates concentrated calcium carbonate, which was incorporated into and eventually reacted with the host sediment. However, other evidence for desiccation (e.g. mud cracks) has not been observed.

2.6 Summary

Within the Upper Psammite and Striped Schist, the presence of occasional bimodal bipolar palaeocurrents, complex sand waves, top-surface gravel lags, thin waning flow sequences, possible lenticular and flaser bedding and the rarity of channelling are consistent with a shallow-marine rather than fluvial origin. The lack of pronounced palaeocurrent variability (e.g. herring-bone cross bedding), or of evidence for exposure, suggests a dominantly subtidal setting. The next section examines the spatial relationships of the various facies and attempts to define a more detailed sedimentary model.

2.6.1 *Facies relationships*

On the outcrop scale, sequences were examined using the technique of Markov chain analysis. No meaningful relationships or cycles were revealed. Significant vertical and lateral changes do occur within the Upper Psammite on a broader regional scale and these are listed in Fig. 2.22*A* and summarized in Fig. 2.23.

Briefly, in the lower parts of the formation, trough-parallel sets (subfacies 1B) are the dominant type of cross-bedding. Thick units (up to several metres) of plane bed (4) occur, but are confined to the basal few hundred metres of the Upper Psammite (Fig. 2.22*A*). Upwards, trough festoon sets (1C) dominate, while thick tabular sets (2) and occasional top-surface lags are developed. Facies 4 is restricted to thin beds. These changes are accompanied by an increase in grain size, and a switch from bidirectional bipolar to unidirectional palaeocurrents. Soft-sediment deformation structures are present throughout the Upper Psammite. Figs 2.12 and 2.15 show graphic logs from the lower and upper parts of the formation respectively.

The main lateral changes along the S to N transport path are a decrease in the grain size of sands accompanied by an increase in the proportion of silt (facies 6 and 7). A change in the style of cross-bedding may occur with subfacies 1C, facies 2 and 3 passing into 1B.

Owing to the tectonic contact between the Upper Psammite and Striped Schist at Ardnamurchan, only vertical changes are definite, any lateral transitions are inferred. However, as the same contact only 16 km to the north at Morar is clearly stratigraphic, a large relative displacement between the two areas is unlikely and lateral continuity is assumed.

2.6.2 *Interpretation: lateral changes*

The inferred lateral changes are, in a shallow-marine context, easiest to explain as changes down a net sand transport–deposition path. Such paths result from distortions of the tidal wave as it propagates from the ocean on to a constricted shelf. The tidal range is amplified and current velocity increased. Inequalities (time–velocity asymmetry) develop between ebb and flood currents and, as the sand and gravel transport rate is approximately proportional to the *cube of the difference* between the current velocity and threshold velocity for a given grade of sediment (Haworth, 1982), even small asymmetries will produce a net sediment transport. Adjacent landmasses impose a 'directionality' on the

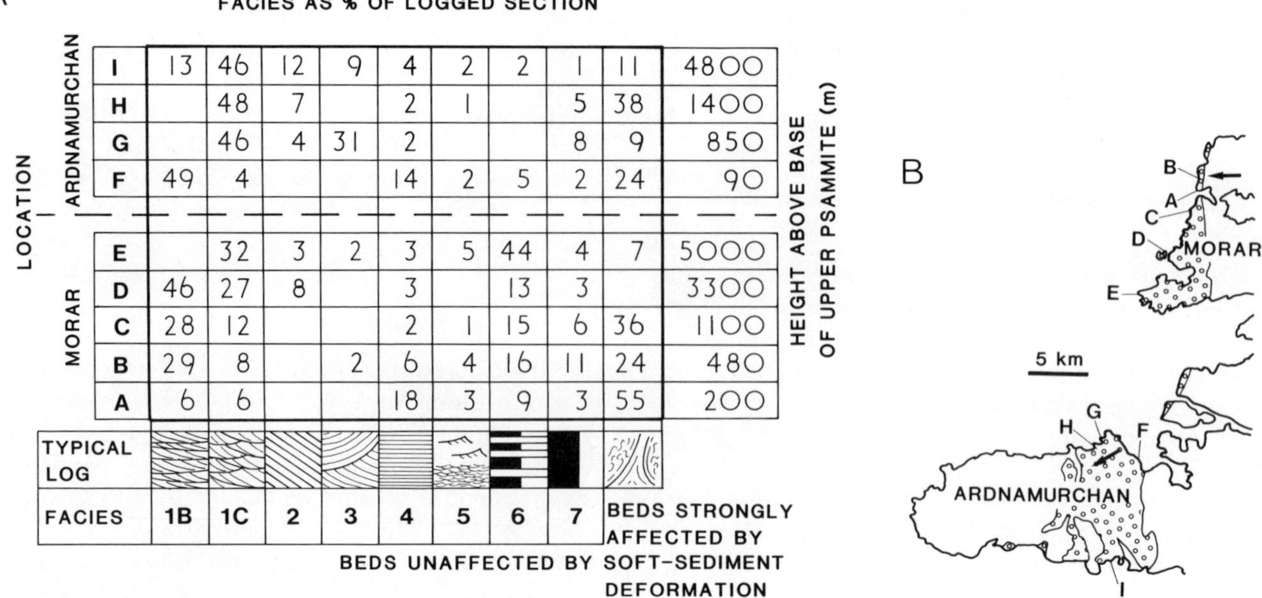

FACIES AS % OF LOGGED SECTION

LOCATION		1B	1C	2	3	4	5	6	7	BEDS STRONGLY AFFECTED BY SOFT-SEDIMENT DEFORMATION	HEIGHT ABOVE BASE OF UPPER PSAMMITE (m)
ARDNAMURCHAN	I	13	46	12	9	4	2	2	1	11	4800
	H		48	7		2	1		5	38	1400
	G		46	4	31	2			8	9	850
	F	49	4			14	2	5	2	24	90
MORAR	E		32	3	2	3	5	44	4	7	5000
	D	46	27	8		3		13	3		3300
	C	28	12			2	1	15	6	36	1100
	B	29	8		2	6	4	16	11	24	480
	A	6	6			18	3	9	3	55	200

Figure 2.22 (*A*) Variations in the proportions of sedimentary facies in logged sections at Ardnamurchan and Morar; (*B*) location of the sections.

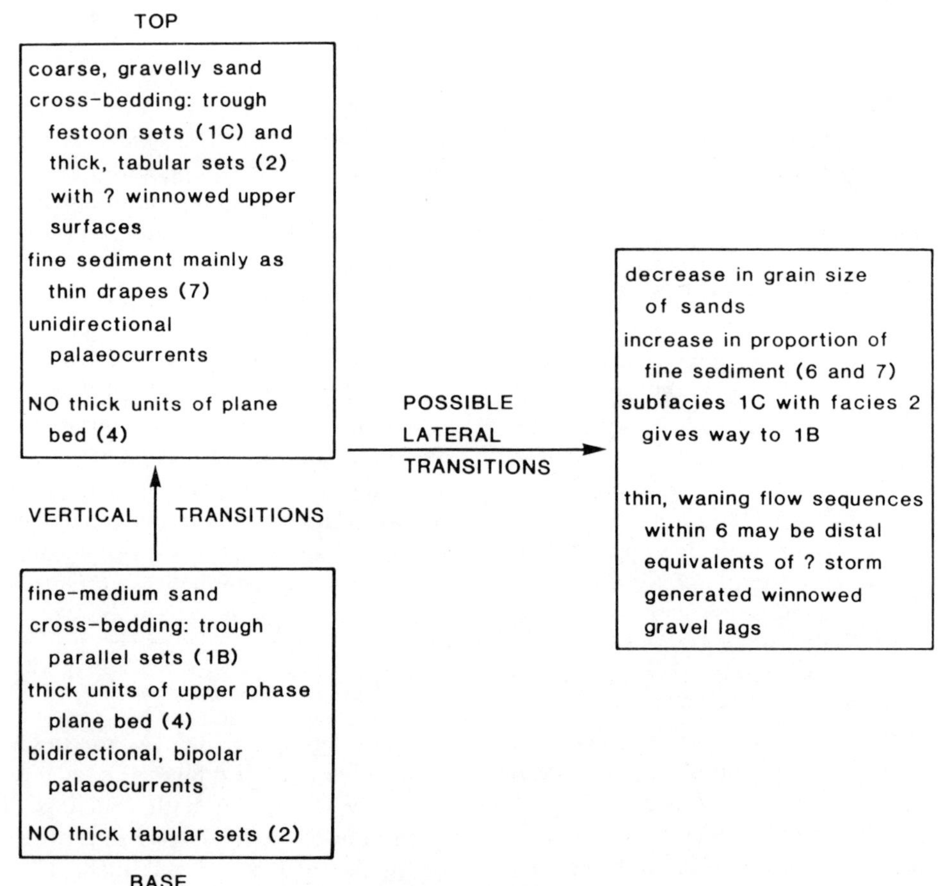

Figure 2.23 Summary of facies relationships.

currents (Spearing, 1975) which tend to parallel the shoreline.

Such effects are greatest in partially enclosed basins or gulfs which co-oscillate, or resonate with the controlling tide. Stride (1963) and Belderson and Stride (1966) described an ebb-dominated transport–deposition path in the Celtic Sea, which orginates from the resonant Bristol Channel. A series of erosional–depositional zones are developed which reflect the progressive reduction in current velocity along the path. Erosion at

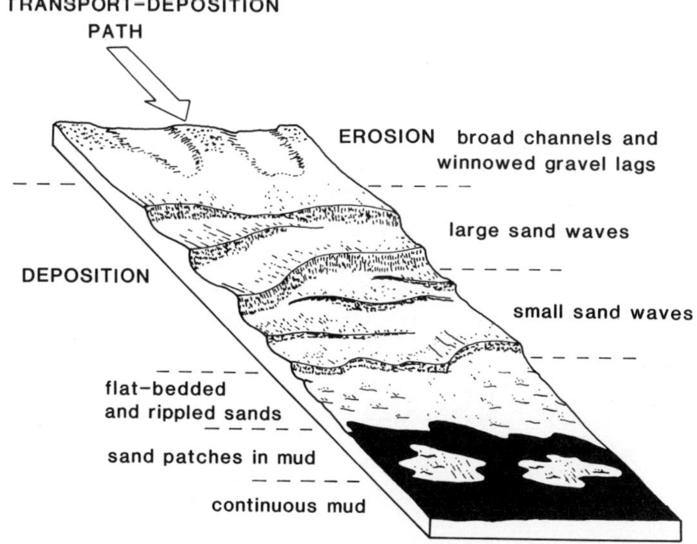

Figure 2.24 Anderton's (1976, Fig. 18) model for tidal shelf sedimentation. Transport/deposition path is tens to hundreds of kilometres long.

the head is successively followed by sand ribbons, sand waves, sand and muddy sand, and finally sand patches on (relict) gravel (Belderson and Stride, 1966).

Similar asymmetrical tides and transport paths linked to resonant gulfs have been invoked to interpret a number of ancient shallow-marine deposits which, like the Upper Psammite, have dominantly unidirectional palaeocurrents. Some examples are the Jura Quartzite (Anderton, 1976), the Lower Sandfjord Formation (Levell, 1980*b*) and the Lower Greensand of SE England (Bridges, 1982). The model devised by Anderton can be applied to the lateral changes in the Upper Psammite. He envisaged the Celtic Sea under equilibrium conditions (i.e. increased sand input) and predicted the depositional zones shown in Fig. 2.24. Storms periodically reinforced the dominant tidal current and caused these zones to migrate along the transport path so that, for instance, the tops of sand waves were eroded and sand waves deposited over distal muds. This could explain the interbedding of facies 1 with 6 and 7, frequently observed in the Morar area (e.g. Fig. 2.12) and the preferential development of winnowed gravel lags on Ardnamurchan. The latter may be the proximal, erosional equivalents of the thin waning flow sequences identified at Morar. The transition from subfacies 1C, 2 and 3 to 1B reflects the reduction in current velocity along the transport path—the bedforms become smaller and less three-dimensional.

2.6.3 *Interpretation: vertical transitions*

The vertical facies changes within the Upper Psammite provide evidence for both transgression and regression.

If an intertidal origin for at least the upper parts of the Striped Schist is accepted, then the Upper Psammite must initially have been transgressive. Furthermore, the presence of thick units of facies 4 near the base of the formation indicates vey shallow water (cm to tens of cm depth), the appearance of thick tabular sets (2) upwards indicates deepening. Observations from natural environments (Allen, 1984) suggest that bedforms 2 m high formed in water depths greater than 5 m. Additionally, in a gulf setting, increasing water depth might also explain the upwards change in palaeocurrent pattern and the shift to higher-energy cross-bedding (i.e. subfacies 1B to 1C). Under deepening water, gulf dimensions might alter to a form resonant with the controlling tide, so that a more fully developed transport–deposition path evolves. As a comparison Amos (1978) described a coarsening-upwards sequence in the Bay of Fundy which reflects a recent 50% increase in tidal amplitude. This, in turn, probably resulted from the change in the Bay's dimensions during the post-glacial sea-level rise.

However, although the vertical changes may relate to increasing water depth, and, taking the lateral transitions into account, the facies relationships within the Upper Psammite were, overall, regressive as the proximal facies (1C and 2) progressively interfinger with and overlie the distal facies (1B, 6 and 7) (Fig. 2.22*A*). The upwards increase in grain size is also more typical of a regressive shallow-marine sequence (Simonson, 1984). This situation is summarized in Fig. 2.25.

2.6.4 *Discussion*

There are several problems associated with this simple regressive depositional model, which are:

(i) The Upper Psammite (at least 5 km thick) is interpreted as a regressive coarsening-upwards sequence. In other ancient shallow-marine deposits such sequences tend to be much thinner, approximately 12 to 50 m (Banks, 1973*b*; Simonson, 1984) and may be stacked in response to repeated transgressive/regressive cycles (e.g. Beaumont 1984; Simonson, 1984). Furthermore, the eventual products of regression would be deltaic or fluvial sediments, but no such incursions have been recognized in the Upper Psammite.

(ii) The lateral transitions inferred within the

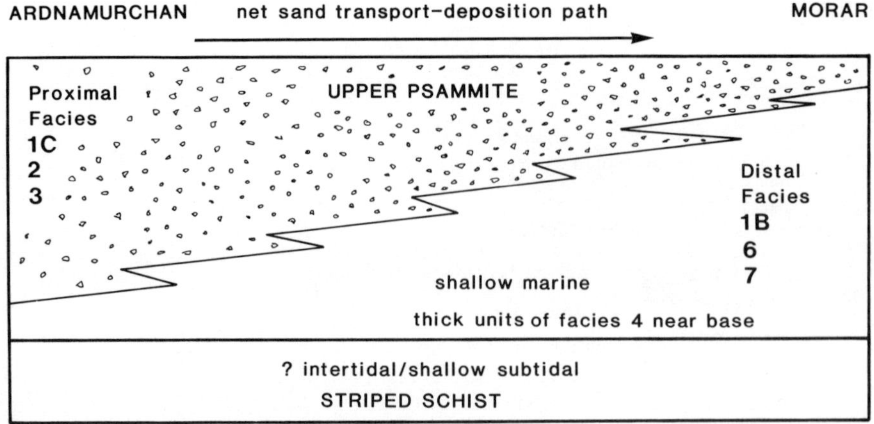

Figure 2.25 Idealized section through Upper Psammite, showing overall regressive relationships of the sedimentary facies.

Upper Psammite, from proximal sequences at Ardnamurchan to distal at Morar, occur over a distance of about 16 km. This appears short in comparison with modern sand transport paths which, around the British Isles, are anything up to 550 km long (Johnson *et al.*, 1982).

(iii) As a result of considerable reworking, ancient tidal sands are typically mature or supermature quartzites. In comparison, the Upper Psammite sands are relatively immature arkoses.

(iv) If Anderton's (1976) model for shallow-marine sedimentation is applied to the Upper Psammite, then the interbedding of sand sheet facies with thick silty units (facies 6 and 7) was produced by the migration of facies belts under storm-reinforced flows. However, evidence for storm activity within the Upper Psammite is rare.

A possible solution to the first three points is that they reflect the tectonic characteristics of the original sedimentary basin. The lack of minor transgressive or regressive cycles superimposed on the overall regressive pattern indicates a delicate balance between sediment input and subsidence. The abundant soft-sediment structures can be interpreted as a response to rapid deposition or to earthquakes generated by repeated movements along (basin margin?) faults. Either mechanism implies both rapid subsidence and rapid sedimentation rates. Therefore the sediments were available for reworking for a comparatively short time, and are consequently less mature. Similarly, they were less likely to be transported long distances down a net sand transport–deposition path. In comparison, sedimentation and subsidence rates in the continental shelf seas around Britain are negligible and sands are available for prolonged reworking down transport paths.

There is no obvious solution to the fourth problem, as the lack of evidence for storms may be a preservational phenomenon. Thin sand units (within facies 6) are more susceptible to deformation, therefore more storm beds may be present but are not recognized.

2.6.5 *The Upper Psammite on the eastern limb of the Morar Antiform; a possible control on basin geometry*

As previously discussed, the original geographical separation of the two limbs is not known. However, if lateral continuity is assumed, then the eastern limb psammites (in low-strain zones) exhibit differences which may reflect the original geometry of the Morar Basin. These are:

(i) The sequence comprises mostly sheets of subfacies 1B (trough-parallel sets) with subordinate silty units

(ii) These sands were originally fine to medium grained. The coarse gravelly bands which are present on the western limb, particularly around Ardnamurchan, are absent

(iii) Large-scale cross-sets (facies 2 and 3) and thick units affected by soft-sediment deformation were not observed

(iv) The Upper Psammite is much thinner than on the western limb. Immediately S of Arisaig (Fig. 2.1) the formation is over 5 km thick; at the same latitude on the eastern limb it has thinned to about 1.5 km (Powell, 1966). This may be partly or wholly due to the effects of the adjacent Sgurr Beag Thrust. Certainly, the rapid northward thinning of the Upper Psammite on the eastern limb (Fig. 2.1) is more compatible with the slide cutting down-section than with an original stratigraphic thinning. However, Powell (1974) suggested that both the W–E and S–N thinning may represent original stratigraphic changes accentuated by sliding.

Therefore the proximal–distal transitions observed on the western limb (Figs 2.23 and 2.25) also run from east to west. If the Upper Psammite accumulated in a gulf setting, then the western shoreline was the main sediment source and, if the thickness variations reflect original values, the basin depocentre also lay to the west. A common style of basin which could display these features is a simple half-graben, in this case trending S–N, with an axial flow system and a controlling fault along its western margin (Fig. 2.26). This interpret-

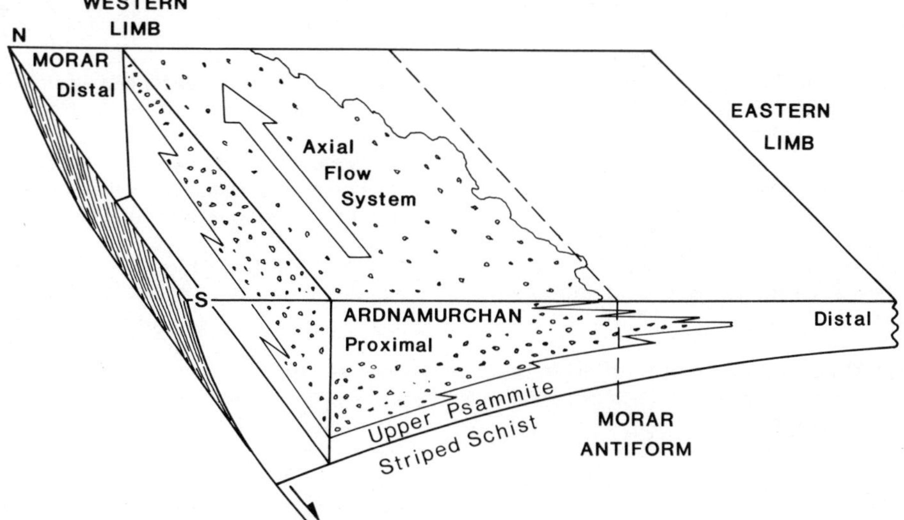

Figure 2.26 Half-graben basin model for the Upper Psammite; based on facies relationships, grain size and possible thickness changes, and the concentration of soft-sediment deformation along the western limb, adjacent to the proposed marginal fault.

ation may also explain the distribution of soft-sediment deformation structures. If a seismic origin is accepted, then their presence along the western margin and absence on the eastern margin relates to the proximity of the original basin margin fault (Leeder, 1987). In this respect, it is worth noting that although intense soft-sediment deformation is uncommon in ancient shallow-marine sands, occurrences are usually linked with seismic activity (Anderton, 1976; Brenchley and Newall, 1977; H. D. Johnson, 1977).

2.7 Conclusions

On the basis of bimodal, bipolar palaeocurrents within subfacies 1B, complex sand waves with silt drapes and reactivation surfaces (facies 2), and winnowed gravel lags and the thin waning-flow sequences (tempestites?) which may be their distal equivalents, the Upper Morar Psammite is considered to have been deposited on a shallow shelf under the influence of strong tidal currents. The role of storms in reinforcing these currents is uncertain.

Lateral facies transitions are interpreted as progressive changes along a net sand transport–deposition path, possibly associated with a large gulf. In such a setting, strongly asymmetrical tides may develop (particularly if the gulf resonates with the controlling tide) and hence the mainly unidirectional S to N palaeocurrents within the Upper Psammite. Taking vertical transitions into account, the sequence is, overall, regressive, as proximal facies overlie more distal facies.

On the basis of contrasts between the Upper Psammite outcrops along the two limbs of the Morar Antiform, it is proposed that the formation was deposited in a simple N–S-trending half-graben in which the depocentre and faulted margin lay to the west.

Finally, some of the constraints imposed upon this interpretation, both by the nature of the sediments themselves and the structural setting, are emphasized. Sedimentologically, the interpretation is based on the exception rather than the rule. 'Diagnostic' structures which indicate a shallow-marine setting are not common; a typical Upper Psammite outcrop consists of sheets of unremarkable, small- to medium-scale cross-bedded sands (facies 1) with unidirectional S to N palaeocurrents and a few intervening silty beds. This could equally be interpreted as a fluvial or a shallow-marine deposit. Correlations, both along the western limb of the Morar Antiform between Ardnamurchan and Morar, and between the two limbs, assume lateral equivalence. This may not be so, but owing to a lack of marker horizons within the formation, evidence for displacements is not forthcoming, therefore the effects of the Gortenfern Disturbance and the Knoydart Slide cannot be assessed.

Acknowledgements

I would like to thank Dr Derek Powell and Prof. Alec Smith for their advice and encouragement during the course of this study. Thanks are also extended to Craig Hildrew and Kevin D'Souza for help with drafting and photography respectively.

This research was supported by a Natural Environment Research Council post-graduate studentship at Royal Holloway and Bedford New College (formerly Bedford College), University of London, UK.

References

Adams, A. E. and Cossey, P. J. (1978) Geological history and significance of a laminated and slumped unit in the Carboniferous Limestone of the Monsal Dale region, Derbyshire. *Geol. J.* **13**, 47–60.

Aigner, T. (1982) Calcareous tempestites: storm dominated stratification in Upper Muschelkalk limestones. In Einsele, G. and Seilacher, A. (eds.), *Cyclic and Event Stratification.* Springer-Verlag, New York, 180–198.

Aigner, T. (1985) Storm depositional systems. In Friedman,

G. M., Horst, J. N. and Seilacher, A. (eds.), *Lecture Notes in Earth Sciences* Vol. 13, Springer-Verlag, New York, 174 pp.

Allen, J. R. L. (1970) *Physical Processes of Sedimentation.* George Allen and Unwin, London, 248 pp.

Allen, J. R. L. (1972) Intensity of deposition from avalanches and the loose packing of avalanche deposits. *Sedimentology* **18**, 105–111.

Allen, J. R. L. (1980) Sand waves; a model of origin and internal structure. *Sedim. Geol.* **26**, 281–328.

Allen, J. R. L. (1981*a*) Lower Cretaceous tides revealed by cross-bedding with mud drapes. *Nature* **289**, 579–581.

Allen, J. R. L. (1981*b*) Palaeotidal speeds and ranges estimated from cross-bedding sets with mud drapes. *Nature* **293**, 394–396.

Allen, J. R. L. (1982) Mud drapes in sand wave deposits: a physical model with application to the Folkestone Beds. *Phil. Trans. R. Soc. London* **A306**, 291–345.

Allen, J. R. L. (1984) Sedimentary structures: their character and physical basis. In *Developments in Sedimentology* **30** (2nd Edition) Elsevier, Amsterdam, 1256 pp.

Allen, J. R. L. and Collinson, J. D. (1974) The superimposition and classification of dunes formed by unidirectional aqueous flows. *Sedim. Geol.* **12**, 169–178.

Allen, J. R. L. and Narayan, J. (1964) Cross-stratified units, some with silt bands, in the Folkestone Beds (Lower Greensand) of southeast England. *Geol. Mijnbouw* **43**, 451–461.

Amos, C. L. (1978) The post-glacial evolution of the Minas Basin, N. S. A sedimentological interpretation. *J. Sed. Pet.* **48**, 965–982.

Anderton, R. (1976) Tidal shelf sedimentation: an example from the Scottish Dalradian. *Sedimentology* **23**, 429–458.

Bailey, E. B., Richey, J. E., Eyles, V. A. and Simpson, J. B. (1922) West Highland District. *Summ. Progr. Geol. Surv. G.B. for 1921*, 89–98.

Banks, N. L. (1973*a*) The origin and significance of some downcurrent-dipping cross-stratified sets. *J. Sed. Pet.* **43**, 423–427.

Banks, N. L. (1973*b*) Tide dominated offshore sedimentation, Lower Cambrian, Northern Norway. *Sedimentology* **20**, 213–228.

Barnes, J. J. and Klein, G. de V. (1975) Tidal deposits in the Zabriskie Quartzite (Cambrian) of eastern California and western Nevada. In Ginsburg, R. N. (ed.), *Tidal Deposits: a Casebook of Recent Examples and Fossil Counterparts.* Springer-Verlag, New York.

Beaumont, E. A. (1984) Retrogradational shelf sedimentation: Lower Cretaceous Viking Formation, Central Alberta. In Tillman, R. W. and Siemers, C. T. (eds.), Siliclastic Shelf Sediments 163–179. *Soc. Econ. Palaeont. Miner. Spec. Publ.* **34**, Tulsa.

Belderson, R. H. and Stride, A. H. (1966) Tidal current fashioning of a basal bed. *Mar. Geol.* **4**, 237–257.

Brenchley, P. J. and Newall, G. (1977) The significance of contorted bedding in Upper Ordovician sediments of the Oslo region, Norway. *J. Sed. Pet.* **47**, 819–833.

Brenchley, P. J., Newall, G. and Stanistreet, I. G. (1979) A storm surge origin for sandstone beds in an epicontinental platform sequence. Ordovician, Norway. *Sedim. Geol.* **22**, 185–217.

Bridges, P. H. (1982) Ancient offshore tidal deposits. In Stride, A. H. (ed.), *Offshore Tidal Sands: Processes and Deposits*, Chapman and Hall, London, 172–192.

Clifton, H. E. (1969) Beach lamination: nature and origin. *Mar. Geol.* **7**, 553–559.

Clifton, H. E., Hunter, R. E. and Philips, P. L. (1971) Depositional structures and processes in the non-barred high energy nearshore. *J. Sed. Pet.* **41**, 651–670.

Collinson, J. D. (1970) Bedforms of the Tana River, Norway. *Geogr. Ann.* **52A**, 31–56.

Collinson, J. D. and Thompson, D. B. (1982) *Sedimentary Structures.* George Allen and Unwin, London, 194 pp.

Cotter, E. (1978) The evolution of fluvial style, with special reference to the central Appalachian Palaeozoic. In Miall, A. D. (ed.), Fluvial Sedimentology. *Mem. Can. Soc. Pet. Geol.* **5**, Calgary, 361–383.

Cotter, E. (1985) Gravel-topped offshore bar sequences in the Lower Carboniferous of Southern Ireland. *Sedimentology* **32**, 195–213.

Dalrymple, R. W., Knight, R. J. and Lambiase, J. J. (1978) Bedforms and their hydraulic stability relationships in a tidal environment, Bay of Fundy, Canada. *Nature* **275**, 100–104.

Goldring, R. and Bridges, P. (1973) Sublittoral sheet sandstones. *J. Sed. Pet.* **43**, 736–747.

Hamblin, W. K. (1961) Micro-cross-lamination in Upper Keweenawan sediments of Northern Michigan. *J. Sed. Pet.* **31**, 390–401.

Haworth, M. J. (1982) Tidal currents of the continental shelf. In Stride, A. H. (ed.), *Offshore Tidal Sands: Processes and Deposits.* Chapman and Hall, London, 10–26.

Holdsworth, R. E., Harris, A. L. and Roberts, A. M. (1987) The stratigraphy, structure and regional significance of the Moine rocks of Mull, Argyllshire, W. Scotland. *Geol. J.* **22**, 83–107.

Johnson, E. W. (1981) A prograding Namurian delta. *Geol. J.* **16**, 93–110.

Johnson, H. D. (1977) Sedimentation and water-escape structures in some Late Precambrian shallow marine sandstones from Finnmark, North Norway. *Sedimentology* **24**, 389–411.

Johnson, M. A., Kenyon, N. H., Belderson, R. H. and Stride, A. H. (1982) Sand transport In Stride, A. H. (ed.), *Offshore Tidal Sands: Processes and Deposits.* Chapman and Hall, London, 58–94.

Johnson, M R. W. (1983) Torridonian–Moine. In Craig, G. Y. (ed.), *Geology of Scotland.* Scottish Academic Press.

Johnstone, G. S., Smith, D. I. and Harris, A. L. Moinian assemblage of Scotland. In Kay, M. (ed.), North Atlantic Geology and Continental Drift: a Symposium. *Mem. Am. Ass. Petrol. Geol.* **12**, 159–180.

Jopling, A. V. (1963) Hydraulic studies on the origin of bedding. *Sedimentology* **2**, 115–12.

Jopling, A. V. (1965) Hydraulic factors and the shape of laminae. *J. Sed. Pet.* **35**, 777–791.

Kelley, S. P. and Powell, D. (1985) Relationships between marginal thrusting and movement on major internal shear zones in the Northern Highland Caledonides, Scotland. *J. Struct. Geol.* **7**, 161–174.

Knight, R. J. and Dalrymple, R. W. (1975) Intertidal sediments from the south shore of Cobequid Bay, Bay of Fundy, Nova Scotia, Canada. In Ginsburg, R. N. (ed.), *Tidal Deposits.* Springer-Verlag, New York, 47–55.

Kumar, N. and Sanders, J. E. (1976) Characteristics of storm deposits; modern and ancient examples. *J. Sed. Pet.* **46**, 145–162.

Lambert, R. StJ. and Poole, A. B. (1964) The relationship of the Moine Schists and Lewisian Gneisses near Mallaigmore, Inverness-shire. *Proc. Geol. Ass.* **75**, 1–14.

Leeder, M. R. (1987) (in press) Sediment deformation structures and the palaeotectonic analysis of ancient sedimentary basins. In Jones, M. E. (ed.) Deformation of Sediments and Sedimentary Rocks. *Spec. Pub. geol. Soc. London.*

Levell, B. K. (1980*a*) Evidence for currents with waves in a late Precambrian shelf deposit from Finnmark, North Norway. *Sedimentology* **27**, 153–166.

Levell, B. K. (1980*b*) A late Precambrian tidal shelf deposit, the lower Sandfjord formation, Finnmark, North Norway. *Sedimentology* **27**, 539–557.

Long, D. G. F. (1978) Proterozoic stream deposits; some problems of recognition and interpretation of ancient sandy fluvial systems. In Miall, A. D. (ed.), Fluvial Sedimentology *Mem. Can. Soc. Pet. Geol.* **5**, Calgary, 313–341.

Lowe, D. R. (1975) Water escape structures in coarse-grained sediments. *Sedimentology* **22**, 157–204.

McKee, E. D., Crosby, E. J. and Berryhill, J. R. (1967) Flood deposits, Bijou Creek, June 1965. *J. Sed. Pet.* **37**, 829–851.

Mayall, M. J. (1983) An earthquake origin for synsedimentary deformation in a late Triassic (Rhaetian) lagoonal sequence, Southwest Britain. *Geol. Mag.* **6**, 613–622.

Michelson, P. C. and Dott, R. H. Jr. (1973), Orientation analysis of trough cross stratification in Upper Cambrian sandstones of Western Wisconsin. *J. Sedim. Pet.* **43**, 784–794.

Nelson, C. H. (1982) Modern shallow-water graded sand layers from storm surges, Bering Shelf: a mimic of Bouma sequences and turbidite systems. *J. Sed. Pet.* **52**, 537–545.

O'Brien, B. H. (1985) The geometry of ductile conjugate fold systems in the Ardnamurchan Moine, Scotland. *Geol. J.* **20**, 91–108.

Poole, A. B. and Spring, J. S. (1974) Major structures in Morar and Knoydart, N. W. Scotland. *J. geol. Soc. London* **130**, 43–53.

Powell, D. (1966) The structure of the south-eastern part of the Morar Antiform. Inverness-shrine. *Proc. Geol. Ass.* **77**, 79–100.

Powell, D. (1974) Stratigraphy and structure of the western Moine and the problem of Moine orogenesis. *J. geol. Soc. London* **130**, 575–593.

Ramsay, J. G. and Spring, J. (1962) Moine stratigraphy in the Western Highlands of Scotland. *Proc. Geol. Ass.* **73**, 295–322.

Reineck, H. E. (1975) German North Sea Tidal Flats. In Ginsburg, R. N. (ed.), *Tidal Deposits*. Springer-Verlag, New York, 5–12.

Reineck, H. E. and Singh, I. B. (1980) *Depositional Sedimentary Environments with Reference to Terrigenous Clastics*. (2nd edition) Springer-Verlag, New York.

Richey, J. E. and Kennedy, W. Q. (1939) The Moine and sub-Moine series at Morar, Inverness-shire. *Bull. Geol. Surv. Gt. Br.* **2**, 26–45.

Roberts, A. M., Strachan, R. A., Harris, A. L., Barr, D. and Holdsworth, R. E. (1987). The Sgurr Beag Nappe; a reassessment of the stratigraphy and structure of the Northern Highland Moine. *Bull. geol. Soc. Am.* **98**, 497–506.

Schumm, S. A. (1968) Speculations concerning the palaeohydrologic controls of terrestrial sedimentation. *Bull. geol. Soc. Am.* **79**, 1573–1588.

Simons, D. B. and Richardson, E. V. (1961) Forms of bed roughness in alluvial channels. *Am. Soc. Civil Eng. Proc.* **87**, No. HY3. Paper **2816**.

Simonson, B. M. (1984) A high energy shelf deposit: early Proterozoic Wishart Formation, Northeastern Canada. In Tillman, R. W. and Siemers, C. T. (eds.), Siliciclastic Shelf Sediments. *Soc. Econ. Palaeont. Miner. Spec. Publ.* **34**, 251–268.

Sims, J. D. (1975) Determining earthquake recurrence intervals from deformational structures in young lacustrine sediments. *Tectonophysics* **29**, 141–152.

Soper, N. J. (1960) A note on current-bedded Moines near Garve, Wester Ross. *Geol. Mag.* **97**, 505–508.

Southard, J. B. (1975) Bed configurations. In Harms, J. C., Southard, J. B., Spearing, D. R. and Walker, R. G., Depositional Environments as Interpreted from Primary Sedimentary Structures and Stratification Sequences. *Soc. Econ. Palaeont. Miner. Short. Course* **2**, Dallas, 5–43.

Southard, J. B. and Boguchwal, L. A. (1973) Flume experiments on the transition from ripples to lower flat bed with increasing sand size. *J. Sed. Pet.* **43**, 1114–1121.

Spearing, D. R. (1975) Shallow marine sands. In Harms, J. C., Southard, J. B., Spearing, D. R. and Walker, R. G., *Depositional environments as interpreted from primary sedimentary structures and stratification sequences. Soc. Econ. Palaeont. Miner. Short Course* **2**, Dallas, 103–132.

Stanley, D. J. (1968) Graded bedding–sole markings–greywacke assemblage and related sedimentary structures in some Carboniferous flood deposits, Eastern Massachusetts. *Spec. Pap. geol. Soc. Am.* **106**, 211–239.

Steele, R. and Aasheim, S. M. (1978) Alluvial sand deposition in a rapidly subsiding basin (Devonian, Norway). In Miall, A. D. (ed.), Fluvial Sedimentology. *Mem. Can. Soc. Pet. Geol.* **5**, Calgary, 385–412.

Strachan, R. A. (1986) Shallow marine sedimentation in the Proterozoic Moine succession, Northern Scotland. *Precambr. Res.* **32**, 17–33.

Stride, A. H. (1963) Current-swept sea floors near the southern half of Great Britain. *Q. J. geol. Soc. London* **119**, 175–199.

Tanner, P. W. G. (1971) The Sgurr Beag Slide; a major tectonic break within the Moinian of the Western Highlands of Scotland. *J. geol. Soc. London* **126**, 435–463.

Tanner, P. W. G. (1976) Progressive regional metamorphism of thin calcareous bands from the Moinian rocks of N.W. Scotland. *J. Pet.* **17**, 100–134.

Terwindt, J. H. J. (1971) Litho-facies of inshore estuarine and tidal inlet deposits. *Geol. Mijnbouw* **3**, 515–526.

Terwindt, J. H. J. (1975) Sequences in inshore subtidal deposits. In Ginsburg, R. N. (ed.), *Tidal Deposits*. Springer-Verlag, New York, 85–89.

Terwindt, J. H. J. (1981) Origin and sequences of sedimentary structures in inshore mesotidal deposits. In Nio, S.-D., Shuttenhelm, R. T. E. and van Weering, Tj. C. E. (eds.), Holocene Marine Sedimentation in the North Sea Basin. *Spec. Publ. Intern. Ass. Sedimentologists* **5**, Blackwell Scientific, Oxford, 4–27.

Thompson, R. W. (1968) Tidal flat sedimentation on the Colorado River Delta, Northwestern Gulf of California. *Geol. Soc. Am. Mem.* **107**.

Thompson, R. W. (1975) Tidal-flat sediments of the Colorado River Delta, Northwestern Gulf of California. In Ginsburg, R. N. (ed.), *Tidal Deposits*. Springer-Verlag, New York, 57–65.

Tucker, M. (1982) Storm surge sandstones and the deposition of interbedded limestone; Late Precambrian, Southern Norway. In Einsele, G. and Seilacher, A. (eds.), Cyclic Event Stratification. Springer-Verlag, New York, 363–370.

Tunbridge, I. P. (1984) Facies models for a sandy ephemeral stream and clay playa complex; the Middle Devonian Trentishoe Formation of North Devon, U.K. *Sedimentology* **31**, 697–715.

Visser, M. J. (1980) Neap–spring cycles reflected in Holocene subtidal large-scale bedform deposits: a preliminary note. *Geology* **8**, 543–546.

Wilson, G. W., Watson, J. and Sutton, J. (1953) Current-bedding in the Moine Series of Northwestern Scotland. *Geol. Mag.* **90**, 377–387.

Wright, M. E. and Walker, R. G. (1981) Cardium Formation (Upper Cretaceous) at Seebe, Alberta; storm transported sandstones and conglomerates in shallow marine depositional environments below fair-weather base. *Can. J. Earth Sci.* **18**, 795–809.

3
The Glenfinnan and Loch Eil Divisions of the Moine Assemblage

R. A. STRACHAN, F. MAY, and D. BARR

3.1 Introduction

The Moine Assemblage is a sequence of Proterozoic metasediments which are thought to have been deformed and metamorphosed at *c.* 1000–750 Ma, and subsequently reworked by Caledonian thrusts and related folds at *c.* 470–420 Ma (Brook *et al.*, 1976, 1977; Brewer *et al.*, 1979; Powell *et al.*, 1981, 1983; Sanders *et al.*, 1984; Barr *et al.*, 1985). They have an unconformable relationship with late Archaean Lewisian basement granulites and gneisses which are incorporated within the Moine Assemblage as infolds and tectonic slices. The Moine Assemblage was emplaced on to the Laurentian foreland across the Caledonian Moine Thrust belt (Fig. 3.1).

Stratigraphic analysis of the Moine Assemblage has been hampered by the monotonous nature of its constituent lithologies and its intense deformation. The Moine Assemblage consists largely of psammites, with subordinate semipelitic and pelitic schists and gneisses, and there are few distinctive horizons which can be mapped easily and correlated from one area to another. This problem is exacerbated by numerous lateral facies variations. Original lithological differences and sedimentary structures have often been obliterated by intense ductile deformation and amphibolite-facies metamorphism. The existing tectonostratigraphic framework for the Moine rocks of western Inverness-shire was erected by Johnstone *et al.* (1969) who subdivided it into the Morar, Glenfinnan and Loch Eil Divisions (Fig. 3.1). Although recognizing that the boundaries of these units might be marked by major tectonic discontinuities, Johnstone *et al.* (1969) considered that in broad terms the sequence Morar–Glenfinnan–Loch Eil probably represented a simple stratigraphic succession, with the oldest rocks in the W and the youngest in the E. More recent work has demonstrated that the Morar and Glenfinnan Divisions are separated by a major Caledonian ductile thrust, the Sgurr Beag Thrust (Tanner *et al.*, 1970; Tanner, 1971; Rathbone and Harris, 1979; Rathbone *et al.*, 1983). Displacement across this thrust is likely to be comparable to that on the Moine Thrust (> 50 km). For this reason the nature of the original relationship between the Morar and Glenfinnan Divisions is unclear. In contrast, the Glenfinnan and Loch Eil Divisions are linked by a sedimentary transition zone. The lithostratigraphy and sedimentology of parts of the Morar Division have been described by Glendinning (this volume). The purpose of this contribution is to describe the lithostratigraphy and sedimentology of the Glenfinnan and Loch Eil Divisions. It is first appropriate to review briefly the geological setting of these rocks and to discuss the evidence relating to their age.

3.2 Geological setting

The Glenfinnan and Loch Eil Divisions of Johnstone *et al.* (1969) collectively form the Sgurr Beag Nappe of Rathbone *et al.* (1983) and Barr *et al.* (1986). The major Lewisian inliers of Scardroy and Monar (Fig. 3.1) rest on the Sgurr Beag Thrust and probably represent slices of the basement on which the Glenfinnan Division was deposited. The Glenfinnan Division comprises a sequence of alternating pelitic and psammitic gneisses, striped schists and quartzites. These largely crop out within the Northern Highland 'steep belt' (Leedal, 1952), where lithologies are commonly folded into a steeply inclined or subvertical attitude (Fig. 3.2). Intense deformation and regional migmatization has largely obliterated sedimentary structures in all but a few areas. The Glenfinnan Division is overlain by the psammites of the Loch Eil Division. The level of tectonic strain within the Loch Eil Division is commonly low, with the result that sedimentary structures are locally abundant. The Loch Eil Division lithologies are mainly flat-lying or gently inclined (Fig. 3.2) and form the Northern Highland 'flat belt' (Leedal, 1952). The boundary between steeply inclined and flat-lying rocks is known as the Loch Quoich Line (Clifford, 1957). It is important to emphasize that the Loch Quoich Line does not coincide with the stratigraphic boundary between the Glenfinnan and Loch Eil Divisions. Brown *et al.* (1970), Roberts and Harris (1983) and Roberts *et al.* (1984) have demonstrated that rocks which can be assigned to the Loch Eil Division crop out W of the Loch Quoich Line as synformal outliers within the steep belt (Fig. 3.2). The Loch Quoich Line is thus a purely tectonic feature, of no major stratigraphic importance.

Holdsworth and Roberts (1984) have suggested that the tectonic history of the Glenfinnan and Loch Eil Divisions can be regarded as having occurred in three distinct episodes. The first of these is expressed as rarely preserved primary folds and fabrics, which appear not to be associated with large-scale structures, but which accompanied the mid- to upper amphibolite-facies peak of regional metamorphism. The second episode involved large- and small-scale recumbent folding and the development of a N–S stretching lineation. The third episode produced upright folds which have NNE–SSW-trending axial planes and are associated

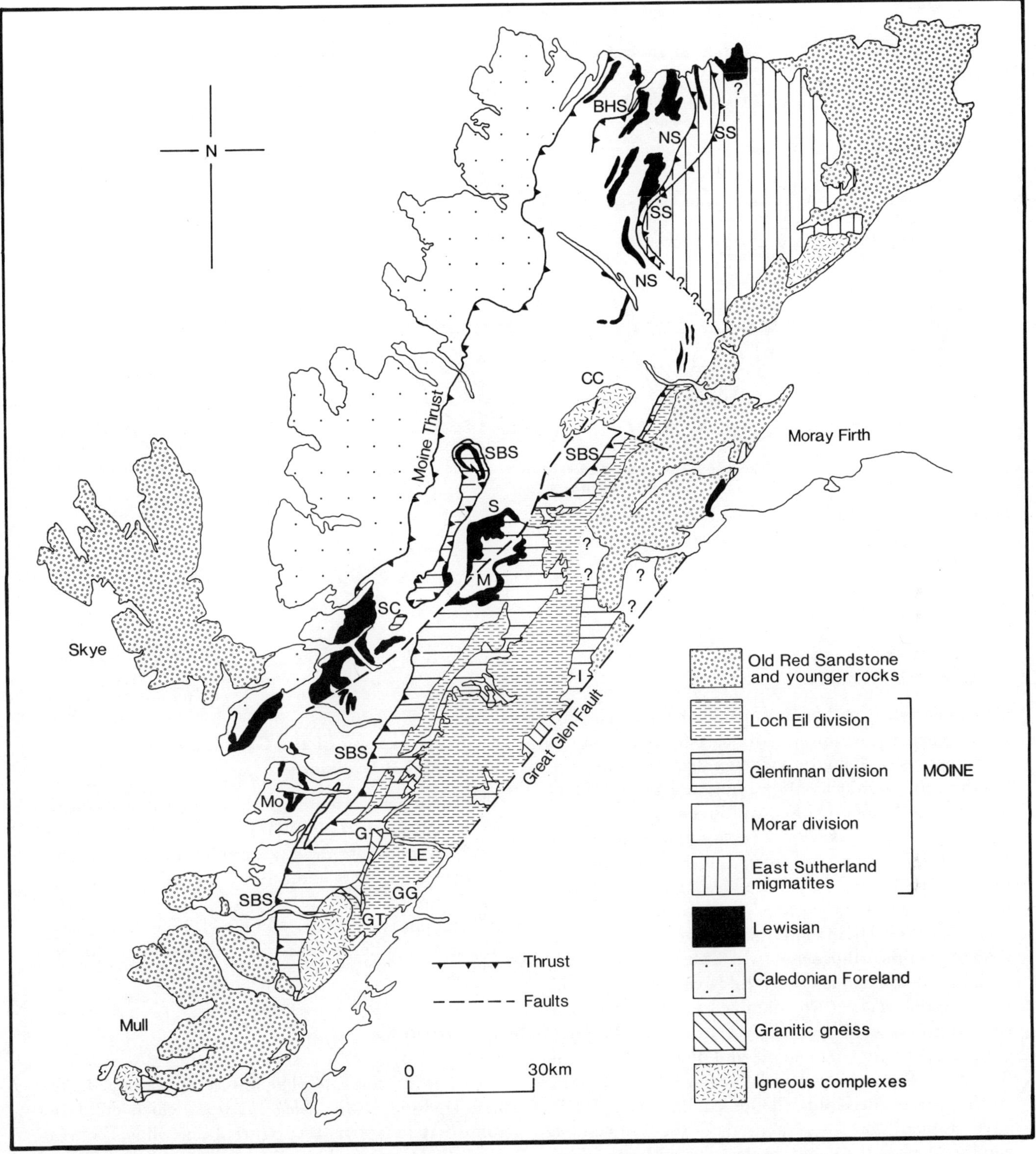

Figure 3.1 Geological map of the Northern Highlands of Scotland to show the context of the Moine divisions. *BHS*—Ben Hope Slide; *CC*—Carn Chuinneag; *G*—Glenfinnan; *GG*—Glen Gour; *GT*—Glen Tarbert; *I*—Invermoriston; *LE*—Loch Eil; *M*—Monar; *Mo*—Morar; *NS*—Naver Slide; *S*—Scardroy; *SS*—Swordly Slide; *SBS*—Sgurr Beag Slide.

with a subvertical stretching lineation. These are poorly developed within the flat belt, but dominate the structure of the steep belt, and it is the boundary between these two domains which forms the Loch Quoich Line (Fig. 3.2). The first two episodes are thought to be Precambrian (Holdsworth and Roberts, 1984), whereas the third episode is Caledonian (Roberts *et al.*, 1984). In certain parts of the Sgurr Beag Nappe the structural history may be either more complex (e.g. Loch Eil, Strachan, 1985) or simpler (e.g. Invermoriston, this chapter) than the sequence described above.

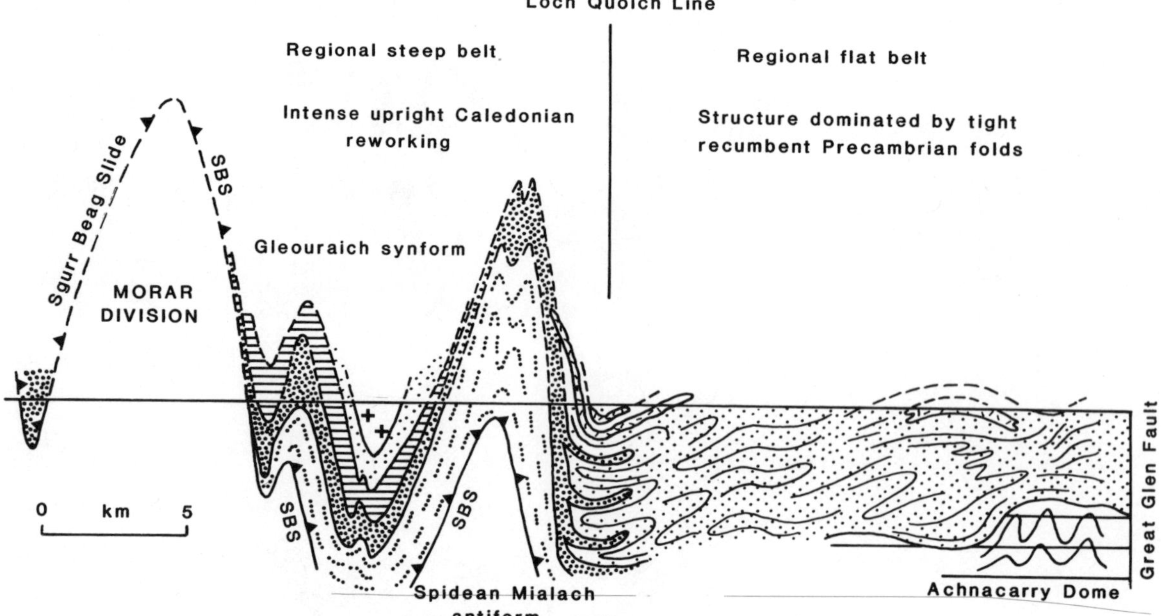

Figure 3.2 Schematic west–east cross-section through the Glenfinnan and Loch Eil Divisions (modified from Roberts *et al.*, 1987).

3.3 Age of the Glenfinnan and Loch Eil Divisions

Caledonian deformation and metamorphism has long been recognized within the Moine Assemblage, but in recent years radiometric evidence has accumulated, pointing to a significant Precambrian tectonothermal history (e.g. Lambert, 1969; van Breemen *et al.*, 1974; Brook *et al.*, 1976, 1977). The absolute age of this Precambrian event rests upon the interpretation of isotopic data obtained from the West Highland Granitic Gneiss and pelitic gneisses of the Glenfinnan Division. The West Highland Granitic Gneiss comprises a suite of granitic orthogneisses which were intruded into both the Glenfinnan and Loch Eil Divisions (Barr *et al.*, 1985). Brook *et al.* (1976) obtained an Rb–Sr whole-rock isochron of 1028 ± 43 Ma from the Ardgour granite gneiss at Glenfinnan, and Piasecki and van Breemen (1979) reported an Rb–Sr age of *c.* 1000 Ma for the granite gneiss at Loch Quoich. Barr *et al.* (1985) have argued that these ages date the syntectonic emplacement of these granite sheets into Moine rocks, which were undergoing high-grade metamorphism. Some support for this interpretation comes from indications of a *c.* 1000 Ma metamorphic event within pelitic gneisses of the Glenfinnan Division (Brewer *et al.*, 1979; Aftalion and van Breemen, 1980). On the basis of this isotopic data, it has been widely argued (e.g. Harris, 1983; Barr *et al.*, 1985) that the Glenfinnan and Loch Eil Divisions were deposited prior to *c.* 1000 Ma when they were deformed and metamorphosed during an orogenic event which has been loosely correlated within the Grenvillian of Canada. It must be acknowledged, however, that the timing of this Precambrian event is relatively poorly constrained, and further isotopic work is clearly a priority.

3.4 Stratigraphy of the Glenfinnan and Loch Eil Divisions

The summary presented here is based largely on the work of Roberts *et al.* (1987). That part of the Sgurr Beag Nappe where the stratigraphy is best understood lies between Glen Scaddle and Cannich (Fig. 3.3). The stratigraphic units are disposed about three major Caledonian folds, the Loch Quoich Line, the Spidean Mialach Antiform and the Gleouraich Synform (Fig. 3.3).

3.4.1 *Glenfinnan Division*

3.4.1.1 *Reidh Psammite*. The lowest unit of the Glenfinnan Division is the *Reidh Psammite* which crops out immediately above the Sgurr Beag Slide, between Kinloch Hourn and the Strathconon Fault (Tanner, 1971; Fig. 3.3). This consists of coarse-grained, often migmatitic psammite, interbanded with subordinate pelite, semi-pelite and quartzite. Sedimentary structures and calc-silicates are generally absent. Tanner (op. cit.) reports the local presence of heavy-mineral bands. The absence of this unit S of Kinloch Hourn may either reflect lateral facies changes or cutting out by a lateral ramp of the Sgurr Beag Thrust. Roberts *et al.* (1987) have correlated the Reidh Psammite with the lithologically similar Quoich Banded Formation which crops out at the E end of Loch Quoich in the core of the Spidean Mialach Antiform (Fig. 3.3).

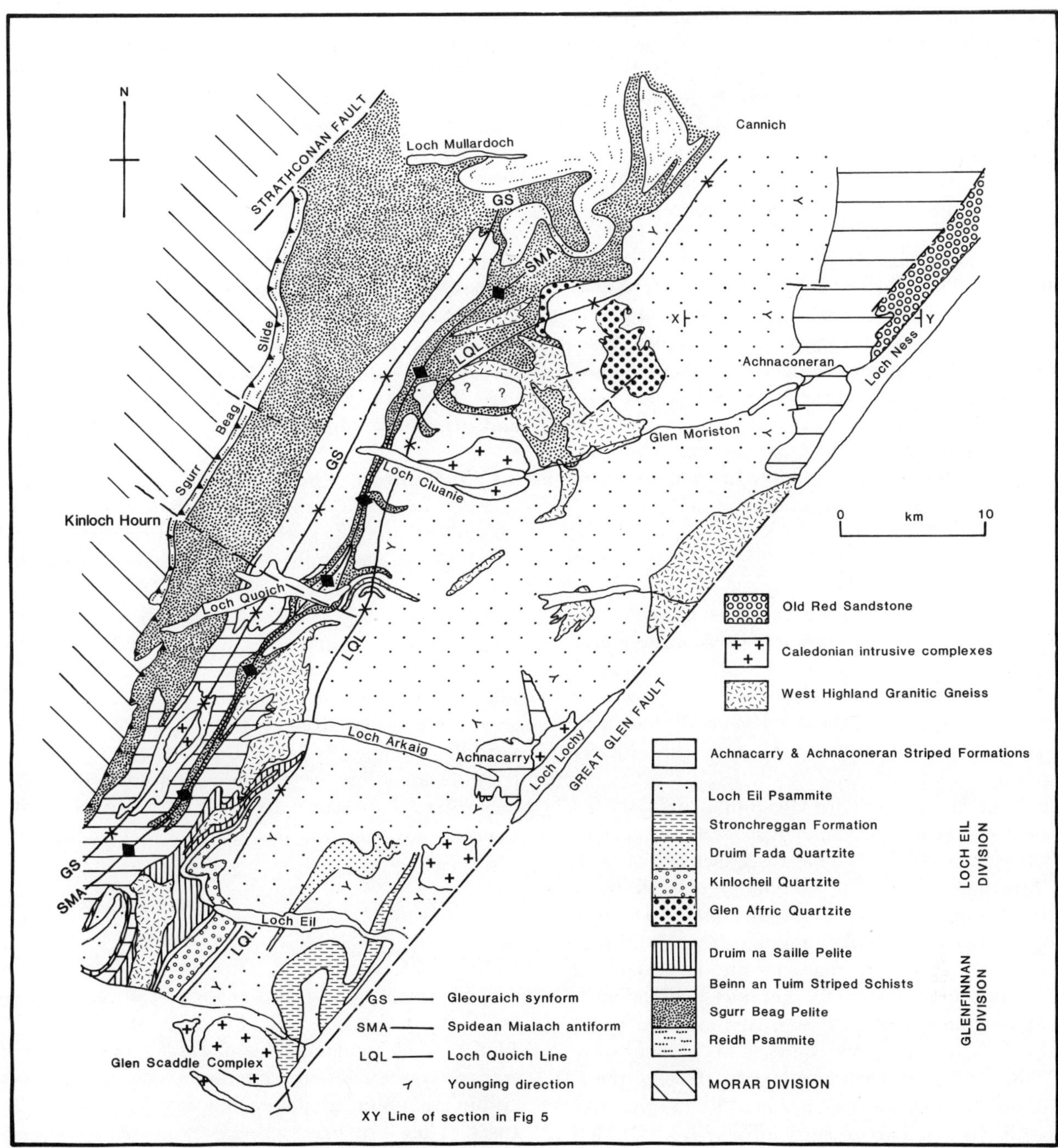

Figure 3.3 Geological map of the Glenfinnan and Loch Eil Division rocks of Inverness-shire (in part from Roberts *et al.*, 1987).

3.4.1.2 *Sgurr Beag Pelite*. Overlying the Reidh Psammite at Kinloch Hourn is the *Sgurr Beag Pelite* (Tanner, 1971). This unit is the most stratigraphically continuous formation within the Sgurr Beag Nappe, and forms a regional marker horizon. To the S it is continuous with the Lochailort Pelite of Powell (1964) (Fig. 3.3). It extends at least as far N as Loch Mullardoch (Fig. 3.3), where it can be demonstrated that it is continuous with the extensive Quoich Pelite which crops out along the eastern limb of the Gleouraich Synform between Glen Dessary and Loch Mullardoch (Fig. 3.3). The Sgurr Beag Pelite is typically represented by a coarse garnetiferous pelitic gneiss with subordinate semi-pelitic and psammitic members, the proportion of which increases northwards (BGS 1:50 000 Scotland Sheet 72 E Glen Affric). Quartzite is locally well developed in the area between Loch Cluanie and Loch Mullardoch. Pods and bands of calc-silicate are common within psammitic bands, and metabasic garnetiferous amphibolites are widespread (Johnstone, 1975).

3.4.1.3 *Beinn an Tuim Striped Schists*. From the Cluanie area southwards the Sgurr Beag Pelite is overlain by a heterogeneous assemblage of psammite, pelite, semi-pelite and quartzite which are commonly interbanded on a metre scale. Sedimentary structures are generally rare. In the Glenfinnan area this formation was called the *Beinn an Tuim Striped Schists* (Dalziel, 1966), and is equivalent to the Strathan Striped Schists and Quartzites in Glen Dessary (Roberts *et al.*, 1984) and the Garry Banded Unit at Loch Quoich (Roberts and Harris, 1983). This formation is thought to be largely absent N of Loch Cluanie, only poorly developed in the Cluanie area itself, but thickens considerably southwards, so that in the Glenfinnan area it is the predominant formation within the steep belt (Fig. 3.3).

3.4.1.4 *Druim na Saille Pelite.* Stratigraphically above the Beinn an Tuim Striped Schists, but present only as far N as Loch Arkaig, lies another formation of pelitic gneiss, the *Druim na Saille Pelite* (Dalziel, 1966). Elsewhere the striped schists are overlain directly by the mainly psammitic rocks of the Loch Eil Division. The Druim na Saille Pelite consist of variably migmatized pelitic gneiss with minor horizons of semi-pelite and psammite. Sedimentary structures are absent; calc-silicate pods are present within psammitic lithologies.

3.4.2 *Loch Eil Division*

The rocks of the Loch Eil Division are less lithologically diverse than those of the Glenfinnan Division, and have not undergone the intense upright folding associated with the development of the steep belt. Hence tectonic strain within the Loch Eil Division is generally low, with the result that sedimentary structures are locally abundant.

The boundary between the Glenfinnan and Loch Eil Divisions is a transitional contact (Roberts and Harris, 1983; Strachan, 1985). In some areas, such as S of Loch Eil (Strachan op.cit), this contact is marked by a relatively rapid passage from Glenfinnan Division pelite into Loch Eil Division psammite over a distance of less than 10 m. More characteristic, however, is a broad transitional zone up to 1 km thick, within which lithologies characteristic of both divisions are interbanded in a stratigraphic sequence. Such zones have been reported by Roberts and Harris, 1983) at the E end of Loch Quoich, and Strachan (1985) to the N of Loch Eil. In some areas, such as S of Loch Arkaig (Fig. 3.3) and at the E end of Glen Affric (BGS 1:50 000 Scotland Sheet 72 E Glen Affric), it can clearly be demonstrated that lithologies assignable to both divisions also pass by lateral and stratigraphic passage into each other.

The base of the Loch Eil Division appears to rest upon different units of the Glenfinnan Division (Fig. 3.4). In the Cluanie area the Loch Eil Division rests on the Sgurr Beag/Quoich Pelite; from the Cluanie area S to Loch Arkaig it rests on the Beinn an Tuim Striped Formation, and S of Loch Arkaig it rests on the Druim na Saille Pelite (Fig. 3.4). The stratigraphically lowest levels of the Loch Eil Division are thus lateral equivalents of the Beinn an Tuim Striped Formation and the Druim na Saille Pelite of the Glenfinnan Division (Fig. 3.4).

3.4.2.1 *Loch Eil Psammite.* Over much of its outcrop the Loch Eil Division comprises a monotonous sequence of psammites, which Johnstone *et al.* (1969) termed the *Loch Eil Psammite*. This unit is laterally equivalent to the Loch Arkaig Psammite of Roberts *et al.* (1984) and the Upper Garry Psammite of Roberts and Harris (1983). The type lithology is a fine- to medium-grained psammite with thin semi-pelitic interbeds. These are locally rhythmically interbebbed on a scale of 1–5 cm. Sedimentary structures are common, and are particularly well developed in the Loch Eil and Glen Moriston areas. Cross-stratification, cross-lamination, small-scale grading, convoluted bedding and rippled surfaces have all been recorded. Calc-silicate rocks are common and represent calcareous concretions formed during diagenesis (Strachan, 1986). Mappable units of semi-pelite and stripped psammite are present locally.

3.4.2.2 *Quartzites.* Strachan (1985) has identified a number of major units of quartzite and siliceous psammite within the Loch Eil Psammite of the Loch Eil area. Three such units have been recognized: the *Kinlocheil* and *Druim Fada Quartzites* and the *Stronchreggan Formation* (Fig. 3.3). These are typically massive, variably feldspathic quartzites with subordinate semi-pelite. Sedimentary structures are common, but calc-silicates are rare. Detailed mapping in the Loch Eil area, and reconnaissance traverses S of the Glen Scaddle Complex, suggest that these units are all laterally discontinuous (Fig. 3.3). Lithologically similar quartzites are present both at the base and within the Loch Eil Division N of Glen Moriston (Fig. 3.3).

3.4.3 *Achnacarry and Achnaconeran Striped Formations*

3.4.3.1 *Achnacarry Striped Formation.* A sequence of striped psammitic and semi-pelitic gneisses crops out at the E end of Loch Arkaig (Fig. 3.3; Johnstone, 1975; BGS 1:50 000 Scotland Sheet 62 E Loch Lochy). These striped rocks contain common calc-silicate bands within psammitic lithologies, but no sedimentary structures. They are here termed the *Achnacarry Striped Formation*, and occupy the core of an upright domal structure (Roberts, 1984). The unit is demonstrably older than the adjacent Loch Eil Psammite which youngs away from their contact to the W, N and S. The boundary between the two units is marked by a broad transition from striped gneisses to psammites over a distance of *c.* 150 m.

3.4.3.2 *Achnaconeran Striped Formation.* Mapping by F. May has demonstrated that lithologically similar rocks are exposed adjacent to the Great Glen between Fort Augustus and Glen Urquhart (Fig. 3.3), where they are known as the *Achnaconeran Striped Formation.* This comprises a sequence of rhythmically interbanded (1–5 cm) psammites and semi-pelites in very variable proportions. Calc-silicate rocks are rare. The unit is divisible into a number of subfacies, based on the relative proportions of psammite and semi-pelite, which grade

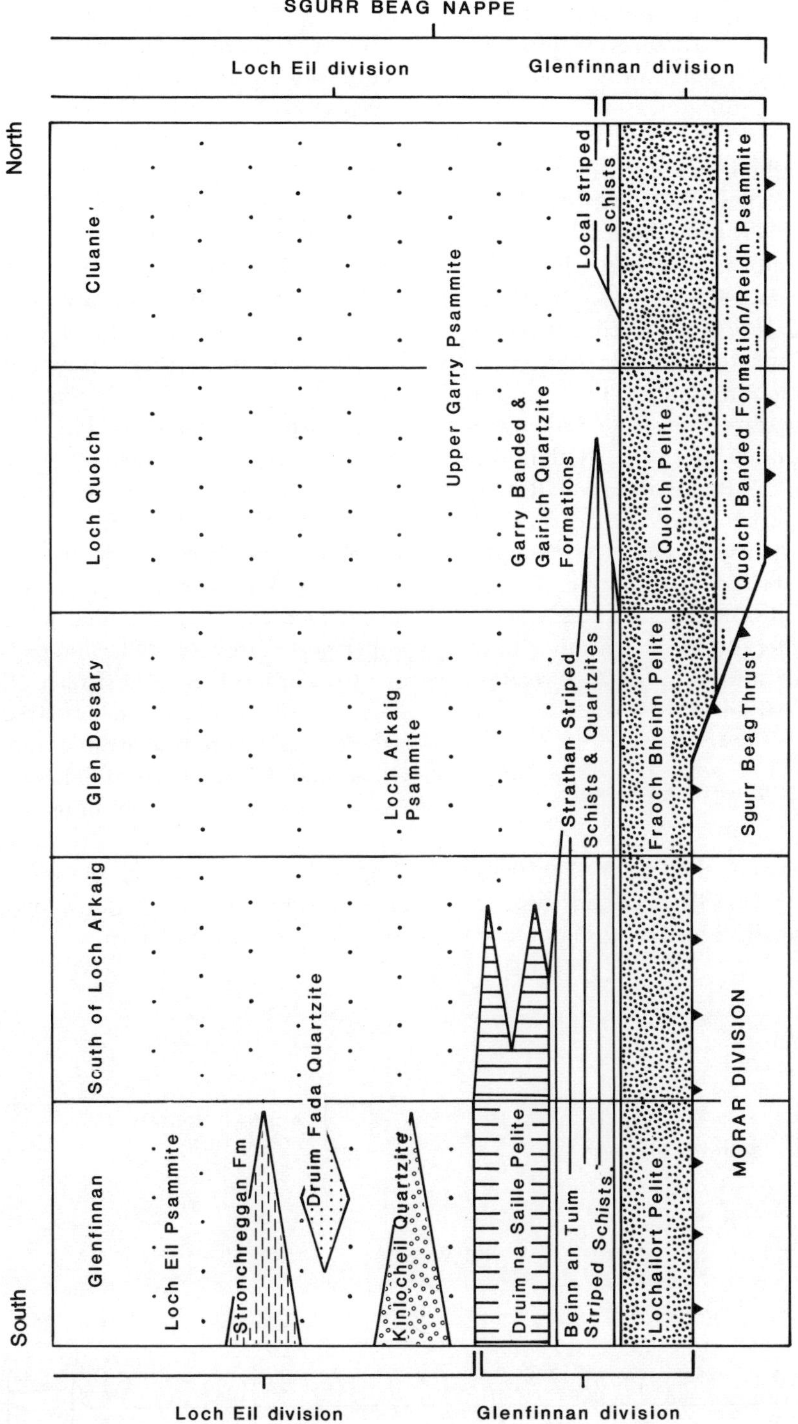

Figure 3.4 Schematic representation of the lateral and vertical changes in lithostratigraphy which occur between Cluanie and Glenfinnan (modified from Roberts *et al.*, 1987).

imperceptibly into each other. The psammitic bands in the thinly striped parts of the sequence locally show cross-lamination and asymmetrical ripples on their upper surfaces. The exposed thickness of the unit is at least 3 km. The junction between the Achnaconeran Striped Formation and the Loch Eil Psammite is transitional over a thickness of 1.5 km, by the large- and small-scale interbanding of striped facies and psammitic facies rocks. Sedimentary structures within the transition zone consistently young away from the Achnaconeran Striped Formation towards the overlying Loch Eil Psammite.

The Achnacarry and Achnaconeran Striped Formations are both clearly older than undoubted Loch Eil Division rocks, and resemble parts of the Glenfinnan Division such as the Beinn an Tuim Striped Schists. On this basis it is reasonable to assign these units to the Glenfinnan Division (Fig. 3.1). Cross-sections drawn across the Glen Moriston area demonstrate, if taken at face value, an extremely rapid thinning of the Loch Eil Psammite in an easterly direction (Fig. 3.5). One possible explanation of this arrangement is that the lower parts of the Loch Eil Psammite are laterally equivalent to the upper parts of the Achnacarry and Achnaconeran Striped Formations. This is represented semi-diagrammatically in the cross-section (Fig. 3.5). It is thus likely that rocks of the Glenfinnan and Loch Eil Divisions not only pass laterally into each other in a NNE–SSW direction, but also in a E–W direction perpendicular to regional strike.

3.4.4 *The northern and southern parts of the Sgurr Beag Nappe*

Examination of published maps (BGS 1:50 000 Scotland Sheet 72 W Glen Affric; 1:63 360 Scotland Sheet 82 Lochcarron) indicates that the Sgurr Beag Pelite may continue for at least 60 km further N into the area to the SE of the Carn Chuinneag granite (Fig. 3.1). Tobisch *et al.* (1970) demonstrate that in the Cannich area the Sgurr Beag Pelite is complexly interfolded with stratigraphically older psammites, possibly equivalent to the Reidh Psammite (Roberts *et al.*, 1987), and younger psammites likely to correlate with the Loch Eil Psammite (Fig. 3.3). The regional structure of this area is, however, extremely complex and incorporates a number of tectonic breaks; further structural analysis is thus necessary before a stratigraphy may be erected. The northern extent of the Loch Eil Division and the Achnaconeran Striped Formation similarly await further mapping. Brown *et al.* (1970) have shown that the Sgurr Beag–Lochailort Pelite, the Beinn an Tuim Striped Schists and the Druim na Saille Pelite are all traceable to the SW of the Glenfinnan area (Fig. 3.1). These units are interfolded with psammites of the Loch Eil Division, and a major psammite termed the Ben Gaire Psammite, which Brown *et al.* (1970) assigned to the Glenfinnan Division. The structure of this area is not, however, fully understood in relation to the area N of Glenfinnan, and further mapping may be necessary before definitive stratigraphic correlations can be made. The exact location of the Glenfinnan–Loch Eil Division boundary S of Glenfinnan is also unclear. This is normally drawn as extending along the eastern and uppermost margin of the Druim na Saille Pelite as far S as Glen Tarbert (Fig. 3.1). Stoker (1983) has, however, shown that the Glen Tarbert area contains a thick sequence of migmatitic striped and pelitic gneisses of Glenfinnan Division aspect. These are in contact with coarse quartzite and psammite of Loch Eil Division aspect in the area between Glen Tarbert and Glen Gour (Fig. 3.1). The boundary between the two divisions may therefore lie some distance to the E of where it has been conventionally drawn.

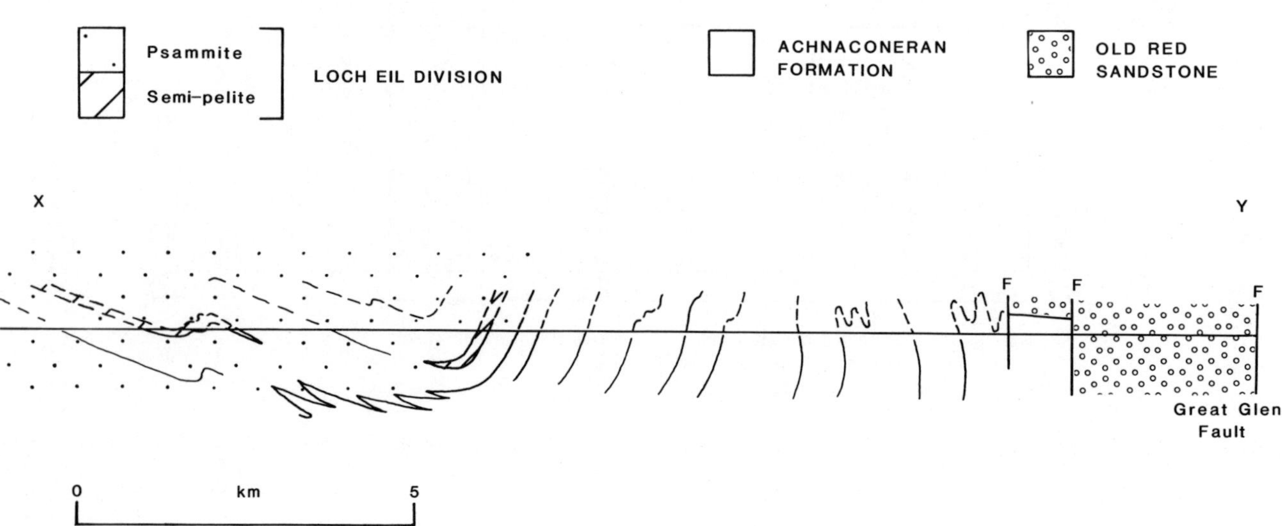

Figure 3.5 Schematic west–east cross-section across the Invermoriston area (see Fig. 3.3 for location).

3.4.5 *Stratigraphic nomenclature*

The existing tectonostratigraphic framework for the Moine Assemblage (Johnstone *et al.*, 1969) was erected at a time when the nature of the relationship between the Glenfinnan and Loch Eil Divisions was unclear. It has been shown that the boundary between the Glenfinnan and Loch Eil Divisions is a stratigraphic transition and may be diachronous. In view of this, there now seems no need to subdivide the rocks above the Sgurr Beag Thrust into two separate tectonostratigraphic units. These rocks may now be assigned to one unit, the Sgurr Beag Nappe (Rathbone *et al.*, 1983). Given that it is now possible to recognize a coherent stratigraphy for the Glenfinnan and Loch Eil Divisions, Roberts *et al.* (1987) have proposed that these two units now be known formally as the *Glenfinnan Group* and the *Loch Eil Group*.

3.5 Regional correlations

3.5.1 *Correlations between the rocks of the Sgurr Beag Nappe and the Morar Division*

The nature of the relationship between the Morar and Glenfinnan Divisions was for many years unclear. Although Tanner (1971) and Tanner *et al.* (1970) demonstrated a tectonic break along much of the length of the Morar/Glenfinnan boundary in Ross-shire and North Inverness-shire, Brown *et al.* (1970) and Powell (1974) argued that in W Inverness-shire the Upper Morar Psammite of the Morar Division passes stratigraphically upwards into the Lochailort Pelite of the Glenfinnan Division. This led Johnstone (1975) to consider that in broad terms the Morar Division was the oldest of the three divisions. Subsequent workers (Rathbone and Harris, 1979; Powell *et al.*, 1981; Barr *et al.*, 1986) have, however, shown that the Sgurr Beag Thrust follows the contact between the Morar and Glenfinnan Divisions throughout W Inverness-shire, and it is thus concluded that the boundary between the Morar and Glenfinnan Divisions is entirely tectonic on the Scottish mainland.

The Morar Division appears to be of broadly the same age as the Glenfinnan Division, although more radiometric dates are required to clarify this. Noting that both the Morar and Glenfinnan Divisions rest unconformably on Lewisian basement, several authors (e.g. Tanner *et al.*, 1970; Harris, 1983; Barr *et al.*, 1986) have suggested that the two divisions are laterally equivalent. Tanner *et al.* (1970) and Rathbone *et al.* (1983) suggested that the Moine basin was partially segmented by a basement high which separated the Morar and Glenfinnan Divisions and is the source of the basement slices carried on the Sgurr Beag Thrust. The Morar Division accumulated in a shallow-marine environment (see Glendinning, this volume), possibly much nearer shore than the Glenfinnan and Loch Eil Divisions.

No sediments with intermediate characteristics to the Morar and Glenfinnan Divisions have been recognized, so it is likely that a significant part of the basin is cut out at the Sgurr Beag Thrust. However, Barr *et al.* (1986) linked the eastward increase in pre-Caledonian metamorphic grade within the Morar Divisions with the fact that westerly outliers of the Glenfinnan Divisions record lower pre-Caledonian grade than the main outcrop. They also suggested that these rocks were located close to one another prior to Caledonian tectonism. These intermediate rocks share one other peculiarity: the Knoydart Pelite (a tectonic slice of eastern Morar Division—Barr *et al.*, 1986) and the Mull and Sguman Coinntich Pelites (outliers of the Glenfinnan Division) are unique among Moine pelites in having high boron contents (Plant *et al.*, 1984).

A true stratigraphic relationship between the Morar and Glenfinnan Divisions may be uniquely preserved on the Ross of Mull. Although Rathbone (1980) and Barr (1983) considered that here a tectonic contact separates rocks of Morar Division aspect from rocks of Glenfinnan Division aspect, Holdsworth *et al.* (1987) have questioned this conclusion and favour a simple stratigraphic passage of Morar Division rocks into younger Glenfinnan Division rocks. The anomalously high metamorphic grade of the Morar Division rocks here (kyanite, as opposed to garnet or biotite on the mainland to the east) indicates that they lie in an outlier of one of the higher, more internal Moine nappes (the Knoydart or Sgurr Beag Nappes of Barr *et al.*, 1986). If the conclusions of Holdsworth *et al.* (1987) are correct, the Morar Division passed both laterally (distally, into the deeper parts of the basin) and vertically (due to basin deepening or marine transgression) into the Glenfinnan Division.

3.5.2 *Correlation between the rocks of the Sgurr Beag Nappe and the Sutherland Moine Assemblage*

A number of authors (e.g. Soper and Barber, 1982; Barr *et al.*, 1986; Butler, 1986) have pointed out that the E Sutherland Moine (discussed in detail by Moorhouse and Moorhouse, this volume) occupies a similar structural position to the Sgurr Beag Nappe, and have speculatively correlated the Sgurr Beag Thrust with the Naver or Swordly Slides. The E Sutherland Moine comprises a poorly-known sequence of pelites and psammites in the west, which passes eastwards into dominantly psammitic lithologies with subordinate semipelite and quartzite. This lithological change invites comparison with the Glenfinnan and Loch Eil Divisions of the Sgurr Beag Nappe, although it would be premature to attempt further correlation at this stage.

3.6 Sedimentology of the Glenfinnan and Loch Eil Divisions

To date, the only systematic sedimentological analysis of any of the rocks of the Sgurr Beag Nappe is that of Strachan (1986) with reference to the Loch Eil Division rocks of the Loch Eil area. Previous workers largely confined themselves either to descriptions of sedimentary structures at particular localities (e.g. Tobisch, 1966) or to generalized statements concerning the probable environments of deposition (e.g. Johnstone, 1975). Sedimentological work is restricted to the relatively undeformed rocks of the Loch Eil Division; environmental interpretation of the Glenfinnan Division rocks must necessarily be tentative owing to their high state of deformation and gneissose nature.

3.6.1 *Loch Eil Division*

3.6.1.1 *Facies analysis of the Loch Eil area.* Strachan (1986) subdivided the Loch Eil Division rocks of the Loch Eil area into three broad lithofacies. In terms of their inferred sedimentary precursors these are: (i) sandstone, (ii) interbedded sandstone and siltstone, (iii) siltstone.

(i) Sandstone facies This constitutes the psammites and quartzites of the Loch Eil area, which total over 90% of the local succession. The psammites (60–80% quartz, 20–40% feldspar, 0–10% mica) probably represent metamorphosed arkosic or feldspathic greywackes (Pettijohn, 1975). The quartzites (80–100% quartz, 0–20% feldspar, 0–10% mica) are likely to represent subarkoses with minor orthoquartzites (Pettijohn, 1975). No coarse clastic material, such as pebble beds or conglomerates, has been found, and it is inferred that these sands were originally well sorted and that grain size was fine to medium.

The sandstone beds are parallel sided and laterally continuous, and range from 1 to 50 cm thick. The bases and tops of individual beds are planar and non-erosive. There is every transition both laterally and vertically from sequences which comprise 100% sand with no interbedded silt, to sequences where thin interbeds and laminae of silt are common.

Cross-lamination and cross-stratification are the most common sedimentary structures. Cross-laminated sets are both tabular and trough-bedded, and ripples may display undulatory tops which are either draped by siltstone or overlain by another current-rippled set. Cross-stratified sets are generally tabular, with parallel sides and non-erosive bases (Fig. 3.6). They range in thickness from 5 to 20 cm, although exceptionally they may attain thickness of up to 1 m. Both cross-laminated and cross-stratified sets form bidirectional 'herringbone' structure at intervals within the succession (Fig. 3.7). Ripple marks are locally preserved on bedding surfaces (Fig. 3.8).

The cross-laminated and cross-stratified sands are commonly interbedded, which suggests that the energy of current flow was extremely variable. The relatively small-scale nature of the sedimentary structures and the inferred grain-size suggest a medium- to low-energy environment of deposition. The presence of bidirectional current-formed structures is particularly important, in that these are diagnostic of tidal environments (Reineck and Singh, 1973). The symmetrical ripple marks observed are identical to wave ripples recorded from modern environments (Reineck and Singh, 1973) and are indicative of reworking of these sands by wave activity after deposition.

Many sandstone beds appear massive and apparently structureless in the field, but polishing of cut slabs frequently reveals internal laminae and current-formed

Figure 3.6 Cross-stratification within the Kinlocheil Quartzite.

Figure 3.7 Cross-laminated sands showing bidirectional herring-bone' structure within the Loch Eil Psammite.

Figure 3.8 Symmetrical ripple marks within the Loch Eil Psammite.

structures. It is thus concluded that a large proportion of these beds also accumulated as dunes and ripples. This interpretation is broadly supported by the interbedding of 'massive' beds with cross-stratified and cross-laminated sands which suggests a common origin. Nevertheless, some massive sands display local grading into overlying silty layers, and it is likely that these may have accumulated from suspension.

Most of the sandstone facies is thought to have accumulated as a result of the migration of small-scale ripples and dunes. Sand-dominated sequence reflect a more or less constant depositional regime; those sequences which are characterized by frequent thin laminae and interbeds of silt indicate only periodic deposition of sand, separated by relatively quiescent periods during which silt accumulated.

(ii) Interbedded sandstone and siltstone facies This constitutes less than 10% of the succession, and is distinguished by the regular alternation of layers of sandstone (both psammites and quarzites) with siltstone. Where typically developed, sandy beds (2–3 cm thick) alternate with thinner silty laminae and beds (0.5–2 cm thick) to impart a regular striped appearance. Most layers are planar and subparallel for up to 10 m, but some may locally wedge out. Cross-lamination and convoluted bedding are only rarely present, and most of the sandstone beds are internally structureless. Sandstone beds have sharp bases and may locally grade into overlying silt laminae and beds.

This facies results from the rhythmic deposition of layers of sand and silt. The general lack of current or wave-formed structures through substantial thicknesses of sediment suggests that depositional processes were relatively transquil. The silt layers were probably deposited from suspension, with regular influxes of sand giving rise to the sandy layers. These may also have accumulated from suspension, in particular those sandy layers which grade into overlying silty layers, or under lower flow regime conditions.

(iii) Siltstone facies This constitutes only 1–2% of the total succession and occurs in two main forms:

(a) As thin laminae or beds separating sandy beds, and occasionally forming 'mud drapes' over ripple forms

(b) As lenticular horizons within major formations. These are up to 30 m thick and may extend for up to 500 m laterally. Exceptionally, a semipelitic unit within the Kinlocheil Quartzite is 250 m thick and 3.5 km long (Strachan, 1982). These units commonly contain thin (< 10 cm thick) parallel-sided cross-laminated sandstones.

This facies is now represented by strongly deformed biotite schists which do not preserve any sedimentary structures. All occurrences of this facies are likely to have accumulated from suspension.

Detailed mapping of these lithofacies has so far failed to reveal any systematic lateral or vertical arrangement or cyclicity.

Figure 3.9 Palaeocurrent directions within the lower part of the Loch Eil Division at Loch Eil, as determined from cross-stratification.

3.6.1.2 *Palaeocurrents.* Cross-bed foresets within the lower parts of the Loch Eil Division at Loch Eil were originally inclined to the NNE (Fig. 3.9) prior to folding (Strachan, 1986). This therefore represents to dominant direction of sediment transport within this part of the succession. A subsidiary mode to the SSW represents occasional reversals in the direction of sediment transport indicated by 'herringbone' bedding.

3.6.1.3 *Environmental interpretation.* Interpretation of thick Proterozoic arenaceous sequences presents a number of difficulties, which arise partly from the lack of faunal indicators and also from the different run-off conditions pertaining during the Proterozoic (Long, 1978). It is likely that thick, dominantly sandstone sequences such as the Loch Eil Division accumulated under either fluviodeltaic or shallow-marine conditions, or a combination of these environments. Discrimination between these two environments is particularly difficult in the absence of faunal indicators, and it is clear that all interpretation must be generalized.

The sandstone facies is volumetrically the most important component of the Loch Eil Division succesion. Any interpretation of these sandstones must account for the following:

(a) A lack of any obvious large-scale cyclicity, relating in particular to the relative proportions of sand and silt

(b) The absence of any coarse clastic material, such as conglomeratic lags or pebble beds

(c) The presence of herring-bone cross-beds and wave-generated ripples

(d) The exclusively tabular nature of cross-stratified sets

(e) The small-scale nature of sedimentation sets

(f) The relatively common occurrence of thin silt suspension laminae and beds.

A fluviodeltaic interpretation for these sandstones seems implausible: it is unlikely that a river or deltaic system could have deposited such a thickness of sandstone with no channel-fill deposits, virtually no fine grained sediment, exclusively tabular bedding and no detectable cyclicity.

The Loch Eil Division sandstones do, however, appear to have ready analogues in both the modern shallow-marine environment (Belderson and Stride 1966; Kenyon, 1970) and in ancient shallow-marine shelf deposits (Johnson, 1978; Levell, 1980). The existence of herring-bone bedding indicates that parts of the succession were deposited in a regime which was subject to tidal activity. The occurrences of wave ripples are clearly compatible with this. The exclusively tabular nature of cross-stratified units may also indicate shallow-marine deposition. Although both tabular and trough cross-stratification may be produced in the fluvial environment, trough sets (channels) are more important than tabular sets (interchannel bars). In the shallow-marine environment, where the energy of deposition is commonly less, tabular sets produced by the migration of laterally extensive dunes are much more common than trough sets. Related to this is the uniformly small-scale nature of individual sedimentation units. This implies deposition by low-velocity currents through considerable thicknesses of sediment, which is consistent with deposition in a shallow-marine environment. Such an interpretation is also favoured by the lack of any coarse clastic material, which implies a distal rather than a proximal environment of deposition. Dominantly sandstone sequences reflect more or less constant deposition, probably under the influence of tidal currents. Those parts of the sandstone facies which have a higher proportion of thin suspension silt laminae and beds were deposited in a setting where the transport of sand was more intermittent and of longer duration than diurnal tidal cycles. In such a setting it is likely that sedimentation will be influenced by storms as well as tidal and oceanic currents.

The interbedded sandstone and siltstone facies may also be integrated into a shallow-marine shelf model. The facies are likely to represent areas of the shelf which are characterized by relatively high rates of silt deposition and starvation of sand. Such lateral heterogeneities in the style of deposition are common in modern shallow-marine shelf environments. Mud may accumulate in a variety of sub-environments from nearshore to deep offshore (McCave, 1972), and thus coexist with areas of active sand deposition. Regular incursions of sand relate to variations in tidal and storm activity. The siltstone facies may thus also be assigned to a shallow-marine shelf setting. The thin, laterally persistent cross-laminated sands within this facies bear a strong resemblance to storm deposits recorded by other workers (e.g. Brenchley *et al.*, 1979).

3.6.1.4 *Comparison with the Glen Moriston area.* Relatively undeformed psammites and semipelites containing sedimentary structures are also present in the Glen Moriston area (Fig. 3.3). A preliminary analysis of the sedimentology of these rocks suggests that

Figure 3.10 Wave-rippled sands within the Loch Eil Psammite.

they are very similar to those of the Loch Eil area. They are dominated by thick sequences of sandstones which contain cross-lamination and small-scale tabular cross-stratification. Wave ripples have been identified on polished slabs of sandstone (Fig. 3.10). Herring-bone structure has not been identified. Obvious channels are absent, but broad, shallow depressions in the tops of some beds may represent the result of penecontemporaneous erosion. Convolute bedding and graded bedding have not been recorded. There is no trace of any coarse clastic material such as pebbly horizons. Thinly interbedded and interlaminated sandstones and siltstones are locally present, and discrete units of laminated sandstone up to 150 m thick have also been mapped. These lithological types are comparable in most respects with those described from the Loch Eil area, and accordingly it is thought that the Loch Eil Division rocks of the Glen Moriston area also accumulated in a general shallow-marine setting. Unfortunately, it has not proved possible to analyse palaeocurrent data for this area, due to a lack of cross-beds suitable for measurement.

3.6.2 *Glenfinnan Division*

Sedimentological information from these rocks is sparse because strong deformation has obliterated much sedimentary detail. Cross-stratification has been recorded from psammitic units within the Glen Cannich area (Tobisch, 1966) and locally at Loch Quoich (Roberts and Harris, 1983). Environmental interpretation of these rocks does, however, have to rest on general observations relating to the broad characteristics of the different lithological units. The association of thick, regionally extensive horizons of pelite (silt–mud) and striped schists (rhythmically layered sands and silts) seems most consistent with deposition in a deep-water marine environment (Strachan, 1986). The rapid alternation of sand and silt layers within the striped schists may reflect deposition from distal turbidites.

3.7 Depositional tectonics

It is generally considered that the boundary between the Glenfinnan Division and the Lewisian basement gneisses which overlie the Sgurr Beag Thrust is a tectonically modified unconformity (Tanner *et al.*, 1970; Tanner, 1971; Rathbone, 1980). This implies that the sediments of the Sgurr Beag Nappe were deposited in a basin floored by continental-type basement rocks. The basin is likely to have been elongate in a NE–SW direction, with sediments derived from the NW (or SW), transported axially along the basin by longshore currents (Strachan, 1986). The relatively uniform cross-strike nature of sedimentary environments within the Loch Eil Division, and the lack of slope or shelf-edge deposits may indicate that the sediments were deposited in an intracontinental rift like the North Sea, rather than on a passive margin. Such an areally restricted setting may explain the absence of close stratigraphic analogues elsewhere in the Proterozoic of NW Europe.

The metasediments of the Sgurr Beag Nappe contain relatively high amounts of feldspar, which implies that they were derived by rapid mechanical weathering. The absence of any distinctive clast types within the metasediments means that it is difficult to directly deduce the composition of the source terrain(s). Plant *et al.* (1984) present geochemical data which suggest that the Glenfinnan and Loch Eil Division Moine Assemblage were derived from the erosion of mid- to upper-crustal complexes intruded by evolved granites. Pb-isotopic zircon systematics (Pidgeon and Aftalion, 1978; Halliday *et al.*, 1979) are consistent with derivation from a terrain *c.* 1700 Ma in age. Accordingly, Plant *et al.* (1984) speculate that the Ketilidian (*c.* 1800 Ma) and adjacent Archaean provinces of southern Greenland acted as source lands for much of the Moine. The great thickness of Moine metasediments implies that these source lands were drained by an active river-deltaic system which transported sediment eastwards into the Moine Basin(s).

The geometry of the basin is difficult to evaluate, given the complexity of the subsequent orogenic deformation and the lack of complete cross-strike sections, The general uniformity of depositional conditions within the Loch Eil Division indicates that subsidence kept pace with sedimentation, but the lenticular (rather than wedge-shaped) form of subunits, and the lack of recorded large-scale growth faulting, argue against active crustal extension as the driving process. However, the variations in the style of sedimentation within the Sgurr Beag Nappe as a whole may relate to changes in the rate of subsidence during the development of the basin. The relatively deep-water sediments of the Glenfinnan Division may represent sediments laid down during and immediately after active stretching. Rapid extension could have given rise to several kilometres of subsidence on a time-scale of the order of 10 Ma (e.g. Jarvis, 1984), producing a deep compartmented basin. Thermal subsidence would have continued at an exponentially decreasing rate for the next 50–100 Ma (McKenzie, 1978). If sediment input rates were maintained, basin would have shallowed as subsidence rates declined through time. The shallower-water sediments of the Loch Eil Division may therefore have been laid down during the thermal subsidence phase, and thus be broadly analogous to the Tertiary sediments of the North Sea or of the North Atlantic continental shelves. The numerous lateral variations in lithofacies and thicknesses of units within the Sgurr Beag Nappe imply differential subsidence along the strike of the basin, probably accommodated by transfer faults (Lister *et al.*, 1986).

Acknowledgements

R. A. Strachan and D. Barr acknowledge receipt of NERC studentships held at the University of Keele and the University of Liverpool respectively; RAS also gratefully acknowledges continued funding from the Research and Advanced Study Committee of Oxford Polytechnic. F. May's contribution is published by permission of the Director of the British Geological Survey. D. Barr's contribution is published by permission of Britoil plc. Ann Mackenzie is thanked for typing the manuscript, and Simon Deadman for his photographic work.

References

Aftalion, M. and van Breemen, O. (1980) U–Pb zircon, monazite and Rb–Sr whole-rock systematics of granitic gneiss and psammitic to semipelitic host gneiss from Glenfinnan, northwestern Scotland. *Contrib. Miner. Pet.* **72**, 87–98.

Barr, D. (1983) Genesis and Structural Relations of Moine Migmatites. Unpublished Ph.D. Thesis, University of Liverpool.

Barr, D., Roberts, A. M., Highton, A. J., Parson, L. M. and Harris, A. L. (1985) Structural setting and geochronological significance of the West Highland Granitic Gneiss, a deformed early granite within Proterozoic, Moine rocks of NW Scotland. *J. geol. Soc. London* **142**, 663–676.

Barr, D., Holdsworth, R. E. and Roberts, A. M. (1986) Caledonian ductile thrusting in a Precambrian metamorphic complex: the Moine of NW Scotland. *Bull. geol. Soc. Am.* **97**, 754–764.

Belderson, R. L. and Stride, A. H. (1966) Tidal fashioning of a basal bed. *Mar. Geol.* **4**, 237–257.

Brenchley, P. J., Newall, G. and Stanistreet, I. G. (1979) A storm surge origin for sandstone beds in an epicontinental platform sequence, Ordovician, Norway. *Sedim. Geol.* **22**, 185–217.

Brewer, M. S., Brook, M. and Powell, D. (1979) Dating of the tectonometamorphic history of the southwestern Moine, Scotland. In Harris, A. L., Holland, C. H. and Leake, B. E. (eds.), The Caledonides of the British Isles—Reviewed. *Spec. Publ. geol. Soc. London* **8**, 129–137.

Brook, M., Powell, D. and Brewer, M. S. (1976) Grenville age for rocks in the Moine of north-western Scotland. *Nature* **260**, 515–517.

Brook, M., Powell, D. and Brewer, M. S. (1977) Grenville events in Moine rocks of the Northern Highlands, Scotland. *J. geol. Soc. London* **133**, 489–496.

Brown, R. L., Dalziel, I. W. D. and Johnson, M. R. W. (1970) A review of the structure and stratigraphy of the Moinian of Ardgour, Moidart and Sunart—Argyll and Inverness-shire. *Scott. J. Geol.* **6**, 309–335.

Butler, R. W. H. (1986) Structural evolution in the Moine of northwest Scotland: a Caledonian linked thrust system? *Geol. Mag.* **123**, 1–11.

Clifford, T. N. (1957) The stratigraphy and structure of part of the Kintail district of southern Ross-shire—its relationship to the Northern Highlands. *Q. J. geol. Soc. London* **113**, 57–92.

Dalziel, I. W. D. (1966) A structural study of the granitic gneiss of western Ardgour, Argyll and Inverness-shire. *Scott. J. Geol.* **2**, 125–152.

Halliday, A. N., Stevens, W. E. and Harmon, R. S. (1979) Petrogenetic significance of Rb–Sr and U–Pb isotopic systems in the *c.* 400 Ma old British Isles granitoids and their hosts. In Harris, A. L., Holland, C. H. and Leake, B. E. (eds.), the Caledonides of the British Isles—Reviewed. *Spec. Publ. geol. Soc. London* **8**, 653–661.

Harris, A. L. (1983) The growth and structure of Scotland. In Craig, G. Y. (ed.), *Geology of Scotland.* Scottish Academic Press, Edinburgh, 1–22.

Holdsworth, R. E. and Roberts, A. M. (1984) A study of early curvilinear fold structures and strain in the Moine of the Glen Garry region, Inverness-shire. *J. geol. Soc. London* **141**, 327–338.

Holdsworth, R. E., Harris, A. L. and Roberts, A. M. (1987) The stratigraphy, structure and regional significance of the Moine rocks of Mull, Argyll-shire, W. Scotland. *Geol. J.* **22**, 83–108.

Jarvis, G. T. (1984) An extensional model of graven subsidence—the first stage of basin evolution. *Sedim. Geol.* **40**, 13–31.

Johnson, H. D. (1978) Shallow siliciclastic seas. In Reading, H. (ed.), *Sedimentary Environments and Facies.* Blackwell Scientific, Oxford, 209–258.

Johnstone, G. S. (1975) The Moine Succession. In Harris, A. L., Shackleton, R. M., Watson, J. V., Downie, C., Harland, W. B. and Moorbath, S. (eds.), A Correlation of the Precambrian Rocks in the British Isles. *Spec. Rep. geol. Soc. London* **6**, 30–42.

Johnstone, G. S., Smith, D. I. and Harris, A. L. (1969) The Moinian Assemblage of Scotland. In Kay, M. (ed.), North Atlantic—Geology and Continental Drift. *Mem. Am. Ass. Pet. Geol.* **12**, 159–180.

Kenyon, N. H. (1970) Sand ribbons of European tidal seas. *Mar. Geol.* **9**, 25–39.

Lambert, R. StJ. (1969) Isotopic studies relating to the Precambrian history of the Moinian of Scotland. *Proc. geol. Soc. London* **1652**, 243–245.

Leedal, G. P. (1952) The Cluanie igneous intrusion, Inverness-shire and Ross-shire. *Q. J. geol. Soc. London* **108**, 35–63.

Levell, B. K. (1980) A late Precambrian tidal shelf deposit, the Lower Sandfjord Formation, Finnmark, North Norway. *Sedimentology* **27**, 539–557.

Lister, G. S., Etheridge, M. A. and Symonds, P. A. (1986) Detachment faulting and the evolution of passive continental margins. *Geology* **14**, 246–250.

Long, D. G. F. (1978) Proterozoic stream deposits: some problems of recognition and interpretation of ancient sandy fluvial systems. In Miall, A. D. (ed.), Fluvial Sedimentology. *Mem. Can. Soc. Pet. Geol.* **5**, 313–341.

McCave, I. N. (1972) Transport and escape of fine-grained sediment from shelf areas. In Swift, D. J. P., Duane, D. B. and Pilkey, O. H. (eds.), *Shelf Sediment Transport: Process and Pattern.* Dowden, Hutchinson and Ross, New York, 225–248.

McKenzie, D. (1978) Some remarks on the development of sedimentary basins. *Earth Planet. Sci. Letters* **40**, 25–32.

Pettijohn, F. J. (1975) *Sedimentary Rocks.* Harper and Row, New York, 3rd Edn.

Piasecki, M. A. J. and van Breemen, O. (1979) The 'Central Highland Granulites': cover–basement tectonics in the Moine. In Harris, A. L., Holland, C. H. and Leake, B. E. (eds.), The Caledonides of the British Isles—Reviewed. *Spec. Publ. geol. Soc. London* **8**, 139–144.

Pidgeon, R. T. and Aftalion, M. (1978) Cogenetic and inherited zircon U–Pb systems in granites of Scotland and England. In Bowes, D. R. and Leake, B. E. (eds.), Crustal Evolution in Northwestern Britian and Adjacent Regions. *Geol. J. Spec. Issue* **10**, 183–220.

Plant, J. A., Green, P. M. and Watson, J. V. (1984) Moine–Dalradian relationships and their palaeotectonic significance. *Proc. R. Soc. London* **A395**, 185–202.

Powell, D. (1964) The stratigraphical succession of the Moine Schists around Lochailort (Inverness-shire) and its regional significance. *Proc. Geol. Ass. London* **75**, 223–250.

Powell, D. (1974) Stratigraphy and structure of the western Moine and the problem of Moine orogenesis. *J. geol. Soc. London* **130**, 575–593.

Powell, D., Baird, A. W., Charnley, N. R. and Jordan, P. J. (1981) The metamorphic environment of the Sgurr Beag Slide; a major crustal displacement zone in Proterozoic, Moine rocks of Scotland. *J. geol. Soc. London* **138**, 661–673.

Powell, D., Brook, M. and Baird, A. W. (1983) Structural dating of a Precambrian pegmatite in Moine rocks of northern Scotland and its bearing on the status of the 'Morarian Orogeny'. *J. geol. Soc. London* **140**, 813–24.

Rathbone, P. A. (1980) Basement–Cover Relationships in the Moine Series of Scotland, with Particular Reference to the Sgurr Beag Slide. Unpublished Ph.D. Thesis, University of Liverpool.

Rathbone, P. A. and Harris, A. L. (1979) Basement–cover relationships at Lewisian inliers in the Moine rocks. In Harris, A. L., Holland, C. H. and Leake, B. E. (eds.), The Caledonides of the British Isles—Reviewed. *Spec. Publ. geol. Soc. London* **8**, 101–107.

Rathbone, P. A., Coward, M. P. and Harris, A. L. (1983) Cover and basement: a contrast in style and fabrics. *Mem. geol. Soc. Am.* **158**, 213–223.

Reineck, H. E. and Singh, I. B. (1973) *Depositional Sedimentary Environment—With Reference to Terrigenous Clastics.* Springer-Verlag, Berlin.

Roberts, A. M. (1984) Stratigraphy and Structure in Moine Rocks along the Loch Quoich Line, Inverness-shire, Scotland. Unpublished Ph.D. Thesis, University of Liverpool.

Roberts, A. M. and Harris, A. L. (1983) The Loch Quoich Line—a limit of early Palaeozoic crustal reworking in the Moine of the Northern Highlands of Scotland. *J. geol. Soc. London* **140**, 883–892.

Roberts, A. M., Smith, D. I. and Harris, A. L. (1984) The structural setting and tectonic significance of the Glen Dessary Syenite, Inverness-shire. *J. geol. Soc. London* **141**, 1033–1042.

Roberts, A. M., Strachan, R. A., Harris, A. L., Barr, D. and Holdsworth, R. E. (1987) The Sgurr Beag Nappe—a reassessment of the stratigraphy and structure of the Northern Highland Moine. *Bull. geol. Soc. Am.* **98**, 497–506.

Sanders, I. S., van Calsteren, P. W. C. and Hawkesworth, C. J. (1984) A Grenville Sm–Nd age for the Glenelg eclogite in north-west Scotland. *Nature* **312**, 439–440.

Soper, N. J. and Barber, A. J. (1982) A model for the deep structure of the Moine thrust zone. *J. geol. Soc. London* **139**, 127–138.

Stoker, M. S. (1983) The stratigraphy and structure of the Moine rocks of Eastern Ardgour. *Scott. J. Geol.* **19**, 369–385.

Strachan, R. A. (1982) Tectonic sliding within the Moinian Loch Eil division near Kinlocheil, West Inverness-shire. *Scott. J. Geol.* **18**, 187–203.

Strachan, R. A. (1985) The stratigraphy and structure of the Moine rocks of the Loch Eil area, West Inverness-shire. *Scott. J. Geol.* **2**, 9–22.

Strachan, R. A. (1986) Shallow-marine sedimentation in the Proterozoic Moine Succession, northern Scotland. *Precambr. Res.* **32**, 17–33.

Tanner, P. W. G. (1971) The Sgurr Beag Slide—a major tectonic break within the Moinian of the Western Highlands of Scotland. *Q. J. geol. Soc. London* **126**, 435–463.

Tanner, P. W. G., Johnstone, G. S., Smith, D. I. and Harris, A. L. (1970) Moinian stratigraphy and the problem of the Central Ross-shire Inliers. *Bull. geol. Soc. Am.* **81**, 299–306.

Tobisch, O. T. (1966) Observations on primary deformed sedimentary structures in some metamorphic rocks from Scotland. *J. Sedim. Pet.* **35**, 415–419.

Tobisch, O. T., Fleuty, M. J., Merh, S. S., Mukhopadhyay, D. and Ramsay, J. G. (1970) Deformational and metamorphic history of Moinian and Lewisian rocks between Strathconon and Glen Affric. *Scott. J. Geol.* **6**, 243–265.

van Breemen, O., Pidgeon, R. T. and Johnson, M. R. W. (1974) Precambrian and palaeozoic pegmatites in the Moines of northern Scotland. *J. geol. Soc. London* **130**, 493–508.

4
The Central Highland Division

M. A. J. PIASECKI and S. TEMPERLEY

4.1 Introduction

The rocks which extend from the Great Glen Fault to the Dalradian (inset in Fig. 4.1) were originally called the 'Central Highland Granulites' (e.g. Hinxman and Anderson, 1915). They were described as siliceous and quartzofeldspathic metasediments in which belts of pelites were present; emphasis was placed upon the presence of gneissose and migmatitic rocks in some areas, and non-migmatitic, flaggy rocks in others. Latter, the psammites became equated with the youngest of the Moine Assemblage rocks to the north of the Great Glen Fault, and were sometimes referred to as the 'younger Moine', whereas the pelitic belts were assigned to the Lower Dalradian (Anderson, 1956; Harris and Pitcher 1975; Johnstone 1975). Before Precambrian isotopic ages were recorded from these rocks, it was assumed that they had all been affected only by early Palaeozoic orogenic events; the gneisses and migmatites represented more intense 'Caledonian' metamorphism than the unmigmatized, flaggy metasediments which retained primary sedimetary structures. Local structural–metamorphic sequences were correlated with those previously established in the Dalradian (Anderson, 1956; Smith, 1968; Piasecki, 1975).

More recent work (Piasecki and van Breemen, 1979*a* and *b*; Piasecki, 1980, 1983, Piasecki and Temperley in press) has shown that the gneissic and migmatitic rocks have a substantially longer structural–metamorphic history than the flaggy metasediments. A major event of ductile shearing has been recognized and dated at *c.* 750 Ma, the mylonitic fabrics of this event severely attenuating the record of an earlier orogenic event in

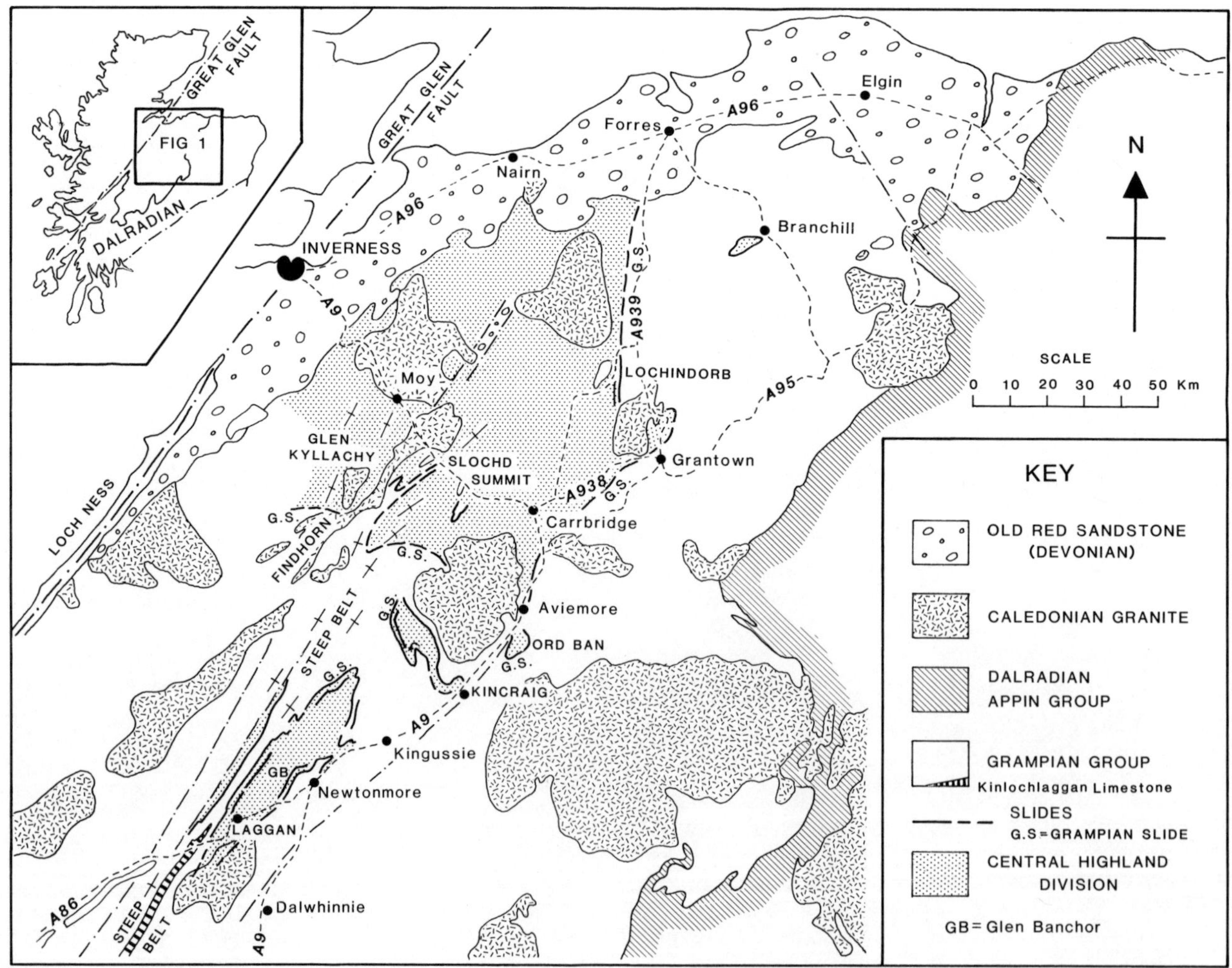

Figure 4.1 Outline geological map of the northern sector of the Scottish Central Highlands, showing the distribution of the Central Highland Division. Inset: location of Fig. 4.1 in relation to the Scottish Highlands.

the gneisses, but forming the earliest fabrics in the flaggy rocks. The latter also appear to have been deposited unconformably on the gneisses. On this basis, the Central Highland Granulites have been separated into a basement assemblage, the Central Highland Division, structurally and stratigraphically overlain by a cover of the flaggy, unmigmatized metasediments, the Grampian Group (Fig. 4.1). The latter appears to pass sedimentologically into the Dalradian Appin Group (Harris *et al.*, 1978).

The Central Highland Division forms a main outcrop of some 1000 km^2 (Fig. 4.1). Outside this, the gneisses reappear in inliers within the Grampian Group at Kincraig and at Laggan, where they may extend southwards into the Ben Alder region. Another inlier may occur at Branchill. Further to the SW, this gneissic core underlies the Grampian Group, its isotopic signature being detectable in late- and post-Caledonian granites (Clayburn, 1981).

4.2 Lithofacies

Where the rocks of the Central Highland Division have been least reworked by the 750 Ma shearing and by early Palaeozoic events, they are seen to be very coarse, banded and dominantly migmatitic. A pervasive compositional banding in the gneisses (Fig. 4.2*a–d*) partly represents the effect of diffusional gneissic differentiation, and partly the effect of stromatic migmatization. Preserved primary sedimentary structures are very rare, but their presence indicates that this metamorphic layering represents a severely modified and variably transposed original sedimentary bedding. Four lithological types comprise most of the Central Highland Division:

(i) Coarse quartzofeldspathic gneisses, banded with alternating feldspathic, siliceous and biotitic layers, commonly with stromatic migmatitic neosome (Fig. 4.1*a*)

(ii) Coarse semi-pelitic gneisses, massive to regularly layered. They are usually migmatized (Fig. 4.2*b*), and commonly develop quartzofeldspathic or siliceous ribs, grading into other gneissic lithologies

(iii) Siliceous gneisses, in which psammitic layers are interbanded with layers richer in biotite and/or feldspar (Fig. 4.2*c*), and which are frequently migmatized (Fig. 4.2*d*). In rare areas of 'low' strain and no migmatization, rhythmic bedding is present, with occasional development of thin, gradded units

(iv) Very coarse, feldspathic 'quartzites' which are usually massive and often form conspicuous migmatitic quartzites (Fig. 4.2*e*). With the development of biotitic layers, the quartzites grade into the siliceous gneisses. A quartzite near Kincraig has yielded one example of a cross-bedded erosional channel, 35 cm wide and 5 cm deep.

In addition, pods and lenticular bodies of gneissic garnet-amphibolite are often developed in the semi-pelitic and quartzofeldspathic gneisses.

Owing to the extensive gneissification, migmatization and the rarity of preserved sedimentary structures, so far only local structural–lithological successions have been recognized in the Central Highland Division. In Glen Banchor, a 1-km thick exposed section of the gneisses contains the best documented succession:

(i) A Lower Siliceous Group occurs at the lowest exposed structural level in the section. It forms a highly quartzose assemblage, characterized by thick units of psammitic gneisses, massive feldspathic quartzites and only minor semi-pelitic units.

(ii) The Middle Pelitic Group consists essentially of coarse, one-mica biotite-gneisses, associated with finer-grained schists rich in muscovite and almandine garnet. It can be demonstrated that the gneisses occur in lensoid areas of 'low' ductile shearing strain, and that the schists are their reworked equivalents. The upper part of the group is particularly highly sheared and contains at least three zones of very high strain.

(iii) The Upper Psammitic Group contains more variable lithologies. At its base is a psammitic unit with discontinuous quartzites. It passes structurally upwards into semi-psammites, then into a second psammitic unit which has no quartzites. This is followed by a semi-pelitic unit in which zones of 'high' strain anastomose around lensoid areas of 'low' strain. It contains numerous pods of garnet-amphibolite. At the structural top of the sequence is a third psammitic unit with thin quartzites, and also with pods of garnet-amphibolite.

4.3 Orogenic history

The shear zones dated at *c.* 750 Ma have contributed substantially to resolving the complex orogenic history of the Central Highland Division. They serve as time markers separating a late Proterozoic gneiss-forming event older than the shearing, from early Palaeozoic events which modify the shear zones.

4.3.1 *Late Proterozoic history*

The late Proterozoic orogenic history is characterized by the formation of ubiquitous, penetrative gneissosity (Fig. 4.2*a–e*), which probably developed subhorizontally and subparallel to the bedding. This fabric was then folded by at least two sets of near-recumbent, isoclinal folds (Fig. 4.2*c*) on north-easterly axial trends. Much of the Kincraig inlier lies on the upper limb of a major recumbent fold of the second fold phase, with an amplitude in excess of 10 km. In this fold, a very coarse migmatitic quartzite, less than 100 m thick, is thickened in the hinge to more than 300 m (Figs 2 and 3–1 in Piasecki, 1980). This fold exposes another structural–lithological succession: it has a core of psammitic and quartzofeldspathic gneisses, which pass structurally upwards into quartzite and then into semipelitic gneiss in the upper limb.

The thermal history related to the development of this structural sequence is complex and little known. An early metamorphism of unknown grade was related to

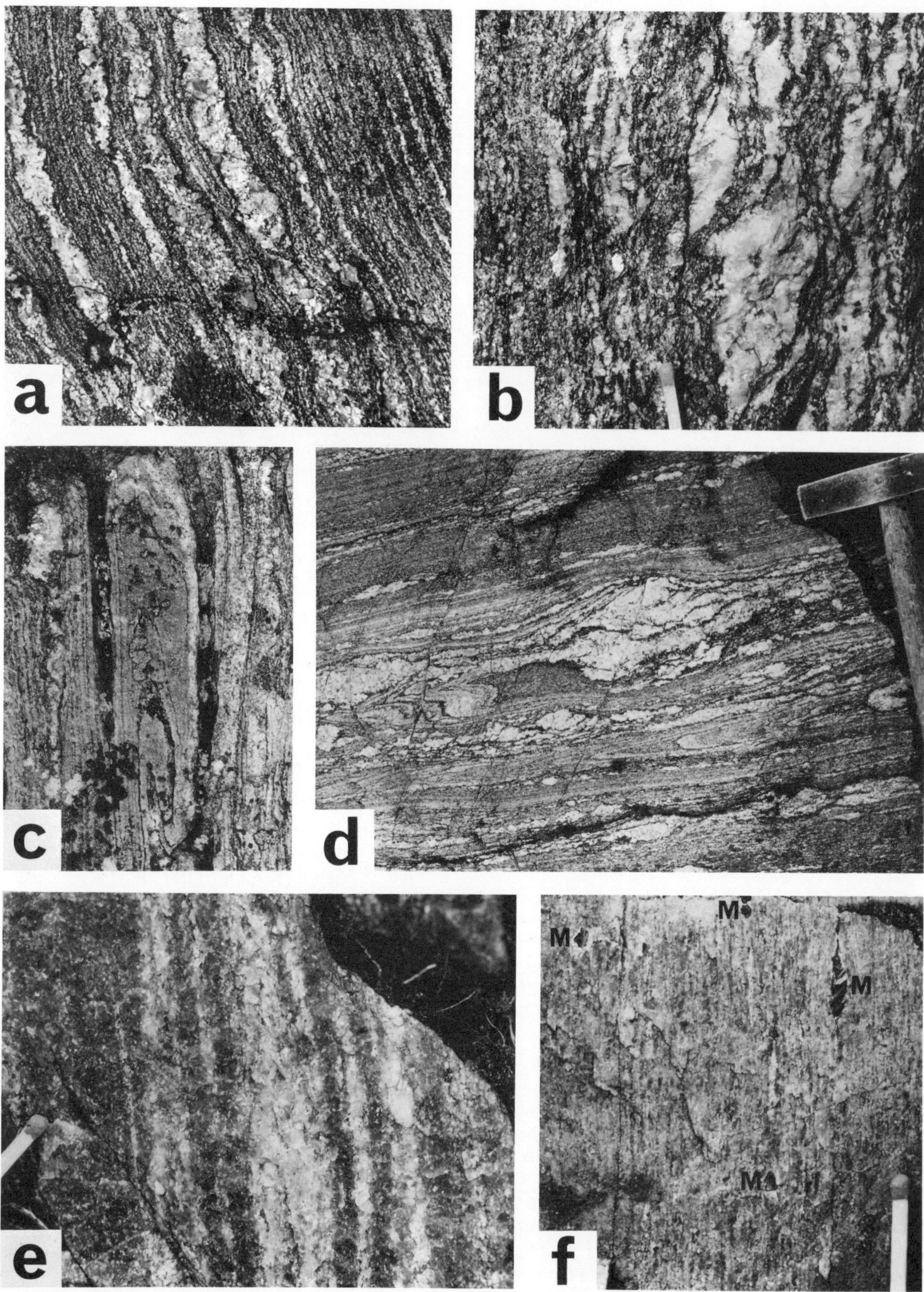

Figure 4.2 Main lithologies and Grenville (?) structures in the Central Highland Division, and the products of Knoydartian shearing. (*a*) Coarse migmatitic, quartzofeldspathic gneiss; Kincraig inlier. (*b*) Typical semi-pelitic biotite-gneiss with coarse quartzofelds pathic neosome; $\frac{1}{2}$ natural size; near Aviemore. (*c*) Siliceous gneiss showing a 'type 3' (Ramsay, 1967) fold interference pattern caused by the superimposition of two sets of isoclinal folds; Glen Banchor. (*d*) Siliceous gneiss with isoclinal folds, showing broadly syntectonic migmatization: the neosome is folded (centre right), but also parallel to the folds' axial surface (centre); Kincraig inlier. (*e*) Typical coarse, migmatitic quartzite, with stromatic, microcline-rich neosome; natural size; Blargie, near Laggan. (*f*) Mylonite from a shear zone, derived from the same quartzite, 4 m from that in Fig. 4.2*e*. Note the attenuation and the comminution in grain size. Muscovite porphyroblasts (*M*) replace the feldspars, yet the rock's chemical composition remains the same as that of the source quartzite; $\frac{1}{2}$ natural size.

the formation of the earliest bedding-parallel foliation. Metamorphism culminated in one or more episodes of widespread migmatization associated with sillimanite growth, which overlapped one, if not both phases, of isoclinal folding (Fig. 4.2*d*). The highest metamorphic grade now seen in the Central Highland Division, as recorded on the zonal map of the Scottish Caledonides (Winchester, 1974) was reached during this late Proterozoic event. It includes the regional migmatization described by Ashworth (1979*a* and *b*), some of the *P–T* values recorded by Wells (1979) and the possible formation of eclogitic assemblages in some amphibolites (Baker, 1986), unless the latter can be shown to have affinities with the Lewisian.

4.3.2 *Latest Proterozoic (Knoydartian) shearing event*

The Central Highland Division, and the lower part of the Grampian Group have been variably modified by a system of shear zones which formed under middle to high amphibolite-facies metamorphic conditions (for description see Piasecki and van Breemen, 1983, Piasecki and Temperley, in press). One of these, the Grampian Slide, has developed along the interface between the Central Highland Division and the Grampian Group, obscuring the original relationship between these assemblages (Fig. 41). Related ductile shear zones developed structurally below, in the Central Highland Division, and above, in the Grampian Group, along the subhorizontal lithological boundaries and within less competent lithologies.

The shearing occurred in zones from 10 to 200 m wide, which appear to anastomose on the regional scale, and within which all stages of transition can be seen from unsheared rocks, through protomylonites into mylonites and ultramylonites (Fig. 4.2*e* and *f*). Stretching lineations related to the shearing movements trend N to NNE; and S–C mylonite fabrics including shear bands (Berthe *et al.*, 1979; White *et al.*, 1980; Lister and Snoke, 1984), indicate a sense of northerly to north-north-easterly directed transport. Folds formed during this shearing event are subrecumbent, asymmetrical 'drag' folds verging towards the N, their hinges orientated at low angles to the stretching lineation. They are intimately related to the numerous stages in the development of the mylonitic fabric in the shear zones. Thus, they fold earlier-formed mylonitic fabrics, whereas that increment of this mylonitic fabric which was forming at the time the folds were developing defines their axial surfaces. The shearing movements were accompanied by the nucleation in the shear zones of new minerals and systems of veins. These include fine-grained muscovite replacing matrix biotite, porphyroblasts of muscovite, feldspar and garnet; and swarms of quartz veins and of small muscovite-bearing pegmatites (Fig. 4.3*a* and *c*; for details of syntectonic mineralization in shear zones see Piasecki and Cliff, in press).

Studies of whole-rock and mineral chemistry across these shear zones show that the composition of the mylonites remains the same as that of their source rocks (cf. Fig. 4.2*e* and *f*): and that the small pegmatites (Fig. 4.3*c*) form by isochemical segregation within the shear zones (E. Hyslop, pers. comm.). Rb–Sr dating of muscovite porphyroblasts from these pegmatites, together with their host mylonites, has yielded *c.* 750 Ma ages for the time of the shearing movements (Piasecki and van Breemen, 1979*a* and *b*; 1983).

The shearing event corresponds with the Knoydartian event (also referred to as Morarian) in the Moine Assemblage of the Northern Highlands (Piasecki and van Breemen, 1983; Piasecki, 1984). It is believed to represent a major regime of northerly movements on gently inclined, deep-seated shear zones associated with amphibolite-facies metamorphism. The compressional, rather than extensional nature of this event is indicated by the presence of tectonic slivers of the Central Highland Division, developed a short distance above the sheared base of the Grampian Group, locally repeating the basement–cover sequence (Piasecki and Temperley, in press).

4.3.3 *Early Palaeozoic orogeny*

Early Palaeozoic structures, most commonly near-upright folds, rework the initially gently inclined gneisses and the Knoydartian shear zones. Several phases of upright folding are involved, most of which have limited distribution—only a NNE-trending phase being developed on the regional scale.

The Central Highlands are crossed by a 'steep belt' (Fig. 4.1), first described by Thomas (1979) as an asymmetrical, orogenic 'root zone' of Grampian age (pre-Arenigian, *sensu* Lambert and McKerrow 1976) that separates nappes verging in opposite directions. In Glen Banchor, this belt forms a single monoclinal, asymmetrical structure, corresponding more closely with a subvertical shear zone, in which the gneisses have been reworked into a steep attitude by the late NNE-trending upright folds. Folds of this phase are developed throughout the region, but as they are traced into the steep belt, so they progressively tighten and become isoclinal. As this belt enters the main outcrop of the Central Highland Division, so it widens, branching out into a network of steep zones. These upright folds can be correlated with the last pervasive Caledonian folds in Glen Kyllachy (Piasecki, 1975), where pegmatites emplaced during the waning phase of this folding have been dated at *c.* 443 Ma (van Breemen and Piasecki, 1983). The steep belt may have provided an important structural control for the emplacement of late-orogenic granites, such as the Glen Kyllachy and Moy granites at *c.* 443 Ma, and the post-orogenic Findhorn Granite at *c.* 413 Ma (van Breemen and Piasecki, 1983; Zaleski, 1985).

The Central Highland Division represents the deepest level of exposure in the Scottish Central Highlands, and at this level there appears to be no evidence for the presence of the early, recumbent fold-nappes so typical of Grampian deformation in the Dalradian Supergroup (Roberts, 1974; Roberts and Treagus,

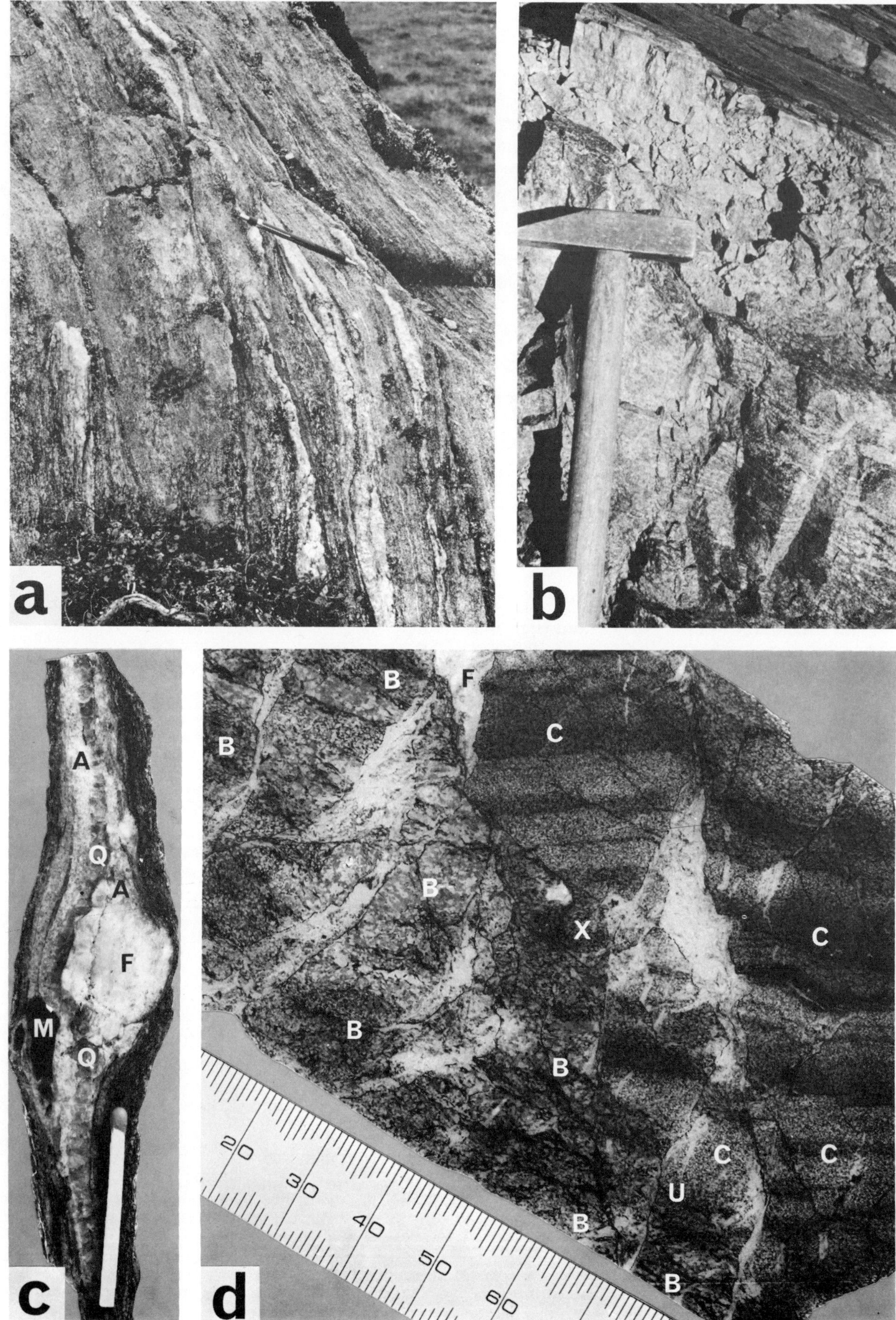
a
b
c
A
Q
A
F
M
Q
d
B
F
B
C
B
X
C
B
B
C
C
U
B
B
20
30
40
50
60

1977; Bradbury *et al.*, 1979; Thomas, 1979). Local developments of gently inclined, minor folds younger than the shearing (the 'flattening' folds of Piasecki, 1980) are not related to fold-nappes; and in the lower part of the Grampian Group, primary graded-bedding structures demonstrate that the cover rocks young upwards, away from their contact with the basement, without any regional inversion. The contribution of the Grampian event to the fabric of the Central Highland Division is difficult to assess; it may be limited to static recrystallization, and local modification of the initial subhorizontal foliation. It may also include local resumption of movements on the Grampian Slide, related to the resetting of some of the Knoydartian ages towards Caledonian values (Piasecki, 1980).

4.4 Relationship between the Central Highland Division and the Grampian Group

Despite the shearing along the interface between these assemblages, their initial contact relationship can be discerned. On the regional scale, the shear zone varies in thickness and locally deviates from the contact, preserving small areas which still retain some of the criteria of a stratigraphic unconformity, albeit deformed and metamorphosed. Thus, near Lochindorb (Fig. 4.1), an unsheared and unmigmatized Grampian Group semipsammite, with little deformed calc-silicate ribs, rests in turn on all the main lithologies of the Central Highland Division, including the highly migmatitic quartzites. Structurally above this contact, the basement–cover sequence is repeated, but here it is intensly sheared (Piasecki and Temperley in press). Other weakly sheared to unsheared contacts are preserved at Slochd Summit (Fig. 4.1 and Fig.4.3*b*). At one of these, a small block downfaulted from the pre-Grampian Group erosion surface, yielded remnants of a preserved original contact (Fig. 4.3*d*). This shows that in the outermost 1 cm of the coarse gneiss, the rock has lost some of its cohesion in a manner reminiscent of erosion ('U' in Fig. 4.3*d*). On this rests unsheared semipsammite with inverse graded bedding (coarsening upwards), suggestive of a high-density flow structure.

4.5 Affinities and age of the Central Highland Division

In terms of lithologies and metamorphic grade, the Central Highland Division closely resembles the high-grade, gneissic metasediments of the Glenfinnan Division of the Moine Assemblage. The two assemblages also share distinct similarities in their regional geochemistry. There are however, lithological differences: for instance, the Central Highland Division is more psammitic; and small calcsilicate ribs and pods regarded as original concretionary bodies, are common in the Glenfinnan Division, but are rare in the Central Highland Division, implying a different diagenetic history. Numerous comparisons can also be drawn with the Sutherland Migmatite Complex.

In the Central Highland Division, the early history of repeated isoclinal folding, regional migmatization and gneissification which predates the Knoydartian shearing, can be most reasonably correlated with the Grenville Orogeny. If this assumption is correct, then the Central Highland Division represents a southerly extension of the Moine Assemblage: and the deposition of the Grampian Group upon a post-Grenville erosion surface in an opening ensialic basin, commenced during the interval between 1000 and 800 Ma. The Knoydartian event may be taken to represent reverse (compressional) movement on earlier-formed listric extensional faults related to the opening of this basin, and probably of proto-Iapetus. Subsequently, during the early Palaeozoic orogeny, the Central Highland Division has been reworked, but to a lesser extent than the Moine Assemblage N of the Great Glen Fault.

4.6 Regional implications

The geology of the Central Highland Division imposes some constraints on the magnitude of movements on the Great Glen Fault which separates the terrains of the Northern and Central Highlands, and on the magnitude of the initial separation of these terrains.

In terms of deeper crustal structure, both terrains appear to contain a similar Moine-type layer underlain by Lewisian-type material (Bamford *et al.*, 1978; Clayburn 1981). Both terrains contain Precambrian records which appear to be of a comparable order of complexity and age. In the Moine Assemblage north of the Great Glen Fault, including the Glenfinnan Division, the effects of the Grenville event have been documented for over a decade (e.g. Brook *et al.*, 1976). The possible presence of the Knoydartian event has been discussed for even longer (Giletti *et al.*, 1961, Long and Lambert, 1963; Bowes, 1968; Lambert, 1969; van Breemen *et al.*,

Figure 4.3 Products of Knoydartian shearing and the cover–basement unconformity. (*a*) Typical mylonite derived from semi-pelitic gneiss, with a swarm of quartz veins subconcordant to the mylonitic C-fabric; this rock also contains S–C fabrics and shear bands; Blargie near Laggan. (*b*) Weakly sheared Grampian Division semi-psammite (top) resting on coarse quartzofeldspathic gneiss of the Central Highland Division; Slochd Summit (Locality F1, Fig. 6 in Piasecki and Temperley, in press). (*c*) A typical, small pegmatitic segregation from a shear zone, formed during the shearing event. Porpyroblasts of muscovite (*M*), and oligoclaseandesine (*F*) granulated–recrystallized to a polygonal aggregate (*A*), both in a quartz vein (Q) subparallel to the mylonitic C-fabric. The match rests on mylonite; $\frac{1}{2}$ natural size; Laggan inlier. (*d*) Preserved remnant of the cover–basement unconformity. Large thin-section (photographed during grinding at 70 μm thickness) of part of a small block (right of *F*) downfaulted from the pre-Grampian Group erosion surface (above photograph). *B*, basement gneiss; *C*, cover; small faults enhanced for clarity. The erosion surface is preserved intact at *U*, slightly slipped at *X*. See also text. The scale is in centimetres; Slochd Summit (locality F7 on Fig. 6 in Piasecki and Temperley, in press).

1974, 1978; Piasecki and van Breemen, 1983). More recently, new Knoydartian ages have been obtained from numerous shear zones in the Glenfinnan and Morar Divisions, including the Sgurr Beag Thrust, indicating that the Knoydartian event of northerly-directed shearing and amphibolite-facies metamorphism has been widespread in the Moine Assemblage (Piasecki, 1984; Piasecki and van Breemen, unpublished work).

However, the Central Highland Division does not provide a *direct* equivalent to the Glenfinnan Division: rather, the two divisions seem to represent broadly equivalent, coeval complexes, in all probability, originally not too distant from each other. There is likewise no *direct* comparability between the early Palaeozoic development of the two terrains. Indeed, there are major differences. The most notable of these are the absence from the Northern Highlands of the equivalents of the Grampian Group, of the Dalradian Supergroup, and of the Grampian tectonometamorphic event: and the contrasting Caledonian plutonism in the two terrains which may be related to differing subduction history (van Breemen and Piasecki, 1983).

On the other hand, the end of the pervasive Caledonian upright folding (reworking) in the Central Highland Division and in the Northern Highlands are coeval at approximately 440 Ma (van Breemen *et al.*, 1974, 1979; van Breemen and Piasecki, 1983). These folds form regional steep belts in both terrains. The 'Loch Quoich Line', or the front of early Palaeozoic reworking in the Northern Highlands (Roberts and Harris, 1983) cannot be directly compared with the steep belt of the Central Highlands, because the two structures have opposing symmetries, but otherwise they appear to represent a broadly coeval pair of major, tectonically related features. Taken together, these considerations appear to preclude substantially larger sinistral (south-westerly) movements on the Great Glen Fault than *c.* 200 km (Winchester, 1973; van Breemen and Piasecki, 1983): they also imply that a NW–SE separation of these terrains across the Great Glen Fault may not have been extensive.

References

Anderson, J. G. C. (1956) The Moinian and Dalradian rocks between Glen Roy and the Monadhliath Mountains, Inverness-shire. *Trans. R. Soc. Edinburgh* **63**, 15–36.

Ashworth, J. R. (1979*a*) Comparative petrography of deformed and undeformed migmatites from the Grampian Highlands of Scotland. *Geol. Mag.* **116**, 445–456.

Ashworth, J. R. (1979*b*) Textural and mineralogical evolution of migmatites. In Harris, A. L., Holland, C. H. and Leake, B. E. (eds.), The Caledonides of the British Isles–Reviewed. *Spec. Publ. geol. Soc. London* **8**, 357–361.

Baker, A. J. (1986) Eclogitic amphibolites from the Grampian Moines. *Miner. Mag.* **50**, 217–221.

Bamford, D., Nunn, D., Prodehl, C. and Jacob, B. (1978) LISPB IV. Crustal structure of northern Britain. *Geophys. J. R. astron. Soc.* **54**, 43–60.

Berthe, D., Choukroune, P. and Jegouzo, P. (1979) Orthogneiss, mylonite and non-coaxial deformation of granites: the example of the South Armorican Shear Zone. *J. Struct. Geol.* **1**, 31–42.

Bowes, D. R. (1968) The absolute time scale and the subdivision of the Precambrian rocks of Scotland. *Geol. Foren. Forh. Stockholm* **90**, 175–188.

Bradbury, H. J., Harris, A. L. and Smith, R. A. (1979) Geometry and emplacement of nappes in the Central Scottish Highlands. In Harris, A. L., Holland, C. H. and Leake, B. E. (eds.), The Caledonides of the British Isles—Reviewed. *Spec. Publ. geol. Soc. London* **8**, 213–220.

Brook, M., Brewer, M. S. and Powell, D. (1976) Grenville age for rocks in the Moine of north-western Scotland. *Nature* **260**, 515–517.

Clayburn, J. A. P. (1981) Age and Petrogenetic Studies of some Magmatic and Metamorphic Rocks in the Grampian Highlands. Ph.D. Thesis, University of Oxford.

Giletti, B. J., Moorbath, S. and Lambert, R. StJ. (1961) A geochronological study of the metamorphic complexes of the Scottish Highlands. *Q. J. geol. Soc. London* **117**, 233–272.

Harris, A. L., Baldwin, C. T., Bradbury, H. J., Johnson, H. D. and Smith, R. A. (1978) Ensialic basin sedimentation: the Dalradian Supergroup. In Bowes, D. R. and Leake, B. E. (eds.), Crustal Evolution in northwestern Britain and adjacent regions. *Geol. J. Spec. Issue* **10**, 115–138.

Harris, A. L. and Pitcher, W. S. (1975) The Dalradian Supergroup. In Harris, A. L., Holland, C. H. and Leake B. E. (eds.), A Correlation of the Precambrian rocks in the British Isles. *Spec. Rep. geol. Soc. London* **6**, 52–75.

Hinxman, L. W. and Anderson, E. M. (1915) The Geology of Mid-Strathspey and Strathdearn. *Mem. geol. Survey, U.K.* Sheet 74.

Johnstone, G. S. (1975) The Moine succession. In Harris, A. L., Shackleton, R. M., Watson, J., Downie, C., Harland, W. B. and Moorbath, S. (eds.), A Correlation of the Precambrian Rocks in the British Isles. *Spec. Rep. geol. Soc. London* **6**, 32–42.

Lambert, R. StJ. (1969). Isotopic studies relating to the Precambrian history of the Moinian of Scotland. *Proc. geol. Soc. London* **1652**, 243–244.

Lambert, R. StJ. and McKerrow, W. S. (1976) The Grampian orogeny. *Scott. J. Geol.* **12**, 271–292.

Lister, G. S and Snoke, A. W. (1984) S-C Mylonites. *J. Struct. Geol.* **6**, 617–638.

Long, L. E. and Lambert, R. StJ. (1963) Rb–Sr isotopic ages from the Moine Series. In Johnson M. R. W. and Stewart F. H. (eds.), *The British Caledonides*, Oliver and Boyd, Edinburgh, 217–247.

Piasecki, M. A. J. (1975) Tectonic and metamorphic history of the Upper Findhorn, Inverness-shire, Scotland. *Scott. J. Geol.* **11**, 87–115.

Piasecki, M. A. J. (1980) New light on the Moine rocks of the Central Highlands of Scotland. *J. geol. Soc. London* **137**, 41–59.

Piasecki, M. A. J. (1984) Ductile thrusts as time markers in orogenic evolution: an example from the Scottish Caledonides. In Galson, D. A. and Mueller, St. (eds.), First European Geotraverse Workshop: the Northern Segment. *Eur. Sci. Found. Publ. Strasbourg* 109–114.

Piasecki, M. A. J. and Cliff, R. A. (in press) Syntectonic mineral growth in shear zones, and Devonian shearing in northern Trondelag. *Norges Geol. unders.*

Piasecki, M. A. J. and Temperley, S. (in press) The northern sector of the Central Highlands. In Allison, I., May, F. and Strachan, R. (eds.), *An Excursion Guide to the Moine of the Scottish Highlands.* Edinburgh and Glasgow geological Societies.

Piasecki, M. A. J. and van Breemen, O. (1979*a*) A Morarian age for the 'younger Moines' of central and western Scotland. *Nature* **78**, 734–736.

Piasecki, M. A. J. and van Breemen, O. (1979*b*) The 'Central Highland Granulites': cover–basement tectonics in the

Moine. In Harris, A. L., Holland, C. H. and Leake, B. E. (eds.), The Caledonides of the British Isles–Reviewed. *Spec. Publ. geol. London* **8**, 139–144.

Piasecki, M. A. J. and van Breemen, O. (1983) Field and isotopic evidence for a *c.* 750 Ma tectonothermal event in the Moine rocks in the Central Highland region of the Scottish Caledonides. *Trans. R. Soc. Edinburgh: Earth Sci.* **73**, 119–134.

Ramsay, J. G. (1967) *Folding and Fracturing of Rocks*. McGraw-Hill, New York, 568 pp.

Roberts, J. L. (1974) The structure of the Dalradian rocks in the SW Highlands of Scotland. *J. geol. Soc. London* **130**, 93–124.

Roberts, A. M. and Harris, A. L. (1983) The Loch Quoich Line—a limit of early Palaeozoic crustal reworking in the Moine of the Northern Highlands of Scotland. *J. geol. Soc. London* **140**, 883–889.

Roberts, J. L. and Treagus, J. E. (1977) Polyphase generation of nappe structures in the Dalradian rocks of the southeast Highlands of Scotland. *Scott. J. Geol.* **13**, 237–254.

Smith, T. E. (1968) Tectonics in Upper Strathspey, Inverness-shire, Scotland. *Scott. J. Geol.* **4**, 68–84.

Thomas, P. R. (1979) New evidence for a Central Highland root zone. In Harris, A. L., Holland, C. H. and Leake, B. E. (eds.), The Caledonides of the British Isles—Reviewed. *Spec. Publ. geol. Soc. London* **8**, 205–212.

Thomas, P. R. (1980) The structure and stratigraphy of the Moine rocks N of the Schiehallion complex, Scotland. *J. geol. Soc. London* **137**, 469–482.

van Breemen, O. and Piasecki, M. A. J. (1983) The Glen Kyllachy Granite and its bearing on the nature of the Caledonian Orogeny in Scotland. *J. geol. Soc. London* **140**, 47–62.

van Breemen, O., Pidgeon, R. T. and Johnson, M. R. W. (1974) Precambrian and Palaeozoic pegmatites in the Moines of northern Scotland. *J. geol. Soc. London* **130**, 493–507.

van Breemen, O., Halliday, A. N., Johnson, M. R. W. and Bowes, D. R. (1978) Crustal additions in Precambrian times. In Bowes, D. R. and Leak, B. E. (eds.), Crustal Evolution in North-western Britain and Adjacent Regions *Geol. J. Spec. Issue* **10**, 81–106.

van Breemen, O., Aftalion, M., Pankhurst, R. J. and Richardson, S. W. (1979) Age of the Glen Dessary Syenite, Inverness-shire: diachronous Palaeozoic metamorphism across the Great Glen. *Scott. J. Geol.* **15**, 49–62.

Wells, P. R. A. (1979) *P–T* conditions in the Moines of the Central Highlands, Scotland. *J. geol. Soc. London* **136**, 663–671.

White, S. H., Burrows, S. E., Carreras, J., N. D. and Humphreys, F. J. (1980) On mylonites in ductile shear zones. *J. Struct. Geol.* **2**, 175–187.

Winchester, J. A. (1973) Pattern of regional metamorphism suggests a sinistral displacement of 160 km along the Great Glen Fault. *Nature* **246**, 81–84.

Winchester, J. A. (1974) The zonal pattern of regional metamorphism in the Scottish Caledonides. *J. geol. Soc. London* **130**, 509–524.

Zaleski, E. (1985) Regional and contact metamorphism within the moy intrusive Complex, Grampian Highlands, Scotland. *Contrib. Miner. Pet.* **89**, 296–306.

5
The Moine Assemblage in Sutherland

S. J. MOORHOUSE and VALERIE E. MOORHOUSE

5.1 Introduction: a history of speculation and controversy

Sutherland is the least well-known area of the Moine Assemblage so it is something of a paradox that the name derives from the A' Mhoine area of the north Sutherland coast. The relative lack of research in the region is even more surprising, as the area has been an enduring source of speculation and controversy for over 100 years. Even the age of deposition and primary deformation of the Moine Assemblage in Sutherland–Proterozoic or Phanerozoic—has remained controversial.

In the middle 19th century, one view in the 'Highland Controversy' held that there was a conformable succession from the Cambro-Ordovician cratonic cover, up into the 'Silurian' age 'Eastern Schists' of the A' Mhoine area (Lapworth, 1885). The earliest Geological Survey view, formulated during late 19th century mapping of the north coast of Sutherland (for which no memoir was ever published), was that the area consisted of a mechanical mixture of Palaeozoic (Moine) and Archaean (Lewisian) produced during Caledonian thrusting (Peach and Horne, 1884). Before the end of that century, Greenly's discovery of sedimentary structures in the Moine led to a return of the view that it was a metasedimentary sequence, with the Lewisian now interpreted as inliers of basement gneiss (Flett, 1906; Peach *et al.*, 1907; Greenly, 1938).

Later Geological Survey work in the Moine Assemblage of Sutherland (Read *et al.*, 1926; Read, 1931) distinguished an apparently inverted west to east increase in metamorphic grade, culminating in the development of the east Sutherland migmatite complexes. His work in Sutherland led Read to attempt a drastic reinterpretation of many of the 'Lewisian inliers' as 'Durcha-type Moine Assemblage' integral sedimentary or volcanic parts of the Moine Assemblage (Read, 1934). This idea, based on work in the Loch Shin area of south Sutherland, was enthusiastically taken up by workers elsewhere, in the erroneous reinterpretation of Lewisian inliers (e.g. Sutton and Watson, 1953). Read's 'Durcha-type Moine Succession' theory was finally laid to rest by the pioneering geochemical correlation work of Winchester and Lambert (1970), which not only showed the 'Durcha type' rocks of the Loch Shin area to be Lewisian, but also scotched any idea that the Moine Assemblage contained major developments of calcareous sediments or volcaniclastics. The reality of the Lewisian origin of other Sutherland basement bodies has been shown by further geochemical studies (Moorhouse, 1976; Moorhouse and Moorhouse, 1977), and the recognition of Scourian pyroxene granulites in the Borgie (O'Reilly, 1971; Moorhouse, 1971, 1976) and Naver (Moorhouse and Harrison, 1976) inliers and of picritic Scourie dykes in the Borgie and Mudale inliers (Moorhouse, 1977).

On the basis of his interpretation of metamorphic fabrics in Sutherland, Read believed that the Moine Assemblage was a polyorogenic Proterozoic sequence (Read, 1934). Some later workers followed Read and recognized a prolonged Precambrian history in the Sutherland Moine Assemblage, including basement–cover interfolding, regional Precambrian migmatization, and Caledonian (*s.l.* throughout this work) overprinting of Precambrian garnets, in amphibolites and semi-pelites, throughout the Sutherland Moine Assemblage from the Moine Thrust zone into the east Sutherland migmatite complexes (Johnson, 1975; Moorhouse and Harrison, 1976; Moorhouse and Moorhouse, 1979*a*, 1979*b*, 1983; Evans and White, 1984; Barr *et al.*, 1986). However, others disagreed with Read, maintaining that all the deformation in the Sutherland Moine Assemblage was Caledonian (Soper, 1971; Soper and Brown, 1971; Soper and Wilkinson, 1975). In this latter view, Sutherland is a special case, in all or part representing a younger (?late Precambrian–early Phanerozoic) sequence, separated from the southern Moine Assemblage by either a Precambrian metamorphic front or a major unconformity (Soper and Barber, 1979). Indeed Soper and Barber (1982) go so far as to correlate the west Sutherland Moine Assemblage (between the Moine Thrust and the Meadie Thrust; see Fig. 5.1) with the foreland Torridonian Group or the probably younger Grampian Group (Harris *et al.*, 1978).

In view of this controversy, and the almost complete absence of reliable geochronological evidence, before considering the relationship of Sutherland to the Moine Assemblage elsewhere it is essential to attempt a clarification of the correlation of structural and metamorphic events, to determine the age of deposition and primary deformation of the Sutherland Moine Assemblage and its most likely correlatives.

5.2 Lithologies, tectonics and age of deposition

5.2.1 *Structure and tectonometamorphic sequence*

5.2.1.1 *Regional structure.* The regional structure of the Sutherland Moine Assemblage (Figs 5.1 and 5.2), is dominantly a stack of ESE to NW translated Caledonian ductile thrust (slide) bounded nappes (S. J. Moorhouse 1977; V. E. Moorhouse, 1979; Soper and Barber, 1982; Moorhouse and Moorhouse, 1983; Barr

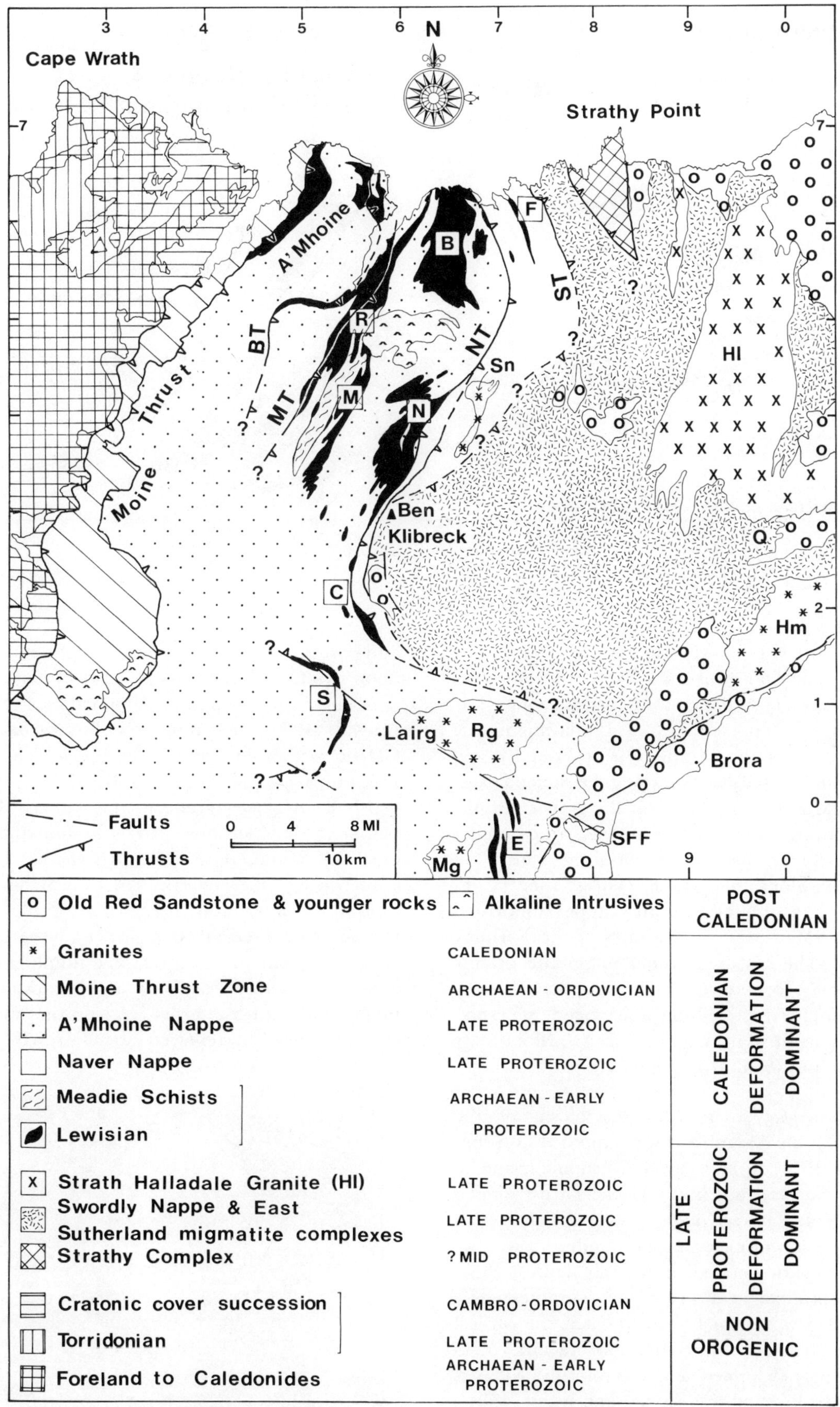

Figure 5.1 Geological map of Sutherland. Thrusts: *BT* Ben Hope; *MT* Meadie; *NT* Naver; *ST* Swordly. Granites: *Hl* Strath Halladale; *Sn* Strathnaver; *Hm* Helmsdale; *Rg* Rogart; *Mg* Migdale. Lewisian inliers: *R* Ribigill; *M* Mudale; *B* Borgie; *N* Naver; *F* Farr; *C* Crask; *S* Shin; *E* Evelix. Q Scaraben Quartzite. *SFF* Strath Fleet Fault.

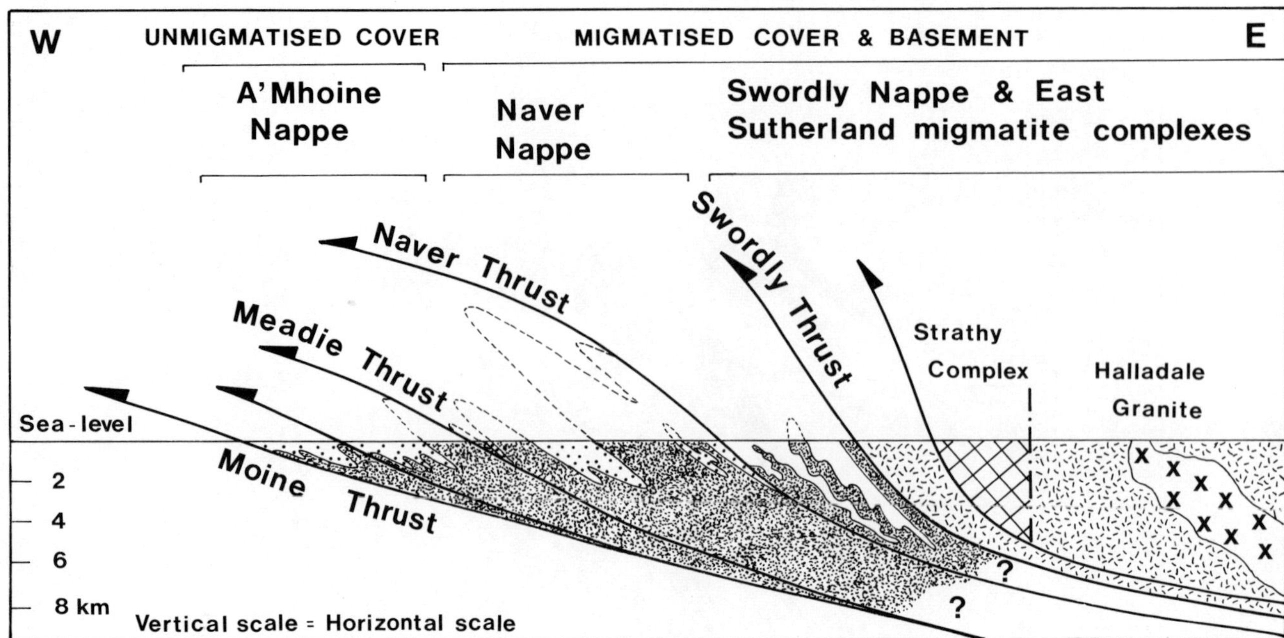

Figure 5.2 Schematic cross-section of the Sutherland Moine (Key as Fig. 5.1; Lewisian basement stippled.)

et al., 1986; Butler, 1986; Moorhouse *et al.*, 1987). The location of these (D2) thrusts may be partly controlled by the disposition of earlier (D1) high-strain zones. The westernmost A' Mhoine Nappe of non-migmatitic psammites overlies the Moine Thrust Zone (Peach *et al.*, 1907) and is itself overlain along the Naver Thrust zone (S. J. Moorhouse, 1977) by the Naver Nappe, consisting of migmatitic metasediments, amphibolites and granites of the Bettyhill Assemblage. This is, in turn, overlain along the Swordly Thrust zone (V. E. Moorhouse, 1979) by the Swordly Nappe, comprising the migmatitic gneisses and granites of the Kirtomy Assemblage. The east Sutherland migmatite granite complexes may form a still higher nappe, with the Strathy Complex representing a basement slice emplaced along its basal thrust (Figs 5.1, 5.2; Moorhouse and Moorhouse, 1983; Moorhouse *et al.*, 1987).

5.2.1.2 *Metamorphic history.* Metamorphism of the Sutherland Moine Assemblage was supposed by Soper and Brown (1971) to be entirely Caledonian, and to range from chlorite grade in the Moine Thrust zone to sillimanite in the Klibreck migmatites (Fig. 5.1), with a primary inversion of metamorphic grades, caused by fluid flow in pelitic horizons in the migmatites. However, Read's original polyorogenic idea is vindicated by recent detailed work which confirms that the earliest Caledonian structures in Sutherland rework rocks already strongly deformed and metamorphosed, presumably in the Precambrian (Evans and White, 1984). In addition, the dating of the Strath Halladale granite as Precambrian (649 ± 30 Ma; Brook, in Pankhurst, 1982), proves the reality of extensive Precambrian deformation in the Sutherland Moine Assemblage, as this little-deformed granite cuts the regional migmatites and post-dates the peak of metamorphism and deformation (McCourt, 1980; Storey and Lintern, 1981).

Clearly, the apparent simplicity of Soper and Brown's (1971) interpretation is illusory, caused by the Caledonian ductile thrust system which telescoped Precambrian metamorphic grades that increased eastwards from garnet grade in the western A' Mhoine Nappe (S. J. Moorhouse, 1977), to sillimanite grade in northeast Sutherland (V. E., Moorhouse, 1979). This sequence was then overprinted by Caledonian metamorphism which also increased in grade eastwards, retrogressing Precambrian garnet in the A' Mhoine Nappe (e.g. in the 'early Moine amphibolites'; S. J. Moorhouse, 1977; Moorhouse and Moorhouse, 1979*a*), maintaining at least garnet grade in the Naver Nappe (e.g. evidenced by renewed garnet growth in the Ard

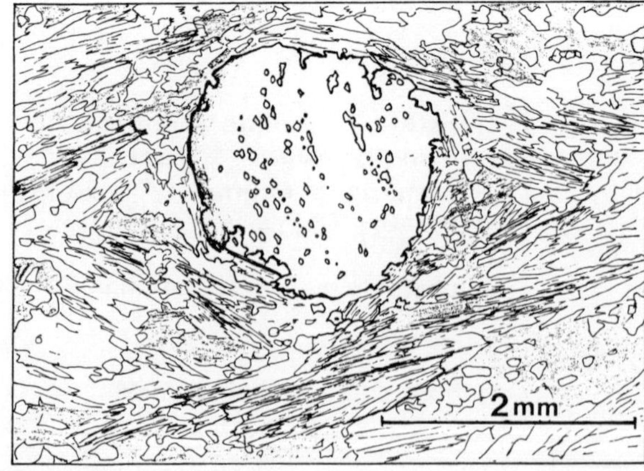

Figure 5.3 Early (Precambrian) garnet with discordant (S1) inclusion trail wrapped by later (Caledonian) mica schistosity (S2), in A' Mhoine Nappe Althaharra semi-pelite. (From Plate 6, S. J. Moorhouse, 1977).

Table 5.1 Tectonometamorphic events in the Sutherland Moine Assemblage.

A' Mhoine Nappe	Naver Nappe	Swordly Nappe
Pre-D1A* Deposition of arenaceous Moine sediments with local basal conglomerates unconformably on Lewisian basement. Early tholeiitic basic intrusives, ?tensional environment.	**Pre-D1N** Deposition of arenaceous/argillaceous Moine sediments on Lewisian basement. Early tholeiitic intrusives, tensional environment?	**Pre-D1S** Deposition of Moine greywacke sediments on Strathy Complex basement. Intrusion of tholeiitic amphibolites ?tensional environment.
		D1Sa Upper amphibolite-facies metamorphism, gneissosity (S1S). Rare siliceous lenses preserve F1Sa isoclinal folding of this S1S gneissosity.
D1A Heterogeneous strain producing locally intense zones of bedding-parallel (S1A) 'flattening' fabrics and N–S mineral extension lineation at greenschist–amphibolite transition facies. Local tight to isoclinal F1A folds.	**D1N** High strain at upper amphibolite facies, producing a strong gneissosity (S1N) and regional migmatization. Pre-D2N emplacement Clerkhill granite–appinite suite (G1N), ?partly mantle-derived.	**D1Sb** Second-phase isoclinal folds F1Sb, very rare pre-D2S isoclinal refolds. Continued upper amphibolite-facies metamorphism produces coarse regional migmatitic banding, enveloping siliceous lenses preserving F1Sa/F1Sb refolds.
D2A Ubiquitous, tight to isoclinal (F2A) folding on all scales; zones of high strain produce sheath folds. SE axes, congrous mineral extension lineation (L2A). Interfolding of basement and cover throughout the nappe. Movement along Meadie, Ben Hope and Moine (ductile) thrusts.	**D2N** Tight to isoclinal NW- and SE-plunging F2N folds of S1N gneissosity and migmatitic lits. K-feldspars in G1N augened. Congruous mineral extension lineation (L2N), variable axial surfaces. Sheath folds and zones of high strain. Movement along Naver Thrust. Amphibolite-facies metamorphism. Formation of (G2N) foliated granites as partial melts of local gneisses.	**D2S** Tight to isoclinal variable NW/SE plunging F2S folds of D1S migmatitic banding. Movement along Swordly and Strathy Thrusts. Upper amphibolite-facies metamorphism. Formation of (G2S) foliated granites as partial melts of local gneisses.
D3A Continuing D2A stress field and thrust movements, local open to tight folding, variable sense overturning, weak axial-surface schistosity and intersection lineation, S3A, and L3A, broadly coaxial with F2A. Structures small scale and spatially associated with zones of D2A high strain. Metamorphic grade similar or slightly lower than D2A event. Post-F2A syn-D2A/D3 A intrusion of the deformed 'Loch a' Mhoid metagabbros' and Strath Vagastie augen granite (G3A) ?partly mantle-derived.	**D3N** Upright, tight SE-plunging F3N folds, steep E axial surfaces. Variable axial extension, intersection and rodding lineation (L3N). Folds D2N, high-strain fabrics and foliation in G2N granites. Coaxial refolds of F2N and L2N. Mid(?)-amphibolite-facies metamorphism Post-F3N unfoliated pegmatite/granite (G3N).	**D3S** Upright, tight NW/SE plunging F3S folds, weak fabrics, coaxial refolds of F2S. Upper amphibolite-facies metamorphism, abundant coarse white granite (G3S), formed as local partial melt, engulfs all earlier structures. Continued strain and pegmatite production in Swordly Thrust Zone.
	Late D3N Variable, generally SE-plunging monoformal folds. Widespread retrogression.	**Late D3S** Variable, generally SE-plunging monoformal folds. Widespread retrogression.
D4A Widespread, late to post-orogenic monoformal, conjugate and kink-band folding and faulting.	**D4N** Local brittle folds. Faulting and tension gashes, emplacement of microgranitic, porphyritic and lamprophyric sheets (G4N). Faulting.	**D4S** Essentially post-orogenic local brittle conjugate and monoformal folds. Late microgranites occupy tension gashes. Continued faulting.

*The exact correlation of events from nappe to nappe is uncertain. D1A events are believed to be Precambrian and D2A Caledonian (*s.l.*), as D2A is correlated with early mylonite formation in the Moine Thrust zone. The early regional migmatites of D1N and D1S are broadly correlated and considered to be Precambrian. This is supported by correlation with the early regional migmatites cut by the Precambrian Strath Halladale Granite. D2N and D2S could be late Precambrian or Caledonian if they directly correlate with D2A. Pegmatites with Caledonian ages (unpubl. data) in the Swordly Thrust zone suggest D3S is Caledonian and comparable with D3N and D3A.

Mor amphibolites of Moorhouse and Moorhouse, 1979*a*) before a widespread late retrogressive phase. Thus, throughout the Sutherland Moine Assemblage there are numerous examples of early (Precambrian) garnets, containing inclusion trails marking an early (Precambrian) S1 schistosity, wrapped by a later (Caledonian) S2 mica schistosity (Fig. 5.3).

5.2.1.3 *Tectonometamorphic correlation.* The structural, metamorphic and igneous history within each nappe can be referred to a polyphase sequence of deformation, but correlation from one nappe to another is not certain (Table 5.1). Also, the precise relationship of the Klibreck and Strath Halladale migmatite complexes (Horne and Greenly, 1896; Read, 1931; Brown, 1971; McCourt, 1980; Storey and Lintern, 1981; Fig. 5.1) to these tectonic units is uncertain and complicated by intervening areas of poor exposure. However, the general sequence of events on the north coast, at Ben Klibreck and in Strath Halladale is remarkably similar, with early (Precambrian) structures, fabrics and regional migmatites being reworked and overprinted by later events and granite emplacement (Table 5.1).

One particular problem is that the structural relationships of the (650 Ma) Strath Halladale granite are ambiguous, as intrusion is said to post-date F3 in the Strath Halladale area (McCourt, 1980), whereas D3N in the Naver Nappe is apparently Caledonian (Table 5.1). A possible explanation is that the structural age of the Strath Halladale granite relates to marginal minor granite sheets, not the main granite body itself (Storey and Lintern, 1981). Clearly, the Strath Halladale granite is post-D1, as it cuts regional migmatites, but it could be that the 650 Ma date (on drill-core samples) relates to a syn- or pre-D2 main granite intrusion. This would enable a simpler correlation with events further west (Table 5.1).

Another problem concerns the pre- or syn-D3A (= Gallaig phase of Moorhouse and Moorhouse, 1979*b*) Strath Vagastie (G3A) augen granite (Table 5.1; Read, 1931; Fergusson, 1978), which is dated at 405 ± 11 Ma (Pidgeon and Aftalion, 1978). We find this date surprisingly young, in view of the clear tectonic fabric in parts of this suite of minor granite bodies (but see opposite view of Fergusson, 1978), and place no reliance upon it. Clearly a programme of structurally constrained radiometric dating is long overdue for the Sutherland Moine Assemblage.

5.2.2 *The A' Mhoine Nappe*

5.2.2.1 *Lithology and structure.* The dominantly psammitic A' Mhoine Nappe lithologies are of Morar Division aspect (Johnstone, 1975), they contain minor developments of semi-pelite and conglomerate (Fig. 5.4; Mendum, 1976), with sedimentary structures locally well preserved in areas of low strain (Fig. 5.5; Wilson *et al.*, 1953; Moorhouse *et al.*, 1987). The extensive development of Lewisian basement within this nappe suggests that the A' Mhoine psammites were never very thick. Neither were they ever regionally migmatized, and there is no good evidence to suggest a metamorphic grade higher than the greenschist–amphibolite-facies transition. This produced the early (D1A) garnet growth found throughout this nappe in both metasediments and early metabasic rocks (Fig. 5.3; Table 5.1), which predates Caledonian deformation, in contradiction of the views of Soper and Brown (1971), Soper and Wilkinson (1975) and Soper and Barber (1982). The kyanite and staurolite referred to by previous writers (Read, 1931; Soper and Brown, 1971) we refer to the sub-Moine basement as it occurs in a pelite within the Meadie Schists (Fig. 5.1; Moorhouse *et al.*, 1987), which appear to be largely derived by shearing of Lewisian feldspathic and biotitic lithologies (see later).

Figure 5.4 High-strain flattened quartzofeldspathic and quartzitic pebbles in A' Mhoine Nappe basal Strathan conglomerate overlying Melness Lewisian inlier. Hammer handle = 350 mm. (Mendum, 1976; Moorhouse *et al.*, 1987.)

Within this nappe are a number of D1A and/or D2A high-strain zones and minor ductile thrusts, the most extensive being the D2A Ben Hope Thrust, while the Meadie zone, above the Meadie Thrust (Figs 5.1 and 5.2), contains a concentration of minor ductile thrusts, highly strained horizons and some brittle low-angle faulting (Moorhouse *et al.*, 1987). Greenschist-facies D2A mylonitization associated with large-scale WNW movements along the (ductile) Moine Thrust, forming the tectonic base of the A' Mhoine Nappe, clearly overprints and locally retrogresses D1A fabrics and garnet-grade mineral assemblages (Table 5.1; Read, 1931; Evans and White, 1984; Barr *et al.*, 1986). Similar D2A high-strain fabrics and retrogressive effects are also seen above the Meadie Thrust (Fig. 5.1; Moorhouse, 1977; Moorhouse *et al.*, 1987).

5.2.2.2 *The Meadie zone.* This zone of Lewisian inliers, Meadie Psammite and Meadie Schists structurally overlies the Meadie Thrust and underlies the Altnaharra Psammites (Fig. 5.1). The Meadie Psammite is variably feldspathic, occasionally pebbly or 'gritty' and arkosic, infrequently preserving cross-bedding with

Figure 5.5 Sedimentary structures in the Altnaharra psammite of the upper A' Mhoine nappe. (*a*) Cross-bedding preserved in low-strain hinge zone of F2A fold, scale ruler = 20 mm wide; (*b*) Heavy-mineral bands, specimen = 150 mm across; (*c*) Soft-sediment deformation, specimen = 150 mm across. (From Plates 24, 43, 44; S. J. Moorhouse, 1977.)

thin heavy-mineral bands and minor epidotic lenses. Pebbly and gritty bands occur along only slightly modified unconformities with Lewisian inliers in the north of the zone (Moorhouse *et al.*, 1987). Within the psammite, thin semi-pelites are occasionally developed.

In addition to these undoubtedly Moine lithologies, there are three groups of 'Meadie Schists'—pelitic, semi-pelitic and psammitic (as defined below) which are mainly derived from sheared Lewisian basement (see later). The 'semi-pelitic' and 'psammitic' schists contain variable amounts of quartz, plagioclase, biotite and muscovite, with or without garnet. Chlorite may be a major constituent. They exhibit high-strain fabrics and commonly contain abundant quartz segregations, veins and lenses. These Meadie Schists occur between the Meadie Moine metasediments and biotite, hornblende and pyroxene gneisses of the Mudale and Ribigill Lewisian bodies (Fig. 5.1).

The Meadie Pelite occurs as a number of discontinuous bands and lenses within these schists, nowhere being in contact with undoubted Moine metasediments. It is a garnet-staurolite-kyanite-biotite-muscovite-chlorite pelitic schist, with variable amounts of quartz segregations and uncommon (?metasedimentary) amphibolite lenses (Moorhouse and Moorhouse, 1979*a*). Pelites of this type are very rare within the Moine (Rock *et al.*, 1986); nevertheless, similar pelites occur in the Strath Evelix area of SE Sutherland, associated with the Evelix Lewisian inliers (Fig. 5.1, 'E'), immediately north of the most northerly point to which we have traced the Sgurr Beag Thrust and Glenfinnan Division of Ross-shire (Johnstone, 1975).

5.2.2.3 *The Altnaharra Group.* This group comprises the Moine metasediments which overlie the Meadie zone and are overthrust by the Naver Nappe (Fig. 5.1). The preponderant lithology is psammite, differing from psammites further west only in being rather less 'gritty', more frequently banded and more often rich in quartz, as opposed to rich in feldspar. Sedimentary structures are less commonly preserved, but in low-strain zones cross-bedding and soft-sediment deformation structures are occasionally seen (Figs 5.5*a–c*). As a response to the overthrusting by the migmatites of the Naver Nappe, D2A feldspar augen are locally developed in psammites in the footwall to the thrust (Moorhouse, 1977). Semi-pelites of two types are sparingly developed. Biotite semi-pelites, rarely containing small garnets, occur as minor bands throughout the psammite. Garnet-muscovite-bearing semi-pelites are most often found closely associated with the large Lewisian inliers which occur below the Naver Thrust (Fig. 5.1).

5.2.2.4 *Amphibolites and age relations.* The A' Mhoine Nappe contains at least two suites of metabasic intrusive whose age of intrusion has a crucial bearing on the age of deposition and deformation of the Sutherland Moine Assemblage. Smith (1979) groups all the Sutherland amphibolites into his 'northern zone of amphibolites of igneous origin', extending from north Ross-shire to the north coast of Sutherland. According to him, this zone 'post-dated the emplacement of Lewisian slices into the Moine of Sutherland' [*sic*] and the amphibolites were intruded between *c.* 550 and 450 Ma. On the basis of Smith's idea, Brown (1983) states that this 'northern zone of amphibolites transgresses the boundary of the Morar and Glenfinnan Divisions [and the] tectonic junction between these Divisions ... predates the amphibolite intrusions' [*sic*].

These ideas are untenable, not least because they overlook the presence of two suites of amphibolites in the Sutherland Moine, the 'early Moine amphibolites' and the 'Loch a' Mhoid amphibolites' of Moorhouse and Moorhouse (1979*a*). Amphibolite sheets of the earlier suite of tholeiitic basic rocks, including the Ben Hope Sill (Read, 1931; Moorhouse and Moorhouse, 1977, 1979*a*; Winchester and Floyd, 1984; Moorhouse *et al.*, 1987), are folded by F2A folds which deform a Precambrian (D1A) garnet hornblende fabric (Table 5.1). Thus this suite was emplaced pre-D1A, manifestly earlier than the D2A development of the ductile thrust system and the F2A interfolding of basement and cover. The later suite of mildly alkaline Loch a' Mhoid metadolerites was emplaced post-F2A folding and frequently preserves relict igneous textures, but the dolerites were variably foliated by the continuous D2A-D3A stress field (Table 5.1; Moorhouse *et al.*, 1987). Thus, far from negating the idea of mid-Proterozoic deposition and Precambrian primary deformation of the Sutherland Moine—as suggested by Smith (1979) and Brown (1983)—the evidence of the amphibolites strongly supports this idea.

5.2.2.5 *Lewisian basement inliers.* In the western part of the nappe the basement bodies tend to lie along ductile thrusts, while in the centre and east of the nappe the basement lies in (F2A) fold cores. Many of the basement bodies appear to be in modified unconformable contact with their Moine cover, often with the development of a gritty or conglomeratic facies at the boundary, and they form the cores of originally NW-overturning F2A folds locally modified by D2A ductile thrusting (Moorhouse *et al.*, 1987).

The basement in the lower part of the nappe is lithologically and geochemically similar to the Laxfordian of the foreland, with variably preserved feldspathic migmatites and pegmatites (Moorhouse and Moorhouse, 1977; Moorhouse *et al.*, 1987). Above the Meadie shear zone, in the upper part of the nappe, the basement is closely analogous to the Scourian of the foreland and contains relict pyroxene granulites (Fig. 5.1; O'Reilly, 1971; S. J. Moorhouse, 1971, 1976, 1977; Moorhouse and Harrison, 1976). The immediate footwall of the Naver Thrust contains numerous small zones of highly strained Lewisian lithologies, some are demonstrably in F2A fold cores within the psammites.

Lewisian inliers are less extensive in south Sutherland, possibly indicating a generally deeper basement–cover boundary. However, F2A infolds of definite Lewisian occur in Straths Evelix and Carnaig (Moorhouse and Moorhouse, unpublished) and in the Loch Shin area (Winchester and Lambert, 1970). In the latter case we have seen no evidence for disposition in F2A fold cores, therefore, in view of the high strain associated with these inliers (Peacock, 1975), they may lie on a D2A thrust, possibly a continuation of the Ben Hope–Meadie Thrust system (Fig. 5.1).

5.2.3 *The Naver Nappe*

5.2.3.1 *The Bettyhill Assemblage.* This assemblage (Fig. 5.1) comprises variably migmatitic, frequently highly strained, thinly banded psammitic and subordinate semi-pelitic gneisses. Meta-igneous amphibolites are locally abundant (Moorhouse and Moorhouse, 1979*a*; Moorhouse *et al.*, 1987). A small number of biotite 'pelite' bands occur (Cheng, 1943), but aluminosilicate minerals are generally absent, as they are somewhat feldspathic semi-pelitic gneisses rather than true pelites. Extensively altered epidotic calc-silicate lenses are sparingly developed. No sedimentary structures are preserved, the rocks of this nappe being in general very highly strained and deformed, containing abundant foliated (G2N) and unfoliated (G3N) granite and pegmatite veins. This we interpret as Precambrian migmatization cut by younger (?late Precambrian and Caledonian) granitic sheets.

Within this nappe there are at least three groups of mafic rocks which have been widely confused in the past (Cheng, 1942; Geological Survey maps). It is essential to distinguish them, as they are important in understanding the age of deposition and deformation of the rocks in this nappe.

5.2.3.2 *Amphibolites.* The widespread amphibolites and their hornblende and biotite schist derivatives are inferred, on geochemical and structural grounds, to be part of the same intrusive phase of 'early Moine amphibolites' recognized in the A' Mhoine Nappe. They were intruded pre-D1N, as in the less highly strained examples a D1N fabric is seen folded by F2N folds (Table 5.1; Moorhouse and Moorhouse, 1979*a*; Moorhouse *et al.*, 1987). On the Geological Survey 1:62 500 North map some of the schistose amphibolites are erroneously shown as Lewisian basement, while on Sheet 5 (1:253 440) the Lewisian basement is shown as amphibolite or hornblende schist of intrusive igneous origin (Moorhouse *et al.*, 1987).

5.2.3.3 *Lewisian inliers.* In contrast to the extensive interleaving of basement and cover in the A' Mhoine Nappe, within the Naver Nappe the basement bodies are more localized. Only the Farr basement body is of a significant size (Fig. 5.1). It still retains geochemical characteristics comparable with the Scourian of the foreland (V. E. Moorhouse, 1979), and appears to form a (?D2N/D3N) fold core (Moorhouse *et al.*, 1987). Other minor bodies, referred to the Lewisian basement on geochemical grounds, occur above the Naver Thrust in the Torrisdale steep belt in Strathnaver and in the footwall to the Swordly Thrust (V. E. Moorhouse 1979; Moorhouse *et al.*, 1987).

5.2.3.4 *Appinites—Ach' uaine hybrids.* The third rock group containing mafic lithologies is the Clerkhill augen granite-gneiss/appinite body (Moorhouse *et al.*, 1987), which occurs immediately east of the Farr Lewisian, separated by a thin zone of migmatitic psammitic gneiss. It ranges in composition from ultramafic hornblende-pyroxene 'appinites' ($\pm$ pink feldspar) to K-feldspar granite-gneiss. Cheng (1942) confused these lithologies with the Farr Lewisian inlier (Fig. 5.1). The origin, age of emplacement, and correlatives of this body are currently under further study. However, geochemical work (Moorhouse and Moorhouse, unpublished) shows similarities with the appinite suite of Argyll (Wright and Bowes, 1979) as the rocks trend towards enrichment in K_2O, Sr, Ba, Ni, Cr, and light rare earths. This, of course, implies no age correlation. Similar rocks in central and south Sutherland are the 'Ach'uaine hybrids' of Read (1931), which occur in association both with younger granites (e.g. Rogart; Soper, 1963) and the Precambrian Strath Halladale granite (McCourt, 1980). This supports the suggestion of Brown (1983) that there may be more than one suite of appinitic rocks in Sutherland, especially as there is abundant evidence of the repeated evolution of alkaline magmas in Sutherland (Parsons, 1979), including the post-orogenic Ben Loyal syenites (Robertson and Parsons, 1974). We believe the Clerkhill body belongs to a Precambrian 'appinite' suite (presumably including the similar Strath Halladale rocks) as its structural relations indicate a pre- to syn-D2N intrusion age. Augen granite veins, apparently related to the Clerkhill body, are also developed in the surrounding gneisses and may post-date the regional migmatization (Table 5.1; Moorhouse *et al.*, 1987).

5.2.3.5 *The deformation sequence.* The deformation sequence within this nappe can also be described in terms of four main phases, but here there is early D1N regional migmatization, especially well developed in plagioclase-rich rocks, with the extensive development of later D2N cross-cutting foliated (G2N) granite veins and syn- to post-F3N (G3N) granite and pegmatite veins (Table 5.1). The later pegmatites extend for a short distance west into the footwall of the Naver Thrust (Read, 1931; S. J. Moorhouse, 1977), as does the upright D3N refolding which also forms the Torrisdale steep belt extending south along Strathnaver (V. E. Moorhouse, 1979; Moorhouse *et al.*, 1987). The state of strain in this nappe is generally very high, but there are some lower strain zones, within which F2N/F3N refolds are common (Fig. 5.6; Moorhouse *et al.*, 1987). In the hanging-wall of the Naver Thrust, D1N migmatitic lits are strongly deformed (Moorhouse *et al.*, 1987) and D1N (Precambrian) garnets are augened (S. J. Moorhouse, 1977).

Figure 5.6 Naver Nappe Bettyhill Assemblage, migmatitic psammitic gneiss in F2N/F3N coaxial refold, compass = 60 mm. (From Plate IV 14B; V. E. Moorhouse, 1979).

5.2.4 *The Swordly Nappe and the granite–migmatite complexes*

The Swordly Nappe comprises the Kirtomy Assemblage of migmatitic gneisses with abundant, D2S-D3S foliated and unfoliated granites, and small bodies of 'early Moine amphibolites' (Table 5.1; Moorhouse and Moorhouse, 1979*a*, 1983; Moorhouse *et al.*, 1987). Immediately overlying the Swordly Thrust, the lowermost member of the Kirtomy Assemblage is the Swordly Pelite, a migmatitic biotite-muscovite gneiss. Above the pelite, the Kirtomy migmatitic biotite gneisses are generally feldspathic but with minor, more pelitic developments. Fibrolitic sillimanite is sparingly developed in the more pelitic lithologies and uncommon pyroxene-bearing calc-silicate lenses are found. No sedimentary structures are preserved in these highly migmatitic lithologies.

There are discontinuous lenses of hornblendic gneisses referred to the Lewisian basement in the footwall or shear zone of the Swordly Thrust (Moorhouse *et al.*, 1987), but it is not clear whether or not they lie in fold cores. The Swordly Nappe appears broadly to share the structural sequence of the Naver Nappe, but with a somewhat more involved D1 history (Table 5.1). The southward continuation of the Swordly Thrust is conjectural. It can only be traced with certainty 8 km from the coast into the Strathy Bog. It may then swing west to join the Naver Thrust with the Klibreck migmatites lying above the Naver/Swordly Thrust zone (Fig. 5.1). In a similar manner to the north coast gneisses, the Klibreck Complex comprises Precambrian regional migmatites, but even more extensively overprinted with later granitic sheets.

At the present stage of research, the eastern boundary of the Swordly Nappe is conjectural. On the north coast, between the Kirtomy gneisses and the migmatites and granites of Strath Halladale (McCourt, 1980) the Strathy Complex (Harrison and Moorhouse, 1976; Moorhouse and Moorhouse, 1983; Moorhouse *et al.*, 1987) occurs as a block of lower crustal origin, bounded on the west by a now near-vertical thrust reactivated as a late brittle fault (Figs 5.1 and 5.2). The eastern boundary is obscured by superficial deposits and Old Red Sandstone cover, but may also be faulted. East of this boundary, the relatively siliceous Portskerra

Assemblage (Moorhouse and Moorhouse, 1983) is the coastal extension of the metasedimentary envelope to the Strath Halladale granite (McCourt, 1980). This may be a separate nappe emplaced along the Strathy Thrust (Figs 5.1 and 5.2), with the Strathy Complex representing a fragment of (?early Proterozoic) basement occurring at the base of this nappe (Fig. 5.1). Although it is shown on Geological Survey maps as part of the Moine Assemblage, we have shown the Strathy Complex to be lithologically, geochemically and structurally, dissimilar to any part of the Moine, and it represents the only known occurrence of rocks of basement origin above the Swordly Thrust zone (see Moorhouse and Moorhouse, 1983). In particular, the Strathy quartz-plagioclase grey gneisses are richer in Na, and poorer in Ti, K, Rb, Zr, Nb, La, Ce, Pb and Th than any group of Moine metasediments (Table 5.2). However, the complex is dissimilar to the Lewisian basement bodies in the lower nappes, with higher Y than any comparable group of Lewisian gneisses. The Y and light REE abundances imply an early Proterozoic rather than Archaean Scourian

Table 5.2 Mean analyses of lithological groups from within the Sutherland Moine Assemblage.

	A	B	C	D	E	F	G	H	I	J	K	L	M	N
SiO_2*	62.99	75.95	71.64	62.05	63.30	63.92	77.23	61.91	78.02	58.53	63.00	80.50	72.88	65.02
TiO_2	0.90	0.56	0.81	0.87	0.73	0.72	0.37	1.05	0.50	1.11	0.91	0.48	0.26	0.59
Al_2O_3	15.92	10.66	11.87	16.93	16.42	14.68	10.42	14.89	10.74	16.41	16.54	8.80	12.34	15.53
Fe_2O_3[2]	6.54	3.23	4.89	7.09	7.82	5.89	2.91	7.38	3.17	7.24	7.09	2.49	3.53	5.05
MnO	0.08	0.04	0.05	0.04	0.05	0.05	0.02	0.07	0.06	0.10	0.11	0.03	0.04	0.09
MgO	2.13	1.14	2.02	2.58	3.14	2.32	1.12	2.78	1.03	2.85	1.89	0.63	1.64	2.45
CaO	2.19	1.40	1.67	2.01	1.54	3.33	1.66	2.09	1.41	2.54	1.89	1.13	1.74	5.56
Na_2O	2.92	2.59	2.93	2.84	1.93	4.02	3.02	2.31	2.26	2.77	2.00	2.25	5.16	5.34
K_2O	3.66	3.30	2.69	3.47	2.94	2.04	2.15	4.17	2.92	4.68	4.38	2.88	1.14	1.11
P_2O_5	0.25	0.11	0.16	0.25	0.17	0.14	0.11	0.20	0.08	0.22	0.25	0.09	0.08	0.11
S[1]	341	307	502	350	417	275	297	320	130	250	182	247	844	—
Ba	853	809	747	690	594	604	621	877	763	888	884	725	347	330
Ce	132	85	104	128	116	72	63	92	61	119	148	80	35	67
Cr	78	55	63	80	98	96	60	85	48	79	83	52	29	50
Cu	28	9	9	26	56	23	19	23	7	7	28	12	27	21
Ga	22	8	11	23	22	18	7	22	2	25	24	5	7	19
La	43	28	37	41	37	26	23	37	20	42	82	28	7	19
Nb	19	11	13	20	11	5	5	14	12	19	20	10	3	5
Ni	54	27	40	64	95	72	37	60	23	65	59	19	12	75
Pb	28	24	22	30	26	15	15	21	15	29	24	21	10	10
Rb	143	95	92	155	110	70	60	151	88	183	172	83	21	32
Sr	340	281	357	266	201	435	252	285	222	354	231	207	199	419
Th	17	13	15	17	15	7	7	12	7	18	20	11	2	10
Y	40	22	29	40	31	14	9	25	16	42	37	17	25	10
Zn	97	38	77	109	103	58	31	95	18	118	27	25	62	39
Zr	329	323	409	267	180	256	165	362	298	380	291	292	138	188
Sr/Y	8.50	12.77	12.31	6.65	6.48	31.07	28.00	11.40	13.87	8.43	6.24	12.18	7.96	41.90
K/Rb	212	288	243	186	222	242	297	229	275	212	211	288	451	288
Rb/Sr	0.42	0.34	0.26	0.58	0.55	0.16	0.24	0.53	0.40	0.52	0.74	0.40	0.11	0.08
Ca/Y	391	455	412	359	355	1701	1319	598	630	432	365	475	498	3975
Number of analyses	9	49	36	11	13	14	38	7	15	6	9	88	52	28

* Oxides as per cent.
[1] Trace elements in ppm.
[2] Fe_2O_3 is total Fe as Fe_2O_3.

A. Naver Nappe: Bettyhill Assemblage Moine semi-pelitic gneiss.
B. Naver Nappe: Bettyhill Assemblage Moine psammitic gneiss.
C. Swordly Nappe: Kirtomy Assemblage Moine biotite gneiss.
D. Swordly Nappe: Kirtomy Assemblage Moine semi-pelitic gneiss (Swordly Pelite).
E. A' Mhoine Nappe: Meadie Lewisian garnet-staurolite-kyanite pelite.
F. A' Mhoine Nappe: Meadie Lewisian biotite schist ('semi-pelitic').
G. A' Mhoine Nappe: Meadie Lewisian quartz-plagioclase schist 'psammitic').
H. A' Mhoine Nappe: Meadie Moine semi-pelite.
I. A' Mhoine Nappe: Meadie Moine psammite.
J. A' Mhoine Nappe: Altnaharra Moine biotite semi-pelite.
K. A' Mhoine Nappe: Altnaharra Moine muscovite-biotite semi-pelite.
L. A' Mhoine Nappe: Altnaharra Moine psammite.
M. Strathy Complex: quartz-plagioclase grey gneiss (Moorhouse and Moorhouse, 1983).
N. Ribigill Lewisian: biotite and hornblende gneisses (Moorhouse and Moorhouse, 1977).

Lewisian origin (Moorhouse and Moorhouse, 1983). This suggests a marked change in the nature of the basement beneath east Sutherland, with Strathy Complex type Proterozoic basement—not Scourian Lewisian—immediately underlying the Moine. It may be that basement of this type represents the ultimate source of the mid-Proterozoic zircons found in many Highland granites (Pidgeon and Aftalion, 1978).

Near the east Sutherland coast in the Scaraben area (Fig. 5.1), quartzite and pelite have been tentatively referred to the Dalradian on the basis of their lithological association and the boron content of stream sediments (Johnstone *et al.*, 1979). We are currently studying the geochemistry of these rocks, but the imprecise nature of stream-sediment studies, with the 'boron anomaly' possibly originating from Moine metasediments, or Devonian 'Old Red Sandstone', pegmatites/granites, renders this correlation highly speculative.

5.2.5 *Conclusions on the age of deposition and deformation*

We conclude that all the Sutherland Moine Assemblage was deposited in the late mid-Proterozoic and has undergone strong Precambrian metamorphism and deformation. Precambrian ('early Moine') amphibolites occur in all units so far investigated, apparently ruling out the possibility that part of the Sutherland Moine Assemblage is younger and has only undergone Caledonian deformation. The Sutherland Moine is thus directly comparable with the Moine Assemblage to the south, both in age and deformation. Above the Naver Thrust, Precambrian regional migmatites are strongly deformed and overprinted by ?late Precambrian and Caledonian granitic vein complexes. The Torrisdale steep belt in the Naver Nappe may have been produced by ramping on a lower thrust plane—as it cut through the interfolded Lewisian and Moine which crops out just west of, and is overridden by, the Naver Thrust

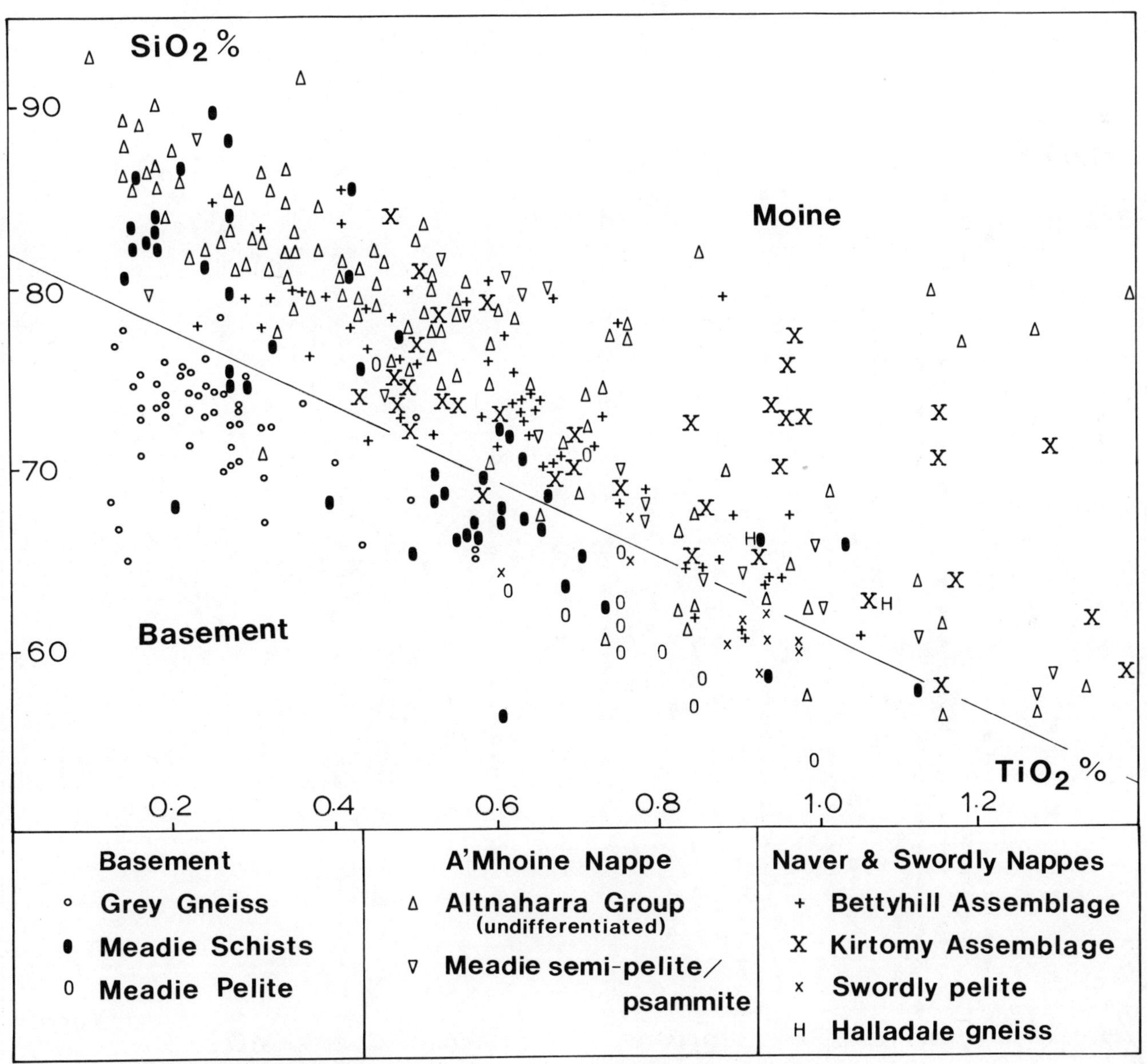

Figure 5.7 SiO_2 *v.* TiO_2 plot for Sutherland Moine metasediments, Meadie Schists and Strathy Complex Grey Gneisses. Boundary from Moorhouse and Moorhouse (1977, 1983).

(Fig. 5.1). Thus, it may be compared to the Caledonian deformation of the Quoich Line steep belt (Barr *et al.*, 1986), albeit on a smaller scale. The eastern limit of strong Caledonian reworking may be marked by the occurrence of the (?early Proterozoic) Strathy Complex on the north coast, possibly marking the base of an upper (Strathy) nappe. East of this, much of the deformation in the Portskerra Assemblage and the Strath Halladale Complex appears to be Precambrian.

5.3 Geochemical characterization of the Sutherland Moine Assemblage

5.3.1 *Analytical method*

Some 296 samples of supposed Moine lithologies from central and northern Sutherland have been analysed, mainly for 26 elements, using Philips PW1212 and PW1450 X-ray fluorescence spectrometers at Birmingham and Hull Universities and The Hatfield Polytechnic (see S. J. Moorhouse, 1977; Moorhouse and Moorhouse, 1979*a*).

The analyses are summarized as fourteen groups in Table 5.2, along with the averages of 28 biotite and hornblende gneisses from the Ribigill Lewisian inlier, and 52 quartz-plagioclase grey gneisses from the Strathy Complex (Moorhouse and Moorhouse, 1977, 1983; full details are available from the authors). The terms pelite, semi-pelite and psammite are used in the sense that 'pelites' contain 70% or more of mica and/or kyanite, staurolite or sillimanite; 'semi-pelite' contains less than 67% SiO_2; and 'psammite' contains more than 67% SiO_2. The latter subdivision is based on the fact that one of the principal minima in a crude SiO_2 histogram of Moine metasediments lies at *c.* 67% SiO_2, both in Sutherland and elsewhere (Winchester *et al.*, 1981).

5.3.2 *Geochemical results*

5.3.2.1 *Geochemistry of the Meadie Schists.* The Lewisian basement affinities of the Meadie Schists are reflected in their tendency to low Ti relative to silica, greater Na

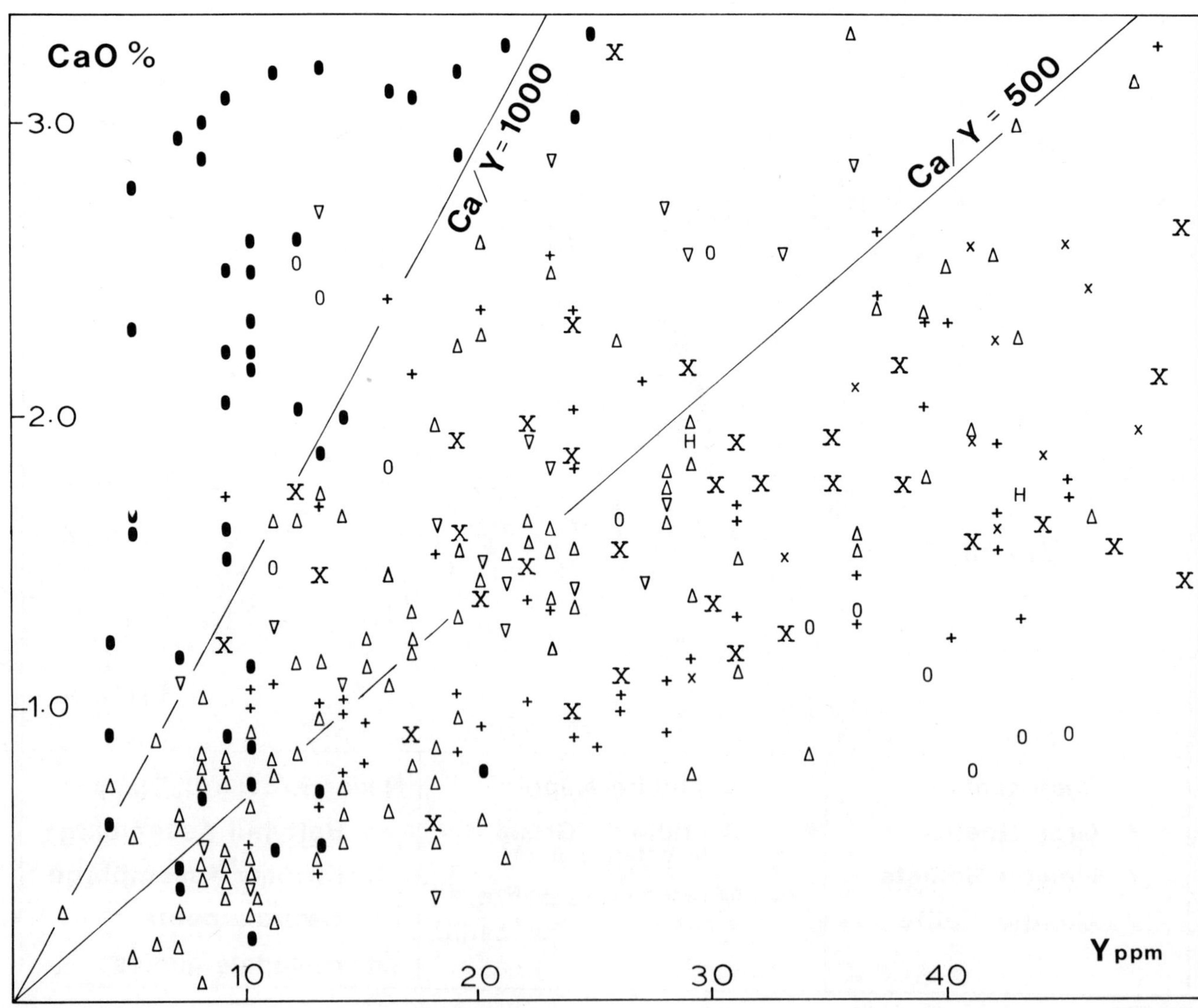

Figure 5.8 CaO *v.* Y, Moine metasediments and Meadie Schists. Key as for Fig. 5.7.

than K, high Ca, Sr and Ca/Y, Ca/La and Sr/Y ratios, with low Y, Zr, Nb, Ce, Pb, Th, and Rb/Sr compared with Moine metasediments (Figs 5.7–5.12, Table 5.2). On Lewisian–Moine discrimination diagrams, such as SiO_2 *v.* TiO_2, CaO *v.* Y, and CaO *v.* La (Figs 5.7–5.9; Moorhouse and Moorhouse, 1977, 1983), they mainly plot in or close to the Lewisian fields, while their relatively low Y distinguishes them from Strathy Complex basement lithologies. On the K_2O *v.* Na_2O plot they mainly fall in a low-K field overlapping Lewisian feldspathic gneisses and the Strathy Complex grey gneisses (Fig. 5.10). The Meadie Schists occupy a similar position to the Morar Basal Pelite on a Sr/Y *v.* SiO_2 diagram (cf. Fig. 4, Winchester *et al.*, 1981), probably because the Basal Pelite is also, at least in part, tectonically derived from Lewisian basement gneiss. The scatter across into the Moine fields on the CaO *v.* Y and CaO *v.* La plots (Figs 5.8 and 5.9) parallels a similar scatter seen in relatively little-deformed feldspathic gneisses from the Ribigill Lewisian inlier (cf. Fig. 4, Moorhouse and Moorhouse, 1977). This is taken as evidence that the Meadie Schists are highly strained derivatives of Lewisian feldspathic and biotitic gneisses.

The Meadie Pelite also shows substantial differences from Moine pelitic and semi-pelitic groups, having lower average K, Rb, Y, Zr and Nb (Table 5.2), and as it lies entirely within Meadie Schists, it is also considered to be of basement origin. The Meadie Pelite fits the criteria of Rock *et al.* (1986) as an 'anomalous pelite' in terms of mineralogy, structural setting, and its association with calcareous clinopyroxene gneisses in the adjacent Mudale Lewisian inlier (S. J. Moorhouse, 1977), which are comparable with calcareous lithologies in the Loch Shin Lewisian (Fig. 5.1). However, in this case the pelite and the enclosing schists are more likely to be tectonically derived from Lewisian, possibly late Scourian or Laxfordian, plagioclase and biotite schists or gneisses.

5.3.2.2 *Psammitic Moine lithologies.* In general, there appears to be very little geochemical difference between the psammitic rocks of the A' Mhoine and Naver

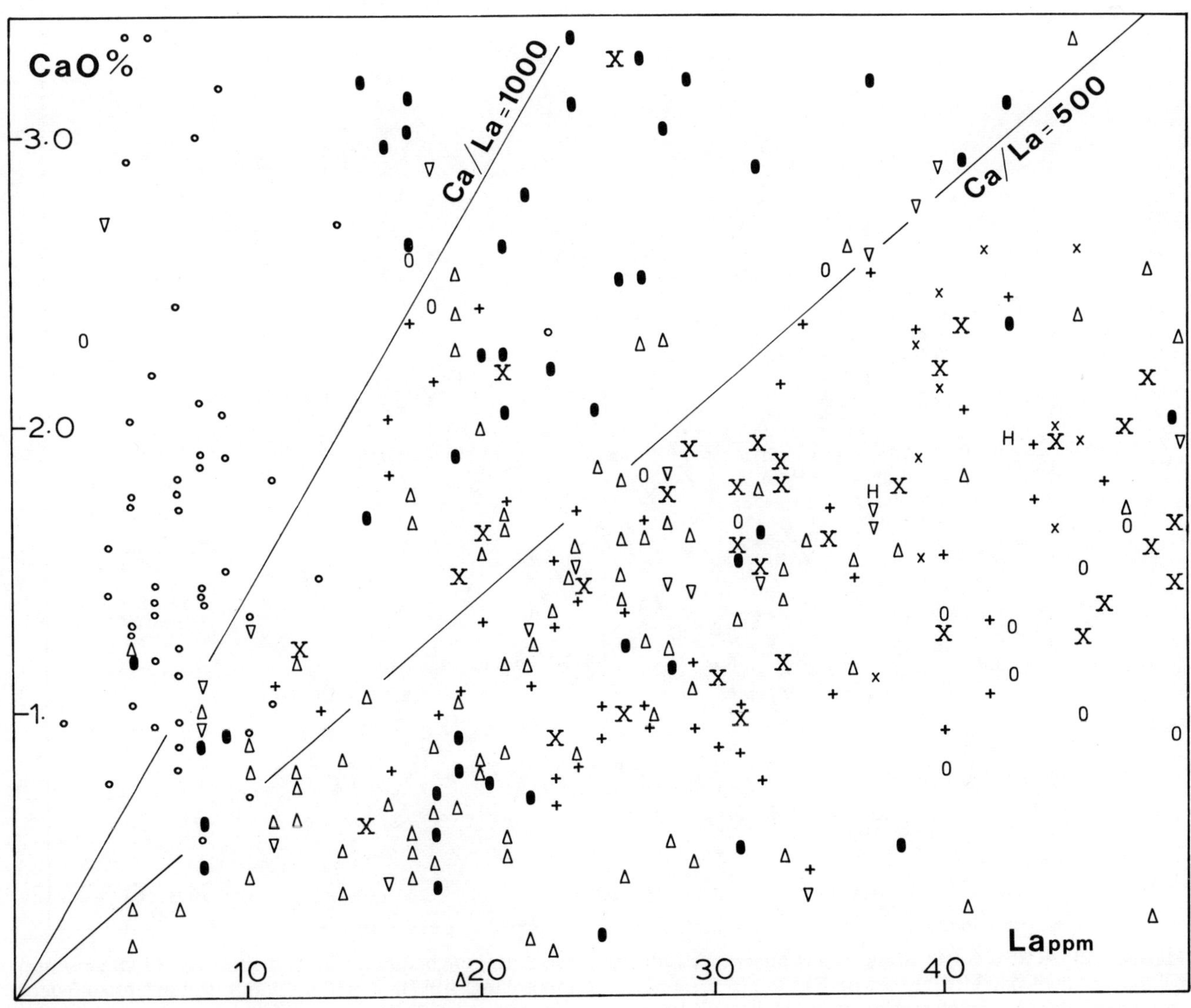

Figure 5.9 CaO *v.* La, Moine metasediments, Meadie Schists, Strathy Complex Grey Gneiss. Key as for Fig. 5.7.

Nappes, except that quartz-rich psammites are more common in the A' Mhoine Nappe (Figs 5.7, 5.11; Table 5.2). However, the Kirtomy Gneiss of the Swordly Nappe has a somewhat more sodic, less potassic composition (Fig. 5.10), which, together with higher average Sr and lower Rb/Sr ratio (Table 5.2), suggests an originally more plagioclase-rich sediment. Thus, on a log Na_2/K_2O *v.* log SiO_2/Al_2O_3 diagram (Pettijohn *et al.*, 1973) the Kirtomy Gneiss crosses the lithic arenite into the greywacke field, while the other psammites fall mainly within the lithic arenite, arkose and subarkose fields (Fig. 5.11). Assuming the Kirtomy Gneiss geochemistry has not been significantly affected by element mobility during migmatization and granite formation, there are two possible explanations of these differences. They may reflect a rather more extended weathering and transport history for the Kirtomy Assemblage, with replacement of feldspars by clays and loss of K in solution. However, as the differences between the Kirtomy Gneiss and the other psammites are not great, the gneisses possibly represent derivation from a more K-feldspar-poor source, perhaps of K-poor, quartz-plagioclase, Strathy Complex-type basement gneisses (Fig. 5.10; Table 5.2; Moorhouse and Moorhouse, 1983). If the psammites of the western nappes were derived from Lewisian gneisses similar to those of the Ribigill inlier—which includes K-feldspar-rich, possibly Laxfordian, pegmatites and migmatites (Moorhouse and Moorhouse, 1977; Moorhouse *et al.* 1987)—this would adequately explain the compositional differences.

5.3.2.3 *Semi-pelitic Moine lithologies.* Using the Nb/Y *v.* Sr, Zr/Nb *v.* Nb/P_2O_5 and Sr *v.* Nb/P_2O_5 discrimination diagrams of Winchester *et al.* (1981, 1983), the sparingly garnetiferous and muscovite-bearing Swordly Pelite consistently plots in the 'Glenfinnan Division field', while the Bettyhill Assemblage biotite semi-pelitic

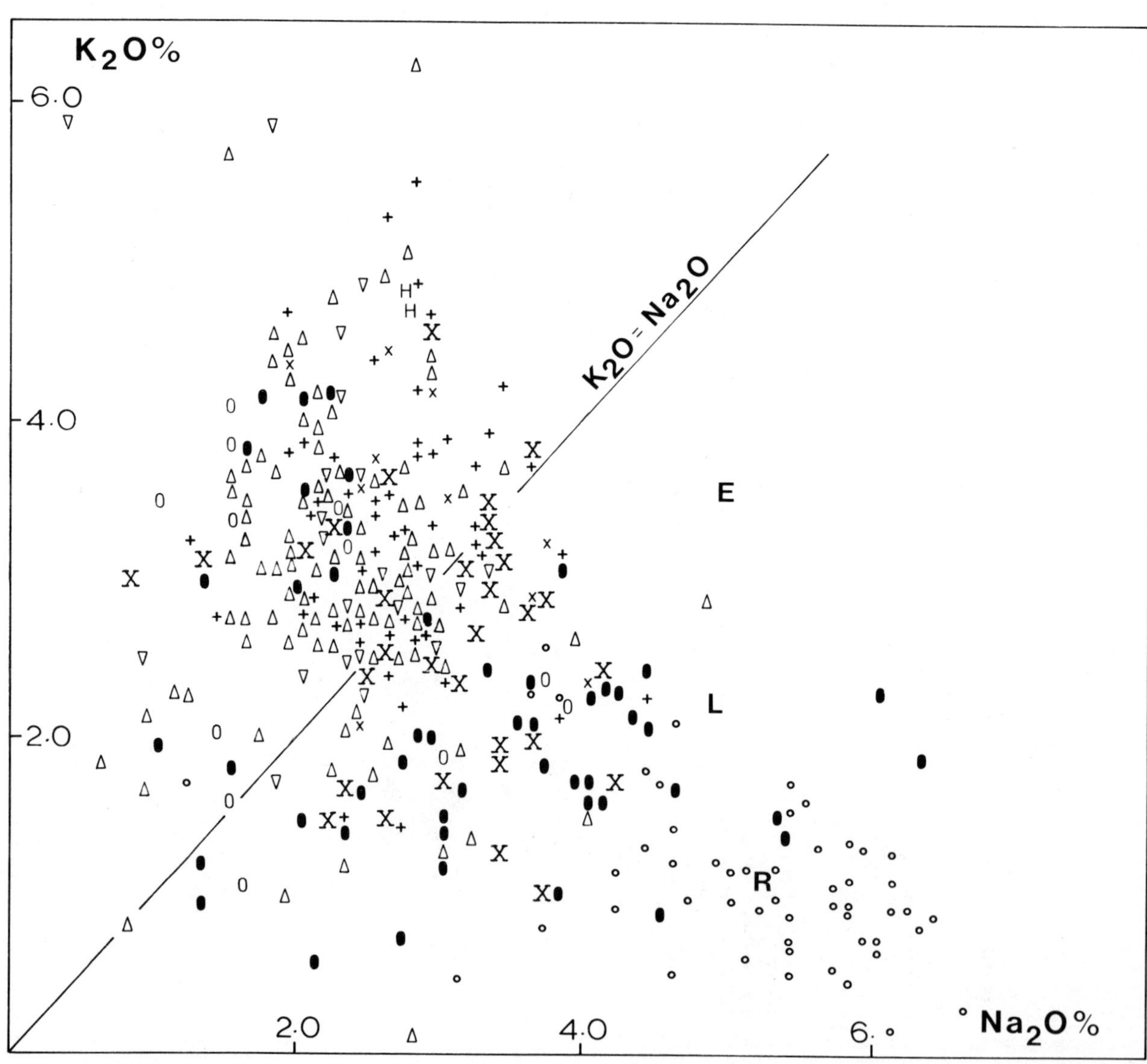

Figure 5.10 K_2O *v.* Na_2O, Moine metasediments, Meadie Schists, Strathy Complex Grey Gneiss. Key as Fig. 5.7, plus: *R* = mean Lewisian hornblende-biotite gneiss, Ribigill Inlier (Moorhouse and Moorhouse, 1977); *E* = mean Lewisian feldspathic migmatitic gneiss, Ribigill Inlier (Moorhouse and Moorhouse, 1977); *L* = mean foreland Laxfordian gneiss (Tarney *et al.*, 1972).

gneisses, and the biotite semi-pelites from the A' Mhoine Nappe, mainly plot in the 'Morar Division field' (Figs 5.13–5.15). This apparently supports the tentative suggestion that the Kirtomy Assemblage of the Swordly Nappe might represent the continuation of the Glenfinnan Division in Sutherland (V. E. Moorhouse, 1979; Barr *et al.*, 1986; Moorhouse *et al.*, 1987). However, the geochemical evidence is not as clear-cut as might be wished. On these diagrams the semi-pelitic Kirtomy Gneisses mainly plot in the 'Morar Division field', while semi-pelites from the A' Mhoine Nappe also fall into two different groups. Biotite semi-pelites mainly occupy the 'Morar Division fields', while garnet–muscovite-bearing semi-pelites fall into the 'Glenfinnan Division fields'. Clearly, these compositional differences are mineralogically controlled, with garnet- and muscovite-bearing pelites and semi-pelites plotting as distinct from biotite semi-pelites in both A' Mhoine and Swordly Nappes. Thus, the use of these diagrams for lithostratigraphic correlation of these Sutherland Moine groups appears to be ambiguous. However, the fact that elsewhere this mineralogical differentiation does not apparently occur, with pelites and more feldspathic semi-pelites (even semi-psammites and psammites) of the Morar Division being clearly distinguishable from the Glenfinnan and Loch Eil Division lithologies (Winchester *et al.*, 1981, 1983), suggests some peculiarity inherent in the formation of some Sutherland semi-pelites.

The field evidence strongly suggests that the A' Mhoine Nappe is the northern continuation of, at least, part of the Morar Division. Therefore, perhaps the muscovite-bearing semi-pelites of the A' Mhoine Nappe do not represent unmodified sedimentary compositions. A clue is given by their field occurrence: they almost all occur adjacent to Lewisian inliers, they invariably display high-strain (D2) fabrics with the muscovite aligned with that fabric. This occurrence of muscovite implies metasomatic transfer of elements during high-strain D2 deformation. If this was accompanied by

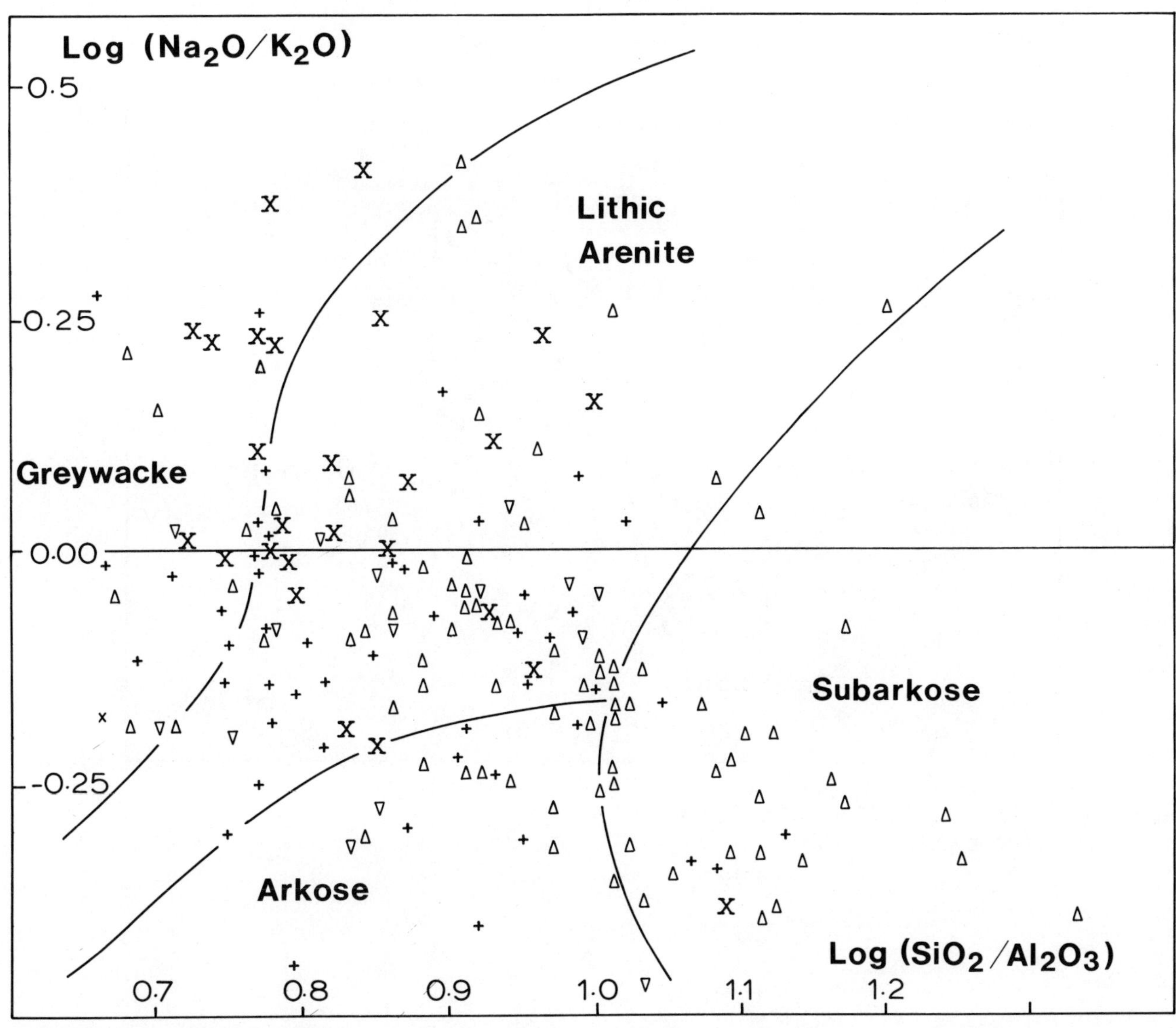

Figure 5.11 Na_2O/K_2O *v.* SiO_2/Al_2O_3 for Moine Psammites and psammitic gneisses. Sandstone fields after Pettijohn *et al.* (1973). Key as for Fig. 5.7.

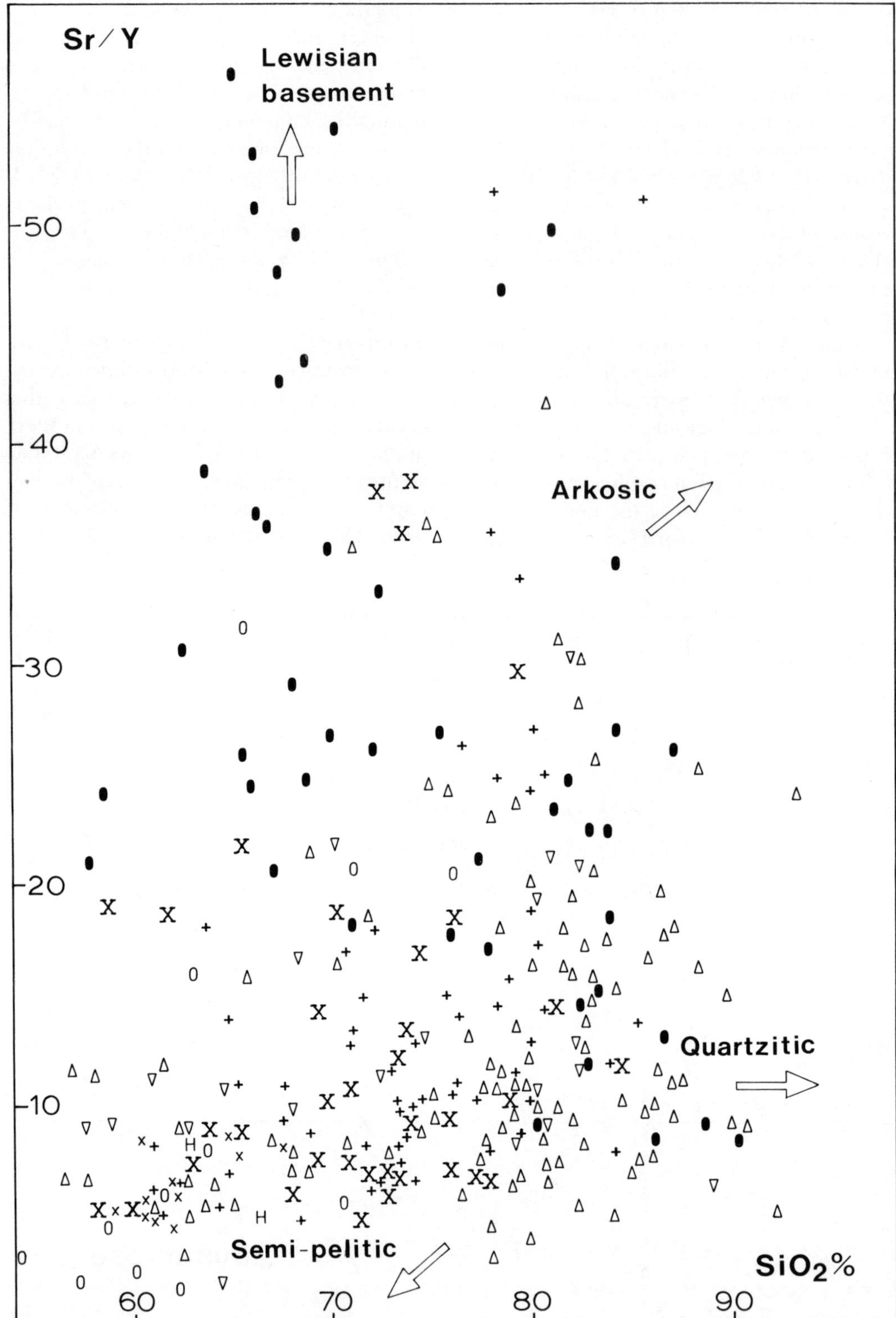

Figure 5.12 Sr/Y *v.* SiO_2 for Moine metasediments and Meadie Schists. Key as for Fig. 5.7.

breakdown of plagioclase, Sr might also be mobilized, resulting in Sr depletion of these rocks. This would cause them to plot in the 'Glenfinnan Division field' on the Nb/Y *v.* Sr and Sr *v.* Nb/P_2O_5 diagrams (Figs 5.13 and 5.15). In fact, the two groups of semi-pelites have similar Nb/Y and Nb/P_2O_5 ratios, with the muscovite-bearing semi-pelites having lower Sr (Table 5.2). They also have lower Zr, and on the Zr/Nb *v.* Nb/P_2O_5 plot (Fig. 5.14) the muscovite semi-pelites fall in a separate low Zr/Nb field from the biotite semi-pelites. This could possibly be explained by loss of Zr from sphene or biotite breaking down during the D2 high strain.

Exactly similar geochemical differences to these are exhibited by the muscovite-bearing Swordly Pelite when compared with the more feldspathic Kirtomy Gneiss and Bettyhill (semi-pelitic) Gneiss. Once again, an examination of the field relations of the Swordly Pelite provides an explanation. The Swordly Pelite lies in the shear zone of the Swordly (ductile) Thrust at the base of the Swordly Nappe. It exhibits a high-strain fabric, and carries abundant muscovite and several generations of muscovite-bearing pegmatite. Closely associated with this shear-zone are small bodies of Lewisian basement gneiss (V. E. Moorhouse, 1979; Moorhouse *et al.*, 1987). Although developed on a significantly larger scale than the smaller muscovite semi-pelites of the A' Mhoine Nappe, the same geochemical arguments of breakdown of plagioclase and sphene or biotite—with the resultant mobility of Sr and Zr—apply to the Swordly Pelite.

The somewhat surprising conclusion is that after excluding the sheared muscovite-bearing semi-pelites and Swordly Pelite, we are left with the other Sutherland semi-pelites plotting as 'Morar Division'. The only exceptions are two garnet-biotite semi-pelites from Strath Halladale (Storey and Lintern, 1981)

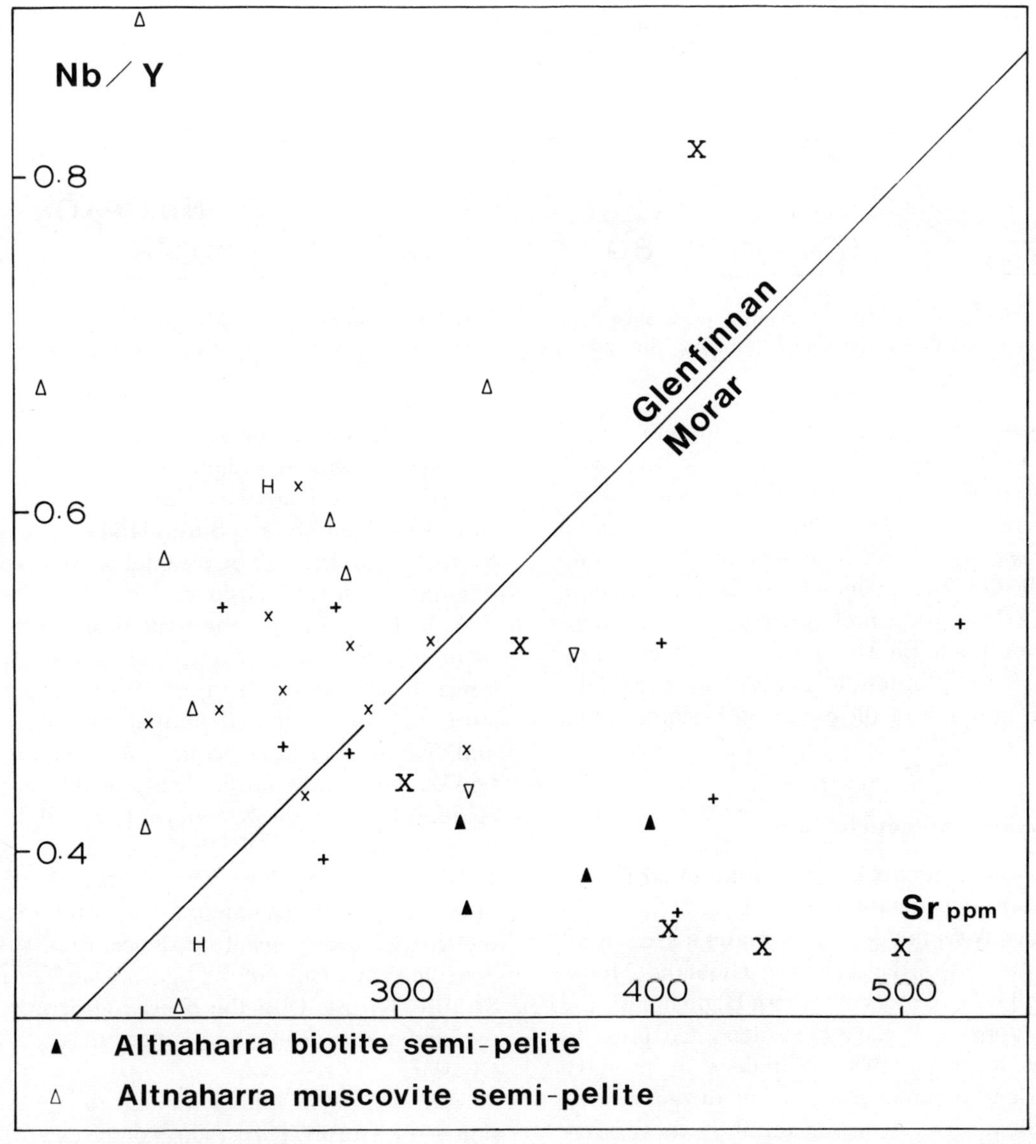

Figure 5.13 Nb/Y *v.* Sr discrimination diagram for pelitic and semi-pelitic Moine metasediments. Boundary between Glenfinnan and Morar Division fields after Winchester *et al.* (1981, 1983). Key as for Fig. 5.7, with Altnaharra Group semi-pelites differentiated into biotite and muscovite bearing.

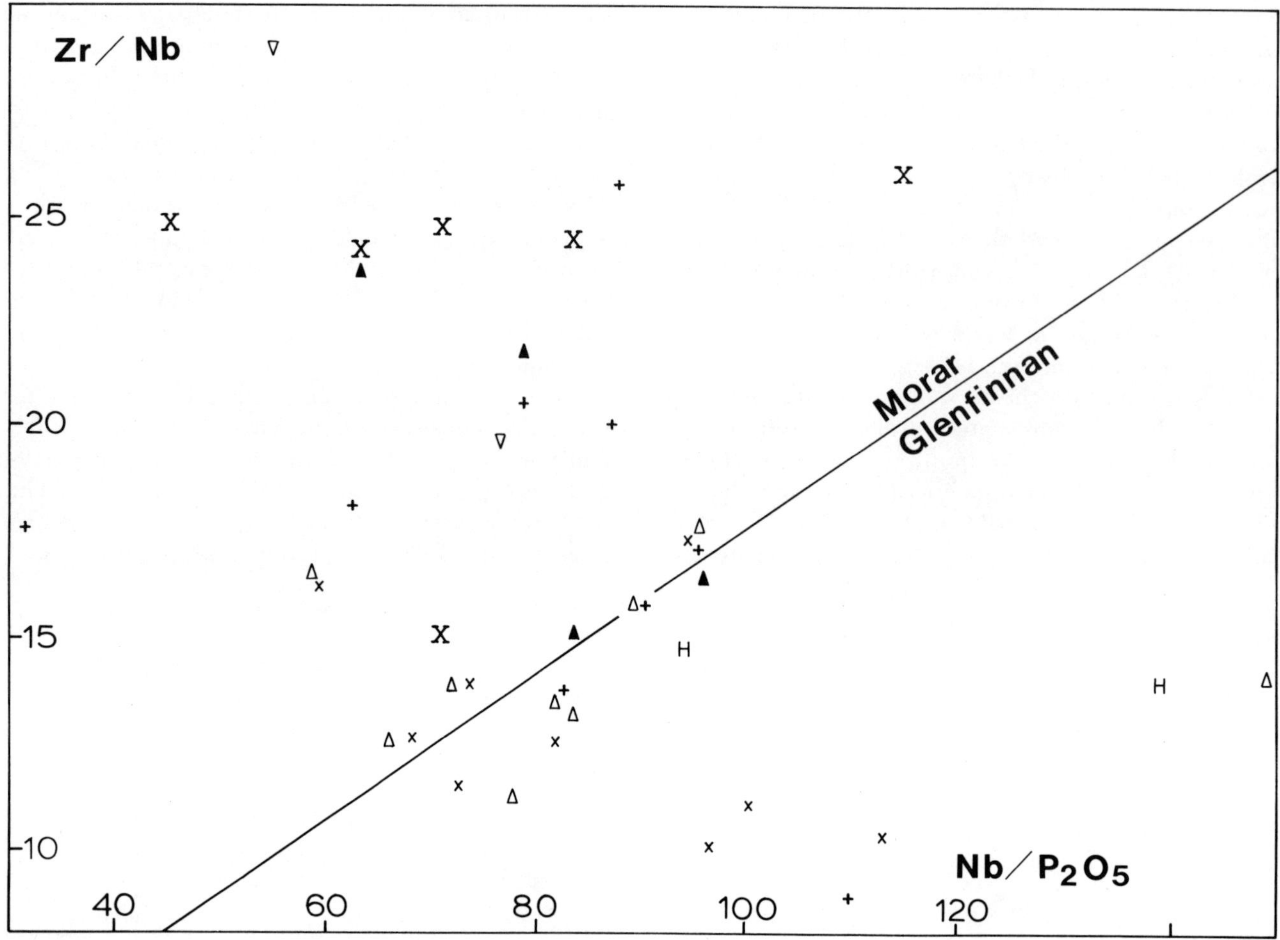

Figure 5.14 Zr/Nb *v.* Nb/P_2O_5 discrimination diagram for pelitic and semi-pelitic Moine metasediments. Boundary between Glenfinnan and Morar Division fields after Winchester *et al.* (1981, 1983). Key as Figs 5.3 and 5.7.

which consistently plot in the 'Glenfinnan fields' (Figs 5.13–5.15). This suggests that there may be a major division of the Sutherland Moine Assemblage, marked by the occurrence of the Strathy Complex on the north coast. Below this boundary the rocks would appear to be the equivalent of the Morar Division, while the overlying rocks probably represent the Glenfinnan and/or Loch Eil Divisions. We are currently undertaking a programme of analysis to clarify the geochemical affinities of these east Sutherland pelitic rocks.

5.4 Discussion and conclusions

The evidence accumulated from studies of structure, metamorphism and metabasic intrusives in the Sutherland Moine Assemblage demonstrates its essential unity with the rest of the Northern Highland Moine Assemblage. Thus, all of the Northern Highland Moine Assemblage appears to have been deposited prior to 1000 Ma, as it presumably all underwent primary deformation and metamorphism at about that time.

The geochemical evidence, as outlined above, suggests that the Sutherland Moine Assemblage, at least as far east as the Strathy Complex, may be the northern continuation of the Morar Division. If this is true, then the Naver and Swordly Nappes may together be comparable with the higher-grade migmatitic Morar Division of the Knoydart Nappe (Powell, 1974; Barr *et al.*, 1986), and the east Sutherland rocks overlying the Strathy Complex may be the equivalent of the Glenfinnan and Loch Eil Divisions.

A further pointer is the structural position of basement bodies in these Sutherland nappes. Unlike the Sgurr Beag Thrust (Tanner, 1970), situated at the structural base of the Glenfinnan Division, the Naver and Swordly Thrusts do not carry major basement bodies. Thus these latter thrusts would appear more comparable with the Knoydart Thrust than the Sgurr Beag Thrust of the W Highland Moine Assemblage. However, the Strathy Thrust does have a major basement body in its hanging-wall, and, together with the sparse geochemical evidence, this suggests the possible equivalence of the Sgurr Beag Thrust and the Strathy Thrust, with the overlying Strath Halladale metasediments being equivalent to the Glenfinnan Division.

At present, little is known about the southern extension of the Strathy Thrust south of the extensive Strathy Bog. Thus the position of the Klibreck/Loch Choire migmatites is uncertain. However, their essentially feldspathic and rather sodic nature (Brown, 1971)—as

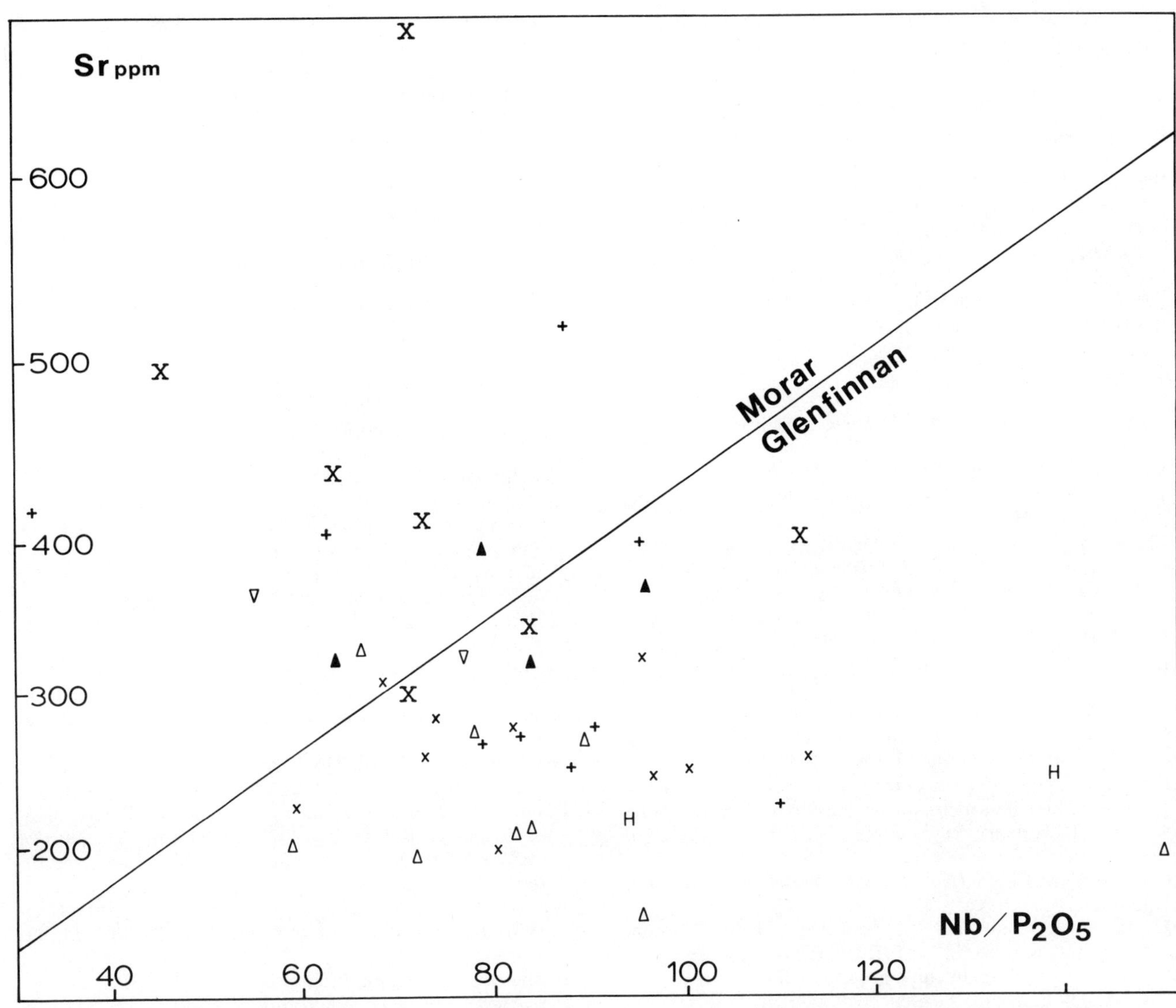

Figure 5.15 Sr *v.* Nb/P_2O_5 discrimination diagram for pelitic and semi-pelitic Moine metasediments. Boundary between Glenfinnan and Morar Division fields after Winchester *et al.* 1981, 1983). Key as for Figs 5.3 and 5.7.

opposed to the generally more pelitic Glenfinnan Division migmatites—invites comparison with the Kirtomy Gneiss of the north coast. Thus the Klibreck migmatites may also represent high-grade Morar Division, and not Glenfinnan Division as has been widely assumed (Barr, 1985; Barr *et al.*, 1986). This implies that the northern extension of the Sgurr Beag Thrust and the Glenfinnan/Loch Eil Divisions may lie east of these central Sutherland migmatites.

This idea explains the minor compositional differences between the psammites of the A' Mhoine Nappe and the feldspathic gneisses of the Naver and Swordly Nappes, which are no greater than the limited geochemical changes found by Winchester *et al.* (1981, 1983) within the Morar Division. The original sandstones of the western A' Mhoine Nappe tended towards pebbly, arkosic compositions, suggesting deposition in fluviatile or near-shore marine conditions. Shallow water is suggested by the common occurrence, in low-strain zones, of cross-bedding and soft-sediment deformation (Fig. 5.5). The sandstones of the Naver Nappe are indistinguishable from those of the A' Mhoine Nappe, while the Kirtomy Gneiss 'greywackes' of the Swordly Nappe may have been deposited in a more rapidly subsiding zone with a more plagioclase-rich source. It is possible that the higher Na_2O/K_2O ratios of the Kirtomy Gneiss 'greywackes' reflect diagenetic albitization of plagioclase which, along with rapid deposition, tend to preserve the Na of detrital feldspar, rather than allow its replacement by K in clay minerals (Blatt *et al.*, 1980). Deposition in such a subsiding zone might partially explain why the metasediments of the Naver and Swordly Nappes were more highly metamorphosed than most of the Morar Division elsewhere.

We have suggested a significant, if necessarily tentative, re-evaluation of the relationship of the Sutherland Moine Assemblage, both in terms of major structure and lithostratigraphic correlation, to the Moine Assemblage elsewhere. Sutherland remains the least known area of the Moine and still subject to speculative interpretation, but we believe that the combination of geochemical and structural evidence reviewed here provides a much firmer base for continuing research in Sutherland.

References

Barr, D., Holdsworth, R. E. and Roberts, A. M. (1986) Caledonian ductile thrusting in a Precambrian metamorphic complex: the Moine of NW Scotland. *Geol. Soc. Am. Bull.* **97**, 754–764.

Blatt, H., Middleton, G. and Murray, R. (1980) *Origin of Sedimentary Rocks* (2nd Edn). Prentice-Hall, New Jersey.

Brown, P. E. (1971) The origin of the granitic sheets and veins in the Loch Coire migmatites, Scotland. *Miner. Mag.* **38**, 446–450.

Brown, P. E. (1983) Caledonian and earlier magmatism. In Craig, G. Y. (ed.). *Geology of Scotland*, Scottish Academic Press, Edinburgh, 167–194.

Butler, R. H. W. (1986) Structural evolution in the Moine of northwestern Scotland: a Caledonian linked thrust system? *Geol. Mag.* **123**, 1–11.

Cheng, Y. C. (1942) A hornblendic complex, including appinitic types, in the migmatite area of north Sutherland, Scotland. *Proc. Geol. Ass.* **53**, 67–85.

Cheng, Y. C. (1943) The migmatite area around Bettyhill, Sutherland. *Q. J. geol. Soc. London* **99**, 107–154.

Evans, D. J. and White, S. H. (1984) Microstructural and fabric studies from the rocks of the Moine Nappe, Eriboll, NW Scotland. *J. struct. Geol.* **6**, 369–389.

Fergusson, I. W. (1978) Structural age of the Vagastie Bridge granite, Sutherland. *Scott. J. Geol.* **14**, 89–92.

Flett, J. S. (1906) On the petrographical characters of the inliers of Lewisian rocks among the Moine gneisses of the north of Scotland. *Summ. Prog. Geol. Surv. Gt. Br. 1905*, 155–167.

Greenly, E. (1938) *A Hand Through Time*, Thomas Murby, London.

Harris, A. L., Baldwin, C., Bradbury, H. J., Johnson, H. D. and Smith, R. A. (1978) Ensialic basin sedimentation: the Dalradian Supergroup. In Bowes, D. R. and Leake, B. E. (eds.), Crustal Evolution in Northwestern Britain and Adjacent Regions. *Geol. J. Spec. Issue* **10**, 115–138.

Harrison, V. E. and Moorhouse, S. J. (1976) A possible early Scourian supracrustal assemblage within the Moine. *J. geol. Soc. London* **132**, 461–466.

Horne, J. and Greenly, E. (1896) On foliated granites and their relations to the crystalline schists in eastern Sutherland. *Q. J. geol. Soc. London* **52**, 633–650.

Johnson, M. R. W. (1975) Morarian orogeny and Grenville belt in Britain. *Nature* **257**, p. 301.

Johnstone, G. S. (1975) The Moine succession. In Harris, A. L., *et al.* (eds.), A Correlation of Precambrian rocks in the British Isles. *Geol. Soc. London, Spec. Rep.* **6**, 30–42.

Lapworth, C. (1885) The Highland controversy in British geology. *Nature* **32**, 558–559.

McCourt, W. J. (1980) The Geology of the Strath Halladale–Altnabreac District. *NERC Report No. ENPU 80–1*, Institute of Geological Sciences.

Mendum, J. (1976) A strain study of the Strathan conglomerate, north Sutherland. *Scott. J. Geol.* **12**, 159–165.

Moorhouse, S. J. (1971) A Geochemical Reconnaissance of some Lewisian 'Inliers' in the Northern Highlands. Unpublished M.Sc. Thesis, University of Birmingham.

Moorhouse, S. J. (1976) The geochemistry of the Lewisian and the Moinian of the Borgie area, north Sutherland. *Scott. J. Geol.* **12**, 159–167.

Moorhouse, S. J. (1977) The Geology and Geochemistry of Central Sutherland. Unpublished Ph.D. Thesis, University of Hull.

Moorhouse, S. J. and Harrison, V. E. (1976) Moinian and Lewisian of Sutherland and the Morarian Orogeny. *Nature* **259**, 249–250.

Moorhouse, S. J. and Moorhouse, V. E. (1977) A Lewisian basement sheet within the Moine at Ribigill, north Sutherland. *Scott. J. Geol.* **13**, 289–300.

Moorhouse, S. J. and Moorhouse, V. E. (1979*a*) The Moine amphibolite suites of central and northern Sutherland, Scotland. *Miner. Mag.* **43**, 211–225.

Moorhouse, S. J. and Moorhouse, V. E. (1979*b*) Structural age of the Vagastie Bridge Granite. *Scott. J. Geol.* **15**, 69–70.

Moorhouse, S. J., Moorhouse, V. E. and Holdsworth, R. E. (1987) Moine excursion 12: North Sutherland. In Allison, I. (ed.), *A Moine Excursion Guide*. Geological Societies of Edinburgh and Glasgow, (in press).

Moorhouse, V. E. (1979) The Geology and Geochemistry of the Bettyhill–Strathy area of north-east Sutherland. Unpublished Ph.D. Thesis, University of Hull.

Moorhouse, V. E. and Moorhouse, S. J. (1983) The geology and geochemistry of the Strathy complex of north-east Sutherland, Scotland. *Miner. Mag.* **47**, 123–137.

O'Reilly, K. J. (1971) Geology and structure of an area around Tongue, north Sutherland. Unpublished Ph.D. Thesis, University of London.

Pankhurst, R. J. (1982) Geochronological tables for British igneous rocks. In Sutherland, D. S. (ed.). *Igneous Rocks of the British Isles*, John Wiley and Sons, Chichester, UK, 575–581.

Parsons, I. (1979) The Assynt alkaline suite. In Harris, A. L., Holland, C. H. and Leake, B. E. (eds.), The Caledonides of the British Isles—Reviewed. *Spec. Publ. geol. Soc. London* **8**, 677–681.

Peach, B. N. and Horne, J. (1884) Report on the geology of the north west of Sutherland. *Nature* **31**, 31–35.

Peach, B. N., Horne, J., Gunn, W., Clough, C. T. Hinxman, L. W. and Teall, J. J. H. (1907) The geological structure of the northwest Highlands of Scotland. *Mem. Geol. Surv. Gt. Br.*

Peacock, J. D. (1975) 'Slide rocks' in the Moine of the Loch Shin area, northern Scotland. *Bull. Geol. Surv. Gt. Br.* **49**, 23–30.

Pettijohn, F. J., Potter, P. E. and Siever, P. (1973) *Sand and Sandstone*. Springer-Verlag, New York.

Pidgeon, R. T. and Aftalion, M. (1978) Cogenetic and inherited zircon U–Pb systems in granites: Palaeozoic granites of Scotland and England. In Bowes, D. R. and Leake, B. E. (eds.), Crustal Evolution in Northwestern Britain and Adjacent Regions. *Geol. J. Spec. Issue* **10**, 183–220.

Powell, D. (1974) Stratigraphy and structure of the western Moine and the problem of Moine orogenesis. *J. geol. Soc. London* **130**, 575–593.

Read, H. H. (1931) The geology of central Sutherland. *Mem. Geol. Surv. Gt. Br.*

Read, H. H. (1934) Age problems in the Moine Series of Scotland. *Geol. Mag.* **71**, 302–317.

Read, H. H., Phemister, J. and Ross, G. (1926) The geology of Strath Oykell and lower Loch Shin. *Mem. Geol. Surv. Gt. Br.*

Robertson, R. C. and Parsons, I. (1974) The Loch Loyal syenites. *Scottish J. Geol.* **10**, 129–146.

Rock, N. M. S., Macdonald, R. Szucs, T. and Bower, J. (1986) The comparative geochemistry of some Highland pelites (Anomalous local limestone–pelite successions within the Moine outcrop; II). *Scott. J. Geol.* **22**, 107–126.

Smith, D. I. (1979) Caledonian minor intrusions of the northern Highlands of Scotland. In Harris, A. L., Holland, C. H. and Leake, B. E. (eds.), The Caledonides of the British Isles—Reviewed. *Spec. Publ. Geol. Soc. London* **8**, 683–697.

Soper, N. J. (1963) The structure of the Rogart igneous complex, Sutherland, Scotland. *Q. J. geol. Soc. London* **119**, 455–478.

Soper, N. J. (1971) The earliest Caledonian structures in the Moine thrust belt. *Scott. J. Geol.* **7**, 241–247.

Soper, N. J. and Barber, A. J. (1979) Proterozoic folds in the northwest Caledonian Foreland. *Scott. J. Geol.* **15**, 1–12.

Soper, N. J. and Barber, A. J. (1982) A model for the deep structure of the Moine thrust zone. *J. geol. Soc. London* **139**, 127–138.

Soper, N. J. and Brown, P. E. (1971) Relationship between metamorphism and migmatisation in the northern part of the Moine Nappe. *Scott. J. Geol.* **7**, 305–325.

Soper, N. J. and Wilkinson, P. (1975) The Moine Thrust and Moine Nappe at Loch Eriboll, Sutherland. *Scott. J. Geol.* **11**, 339–359.

Storey, B. C. and Lintern, B. C. (1981) Geochemistry of the rocks of the Strath Halladale–Altnabreac district. *NERC Report No. ENPU 81–12,* Institute of Geological Sciences.

Sutton, J. and Watson, J. V. (1953) The supposed Lewisian inlier of Scardroy, central Ross-shire and its relations with the surrounding Moine rocks. *Q. J. geol. Soc. London* **108**, 99–126.

Tanner, P. W. G. (1970) The Sgurr Beag Slide—a major tectonic break within the Moinian of the western Highlands of Scotland. *Q. J. geol. Soc. London* **126**, 435–463.

Tarney, J., Skinner, A. C. and Sheraton, J. W. (1972) A geochemical comparison of major Archean gneiss units from Northwest Scotland and East Greenland. *Proc. XXIV Internat. Geol. Congr. (Montreal)* **1**, 162–174.

Wilson, G., Watson, J. V. and Sutton, J. (1953) Current-bedding in the Moine series of northwestern Scotland. *Geol. Mag.* **90**, 377–386.

Winchester, J. A. and Floyd, P. A. (1984) The geochemistry of the Ben Hope Sill suite, northern Scotland *Chem. Geol.* **43**, 49–75.

Winchester, J. A. and Lambert, R. StJ. (1970) Geochemical distinctions between the Lewisian of Cassley, Durcha and Loch Shin, Sutherland and the surrounding Moinian. *Proc. Geol. Ass.* **81**, 275–302.

Winchester, J. A., Lambert, R. StJ. and Holland, J. G. (1981) Geochemistry of the western part of the Moinian assemblage. *Scott. J. Geol.* **17**, 281–294.

Winchester, J. A., Lambert, R. StJ. and Holland, J. G. (1983) Chemical correlation of Moinian rocks in the northern Highlands of Scotland. *Geol. Mag.* **120**, 165–174.

Wright, A. E. and Bowes, D. R. (1979) Geochemistry of the Appinite suite. In Harris, A. L., Holland, C. H. and Leake, B. E. (eds.), The Caledonides of the British Isles—Reviewed. *Spec. Publ. geol. Soc. London* **8**, 699–704.

6
The Moine rocks of Shetland

D. FLINN

In Shetland there are two separate occurrences of Moine-like rocks occurring on either side of the Walls Boundary Fault, a Mesozoic dextral transcurrent fault (Fig. 6.1). On shore, in Shetland, this fault follows the line of an earlier fault (the Great Glen Fault) and the resultant displacement on the two faults is sinistral and of a magnitude that allows no correlation across these two faults within Shetland. The two occurrences of Moine rocks are very different and before faulting were separated by up to 200 km (Flinn, 1985).

6.1 Moine rocks west of the Walls Boundary Fault

The following account of the rocks west of the Walls Boundary Fault is based largely on the work of Pringle (1970) and Robinson (1983).

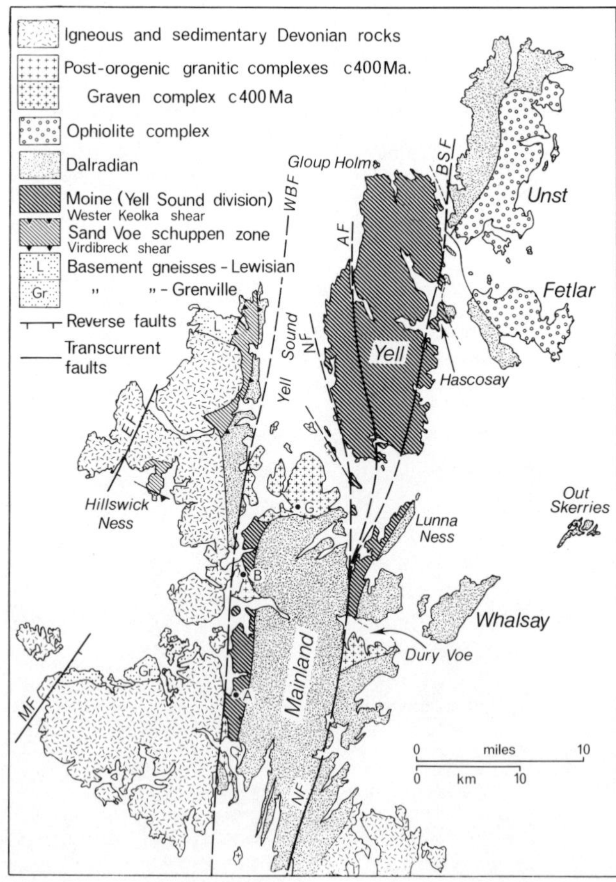

Figure 6.1 The distribution of Moine-type rocks in Shetland. *A*—Aith; *B*—Brae; *G*—Graven. *WBF*—Walls Boundary Fault; *NF*—Nesting Fault; *AF*—Arisdale Fault; *BSF*—Bluemull Sound Fault. All are dextral transcurrent faults. *MF*—Melby Fault; *EF*—Eshaness Fault. These are two components of the St Magnus Bay Fault—a system of *en échelon* reverse faults.

A broad band of psammitic metasediments occurs interleaved with bands of gneiss to form an easterly dipping and northerly striking zone, the Sand Voe schuppen zone (Fig. 6.1). To the west, beyond this zone are quartzofeldspathic orthogneisses of Lewisian age. Brown hornblendes obtained from these gneisses have given K–Ar and Ar–Ar ages ranging 2500–2900 Ma (Flinn *et al.*, 1979; Robinson, 1983). In this account they will be called the Western Gneisses to distinguish them from other gneisses of similar age called the Eastern Gneisses. To the east of the schuppen zone are a series of calcareous and non-calcareous metasedimentary schists and metavolcanic greenstones affected by greenschist-facies metamorphism. These are the Queyfirth Group (Pringle, 1970), and they have been provisionally correlated with the Dalradian of Scotland. Green hornblendes from them have given Ar–Ar ages of 420–440 Ma (Robinson, 1983).

Within the schuppen zone some of the bands of gneiss are composed of Western Gneiss but most are hornblende-banded gneisses often containing ultrabasic inclusions, ranging from fist sized to a kilometre across. These gneisses are similar to the hornblende gneisses found in the Lewisian inliers of Scotland, and they will be called the Eastern Gneisses. Hornblendes from them have given K–Ar and Ar–Ar ages of 1000–2300 Ma (Flinn *et al.*, 1979; Robinson, 1983).

The psammitic metasediments, within the schuppen zone on its western side, are in contact with the Western Gneisses. The junction dips about 45° or more to the east, and is heavily overprinted by a blastomylonitic platy schistosity, which dies out downwards to the west in the gneisses within a few hundred metres, and decreases upwards into the psammites. The term

Figure 6.2 Wester Keolka shear. The hammer head rests on the junction between the Lewisian orthogneiss below and the Moine psammites above. Fethaland, map ref. HU/355 818.

'blastomylonite' will be used in this account for rocks which in thin-section appear to be intermediate between mylonites and schists. The actual contact between the gneisses and the psammites can be recognized with some difficulty as a sharply defined line across which the feldspar content suddenly increases to the west. The line is called the Wester Keolka shear (Pringle, 1970) (Fig. 6.2). On its east side the schuppen zone is limited by the Virdibreck shear (Pringle, 1970) dipping 30–40° to the east, which separates overlying schists of the Queyfirth Group from the garnet-mica schists which form the eastern edge of the schuppen zone. A strongly developed late schistosity occurs adjacent to the Virdibreck shear. The relationships along the Virdibreck shear have been complicated by later cataclastic movements, which are probably associated with movement on the Walls Boundary Fault.

6.1.1 *The Sand Voe schuppen zone*

The Sand Voe schuppen zone at Fethaland is composed of three main units, a psammitic unit, a gneiss unit and a garnetiferous mica schist unit. The western and

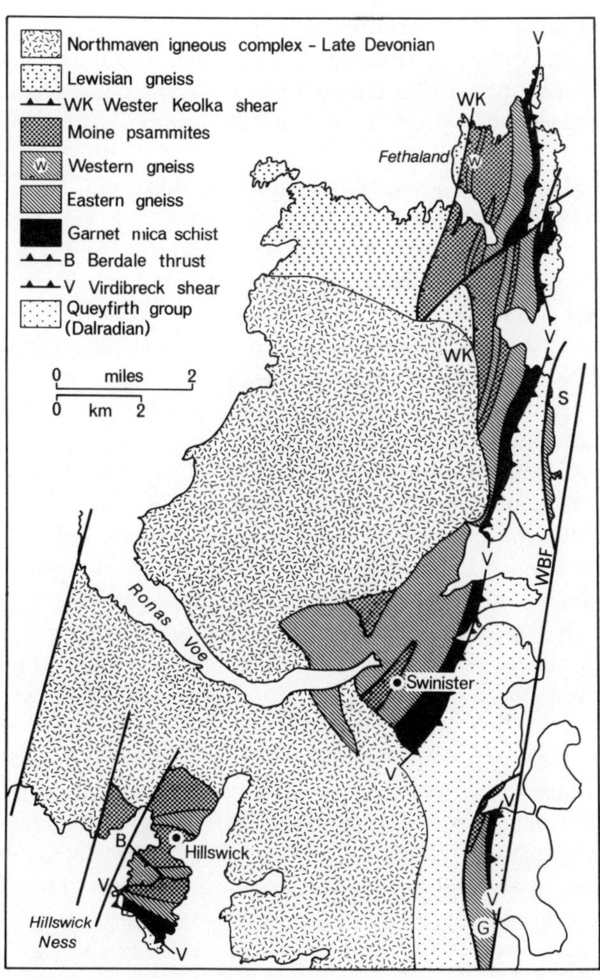

Figure 6.3 The Sand Voe schuppen zone. After Pringle (1970) and Robinson (1983). *G*—Gluss block; *S*—Skea block. WBF—Walls Boundary Fault. Eastern Gneiss in Hillswick Ness: *U*—Upper, *M*—Middle, *L*—Lower.

lowermost unit is formed of more than 800 m of siliceous psammite containing minor bands of both Western and Eastern Gneiss. In Fig. 6.3 only the most prominent band of each is shown. To the east of the psammites lies the Main band of Eastern Gneiss about 250 m thick, and between the Main Gneiss and the Virdibreck shear lie the garnet-mica schists of about 300 m thickness. The junctions between the metasediments and the gneisses within the schuppen zone in the Fethaland area are apparently concordant, and are usually marked by a platy blastomylonitic schistosity. A similar schistosity also occurs, but is variably developed, within the gneisses and the metasediments.

The zone thus appears to be a packet of slices of psammite, of pelite, of basement gneiss (Western Gneiss), and of Lewisian-inlier type gneisses (Eastern Gneiss) brought together by thrusting, for which reason it has been named the Sand Voe schuppen zone. It can be followed from Fethaland to Ronas Voe, where it is truncated by the late Devonian Northmaven igneous complex (Figs 6.1 and 6.3). However, the same lithology also occurs in a large enclave within the complex in the Hillswick area (Robinson, 1984). Here, typical Queyfirth rocks form the south tip of Hillswick Ness (Fig. 6.3); a northwest striking vertical Virdibreck shear follows to the north and further north, beyond a band of garnet-mica schist, three zones of Eastern Gneiss occur, which alternate with three zones of siliceous psammite all with an east–west strike and a steep southerly dip. The succession is truncated to the north by the granite complex. Thus, in the Hillswick area, this fragment of the schuppen zone is oriented approximately normal to its northern continuation. This may be due to rotation of the Hillswick enclave during the emplacement of the Northmaven Complex.

Two further fragments of the Sand Voe schuppen zone occur along the western side of the Walls Boundary Fault (Fig. 6.1). These, the Skea block and the Gluss block, are probably fault blocks bounded to the east by the Walls Boundary Fault and to the west by an older fault. They have been displaced to the south during faulting.

The occurrence of such a tectonically assembled and easterly dipping packet of psammites and Lewisian-inlier type gneisses, lying on a basement of Lewisian gneiss, has led to its correlation with the Moine Thrust zone of northwest Scotland (Pringle, 1970; Flinn *et al.*, 1979). The results of drilling and seismic work in the sea area to the west of Shetland, Orkney and mainland Scotland tend to support this correlation (Andrews, 1985). The most obvious differences between northwest Shetland and northwest Scotland are the steeper dip in Shetland—45° or more but varying between 90° and 35°E—and the higher grade of metamorphism accompanying the formation of the schuppen zone in Shetland. Blastomylonitic chloritoid-mica schists, garnet-mica schists and garnet-hornblende schists are apparently stable products of the creation and emplacement of the schuppen zone which, thus, seems to have formed at a greater depth than the Moine Thrust zone.

6.1.2 *The Eastern Gneisses*

These occur only within the schuppen zone, where they form the 'Main Gneiss' band in the Fethaland–Ronas Voe area and a number of lesser bands to the west, and three major bands in the Hillswick area. The latter will be called the Upper, Middle and Lower bands respectively.

The Main Gneiss is 0.25 km thick in the north, but thickens to the south, where exposure is much poorer especially in the area at the head of Ronas Voe. Beneath it are at least six less substantial slices within the psammite layer. The three bands in the Hillswick enclave are up to 0.5 km thick. All are bounded by zones of strong platy schistosity and blastomylonitization which tends to destroy discordancy at the boundaries and to cause mixing of the rocks at the junctions.

The Eastern Gneisses are dominantly hornblende-banded, with alternating layers and laminae of hornblende-rich and hornblende-poor rock. The earliest minerals present are brown hornblende, garnet and diopsidic pyroxene. The gneisses contain frequent small lenses of hornblendite, actinolite rock, and steatite-serpentinite. Larger masses of ultrabasic rocks include orthopyroxenites and anthophyllite rock. The largest mass found, that at Swinister, is several hundred metres across and has yielded ages up to 2300 Ma (Flinn *et al.*, 1979). An intrusive metagabbro occurs within the schuppen zone at Fethaland. Pegmatites occur in the Main Gneiss and in the Upper Gneiss in the Hillswick area. These appear to predate the formation of the schuppen zone, although only some of them are blastomylonitized. A minor component of the Eastern Gneisses are hornblende-poor quartzofeldspathic gneisses. A particularly remarkable example of such a gneiss is the Setter Quartzite (Pringle, 1970) which forms a band not more than 25 m thick, which can be traced from Fethaland along the Main Gneiss and into the Upper Gneiss on Hillswick Ness. It is easily recognizable because of its richness in quartz (60%) and the presence of small augen of microcline. It is the only rock in the schuppen zone to contain microcline other than the pegmatites. It may be a paragneiss derived from an arkose or it may be a blastomylonitized pegmatite or granite.

The blastomylonite foliation tends to be parallel to the original banding. Where the foliation is well developed, the brown hornblende has been transformed to blue-green hornblende. In places the foliation has been thrown into intrafolial folds, and the closures thus produced then resisted further shearing, so that they form the nucleus of lenses lying within the foliation. In the Fethaland area, fold axes and lineations are concentrated into a partial girdle within the plane of the foliation, with a weak tendency to be parallel to the dip in earlier folds, and parallel to the strike in the later folds (Pringle, 1970). In the Hillswick area structural relations are more complicated, but in general the fold axes of the intrafolial folds and the associated lineations are oriented more nearly parallel to the strike of the foliation than the dip (Robinson, 1983).

6.1.3 *The metasediments—psammites*

The psammitic rocks occur on the west side of the schuppen zone in the Fethaland area, and below each of the three gneiss zones in the Hillswick area. At Fethaland they form a band more than 800 m thick (including the minor Eastern and Western Gneiss bands) which Pringle (1970) has divided into three broad zones, the Western impure quartzite, the garnetiferous semi-pelite and the Main impure quartzite. The impure quartzites or siliceous psammites contain from 60% quartz and 30% oligoclase to more equal amounts of these two minerals, together with about 10% mica and only accessory garnet. Characteristically they contain scattered ex-clastic quartz and sodic plagioclase grains several times larger than the grains around them. Several conglomeratic bands composed of pebbles of Western Gneiss occur. Cross-bedding occurs rarely but has not revealed a 'way up'. The semi-pelitic varieties of the psammites contain some 10% less quartz and oligoclase, and proportionately more mica and garnet.

In the Hillswick area, the same two rock types form the bulk of the metasedimentary succession, but laminated, graphitic, conglomeratic and hornblendic varieties occur in substantial amounts (Robinson, 1983). Although no graphitic rocks have been found in the Fethaland area, they do occur in the poorly exposed area at the head of Ronas Voe and in the Gluss block. Beneath the Lower Gneiss are some hundreds of metres of siliceous psammite with a 40 m band rich in pyrite and graphite. Between the Middle and Lower Gneisses are some 700–800 m of metasediment. The lower third is composed of the usual siliceous psammite and semi-pelites but here these are interlaminated, the former being in layers up to 5 cm thick and the latter in layers up to 1 cm thick. Above this, there is a thick layer of slightly graphitic siliceous psammite with a 65 m graphitic-pyritiferous band. The top third is siliceous psammite. Thin bands of blastomylonitic hornblendic semi-pelite occur in the psammites adjacent to the hornblendic gneisses. One band occurs also in the middle of the psammites, apparently unconnected with the gneisses, but probably marking the site of a shear which contains a slice of Eastern Gneiss elsewhere.

Between the Upper and Middle gneisses lie more than 500 m of siliceous psammite and semi-pelite. The lower 100 m is slightly graphitic and locally laminated as described above. In the middle is a 90 m layer of coarsely gritty to conglomeratic psammite. The grit grains are of quartz and sodic plagioclase (and only occasionally of K-feldspar) and one to two millimetres in diameter, while the pebbles are of quartz and sodic plagioclase and up to a centimetre in diameter. Hornblende and hornblende-garnet semi-pelitic blastomylonites occur adjacent to the hornblende gneisses and within the succession.

Small-scale cross-bedding was found in a graphitic layer, indicating upwards younging in the succession, while large-scale cross-bedding was found in a loose block.

Schistosity, foliation and layering all tend to parallelism throughout the schuppen zone. However, in Hillswick Ness the metasediments can be observed truncating banding in the gneisses, despite the strong development of blastomylonitization at the junctions. Intrafolial folds occur in the metasediments from Fethaland to Hillswick Ness, and like those in the Eastern Gneiss, have the fold axes, the parallel lineation and the directions of elongation of the pebbles widely dispersed in the plane of the foliation—though tending to be more nearly parallel to strike than to dip in the Fethaland area. In the Hillswick area, although the structural relations are less clear than in the Fethaland area, the folds and lineations seem to have the same type of preferred orientation as in the gneisses. That is, the fold axes and lineations lie within the foliation, plunging more nearly horizontally than down-dip (Robinson, 1983).

6.1.4 *The metasediments—the garnet-mica schists*

The garnet-mica schists occur between the Main Gneiss and the Virdibreck shear from Fethaland to the area at the head of Ronas Voe, and between the Upper Gneiss and the Virdibreck shear on Hillswick Ness. They also occur as a 3 m band within the Main Gneiss and immediately beneath it at Fethaland, and can be traced up to 6 km to the south. These schists include both pelitic and semi-pelitic varieties and are very rich in muscovite. The pelites contain chloritoid, kyanite and garnet. The chloritoid tends to enclose garnet and sometimes kyanite, and is best developed in the blastomylonite zone adjacent to the Virdibreck shear. Coarse kyanite-chloritoid segregation veins also occur there. The semi-pelite varieties of the schists are garnet-rich and lack the other minerals, though the garnets tend to be much larger than in the pelitic schists, reaching several centimetres in diameter in some occurrences. The thin bands within and beneath the Main Gneiss are of semi-pelitic garnet-rich type.

6.1.5 *The correlation with Scotland*

The correlation of the metasedimentary components of the Sand Voe schuppen zone with the Moine of Scotland has been based on their dominantly psammitic character, their occurrence as slices in a schuppen zone lying on a Lewisian basement and their association with banded hornblendic gneisses of Lewisian-inlier type. As is to be expected from their mineralogical composition, they have a general similarity to the Moine rocks in their major- and trace-element compositions but with some noticeable differences (Robinson, 1983), in particular they are more siliceous, more mature and contain lower Ti and Zr concentrations. They lack the garnet-studded hornblende schist bands, the heavy-mineral bands and the calc-silicate bands which characterize various parts of the Moine succession in Scotland. The paucity of sedimentary structures may be due to the intense deformation. The graphitic varieties are not matched in Scotland.

6.2 Moine rocks east of the Walls Boundary Fault

The occurrence of Moine rocks on the east side of the Walls Boundary Fault is very different, both lithologically and in its tectonic environment, from that on the west side. Here it forms the western part of what at first appeared to be a continuous succession of metasedimentary rocks about 20 km thick, the East Mainland succession (Flinn *et al.*, 1972). This succession has been divided into four divisions, of which the westernmost, and probably the stratigraphically lowest division, is composed of Moine-type rocks and has been named the Yell Sound Division. The other three divisions are of Dalradian type, the easternmost or Clift Hills Division having been correlated with the Upper Dalradian of Scotland (Flinn *et al.*, 1972; Flinn and Moffat, 1985).

The rocks of the Yell Sound Division are intersected and offset by the Nesting Fault and its splays. To the east of that fault they underlie most of Yell, and the western side of Lunna Ness as far south as Dury Voe (Figs 6.1 and 6.8). To the west of the Nesting Fault they occur in a number of small islands in Yell Sound. They occur in the Graven area of Mainland and continue down the east side of the Walls Boundary Fault to the South of Aith (Figs 6.1 and 6.8).

The outcrop of the Yell Sound Division is interrupted by the Brae Complex, a post-orogenic dioritic complex lithologically very similar to the Garabal Hill Complex of Scotland. It is partially interrupted by the Graven Complex, a post-orogenic granodioritic complex of appinitic type. This complex contains innumerable xenoliths of its country rock, varying in size from a centimetre to a kilometre or more across. Xenoliths larger than a metre across appear to have retained their orientation and position relative to each other, so that it is possible to trace schistosity, compositional banding and lithology of the rocks of the Yell Sound Division through the complex (Figs 6.8 and 6.9).

The outcrop of the division is truncated by the sea to the north of Yell, and it wedges out to the south against the Walls Boundary Fault—which is also its western limit. The eastern boundary of the division was first defined as the eastern limit of the outcrop of the psammitic Moine-like lithology (Flinn *et al.*, 1972) and later, because this is not a precisely defined line on the ground, as the Valayre Gneiss (Flinn *et al.*, 1981). The Valayre Gneiss is a remarkable band of single-crystal, microcline augen, (each of which is several centimetres in diameter), and which can be traced intermittently from Aith to the north coast of Yell (Figs 6.5, 6.8, 6.9).

The Valayre Gneiss provides a good lithological boundary between the Yell Sound Division and the rest of the East Mainland succession to the east. However, it is not necessarily the dividing line between the Moine and the Dalradian rocks of the area.

Another remarkable gneiss band, the Skelladale Gneiss (Flinn, 1954), can be followed from the Aith area for a considerable distance to the north, parallel to the Valayre Gneiss and about 1 kilometre to the east of it (Figs 6.8 and 6.9). The rocks to the east of the

Skelladale Gneiss and forming the remainder of the East Mainland succession appear to be Dalradian, while the affinities of the rocks between the two gneiss bands are not so obvious. The boundary between the Moine and the Dalradian thus lies within, or is formed by, the zone contained between these two gneiss bands. It will be called the Boundary zone. It was previously called the Scatsta pelitic 'group' of the Scatsta Division (Flinn *et al.*, 1972).

In Mainland, between Aith and the Nesting Fault, the Boundary zone is poorly exposed, though well defined. To the east of the Nesting Fault, the Boundary zone is well exposed in the Lunna Ness area, but to the north it lies mostly under the sea, with its western edge overlapping the coast. Along the north-east coast of Yell, the East Mainland succession is truncated by the Hascosay Slide. This is a 0.25-km thick, westerly dipping, partially blastomylonitized tectonic *mélange* of hornblendic and banded hornblendic gneisses with Lewisian-inlier affinities. In north-east Yell, it outcrops parallel to and immediately east of the Valayre Gneiss and appears to truncate the Boundary zone (Figs 6.5, 6.9).

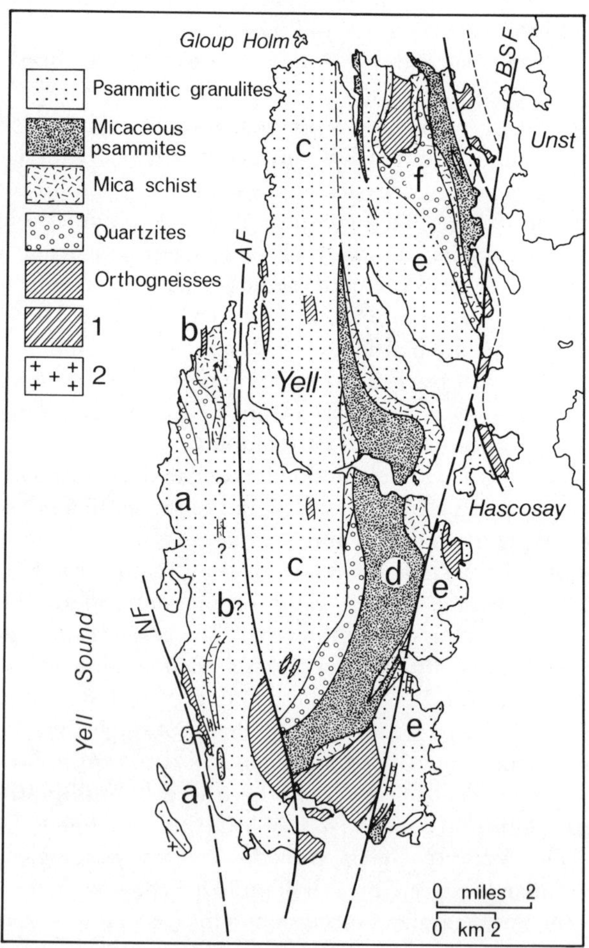

Figure 6.4 The metasediments of the Yell Sound Division in Yell. *NF*—Nesting Fault; *AF*—Arisdale Fault; *BSF*—Bluemull Sound Fault. ?—critical areas lacking exposures. *a*—Yell Sound Psammites; *b*—Graveland lens; *c*—Herra Psammites; *d*—Mid-Yell lens; *e*—Basta psammites; *f*—Cullivoe lens. 1. Hascosay Slide; 2. Graven Complex.

6.2.1 *Lithology of the Yell Sound Division*

The Yell Sound Division is dominantly composed of psammites, together with lesser quantities of mica schists and quartzites. In places, the first two are partially to completely transformed into gneiss. Also present are granitic orthogneisses and hornblende gneisses forming Lewisian inliers.

6.2.1.1 *The metasediments.* The metasedimentary rocks forming the Yell Sound Division are displayed in Figs 6.4 and 6.8. The largest outcrop occurs in Yell (Fig. 6.4) where the splays of the Nesting Fault divide it into four blocks, giving rise to some repetition of the succession. Coastal sections are magnificently well exposed and continuous, but inland exposure is poor over large areas. The rocks are dominantly a monotonous succession of psammitic granulites, with partial to complete gneissification, together with several easily distinguished major occurrences of quartzite and schist.

At first sight the psammites appear to be a monotonous uniform succession, made interesting only by the

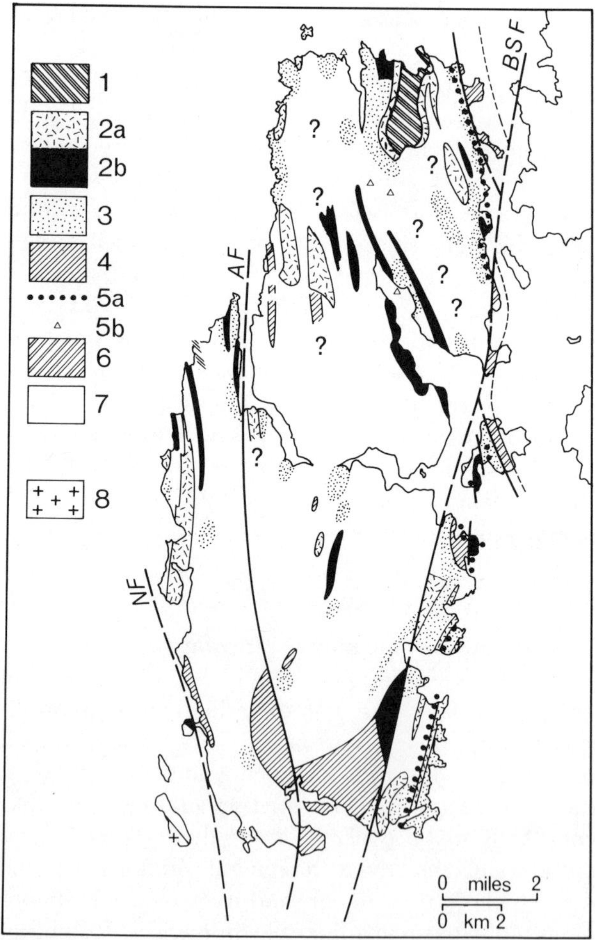

Figure 6.5 The gneisses of Yell. *NF*—Nesting Fault; *AF*—Arisdale Fault; *BSF*—Bluemull Sound Fault. 1—granitic orthogneisses; 2a—paragneisses containing no microcline; 2b—microcline-rich paragneisses; 3—partially gneissose psammites; 4—Lewisian inliers; 5a—Valayre Gneiss; 5b—isolated exposures of Valayre-type gneiss; 6—Hascosay Slide; 7—non-gneissose metasediments; 8—Graven Complex. ?—critical areas lacking exposures.

presence of zones showing varied degrees of gneissification. Closer inspection reveals compositional banding on all scales from millimetres to hundreds of metres. It is seen in exposure as parallel-sided and continuous bands several centimetres thick, or as continuous parallel laminations a millimetre or two thick, or very commonly as both together. No signs of cross-bedding have been found, and any pinch-and-swell structures or other interruptions to the regularity of the banding appear to be tectonic in origin (Figs 6.11, 6.12, 6.13, 6.14). No gritty or conglomeratic varieties occur. The average grain diameter as seen in thin-section is 0.3 mm.

The mica content of the psammites varies between 10% and 40% and it is sudden changes in the mica content which make the banding and lamination visible. Most psammites contain both biotite and muscovite, but a significant proportion (up to a quarter) of them are completely lacking in muscovite. Quartz and sodic plagioclase tend to occur in equal amounts. Microcline is absent, except locally as interstitial grains bounded by linked, outwards concave segments of grain boundary, giving the impression that these grains are a late growth in the rock (see below). Garnet occurs throughout the psammites and appears in about half the thin-sectioned rocks in amounts up to about 20%. Rutile-sieved garnets are a characteristic feature of several widely scattered areas, and occur less commonly throughout the division. The rutiles occur as minute crystals confined to the outer zone of zoned garnets. Kyanite is widely, but very sparsely, distributed, occurring sometimes as inclusions in the garnet and sometimes in the matrix. Staurolite has a similar occurrence, but is much rarer, except in certain schist bands mentioned below. No sillimanite has been found, except within the aureoles of the Gloup Holm diorite, the Graven Complex, the Brae Complex and the Aith Complex.

Mica schists occur in places in the succession as layers varying from metres to hundreds of metres in thickness. They differ most obviously from the more micaceous psammites in containing much larger mica flakes than the psammites, and sometimes contain very large garnets. They rarely contain more than 50% mica, and often very much less. They have a tendency to contain more quartz than plagioclase. Some are composed entirely of biotite, quartz and kyanite, but most contain muscovite. Several bands of staurolite-mica schist occur.

Pure quartzites occur as bands from a metre to hundreds of metres in thickness, often closely associated and even interbanded with the mica schists.

The nature of the banding, the narrow range of variation shown by the psammites, and the restriction of good exposure to the cliffs makes it very difficult to subdivide the Yell Sound Division and to trace subdivision boundaries across its outcrop.

In Fig. 6.9, the Mesozoic dextral movements on the Nesting Fault and its splays have been restored, so as to provide maximum continuity in the succession across all the fault traces. This restoration shows that the Yell

Figure 6.6 Kyanite pseudomorph of chiastolite. Pseudomorph *c.* 0.5 cm across. Gerherda, west coast of Yell. Map ref. HU/475 208.

Figure 6.7 Valayre Gneiss with schist matrix. Gonfirth. Map ref. HU/367 619.

Sound Division is formed of at least 10 km (allowing for dip) of psammites, within which are included three lenticular masses up to several kilometres thick and about ten times as long, composed dominantly of quartzite, mica schist and micaceous psammite. The Mid Yell lens outcrops on land, but the outcrops of the Graveland and Cullivoe lenses are truncated to the north by the sea (Figs 6.4 and 6.9).

The Graveland, Mid Yell and Cullivoe lenses can be used to provide a crude subdivision of the Moine outcrop in Yell, but the correlation of these subdivisions with the Moine in Mainland can only be guessed at. The Yell Sound Psammites lie west of the Graveland lens and west of the Nesting Fault. They are variably banded and laminated psammites, with a few bands several metres thick of pure quartzite which is usually (and characteristically) intensely colour striped with perfectly parallel and parallel-edged laminations and bands. A band of large garnet mica schist several metres thick occurs to the north of Aith (Fig. 6.8). A large Lewisian inlier is well exposed in the cliffs of East Yell.

The Graveland lens is composed of great thicknesses of bedded pure quartzite and masses of coarse mica schist. It may continue southwards through Yell as a several metre thick band of staurolite schist, though no staurolite occurs in the mica schists of the lens.

Between the Graveland lens and the Mid Yell lens are the Herra Psammites, distinguished by having a relatively high proportion of muscovite-free psammites. They contain several small Lewisian inliers.

The Mid Yell lens is formed by the Great Arisdale Quartzite on the west and the Kaywick Mica Schists on the east. Between them lie the Mid Yell Schists. These are micaceous psammites, unusual in the paucity of interbanded non-micaceous psammites and of any signs of gneissification. They contain sparsely distributed, fist-sized calcareous nodules. Such nodules are exposed elsewhere in the succession only on the north coast of Yell along the strike of the Mid Yell lens. It is not clear whether the great Houlland Lewisian inlier lies within the Mid Yell lens or beside it.

The Basta Psammites between the Mid Yell lens and the Cullivoe lens are truncated to the south, at least in part, by the Boundary zone, but the psammites lying west of the Valayre Gneiss in south-east Yell and Lunna Ness probably belong to this subdivision.

The Cullivoe lens is composed of pure quartzite, coarse mica schists and micaceous psammite, but poor exposure and the complex distribution of these rocks around the Breckin orthogneiss, make the lens difficult to define. It pinches out to the south against the Blue Mull Sound Fault, where it includes a band of rapidly alternating beds of pure quartzite and coarse kyanite-biotite-quartz schists.

The psammites appear originally to have been arkoses and arkosic wackes of medium-sand grain size. The ubiquitous parallel banding and lamination in them is of sedimentary origin, though the thickness of the bands and laminae has probably been reduced by deformation. This banding and lamination differs significantly from the much sharper and more continuous interbanding of different rock types due to blastomylonitization in slide zones (e.g. the Hascosay Slide—see below). The mineral fabrics are much stronger in the slide zones than in the psammites, and the blastomylonitization has the effect of destroying the compositional banding in the psammites adjacent to the slide and not of enhancing it.

The considerable thicknesses of seemingly constant sedimentary facies seem to suggest that deposition occurred in a deep basin out of range of shallow-water processes. Further evidence in support of this interpretation is provided by the lack of cross-bedding and of original pinching and swelling of the layers—features associated with shallow-water deposition. The fine grain size (*c.* 0.3 mm) probably reflecting a similar original grain size, suggests that they may have been deposited by sand-rich turbidity currents (T. Elliott, pers. comm.). The conditions under which the lenticular masses of quartzite and mica schist were deposited are more difficult to deduce. Both have suffered such a coarse recrystallization that any sedimentary features have been destroyed. The fact they lie within the psammite succession probably indicates that they differ from the psammites only in the manner in which the sediment supplied to the basin was sorted before delivery.

Hornblende schist bands and lenses, mostly 10 or 20 centimetres wide, occur frequently but irregularly throughout the division. A high proportion of them are studded with garnets from several millimetres to a centimetre in diameter. In several places, including the south-west tip of Yell, they alternate with psammitic bands of equal thickness so as to form 50% by volume of the exposures, but mostly they are a minor constituent volumetrically. They always appear to be conformable with the psammite banding.

6.2.1.2 *Paragneisses.* In describing the metasedimentary succession (and in Fig. 6.4) the paragneisses have been omitted and been replaced by their protoliths. The determination of the nature of the protoliths was based on a study of the partially gneissified relics in the gneiss bands, and on adjacent metasediments.

A number of recognizably distinct types of paragneiss occur within the Yell Sound Division. In Fig. 6.5 they have been grouped into only two types, those in which K-feldspar is an essential constituent and those in which it is absent or nearly so.

Some of the K-feldspar-free gneisses appear to be the result of a simple recrystallization which triples the grain size and shortens the plagioclase grain boundaries at the expense of the quartz grains. In others the recrystallization has been confined to the plagioclase grains, which have grown to spherical single-crystal microaugen two or three millimetres in diameter. Coarsened psammites and coarse micaceous gneisses, richly streaked with quartz-plagioclase irregular lenticular augen a centimetre or so across, occur very commonly.

The K-feldspar gneisses include psammitic granulites which have been partially replaced by interstitial microcline without otherwise altering the rock. The same interstitial replacement occurs in the recrystallized psammites mentioned above. In some of these gneisses the replacement produces lenticular laminae (less than a centimetre thick) of feldspar-rich rock alternating with less feldspathic rock to give the whole a *lit-par-lit* appearance. Another group of K-feldspar gneisses are those containing well-formed, discrete, single-crystal augen of microcline. These augen vary from one centimetre to several centimetres in diameter in different occurrences. The Valayre Gneiss is the coarsest and most spectacular (Figs 6.7, 6.10).

These different types of gneiss occur throughout the Yell Sound Division and (except for the Valayre Gneiss) tend to form lenticular zones composed almost entirely of gneiss of a particular type, and changing outwards into ungneissified psammite by gradational change and increasingly frequent interbanding with psammites. It is apparent from a study of the gradational changes and the interbanding that some of the differences between the types of gneiss arise from differing protoliths, but others must be the result of different processes operating on the same type of metasediment.

6.2.1.3 *Orthogneisses.* A large mass of orthogneiss, the

Breckin Gneiss, occurs in the Cullivoe lens, and several smaller sheets (the Graveland Gneiss) in the Graveland lens (Fig. 6.5). All are tectonized granite, and are associated with otherwise unusually large-scale and complex folds of very irregular type in the adjacent rocks. No signs of an aureole were detected adjacent to them. However, several kilometres west of the Breckin Gneiss, a small band of pelitic schist in the west cliffs of Yell contains many kyanite pseudomorphs of chiastolite, the pattern of inclusions characteristic of chiastolite being easily visible in the field (Fig. 6.6). The chiastolite must have formed as the result of thermal metamorphism of the Moine sediments prior to the regional metamorphism, possibly due to an intrusion of granite of the type forming the protolith for the Breckin and Graveland Gneisses (which now lies beneath the sea immediately west of the cliffs). No other rocks are known to occur which could have produced this effect.

6.2.1.4 *Lewisian inliers.* Seven Lewisian inliers are shown in Fig. 6.5, many of them being ideally exposed in cliff sections. Two main rock types occur in the inliers—hornblende gneisses and hornblende-poor to hornblende-free gneisses. The hornblende gneisses are the most prominent. Similar rocks occur in the Sand Voe schuppen zone and in the Hascosay Slide. The latter occurrence will be referred to later. The hornblende gneisses are very rich in hornblende and are not often interbanded with hornblende-poor gneiss. Such banding as does occur is due to only small variations in hornblende content and to the presence or absence of epidote. They differ from the hornblende schist bands in the psammites in being coarser and in rarely containing garnet. The most commonly associated hornblende-poor gneiss has been called the pegmatite-blebbed gneiss, because its most obvious characteristic is the regular but sparse distribution through it of 'fragments' or blebs of quartz-albite pegmatite several centimetres across. It also contains widely scattered garnets several centimetres in diameter. It is a coarsely crystalline, granulitic, quartz-sodic plagioclase gneiss with some mica and a tendency to contain thin biotite-garnet-hornblende laminae. The hornblende gneisses occur within the pegmatite-blebbed gneiss, both as small xenolith-like lumps and as much larger masses. The hornblende-poor gneisses associated with the hornblendic inliers do not always show these distinguishing features, in which case it becomes difficult to distinguish them from some of the more extreme types of paragneiss, so that the exact limits of the inliers are sometimes difficult to determine.

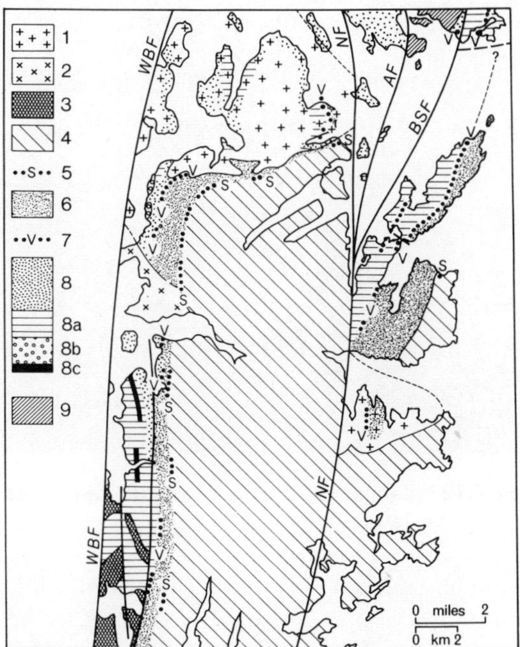

Figure 6.8 Rocks of the Yell Sound Division in Mainland. *NF*—Nesting Fault; *AF*—Arisdale Fault; *BSF*—Bluemull Sound Fault; *WBF*—Walls Boundary Fault. 1.—Graven Complex; 2.—Brae Complex; 3.—Aith Complex; 4.—Dalradian divisions of the East Mainland Succession; 5.—Skelladale Gneiss; 6.—Boundary zone; 7.—Valayre Gneiss; 8.—Yell Sound Division psammites; 8a.—paragneisses; 8b.—quartzite, 8c.—garnet-mica schist; 9—Lewisian inlier.

6.2.1.5 *Pegmatites.* Pegmatites form a major component of the Yell Sound Division, although they are very unevenly distributed. In places, for example the island of Linga to the south of the Brae Complex (Fig. 6.8), they can be seen to form 50% by volume of the rocks, occurring as innumerable sheets injected into the psammites. Spectacular pegmatite veining can be seen along the cliffs of north-west Yell. Some enclaves of pegmatite in the Graven Complex are so large that they were once thought to have been emplaced as an early phase of the Graven Complex (Flinn, 1954). The pegmatites are very variable in type. Some have large muscovite or biotite books, some lack K-feldspar, while others have single-crystal microclines tens of centimetres across. They are associated with a much smaller number of aplites. They vary from masses tens of metres across, to veins a centimetre or two thick. They show some signs of sequential emplacement. Only some pegmatites within the Lewisian inliers and aplites associated with the Hascosay Slide are tectonized. They all cut metamorphic rocks, but except for some thin stringers they are cut by the Graven Complex and the ubiquitous microdioritic lamprophyre dykes, of which about two hundred are exposed in Yell.

6.2.1.6 *Structure of the Yell Sound Division.* The banding, foliation and schistosity in the Yell Sound Division is mostly very steeply inclined, with a northerly strike. In the region of the Graven Complex there is a major vertical-axis kink in the strike, which involves also the Boundary zone and the Dalradian rocks to the east (Figs 6.1, 6.8 and 6.9). It is confined to the block between the Walls Boundary Fault and the Nesting Fault and may be due to dextral drag on those faults. Within Yell, the dip of the banding and foliation changes gradually from vertical in the west to about 30°W in the east, and locally to horizontal at the edge of the Boundary zone in the east.

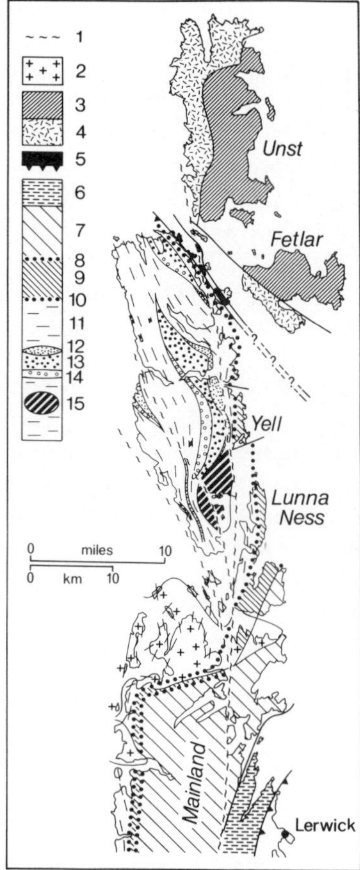

Figure 6.9 Restoration of movement on faults offsetting the outcrop of the Yell Sound Division. 1.—trace of the restored faults; 2.—Graven Complex; 3.—Shetland ophiolite; 4.—metasediments of Unst and Fetlar; 5.—Hascosay Slide; 6.—Clift Hills Division = Upper Dalradian; 7.—Whiteness and Scatsta Divisions = Middle ± Lower Dalradian; 8.—Skelladale Gneiss; 9.—Boundary zone; 10.—Valayre Gneiss; 11.—psammites of the Yell Sound Division = Moine, showing strike of the banding; 12.—mica schists; 13.—mica-rich psammites, 14.—quartzites; 15.—Lewisian inliers.

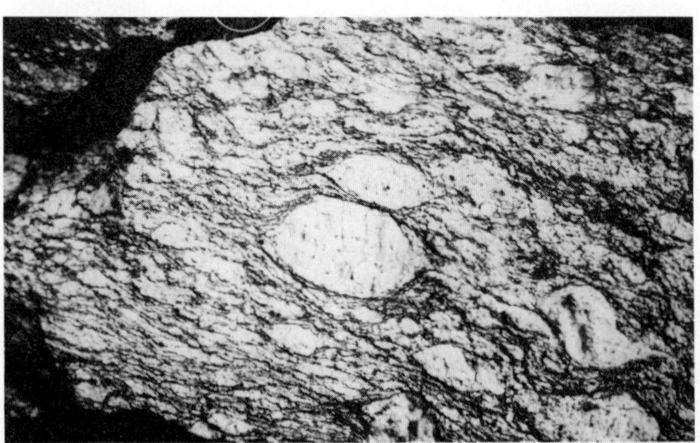

Figure 6.10 Valayre Gneiss with gneissose matrix. Grunna Voe. Map ref. HU/467 671.

Figure 6.11 Laminated and banded psammite. Westsandwick, Yell. Map ref. HU/442901.

Figure 6.12 Laminated and banded psammite. The Hulk, Yell. Map ref. HU/444937.

Large-scale folding is restricted to cliff-scale non-cylindrical contortions in the Graveland area, perhaps associated with the presence of the Graveland orthogneiss sheets and to large-scale diversions of the lithologies about the Breckin orthogneiss. Well-formed near-isoclinal folds, with amplitudes up to ten metres and with horizontal axes, occur in north Yell. These are upright in the west and become overturned to the east with a Z-form profile (looking north) as the Hascosay Slide is approached. A widespread, late, spaced cleavage, with strike rotated 20° clockwise to the earlier schistosity in unfolded areas, is axial planar to at least some of these folds. All these folds fold both the banding and the main schistosity: which is everywhere parallel to the banding. Smaller folds can be found through the division but are lacking in any regularity or simple relationships with each other. Many are intrafolial folds, among which only a small number have the main schistosity as an axial-planar schistosity.

6.2.1.7 *Correlation.* The Yell Sound Division was originally correlated with the Moine (Flinn *et al.*, 1972), on the basis of a study of that part of Mainland lying between the Walls Boundary Fault and the Nesting Fault, and including the Graven Complex and the ground to the south. Here the contrast between the monotonous psammites of the Yell Sound Division and the succession to the south-east, in which psammites are interbedded with semi-pelites, pelites and calc-silicates, and alternate on a larger scale with massive layers of quartzite or marble which can be followed from coast to coast, made it seem credible that the former were Moine and the latter Dalradian. Further work has emphasized these differences and revealed more Moine-type features. Quartzite, and mica schists have now been found to occur in the Yell Sound Division in considerable volume, but form lenticular masses of little use for long-distance stratigraphic division—unlike their mode of occurrence to the east. Lewisian inliers have been found to occur only in the Yell Sound Division and adjacent to it, and garnet-studded hornblende schists are common in the division and confined to it. Gneisses and partially gneissified rocks are very common in the division and in the Boundary zone, but east of the Skelladale Gneiss they are confined to the well-defined Colla Firth Permeation belt (Flinn, 1954; Flinn *et al.*, 1981). Garnet is profusely developed throughout the Yell Sound Division, being easily visible in most exposures and prominent as red streaks on the sandy beaches. In the succession to the east, garnet is difficult to find in thin-section, and in exposures and beach sands, since it only occurs as microscopic grains and is confined to certain areas and to particular rock types. Ptygmatically folded pegmatite stringers occur throughout the Yell Sound Division, though only very sparsely. They have not been seen east of the Valayre Gneiss.

The association of psammitic lithology, garnet-studded hornblende schists and Lewisian inliers serves to confirm the correlation of the Yell Sound Division with the Moine Assemblage (Flinn *et al.*, 1972) and possibly with the Glenfinnan Division of the Moine (Flinn *et al.*, 1979). All the features listed, when considered together give the impression that the Yell Sound Division suffered not only a higher-grade metamorphism, but a more complex history of high-grade metamorphic development than the Dalradian part of the East Mainland succession, so that the Yell Sound Division may be the basement for a cover sequence formed by the other divisions. The alternative interpretation, based on the absence of any visible thrust or slide, that the East Mainland succession is an originally continuous sedimentary succession, seems much less likely. In the absence of suitable radiometric dates it is necessary to look to the Boundary zone for confirmation.

Figure 6.13 Thin banded psammite. The Eigg, Yell. Map ref. HU/452 957.

Figure 6.14 Laminated and thin banded psammite. Gloup Voe, Yell. Map ref. HU/501 048.

6.3 The Boundary zone

The Boundary zone is bounded to the west by the Valayre Gneiss and to the east by the Skelladale Gneiss.

The Valayre Gneiss is a continuous zone, generally a few metres wide, of single-crystal augen of microcline several centimetres in diameter. In gneiss areas, the matrix in which the augen occur is gneissic, and in schist areas the matrix is a schist. In thin-section the matrix is usually seen to be fine grained (*c.* 0.25 mm) and blastomylonitic. To the north of Aith the matrix in places is a micaceous schist, in Lunna Ness it is a granulitic anatectic quartzo-plagioclase gneiss, and in North Yell it is psammite. In the field in some places the gneiss appears to be a blastomylonite, and in others a porphyroblast gneiss. Within the Yell Sound Division, to the west of the Valayre Gneiss, the same type of gneiss occurs in small isolated exposures in the peat bogs of Yell and with more continuity in Lunna Ness (Figs 6.5 and 6.8).

The Valayre Gneiss is very similar to the Tännäs Gneiss of Scandinavia (D. G. Gee pers. comm.) in both appearance and in its occurrence as thin sheets traceable for tens of kilometres or more. However, it is generally accepted in Scandinavia that the Tännäs Augen Gneiss is a mylonitic (probably blastomylonitic in the context of this account) derivative of a Precambrian granite sheet forming one of the Caledonian nappes of Scandinavia (Krill, 1983 and references therein). The Valayre Gneiss has never been a granite, and is a zone of augen, and not a layer of rock derived from a different protolith than the rocks on either side.

The Skelladale Gneiss, on the other hand, is a layer of rock, up to several metres thick, composed of parallel-oriented, lenticular to oval single-crystal microcline augen about one centimetre long and closely packed in a quartzofeldspathic matrix. To the west of the Nesting Fault it forms a continuous layer from south of Aith to the Nesting Fault (Figs 6.8, 6.9). To the east of the fault it is seen only in widely separated exposures south of Lunna Ness, and along the north coast of Whalsay. It appears to be a highly tectonized orthogneiss of granitic origin.

The rocks between the two gneiss bands are in general foreign to the succession on either side. This is least so in the southern sector between Aith and the Nesting Fault. Here the rocks are dominantly very micaceous gneisses and partially gneissified micaceous schists with some calc-silicate and marble streaks (Flinn *et al.*, 1981). They differ sufficiently, individually and as a group, from the rocks to the east in the Scatsta Division to have been made into a subdivision; they also differ from the rocks to the west. These differences, even when considered in conjunction with the blastomylonitic nature of the Valayre and Skelladale Gneisses, are not sufficient to indicate a tectonic break. However, within the Graven Complex to the east of Graven, more exotic rocks are included in the Boundary zone. These occur in considerably greater amounts, across the Nesting Fault, in the Lunna Ness sector of the Boundary zone and include anatectic plagioclase gneisses rich in hornblendite-fish and bands.

In the Lunna Ness sector, the Boundary zone has a three-fold division. In the west, against the Valayre Gneiss, are coarse quartzofeldspathic (sodic plagioclase) anatectic gneisses rich in 10–20 cm bands and fish of hornblendite. To the east of these is a broad band of calc-silicate gneiss, together with some marble and a Cu-Pb-Zn-pyrrhotite strata-bound deposit. To the east again is a broad zone of pelitic gneisses and schists (Flinn *et al.*, 1981).

In the Yell sector of the Boundary zone, the rocks immediately east of the Valayre Gneiss form the south-eastern coast of Yell and are psammitic granulites, many of them partially gneissified by the addition of microcline. Although they are very similar to psammites west of the Valayre Gneiss in Lunna Ness, they differ in the complete absence of garnet-studded hornblende schist bands and are probably part of the boundary zone.

There are further features marking out the Boundary zone from the rocks on either side of it. In the Lunna Ness area, and extending up the south-east coast of Yell, all hornblendes which have not suffered a late alteration are red-brown in colour (Flinn *et al.*, 1981). This indicates that a higher grade of metamorphism has taken place in the Boundary zone in this area than in the rocks on either side, where fresh hornblendes are olive-green in Yell, and green or blue-green in Mainland to the east.

Within the Boundary zone from Hascosay to the south-east tip of Yell, and nowhere else in Shetland, there are epidiorites in the form of very fat lenses or globules up to 20 m in diameter, occurring closely packed in conformable zones. In Hascosay, lenticular globules with cores of ophitic augite-labradorite dolerite can be found, but generally the ophitic texture has been pseudomorphosed by red-brown hornblende and garnet, and the envelopes have been tectonized to hornblende-garnet schists. The globular form of the dolerites seems to indicate that they were intruded at a time when the country rock had the same viscosity as the magma, possibly during the metamorphism which produced the red-brown hornblende in the Boundary zone.

To the east of the Nesting Fault, small serpentinite and steatite-serpentinite masses occur very closely associated with the exposures of Skelladale Gneiss scattered south of Lunna Ness (Fig. 6.1), and along the north side of Whalsay. Within the East Mainland succession and outside the Boundary zone, serpentinite is known only in the form of komatiite in south-east Shetland (Flinn and Moffat, 1985) and as small masses associated with the Brae and Aith Complexes (Flinn *et al.*, 1981).

In south-east Yell, schists more mylonitic than blastomylonitic occur in several places east of the Valayre Gneiss, though they do not form any noticeably continuous features.

All these features considered together make it likely that in Shetland, the Moine and Dalradian sequences are separated by slices of rock derived from elsewhere and assembled tectonically into a schuppen-type boundary zone, either during or prior to the latest regional metamorphism of the area. Most signs of movement recorded in fabrics during the emplacement have been removed or prevented by strong annealing recrystallization during the regional metamorphism.

6.4 The Hascosay Slide

The Hascosay Slide outcrops along the north-east coast of Yell, as far south as Hascosay where it finally passes out to sea (Figs 6.4, 6.5 and 6.9). It lies east of the Valayre Gneiss and thus within the Boundary zone. The slide is narrower than the Boundary zone and characterized by the presence of a relatively fine-grained, finely laminated, blastomylonitic granulite. This appears to have been derived by shearing under epidote–amphibolite-facies metamorphic conditions from hornblendic gneisses, quartzofeldspathic gneisses, psammites and aplites.

Closely spaced aplite sheets up to 10 cm thick, conformable with the schistosity and banding in the psammites, and lying parallel to the slide plane, are a characteristic feature of the edges of the slide zone. Residual masses of gneiss occur within the blastomylonite foliation plane, ranging in size from 10 metres or so in diameter to masses almost filling the slide zone. These include Lewisian-inlier type banded hornblendic gneisses, coarse hornblendites, hornblende-pyroxene-garnet gneisses and quartzofeldspathic paragneisses.

The banding within some of the gneiss masses is in places sharply deflected in their envelopes—into conformity with the foliation in the enclosing blastomylonites. In some places the foliation is thrown into intrafolial folds, some of which form the cores of resistant lenses lying within the foliation. A fabric lineation found throughout the blastomylonite matrix extends for half a kilometre into the psammites to the west. The fold axes and lineations tend to be parallel and to plunge at 0–20° to the north-north-west.

It is apparent that the rocks of Yell have been thrust strongly to the east-north-east on this slide zone. Neither the relict gneiss masses within the slide zone nor the blastomylonitic fabrics are matched in the Boundary zone to the south, so that the Hascosay Slide must be younger than the Boundary zone, even though the former also formed under conditions of active recrystallization, that is before the end of the regional metamorphism. The south-eastward continuation of the Hascosay Slide lies under the sea between Fetlar and the Out Skerries (Fig. 6.1).

6.5 Conclusions

To the west of the Walls Boundary Fault in Shetland, up to several kilometres of psammitic rock occurs interleaved with Lewisian-inlier type hornblendic gneiss and rests on a foreland basement of Lewisian gneiss. This juxtaposition of lithologies is reminiscent of the Moine Thrust zone in north-west Scotland, differing chiefly in the greater angle of dip of the structure and the higher grade of metamorphism at which it was assembled. It seems likely that the psammites are a part of the Moine Assemblage. Despite the intensive deformation, the rocks exhibit some cross-bedding and conglomerates, and relict coarse clastic grains have been preserved. Furthermore, there is a complete absence of any garnet-studded hornblende schists. In the absence of any other evidence, a shallow-water origin and a tentative correlation with the Morar Division of the Moine Assemblage is all that can be proposed.

To the east of the Walls Boundary Fault, the Yell Sound Division of the East Mainland succession appears to be a slice of Moine-type rocks 10 kilometres or more thick, largely composed of psammites characterized by thin, parallel compositional banding and lamination. Within the psammite succession are lenticular masses of quartzite and mica schist, together with Lewisian inliers. The occurrence of calcareous nodules, though rare, together with the ubiquitous occurrence of garnet-studded hornblende schists, suggests a correlation of these rocks with the Glenfinnan Division of the Moine Assemblage. Deep-water deposition is probably indicated by the lack of all signs of shallow-water sedimentary features and the ubiquitous occurrence of parallel and uniform compositional banding and lamination throughout a succession many kilometres thick.

This sequence of Moine rocks is separated from the Dalradian rocks to the east by the Boundary zone, a zone a kilometre or more wide, containing rocks not found in the East Mainland succession on either side, together with some signs of partially annealed mylonitic textures, especially in the narrow gneiss bands on either side of the zone. The zone has also been intruded by dolerite magma and suffered a local high-grade metamorphism. The junction of the Moine and Dalradian in Shetland is a tectonic one, formed during or before the regional metamorphism of the area.

At a later date, but before the end of the regional metamorphism the whole East Mainland succession block has been thrust to the east-north-east, on the west-dipping Hascosay Slide composed of several hundred metres of blastomylonitic hornblende granulite containing relict masses of hornblende gneiss. The metamorphic grade of the blastomylonites in the Hascosay Slide is greater than that of the rocks in the thrust zones beneath the Shetland ophiolite to the east, so that the slide was active before the emplacement of the ophiolite.

The metasediments between the Hascosay Slide and the ophiolite (Fig. 6.9) have previously been correlated tentatively with the Dalradian, on the grounds of similarities with some of the rocks of the Scatsta Division in Mainland. However, these particular rocks in the Scatsta Division have now been allocated to the Boundary zone, which weakens the argument supporting this correlation with the Dalradian.

Acknowledgements

The author is grateful to Prof. T. Elliott and Dr A. L. Harris for pointing out a number of errors and making useful suggestions for alterations and corrections to an earlier version of this paper.

References

Andrews, I. J. (1985) The deep structure of the Moine thrust, southwest of Shetland. *Scott. J. Geol.* **21**, 213–217.

Flinn, D. (1954) On the time relations between regional metamorphism and permeation in Delting, Shetland. *Q. J. geol. Soc. London* **110**, 177–202.

Flinn, D. (1981) Central Shetland, one-inch series, sheet 128, solid edition, IGS, OHMS.

Flinn, D. (1985) The Caledonides of Shetland. In Gee, D. G. and Sturt, B. A. (eds.), The Caledonide Orogen—Scandinavia and Related Areas. *Spec. Publ. geol. Soc. London* **8**, 1158–1171.

Flinn, D. and Moffat, D. T. (1985) A peridotitic komatiite from the Dalradian of Shetland. *Geol. J.* **20**, 287–292.

Flinn, D., May, F., Roberts, J. L. and Treagus, J. E. (1972) A review of the stratigraphic succession of the East Mainland of Shetland. *Scott. J. Geol.* **8**, 335–343.

Flinn, D., Frank, P. L., Brooke, M., and Pringle, I. R. (1979) Basement–cover relations in Shetland. Harris, A. L. and Leake, B. E. (eds.), The Caledonides of the British Isles—Reviewed. *Spec. Publ. geol. Soc. London* **8**, 109–115.

Krill, A. G. (1983) Rb–Sr study of the Rapakivi granite and augen gneiss of the Risbergit nappe, Oppdal, Norway. *Norges geol. Unders.* **380**, 51–65.

Pringle, I. R. (1970) The structural geology of the North Roe area, Shetland. *Geol. J.* **7**, 147–170.

Robinson, T. (1983) Basement/Cover Relations in West Shetland. Unpublished Ph.D. Thesis, University of Liverpool.

7
The Krummedal supracrustal sequence in East Greenland

A. K. HIGGINS

7.1 Introduction

The East Greenland Caledonides extend for 1200 km between latitudes 70–82°N in a coast-parallel belt up to 250 km wide (Figs 7.1 and 7.2). The greater part comprises crystalline complexes, made up of a variety of gneissic and granitic rocks together with metasedimentary rock types. Late Proterozoic (Eleonore Bay Group) and Lower Palaeozoic sequences outcrop mainly between latitudes 72–74°N, while a broad zone of post-Caledonian sediments (Devonian–Cretaceous) covers eastern coastal areas (70–75°N). Tertiary basalts hide the continuation of the Caledonides south of 70°N.

Some of the earliest geologists to visit the crystalline complexes of the inner fjords had considered the gneissic rocks to be Archaean in age, only superficially disturbed by Caledonian events (Parkinson and Whittard, 1931; Odell, 1939, 1944). However, geologists attached to Lauge Koch's geological expeditions at about the same time introduced the concept of a deep-seated Caledonian orogeny, in which the infracrustal rock units were viewed as entirely Caledonian in origin, and the supracrustal rocks were all viewed as metamorphic representatives of the late Proterozoic Eleonore Bay Group (Backlund, 1930; Wegmann, 1935). Subsequent detailed fieldwork, mainly in the southern half of the Caledonides, appeared to confirm the basic concept (Wenk and Haller, 1953; Haller, 1955, 1958), most clearly described in terms of 'stockwerke' tectonics, with a complex deep-seated infrastructure and an overlying simpler suprastructure (Haller, 1953, 1971). The presence of former basement rocks within the infrastructure was recognized in southern areas (Haller, 1971, his Fig. 15), although they were considered to be petrogenetically rejuvenated during the Caledonian orogeny. These viewpoints were largely based on field observations, as the first K–Ar isotopic ages became available only after the conclusion of Lauge Koch's expeditions in 1958 (Haller and Kulp, 1962).

The Geological Survey of Greenland (GGU) carried out systematic geological mapping in the Scoresby Sund region (70–72°N) from 1968 to 1972, and a variety of special studies farther north (72–75°N) from 1973 to 1978. During the Scoresby Sund expeditions, it soon became clear that many of the infracrustal rock units within the Caledonian fold belt exhibited the same geological characteristics as the Archaean and Proterozoic basement complexes elsewhere in Greenland. Rb–Sr whole-rock and zircon isotopic studies have subsequently confirmed their original Archaean or Proterozoic age of formation, although K–Ar and Rb–Sr mineral ages everywhere bear witness to a strong Caledonian metamorphic overprint. North of 72°N, in the classic areas studies by John Haller, Rb–Sr isotopic studies have demonstrated an important early Proterozoic episode of granite intrusion. A variety of isotopic studies in a broad migmatite and granite zone suggest that both middle Proterozoic and Caledonian phases of formation were important. References to the more important isotopic studies are given in the following section.

The age of the supracrustal elements of the crystalline complexes, formerly assumed to be part of the Eleonore Bay Group, was called into question by Grenville (*c.* 1000 Ma) Rb–Sr whole rock ages on some of the metasediments. Subsequently, Grenville ages were yielded by Rb–Sr and zircon studies on granites which intruded migmatized supracrustal rocks. Other supracrustal sequences have proved to be early Proterozoic in age. One of the most widespread of the metasedimentary

Figure 7.1 The East Greenland Caledonides in one of the several possible pre-drift configurations. The frame shows location of Fig. 7.2.

sequences is known in the Scoresby Sund region as the Krummedal supracrustal sequence, and it and equivalent supracrustal sequences are the main subject of this paper. The isotopic evidence for a Grenville orogenic episode affecting them is reviewed in a special section below.

General reviews of the East Greenland Caledonides are given by Haller (1971), Henriksen and Higgins (1976), Higgins and Phillips (1979) and Henriksen (1985). There has been little work in the crystalline areas of East Greenland since 1978, recent fieldwork being largely concerned with the oil prospects of the post-Caledonian basins. However, the northern extremity of the Caledonides was reached in 1980 during the GGU North Greenland expeditions (Hurst and McKerrow, 1985).

7.2 General structure of the Caledonides

The East Greenland Caledonides are best described in terms of a series of N–S-trending zones (Fig. 7.2). These zones are the present-day surface reflection of a series of thick thrust wedges of infracrustal and supracrustal rocks. It is envisaged that during the Caledonian orogeny, complexes of migmatites and granites strongly affected by Caledonian events were thrust westwards over regions less disturbed by Caledonian metamorphism and deformation, which in their turn were thrust westwards over the Caledonian foreland.

The Caledonian foreland is exposed in two tectonic windows along the border of the Inland Ice in the Scoresby Sund region. The Gåseland window, first explored by Wenk in 1958 (Wenk, 1961), exposes a Precambrian gneissic basement overlain by a sequence of marbles and chlorite schists with tillites at the base. If the tillites are late Precambrian (Vendian) as has been suggested (Phillips and Friderichsen, 1981) then the metasediments are younger, and not part of the Eleonore Bay Group as Wenk had assumed. The Charcot Land window exposes an Archaean–early Proterozoic gneiss and granite complex overlain by the Charcot Land supracrustal sequence, a 2000 m succession of marbles, metavolcanics and mica schists (Steck, 1971). The granites, which vein the supracrustals, have given early Proterozoic ages, while the metasediments themselves show isotopic evidence for a metamorphic event 1850–1900 Ma ago (Hansen *et al.*, 1981). A local tillite within the Charcot Land window is correlated with the Vendian tillites (Henriksen, 1981). If the tillites in both windows are Vendian this would confirm the Caledonian age of the thrust episode. Undoubted Lower Palaeozoic rocks are not exposed, but the widespread occurrence of glacial erratics of Cambrian pipe rock (with *Skolithus* trace fossils) points to their presence beneath the Inland Ice to the west (Haller, 1971, Fig. 48).

A broad gneiss and schist zone borders the foreland windows and makes up the thrust units above them; it extends from Gåseland through Hinks Land and Gletscher Land to Isfjord between 70–74°N (Fig. 7.2). The basement gneiss complexes include the Flyverfjord infracrustal complex of the Scoresby Sund region, of which several gneiss and granite units have yielded Archaean Rb–Sr whole-rock and zircon U–Pb ages (2935 Ma, Rex and Gledhill, 1974; 2300–2520 Ma, Steiger *et al.*, 1979). Further north two of Haller's (1971) three infrastructural units have yielded Precambrian ages; the Gletscherland Complex has given a Rb–Sr whole-rock age of 2450 Ma, and the Hagar sheet five Rb–Sr whole-rock ages of 1725–1980 Ma (Rex and Gledhill, 1981). Metasedimentary rocks overlying and interfolded with these infracrustal rocks are referred to the Krummedal supracrustal sequence in the Scoresby Sund region, and equivalent unnamed sequences farther north. The isotopic evidence for Grenville metamorphism affecting these sediments, and the nature of the sedimentary sequences in different areas, are given in special sections below. Both infracrustal and supracrustal rock units have yielded Caledonian Rb–Sr and K–Ar mineral ages, testifying to a widespread thermal event (Rex and Higgins, 1985).

A broad zone dominated by migmatites and granitic rocks has a tectonic contact with the gneiss and schist zone to the west. The main zone extends from Gåsefjord through Milne Land and Renland to the Stauning Alper between 70° and 72°30′N (Fig. 7.2). Palaeosome remnants within the migmatites are almost entirely metasedimentary rock types, except in parts of the southern extremity of the zone. Locally thick, relatively non-migmatitic supracrustal sequences occur as enclaves within the migmatites. The genesis of the migmatite and granite zone is complex, and appears to be in part middle Proterozoic and in part Caledonian. An early phase of augen granite intrusions, associated with migmatite formation, has yielded Grenville ages (Steiger *et al.*, 1979). A varied suite of late monzonite to granite sheets and plutons, associated with a second phase of migmatite development, has largely yielded Caledonian ages (Hansen and Tembusch, 1979; Rex and Gledhill, 1981). The migmatitic sedimentary rocks, which are invaded by the early augen granites, are viewed as remnants of the Krummedal supracrustal sequence.

Liverpool Land is a horst-like region of crystalline rocks, isolated by the post-Caledonian Jameson Land Basin from the crystalline complexes of the inner fjords. It comprises gneissic complexes, many large intrusions, and in the north, high-grade, partly migmatized, metasedimentary rocks. Liverpool Land was the crystalline region for which Lauge Koch had first proposed a deep-seated Caledonian origin (Koch, 1929). Recent isotopic work has demonstrated that most of the granite bodies are Caledonian (Hansen and Friderichsen, 1987).

Crystalline rocks extend north of latitude 74°N in a broad coast-parallel zone. They include infracrustal gneisses and interfolded supracrustal rocks, but little modern fieldwork has been carried out and no isotopic ages are available. While they have been interpreted as largely Caledonian (Haller, 1971), a Proterozoic or

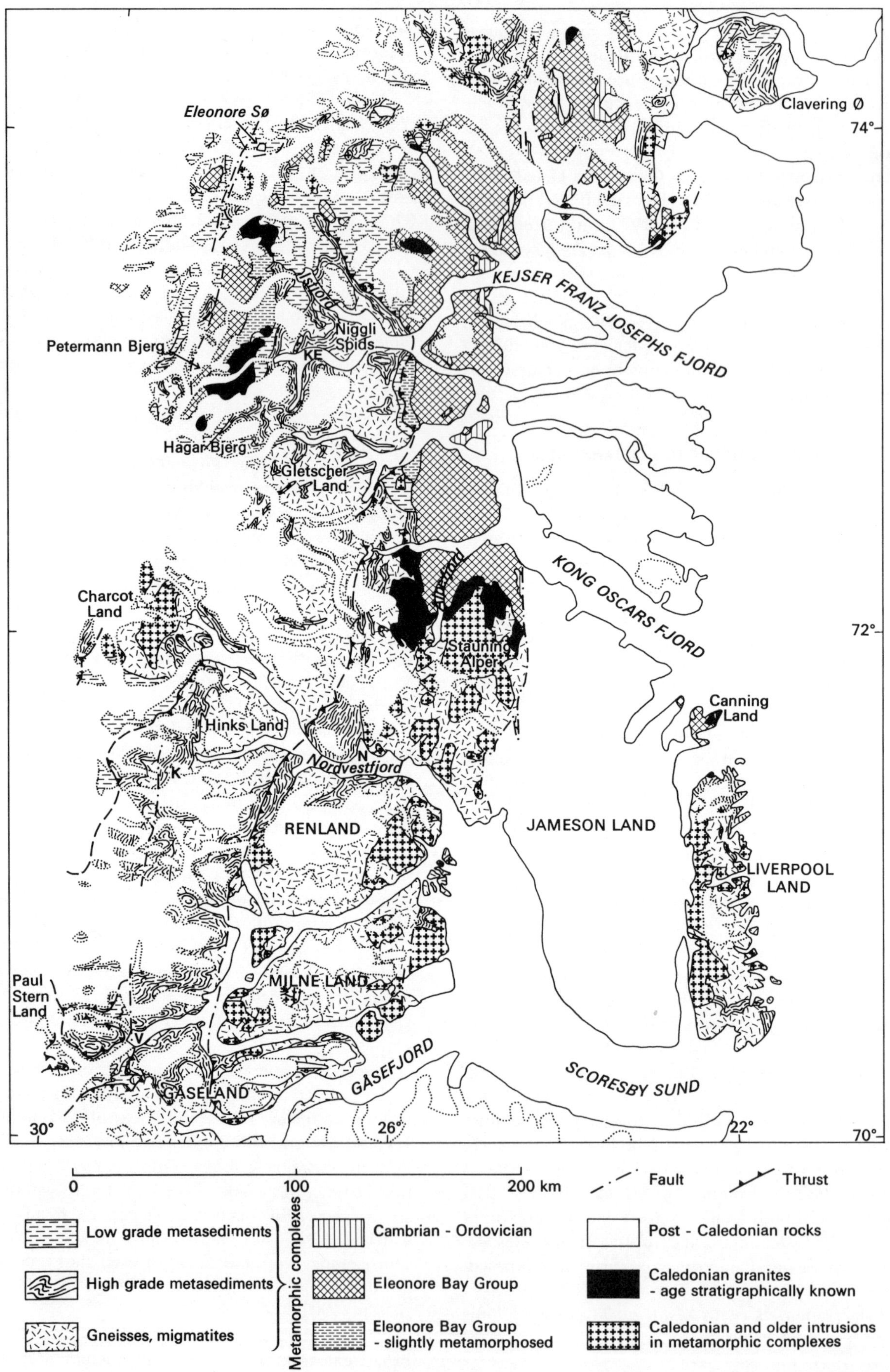

Figure 7.2 Simplified geological map of the southern part of the East Greenland Caledonides. The high-grade metasediments within the metamorphic complexes largely correspond with the Krummedal supracrustal sequence and equivalent sequences of this chapter. *K*: Krummedal, *KE*: Kejser Franz Joseph Fjord, *N*: Nordbugt, *V*: Vestfjord. Parts of the region are shown in greater detail in Figs 7.3, 7.6, 7.7, 7.8, 7.11 and 7.13.

Archaean age has been suggested for the gneissic basement of some areas (Leedal, 1952), and the occurrence of middle Proterozoic metasedimentary rocks cannot be excluded (Henriksen and Higgins, 1976).

The Eleonore Bay Group (late Proterozoic) and Lower Palaeozoic sequence has a total thickness of up to 17 000 m, although the great thicknesses of the lower Eleonore Bay Group are restricted to a few areas. The main outcrops occur in a broad zone between 72° and 74°N, with further outcrops to the west in the nunatak region (Petermann Series) (Fig. 7.2) and farther north around Ardencaple Fjord. These sequences have suffered only the effects of Caledonian deformation and metamorphism, and exhibit a pattern of mainly simple N–S-trending folds, in contrast to the complex deformation patterns shown by the older metasedimentary successions.

In a few areas, metamorphic grade in established lower Eleonore Bay Group rocks can be shown to increase rapidly westwards, and there is an apparent transition into complexly folded migmatites. Such transitions can be interpreted in two ways: that Caledonian migmatitic developments invade and transform the lower parts of the Eleonore Bay Group in the manner suggested by John Haller and co-workers: an interpretation favoured by Caby (1976) and Peucat *et al.* (1985); or that the intense Caledonian migmatitic activity and metamorphism overprints and obscures an original unconformity or *décollement* plane between the lower Eleonore Bay Group and an older migmatite and granite complex (Higgins *et al.*, 1981; Rex and Higgins, 1981). Modern isotopic work is providing increasing evidence of the intensity of Caledonian migmatitic activity and granite emplacement in the migmatite and granite zone, and transformations of the type suggested by Peucat *et al.* (1985) cannot be excluded. However, isotopic studies in the same structural zone also indicate important Grenville orogenic events (see below).

7.3 Isotopic evidence for Grenville orogenesis

Evidence for Grenville orogenic activity in East Greenland comes from isotopic analyses on two rock groups: metasedimentary rocks from the gneiss and schist zone referred to the Krummedal supracrustal sequence or equivalent sequences; and an early granite suite in the migmatite and granite zone which intrudes migmatitic supracrustal rocks.

From the southern part of the gneiss and schist zone in the Scoresby Sund region, Hansen *et al.* (1978) report a Rb–Sr whole-rock isochron age of 1146 ± 93 Ma (1122 Ma when recalculated using the standards recommended by Steiger and Jäger, 1976). The age is based on analyses of seven high-grade metasedimentary samples, and is interpreted as representing an early metamorphism of the Krummedal supracrustal sequence. U–Pb analyses of three monazite fractions from samples in the same region define a discordia with an upper intersection of 1010 ± 10 Ma and a lower intersection at 418 ± 5 Ma (Hansen *et al.*, 1978), suggesting that Caledonian metamorphism has not completely reset the isotopic system of Grenville age monazites.

Farther north (72–74°N), Rb–Sr whole-rock analyses have been undertaken on seven suites of metasediments believed to be northern equivalents of the Krummedal supracrustal sequence (Rex and Gledhill, 1981); none of these have yielded an isochron. However, one suite yielded an 'errorchron' of 1245 ± 100 Ma, and several suites plot as scatters about 1000 Ma reference lines. A plot of all data from all seven collections shows a broad scatter about a 1250 Ma reference line, and illustrates inhomogeneous Rb–Sr geochemistry between different collections and also within some collections (Rex and Gledhill, 1981, their Fig. 3). It is considered that a metamorphic event about 1000 Ma ago was responsible for a partial rehomogenization of the Sr isotopes. A Caledonian metamorphic overprint may have contributed to the scatter, but has not reset the whole-rock isotopic systems.

In the migmatite and granite zone of the Scoresby Sund region, a suite of extensive sheets and massive bodies of foliated garnetiferous augen granites has been emplaced into a sequence of banded paragneisses, which were extensively migmatized during the same orogenic event (Chadwick, 1975). These rusty brown weathering, garnetiferous paragneisses are regarded as equivalents of the Krummedal supracrustal sequence. The augen granites have yielded a Rb–Sr whole-rock isochron age of 987 ± 23 Ma, and a U–Pb zircon age of 1053 ± 40 Ma (Steiger *et al.*, 1979). The augen granites, together with the enveloping migmatized metasedimentary rocks were deformed into a series of large-scale nappe-like structures, which predate intrusion of Caledonian granitic rocks in the same region.

Farther north in the migmatite and granite zone, north of latitude 72°N, a foliated biotite granite forming thin sheet-like bodies in granitic migmatites has yielded a Rb–Sr whole-rock 'errorchron' of 750 ± 130 Ma (Rex and Gledhill, 1981). This result suggests that at least part of the migmatitic developments here are pre-Caledonian, despite the arguments of Peucat *et al.* (1985) that they are Caledonian.

Two granite bodies in the migmatite and granite zone which extends from 73°30′ to 74°30′N have yielded Grenville Rb–Sr whole-rock ages. Both are foliated muscovite granites emplaced into metasedimentary rocks, and have yielded an isochron age of 1000 ± 70 Ma and an 'errorchron' age of 1080 ± 200 Ma (Rex and Gledhill, 1981). Similar granites in the same region have yielded Caledonian ages, but the Caledonian bodies have significantly higher, somewhat variable, initial ratios.

The U–Pb studies of Peucat *et al.* (1985), on zircons from the lower Eleonore Bay Group of the Forsblad Fjord area, are of interest in that detrital zircons gave a lower intercept age of about 1100 Ma. It is deduced that a major metamorphic event occurred at about 1100 Ma ago in the source area of the lower Eleonore Bay Group.

There is thus a substantial body of isotopic evidence

supporting the concept of Grenville orogenic activity in the southern half of the East Greenland Caledonides. Complex deformation patterns, and in particular the E–W and NE–SW-trending folds which differ widely from the superimposed N–S-trending Caledonian folds, may relate to the trend of the Grenville orogenic belt.

7.4 Krummedal supracrustal sequence

The Krummedal supracrustal sequence, and equivalent sequences in East Greenland, show the effects of both Grenville and Caledonian orogenic events. The rocks nearly everywhere have undergone amphibolite-facies metamorphism and show the effects of several phases of superimposed deformation; they are distinguished on Fig. 7.2 as high-grade metasediments. Sedimentary structures are rarely preserved, and no special studies of their sedimentological development have been carried out. Nevertheless, in many areas, general observations can be made on the nature and thickness of the sequence.

7.4.1. *Krummedal*

The sequence is well exposed in the western parts of Krummedal and Rencontredal (Fig. 7.3) in the gneiss and schist zone (Higgins, 1974). Contact with the strongly foliated gneisses below is, due to shearing and reworking, apparently conformable and no angular discordance is preserved.

The lower 2000 m of the sequence generally comprise red-brown weathering, regularly bedded psammitic schist and mica schist. The psammite beds are generally 20–50 cm thick, with thinner mica schist interbeds (Fig. 7.4). Brown-weathering carbonate-bearing lenses are often seen in the psammitic beds. In some localities the basal few hundred metres are fairly massive psammitic rocks with only a small amount of mica schist material, but at other localities mica schists are exclusively developed.

Higher parts of the sequence comprise grey-black weathering semi-pelitic schists. This schist unit is at least 500 m thick, and passes rapidly upwards into gneissic and migmatitic developments, which may represent distinct thrust sheets.

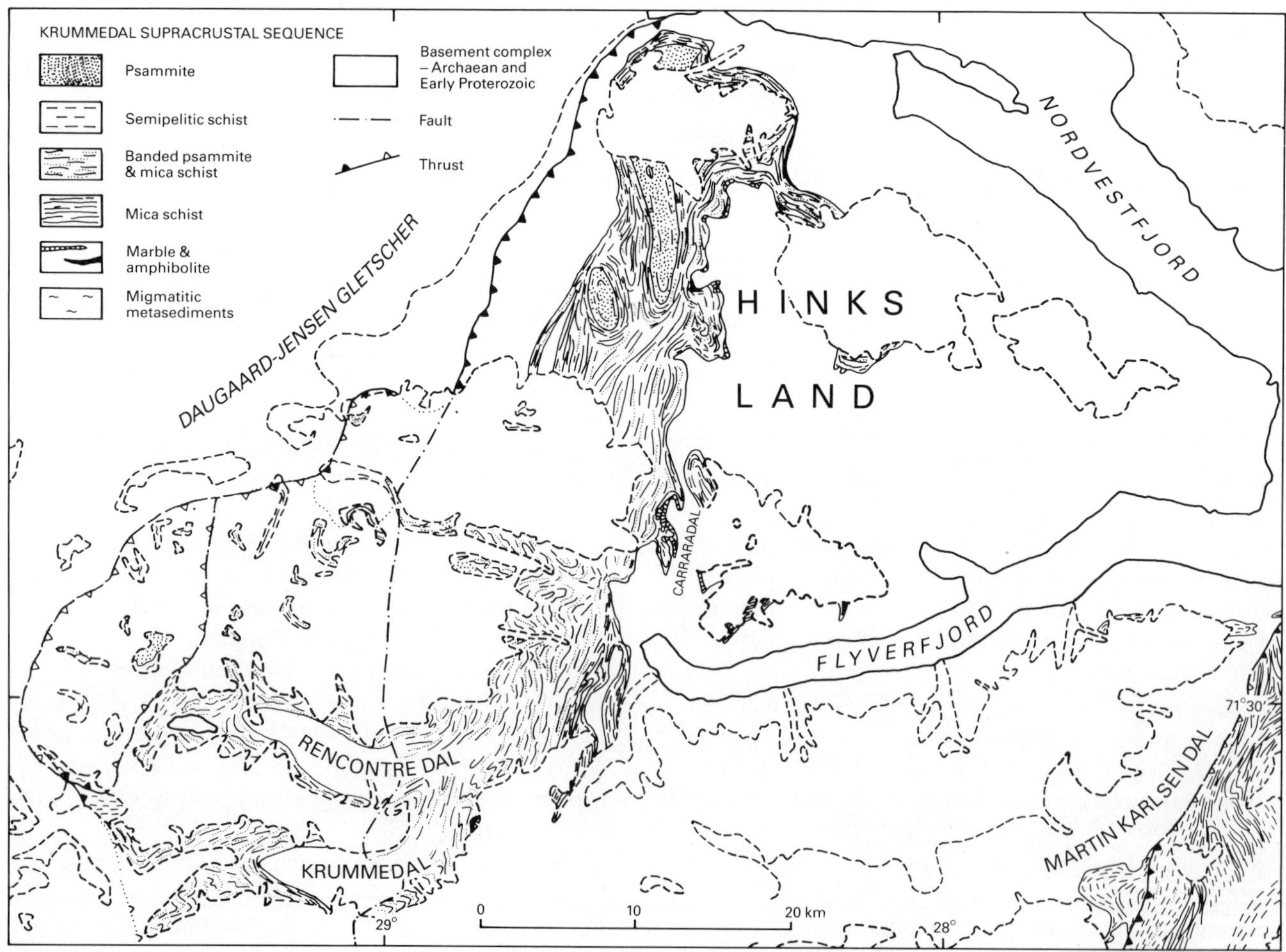

Figure 7.3 Geological map of the Krummedal–Hinks Land region. The Archaean and early Proterozoic basement rocks are here known as the Flyverfjord infracrustal complex.

Parageneses in the mica schists reflect high amphibolite-facies grade metamorphism, with abundant garnet, and occasional kyanite and staurolite.

7.4.2 *Hinks Land*

Parts of central Hinks Land preserve well-exposed metasedimentary sequences overlying complexly folded Archaean gneiss complexes (Fig. 7.3). Vogt (1965), who provided the first description of the region, described the contact as conformable, and grouped the metasediments into his 'Hüllserie' (cover series), estimated at about 1500 m thick. Although only a short distance from the Krummedal area, the supracrustal sequence in Hinks Land displays a number of differences (Higgins, 1974).

A basal unit of white marbles, with associated amphibolite, occurs in many parts of Hinks Land (Fig. 7.5). It is of variable thickness, often with a lens-like outcrop pattern. The best exposures are in Carraradal, where the marbles are at least 100 m thick and strongly folded (Vogt, 1965).

The overlying 600–700 m unit comprises in many areas entirely grey or red-brown mica schists, while in other areas a few psammite beds are present. This unit is overlain abruptly by several hundred metres of massive psammitic rocks developed in units 20–50 m thick, with thinner mica schist intervals.

Figure 7.4 Psammite beds alternating with thin mica schist layers. W end of Krummedal.

Figure 7.5 Thick white basal marble unit in Hinks Land, overlain by dark mica schists. The marbles overlie banded gneisses of the Flyverfjord infracrustal complex, seen in the foreground. Dark object at left (arrowed) is a musk-ox.

7.4.3 *Vestfjord*

In the areas on both sides of Vestfjord (Fig. 7.6), Phillips *et al.* (1973) estimated the exposed maximum thickness of the Krummedal sequence at 1400 m; Wenk (1961) estimated 1500 m for his equivalent 'Phyllite Mica-schist Series' in nearby areas of Paul Stern Land. The base of the sequence is recorded as tectonic in some areas, and at others as a conformable and undisturbed contact with gneissic basement.

The lowest units south of innermost Vestfjord include up to 200 m of amphibolite in layers and lenses, associated with impure quartzites; the mafic rocks are thought to represent metavolcanics. North-west of Vestfjord, in the nunatak region, a lenticular outcrop of metavolcanics at the same level includes breccias and agglomerates associated with gabbros and tuffs.

The bulk of the Krummedal sequence comprises a sequence of fairly uniform mica schists, distinctive for the lack of compositional banding and the intense augening and lineation of feldspars overgrown by garnet and kyanite. These schists grade upwards into more gneissic rocks of similar mineralogy, in which quartzofeldspathic aggregates form lenses and bands. This upwards increase in metamorphic grade, producing an inverted succession of metamorphic zones (first noted by Wenk, 1961), is today interpreted as a result of thrusting of more internal higher-grade rocks on to more external lower-grade rock units (Henriksen, 1986).

7.4.4 *Nordbugt*

North of central Nordvestfjord, at the west edge of the migmatite and granite zone, an approximately 8-km thick sequence of supracrustal rocks outcrops between

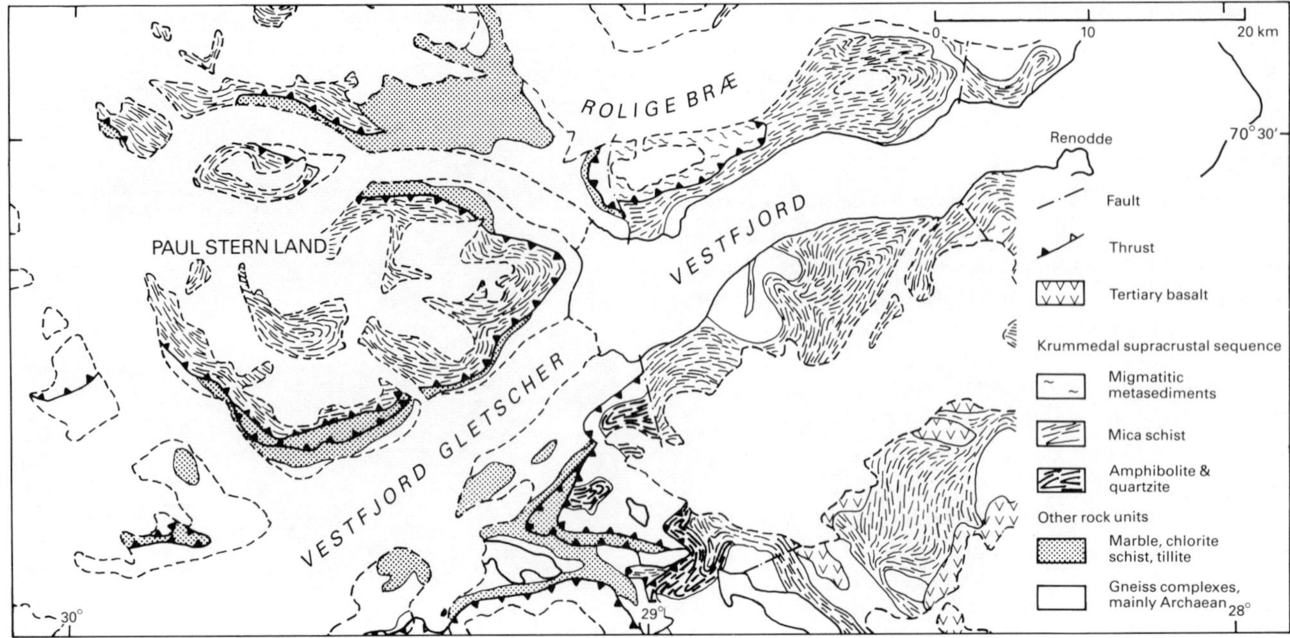

Figure 7.6 Geological map of the area around Vestfjord, including part of western Gåseland. The rock units beneath the thrusts on both sides of Vestfjord Gletscher are foreland rocks of the Gåseland window.

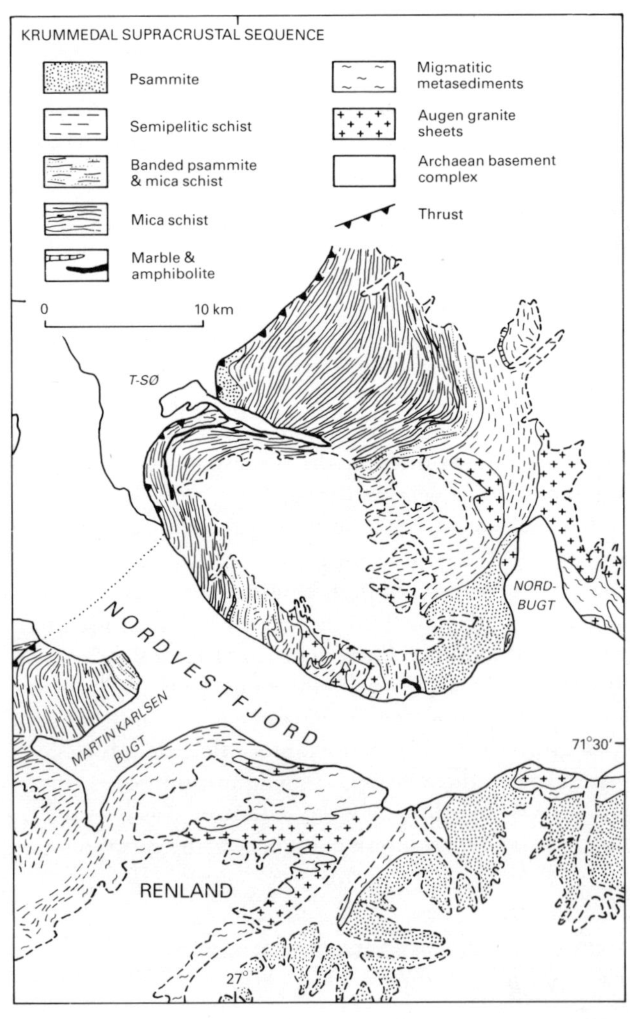

Figure 7.7 Geological map of central Nordvestfjord, showing the thick sedimentary sequence between T-Sϕ and Nordbugt.

Nordbugt and T-Sϕ (Fig. 7.7). The sequence dips eastwards, and in some areas exhibits large-scale folds. Parts of the sequence display diffuse migmatization, and at other levels there are concentrations of granite intrusions (Higgins, 1974; Henriksen *et al.*, 1980).

The western limit of exposure is an eastwards-dipping thrust associated with up to 300-m thick mylonitic developments. The lowest unit above the thrust is a wedge-shaped sequence of psammitic biotite gneiss up to 500 m thick, which may represent an original sandstone or siltstone division; it is rich in sillimanite and kyanite.

The succeeding division, up to 3000 m thick, comprises a sequence of dark-coloured, high-grade, pelitic or semi-pelitic schistose gneisses. The lower part is largely mica schist rich in garnet and sillimanite, with local calc-silicate layers and lenses. The upper part is more homogeneous and more psammitic, with numerous, leucocratic garnet-feldspar lenses.

A red to yellow-brown weathering sequence 1200–1500 m thick follows, made up of alternating psammitic gneiss bands on a 10 cm to 10 m scale, with thin mica schist interbeds. Incipient migmatization is seen in the form of thin garnet-bearing granitic veins. Occasional marble layers occur within the sequence, and there is often a mappable quartzite bed at the top.

The next division comprises up to 3000 m of semi-pelitic to psammitic garnet biotite gneisses, and is a partly migmatitic division with abundant thick sheets of leucocratic augen granite. The highest division in the sequence, about 2000 m thick, is made up of massive banded psammitic gneisses. These pass upwards and eastwards into rock sequences dominated by migmatitic developments and intruded by extensive granites (Henriksen *et al.*, 1980).

7.4.5 *East Milne Land*

At the eastern margin of the migmatite and granite zone in east Milne Land, outcrops of high-grade, relatively non-migmatitic metasedimentary sequences occur (Fig. 7.8). The most important outcrop preserves a sequence about 4000 m thick, in which the occasional preservation of cross-bedding and ripple marks indicates the way up. The outcrop is bounded on all sides by Caledonian intrusions, and is provisionally correlated with the Krummedal supracrustal sequence. The entire sequence is in high amphibolite-facies metamorphic grade (Henriksen and Higgins, 1987).

A conspicuous marble unit, exhibiting strong internal folding, borders the main outcrop. Its stratigraphic position is uncertain. It may relate to the trails of marble lenses of variable size, which can be traced throughout the migmatite and granite zone, and evidently represent remnants of a particular level in the now migmatized sedimentary sequence (Henriksen *et al.*, 1980).

The lower 2000 m of the main sequence comprises thickly bedded semi-pelitic rocks, with occasional units of mica schist and psammite (Fig. 7.9). The next 1000 m consists of regular alternations of massive quartzite units 10–100 m thick (Fig. 7.10), and semi-pelitic schist or mica schist units. Most of the observations of cross-bedding and ripple marks are in this sequence. The uppermost division comprises about 1000 m of very homogeneous semi-pelitic rocks, locally affected by diffuse migmatization.

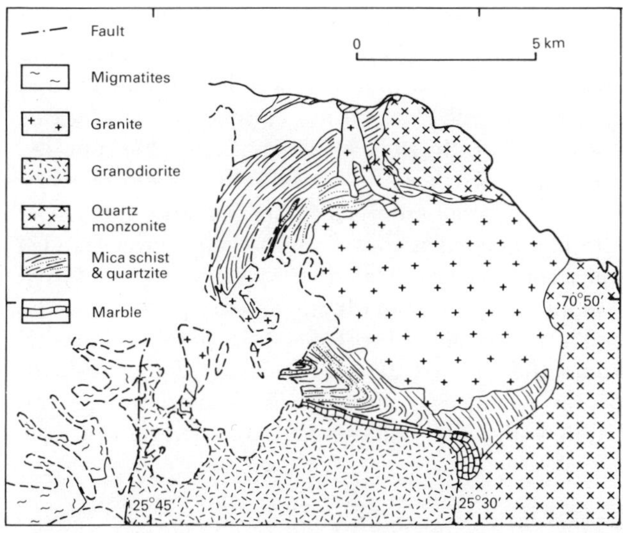

Figure 7.8 Part of NE Milne Land (due S of the group of islands at the N point of Milne Land in Fig. 7.2).

Figure 7.9 Alternating thin layers of quartzite and mica schist, NE Milne Land.

Figure 7.10 Massive white quartzite units up to 100 m wide, alternating with dark-coloured mica schist or semi-pelitic schist units, NE Milne Land.

7.4.6 *North Liverpool Land*

The northernmost areas of Liverpool Land are largely made up of highly deformed, high-grade metasedimentary rocks, bordered to the north and east by the sea, and to the south by migmatites and granites (Fig. 7.11). Crude thickness estimates are of the order of 5–10 km, but should be viewed with reservation as the sequence is disturbed by at least two phases of isoclinal folding and no original sedimentary structures are preserved.

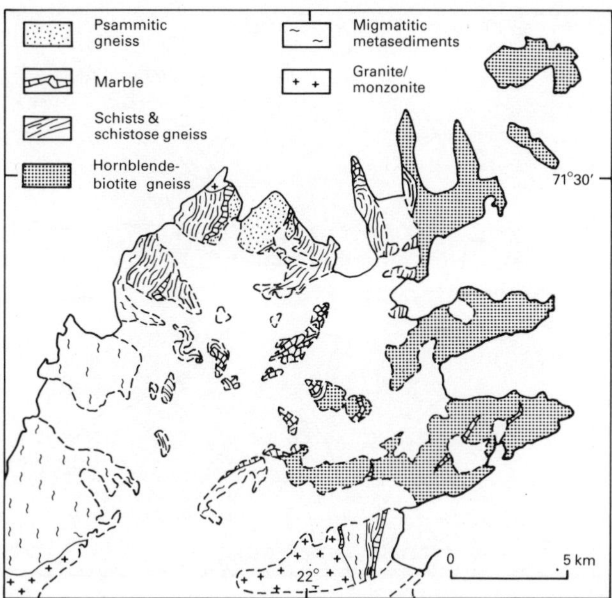

Figure 7.11 Geological map of northern Liverpool Land.

Figure 7.12 Part of northern Liverpool Land, looking NE. The nunatak in the centre of the photograph is made up of a thick yellowish white marble unit, flanked to the right by dark mica schist and hornblende-biotite paragneisses.

The structually lowest unit, about half the thickness, comprises dark-coloured hornblende-biotite paragneisses found in the east; garnet and sillimanite are common. This unit is overlain structually by a variety of semi-pelitic and pelitic schists and schistose gneisses, folded together with units of yellow and white marbles, psammitic rocks and basic gneisses (Fig.7.12). The pelitic rocks often have a rusty red coloration, and contain garnet, sillimanite, and occasionally staurolite. Marbles occur in sequences up to several hundred metres thick. Some are pure marbles often with strongly deformed compositional banding, while others comprise marble bands intercalated with schistose gneiss or quartzite. The structurally highest unit in the sequence consists of 400–500 m of psammitic gneisses intercalated with marble-rich levels.

The age of this sequence is uncertain. To some extent it resembles the Krummedal supracrustal sequence of the western parts of the Scoresby Sund region, but contains a far higher proportion of marbles. Comparisons might also be made with the Eleonore Bay Group, of which unmetamorphosed representatives are known only 20 km to the north in Canning Land (Caby, 1972, 1976), but the two regions exhibit widely different structural and metamorphic histories. Isotopic studies indicate that most of the plutonic rocks of Liverpool Land are Caledonian in age, but where these cut metasedimentary rocks they post-date at least two deformation phases and an episode of high-grade metamorphism.

7.4.7 *Kejser Franz Joseph Fjord*

Thick sedimentary sequences, broadly equated with the Krummedal supracrustal sequence, overlie and are interfolded with the infracrustal gneiss complexes exposed around inner Kejser Franz Joseph Fjord in Frænkel Land and Suess Land (Fig. 7.13); these outcrops have been described by Haller (1953, 1955, 1971), although he identified them as metamorphic representatives of the Eleonore Bay Group. The contact between infracrustal and supracrustal units appears in most places to be conformable, although discordant relationships have been recorded (Higgins *et al.*, 1981).

Basal units of the supracrustal sequence include, in many areas, yellow to orange weathering marbles and calcareous schists, which are conspicuous around the margin of the Niggli Spids Dome (Fig. 7.14). Locally, north-west of Isfjord, the basal unit is a red or white weathering psammite, succeeded by about 400 m of alternating calcareous schists and well-banded micaceous marbles (Haller, 1953; Higgins *et al.*, 1981). Amphibolite lenses and layers are conspicuous locally.

The major part of the sequence in most areas, several kilometres in thickness (Fig. 7.15), comprises red-brown weathering alternations of psammitic, semi-pelitic and pelitic units (Fig. 7.16); banding is usually

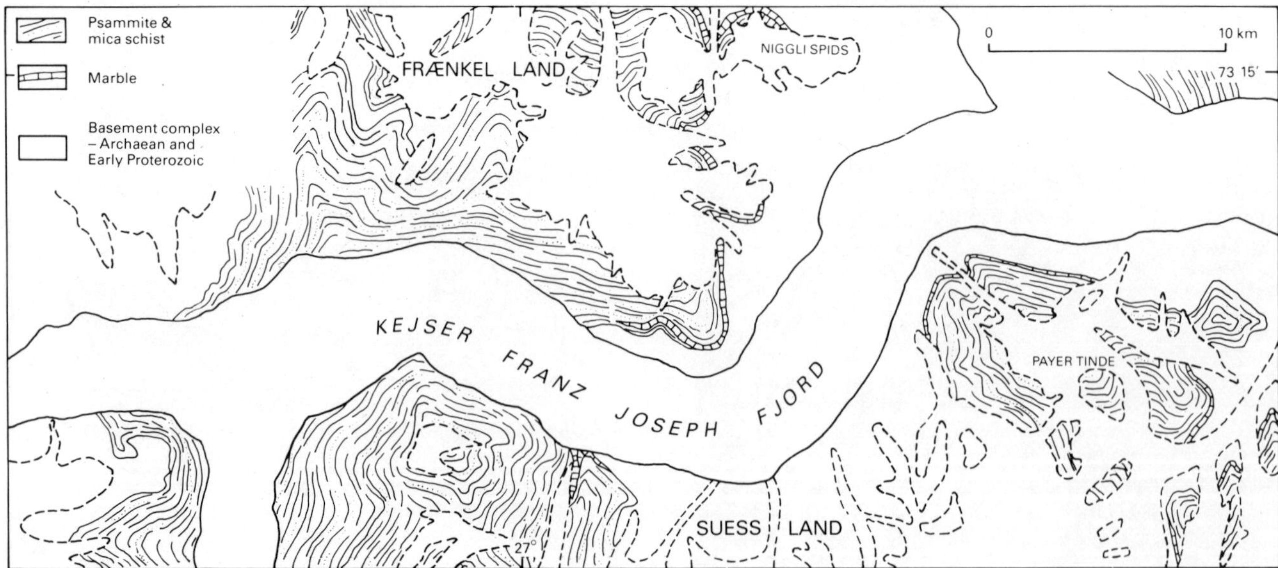

Figure 7.13 Geological map of innermost Kejser Franz Joseph Fjord.

Figure 7.14 Looking NE across Kejser Franz Joseph Fjord, with Frænkel Land and Niggli Spids on the left. The dark sedimentary sequence contains distinct light-coloured marbles towards the base, and overlies massive, cliff-forming augen gneisses of the Niggli Spids Dome.

Figure 7.15 Looking south-east across Kejser Franz Joseph Fjord to Payer Tinde (2310 m high) on Suess Land. The very thick and uniform sequence of metasediments overlies banded gneisses of the Niggli Spids Dome.

on a scale of 10–50 cm. Massive psammitic gneiss units may exceed 50 m in thickness. Haller (1953) presents a series of profiles of this part of the sequence in Andrée Land and Frænkel Land (Haller, 1953, pp. 41–45), although he refers them to the Eremitdal Series of the Eleonore Bay Group.

Metamorphic grade in the region varies between low and high amphibolite facies. Garnet is seen everywhere, while kyanite occurs mainly in the eastern and western margins of the outcrop.

Figure 7.16 Sequence of alternating psammite and mica schist beds, the mica schist generally deeply weathered. N of innermost Kejser Franz Joseph Fjord.

7.5 Conclusions

Metasedimentary sequences deposited prior to a Grenville orogenic episode are widespread in East Greenland between latitudes 70° and 74°N, and probably continue farther north. As a consequence of Grenville, and superimposed Caledonian, deformation and metamorphism, sequences everywhere are tightly folded and at high metamorphic grade. Few sedimentary structures have survived, and no current directions have been measured. South of 72°N these successions are referred to the Krummedal supracrustal sequence.

The sequences in different areas exhibit a comparable development and are several kilometres thick. The basal units in many areas were originally limestones or calcareous shales. Most of the sequences were derived from original sandstone, siltstone and mudstone, often in alternating sandstone and mudstone developments. However, perhaps because we have insufficient data, it is not generally possible to divide the sequences into mappable formations, or to make correlations of any value between different areas. The sometimes great differences between adjacent sections suggest rapid lateral facies variations.

Interpretation of the sedimentary environment is, of necessity, provisional and speculative. However, the general development of the sedimentary facies, their wide distribution and great thickness, are reminiscent of the deposits of deep-water turbidite basins.

Acknowledgements

Publication of this manuscript is authorized by the director of the Geological Survey of Greenland. The technical assistance of Ulla Johansen and Gurli Hansen is gratefully acknowledged.

References

Backlund, H. G. (1930) Contributions to the geology of Northeast Greenland. *Meddr Grønland* **74**, 207–296.

Caby, R. (1972) Preliminary results of mapping in the Caledonian rocks of Canning Land and Wegener Halvø, East Greenland. *Grønlands Geol. Unders. Rapport* **48**, 21–38.

Caby, R. (1976) Investigations on the Lower Eleonore Bay Group in the Alpefjord region, central East Greenland. *Grønlands Geol. Unders. Rapport* **80**, 102–106.

Chadwick, B. (1975) The structure of south Renland, Scoresby Sund, with special reference to the tectonometamorphic evolution of a southern internal part of the Caledonides of East Greenland. *Bull. Grønlands Geol. Unders.* **112** (also *Meddr Grønland* **201**, 2) 67 pp.

Haller, J. (1953) Geologie und Petrographie von West-Andrées Land und Ost-Frænkels Land (NE-Grönland). *Meddr Grønland* **113** (5), 196 pp.

Haller, J. (1955) Der 'Zentrale Metamorphe Komplex' von NE-Grönland. Teil I. Die geologische Karte von Suess Land, Gletscherland und Goodenoughs Land. *Meddr Grønland* **73**, I (3), 174 pp.

Haller, J. (1958) Der 'Zentrale Metamorphe Komplex' von NE-Grönland. II. Die geologische Karte der Staunings Alper und des Forsblads Fjordes. *Meddr. Grønland* **154** (3), 153 pp.

Haller, J. (1971) *Geology of the East Greenland Caledonides.* Interscience Publishers, New York, 413 pp.

Haller, J. and Kulp, J. L. (1962) Absolute age determinations in East Greenland. *Meddr Grønland* **171**, 77 pp.

Hansen, B. T. and Friderichsen, J. D. (1987) Isotopic age dating in Liverpool Land. *Grønlands Geol. Unders. Rapport* in press.

Hansen, B. T. and Tembusch, H. (1979) Rb–Sr isochron ages from east Milne Land, Scoresby Sund, East Greenland. *Grønlands Geol. Unders. Rapport* **95**, 96–101.

Hansen, B. T., Higgins, A. K. and Bär, M.-T. (1978) Rb–Sr and U–Pb age patterns in polymetamorphic sediments from the southern part of the East Greenland Caledonides. *Bull. geol. Soc. Denmark* **27**, 55–62.

Hansen, B. T., Steiger, R. H. and Higgins, A. K. (1981) Isotopic evidence for a Precambrian metamorphic event within the Charcot Land window, East Greenland Caledonian fold belt. *Bull. geol. Soc. Denmark* **29**, 151–160.

Henriksen, N. (1981) The Charcot Land tillite, Scoresby Sund, East Greenland. In Hambrey, M. J. and Harland, W. B. (eds.), *Earth's Pre-Pleistocene Glacial Record.* Cambridge University Press, 776–777.

Henriksen, N. (1985) The Caledonides of central East Greenland 70°–76°N In Gee, D. G. and Sturt, B. A. (eds.), *The Caledonide Orogen: Scandinavia and Related Areas.* Wiley, London, 1095–1113.

Henriksen, N. (1986) Descriptive text to 1:500 000 sheet Scoresby Sund, sheet 12. *Grønlands geol. Unders.* 27 pp.

Henriksen, N. and Higgins, A. K. (1976) East Greenland Caledonian fold belt. In Escher, A. and Watt, W. S. (eds.), *Geology of Greenland.* Geological Survey of Greenland, Copenhagen, 182–246.

Henriksen, N. and Higgins, A. K. (1987) Descriptive text to 1:100 000 sheet Kap Leslie 70 Ø·2 N and Rødefjord 70 Ø·3N. *Grønlands Geol. Unders.* in press.

Henriksen, N., Perch-Nielsen, K. and Andersen, C. (1980) Descriptive text to 1:100 000 sheets Sydlige Stauning Alper 71 Ø·N and Frederiksdal 71Ø·3 N. *Grønlands geol. Unders.* 46 pp.

Higgins, A. K. (1974) The Krummedal supracrustal sequence around inner Nordvestfjord, Scoresby Sund, East Greenland. *Grønlands Geol. Unders. Rapport* **67**, 34 pp.

Higgins, A. K. and Phillips, W. E. A. (1979) East Greenland Caledonides—a continuation of the British Caledonides. In Harris, A. L., Holland, C. H. and Leake, B. E. (eds.), The Caledonides of the British Isles—Reviewed. *Geol. Soc. London. Spec. Publ.* **8**, 19–32.

Higgins, A. K., Friderichsen, J. D. and Thyrsted, T. (1981) Precambrian metamorphic complexes in the East Greenland Caledonides (72°–74°N)—their relationships to the Eleonore Bay Group, and Caledonian orogenesis. *Grønlands Geol. Unders. Rapport* **104**, 46 pp.

Hurst, J. M. and McKerrow, W. S. (1985) Origin of the Caledonian nappes of eastern North Greenland. In Gee, D. G. and Sturt, B. A. (eds.) *The Caledonide Orogen—Scandinavia and Related Areas.* Wiley, London, 1065–1069.

Koch, L. (1929) The geology of East Greenland. *Meddr Grønland* **73**, II (1), 204 pp.

Leedal, G. P. (1952) The crystalline rocks of East Greenland between latitudes 74°30′ and 75°N. *Meddr Grønland* **142** (6), 80 pp.

Odell, N. (1939) The structure of the Kejser Franz Josephs Fjord region, north-east Greenland. *Meddr Grønland* **119** (6), 54 pp.

Odell, N. E. (1944) The petrography of the Franz Josef Fjord region, North-East Greenland in relation to its structures. *Trans. R. Soc. Edinburgh: Earth Sci.* **61**, 221–246.

Parkinson, M. M. L. and Whittard, W. F. (1931) The geological work of the Cambridge expedition to East Greenland in 1929. *Q. J. geol. Soc. London* **87**, 650–674.

Peucat, J. J., Tisserant, D., Caby, R. and Clauer, N. (1985) Resistance of zircon to U–Pb resetting in a prograde metamorphic sequence of Caledonian age in East Greenland. *Can. J. Earth Sci.* **22**, 330–338.

Phillips, W. E. A. and Friderichsen, J. D. (1981) The late Precambrian Gåseland tillite, Scoresby Sund, East Greenland. In Hambrey, M. J. and Harland, W. B. (eds.) *Earth's Pre-Pleistocene Glacial Record.* Cambridge University Press, 773–775.

Phillips, W. E. A., Stillman, C. J., Friderichsen, J. D. and Jemelin, L. (1973) Preliminary results of mapping in the western gneiss and schist zone around Vestfjord and inner Gåsefjord, south-west Scoresby Sund. *Grønlands Geol. Unders. Rapport* **58**, 17–32.

Rex, D. C. and Gledhill, A. (1974) Reconnaissance geochronology of the infracrustal rocks of Flyverfjord, Scoresby Sund, East Greenland. *Bull. geol. Soc. Denmark* **23**, 49–54.

Rex, D. C. and Gledhill, A. R. (1981) Isotopic studies in the East Greenland Caledonides (72°–74°N)—Precambrian and Caledonian ages. *Grønlands Geol. Unders. Rapport* **104**, 47–72.

Rex, D. C. and Higgins, A. K. (1985) Potassium–argon mineral ages from the East Greenland Caledonides between 72° and 74°N. In Gee, D. G. and Sturt, B. A. (eds.) *The Caledonide Orogen—Scandinavia and Related Areas.* London, Wiley, 1115–1124.

Steck, A. (1971) Kaledonische metamorphose der praekambrischen Charcot Land Serie, Scoresby Sund, Ost-Grönland. *Bull. Grønlands Geol. Unders.* **97**, 69 pp.

Steiger, R. H. and Jäger, E. (1977) Subcommission on geochronology: convention on the use of decay constants in geo- and cosmochronology. *Earth Planet. Sci. Letters* **36**, 359–362.

Steiger, R. H. Hansen, B. T., Schuler, C., Bär, M. T. and Henriksen, N. (1979) Polyorogenic nature of the southern Caledonian fold belt in East Greenland. *J. Geol.* **87**, 475–495.

Vogt, P. (1965) Zur Geologie von Südwest-Hinks Land (Ost-grönland). *Meddr Grønland* **154** (5), 24 pp.

Wegmann, C. E. (1935) Preliminary report on the Caledonian orogeny in Christian X's Land (North-East Greenland). *Meddr Grønland* **103** (3), 59 pp.

Wenk, E. (1961) On the crystalline basement and the basal part of the pre-Cambrian Eleonore Bay Group in the southwestern part of Scoresby Sund. *Meddr Grønland* **168** (7), 54 pp.

Wenk, E. and Haller, J. (1953) Geological explorations in the Petermann region, western part of Frænkels Land, East Greenland. *Meddr Grønland* **111** (3), 48 pp.

8
The Stoer Group, Scotland

A. D. STEWART

8.1 Introduction

The red sandstone mountains of north-west Scotland (Fig. 8.1) are remnants of deposits which once filled late Proterozoic rifts on the eastern margin of Laurentia. These rifts may have been connected with a prolonged phase of crustal extension which preceded the opening of the Iapetus ocean. Closure of Iapetus in the Palaeozoic seems to have transformed some of the old normal faults into thrusts, such as the Ben More Thrust and the Outer Isles Thrust.

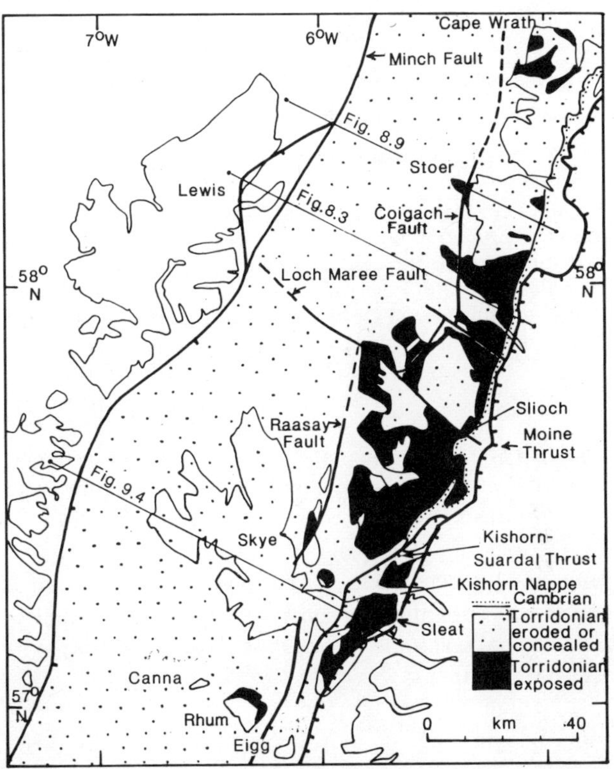

Figure 8.2 Sketch map of NW Scotland, showing the present and former extent of the 'Torridonian', together with the more important pre-Palaeozoic faults which controlled sedimentation.

By their very nature these sandstones are unlikely to have precise lithostratigraphic correlatives in other parts of Scotland. In particular, the Torridon Group and the Moines, though often equated (Geikie, 1895; Kennedy, 1951), had different source areas. This is shown by their diverse detrital zircon suites (MacKie, 1923), and their very different tourmaline contents—reflected in the boron content of stream sediment (Plant, 1984). Time correlation with part of the Moine is conceivable, but if the Moines originated far from their present position and later accreted on to the edge of Laurentia, then such correlation, even if true, might not be very significant.

The outcrop of the sandstones, mainly based on Geological Survey mapping, is shown in Fig. 8.2. The subcrop, however, is considerably more extensive. Westwards as far as the Minch Fault, seismic data show that 'Torridonian' underlies the Triassic (Smythe *et al.*, 1972; Chesher *et al.*, 1983). The Tertiary rocks of Canna, west of Rhum, include subangular blocks of red sandstone and metamorphic rocks from the underlying basement. Tertiary agglomerate on Eigg also contains blocks of red sandstone from the basement (Harker, 1908). To the south-west of Rhum, 'Torridonian' rocks form the sea floor for 125 km, reaching as far south as the latitude of Colonsay (McQuillan and Binns, 1973; Evans *et al.*, 1982, Fig. 1). This explains the occurrence of 'red Torridonian sandstone' and fossiliferous Durness Limestone in the Triassic conglomerates of Mull (Rast *et al.*, 1968). The 500 m thick sequence of breccias and sandstone unconformably overlying Lewisian basement in Iona (Stewart, 1962) may also belong to the 'Torridonian', but definite proof is lacking. (See Bentley, this volume.)

Stratigraphically the 'Torridonian' can be divided into two parts (Fig. 8.3). Red beds up to 2 km thick, outcropping at Stoer and along the coast to the south (Stoer Group) have been shown to be much older than the rest, from which they are separated by a clear angular unconformity and a 90° change in direction of magnetization. The beds above (Torridon Group), up to 5 km thick, are responsible for the spectacular scenery

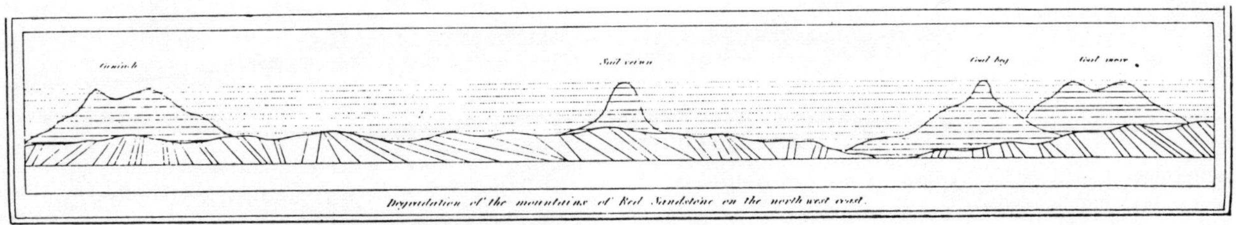

Figure 8.1 The mountains of NW Scotland drawn by Dr John MacCulloch (1819). The sketch shows the unconformable relationship of the strata to the gneisses beneath, which MacCulloch was the first to recognize. Quinag is on the left and Cul More on the right.

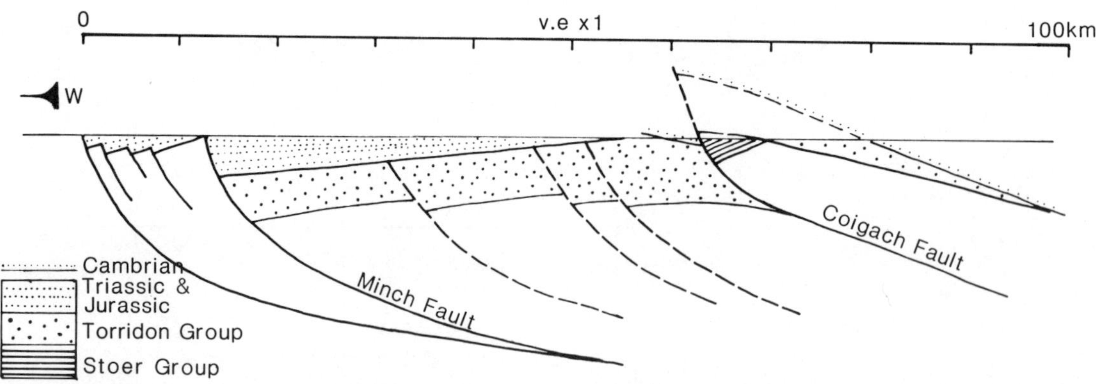

Figure 8.3 Cross-section of the North Minch Basin, showing the relationship between the Stoer and Torridon Groups and the overlying Triassic and Jurassic. There is no definite evidence that the Stoer Group exists west of the Coigach Fault. All faults shown are believed to be Proterozoic in origin. The Minch Fault was reactivated during the Mesozoic. The Coigach Fault was reactivated as a thrust during the Palaeozoic, and identifies with the easterly dipping reflector on the MOIST profile reaching the surface near shot point 2000 (Blundell *et al.*, 1985, Fig. 4). Most of the faults in the MOIST profile, despite showing normal movement in the basement, fail to cut either Mesozoic or Devonian sediments, suggesting that they are all earlier, probably Proterozoic.

of north-west Scotland, and especially the mountains around Torridon, which Nicol (1866) took as the type area for his 'Torridon Sandstone'. Clastic sediments 3 km thick in the Kishorn Nappe (Sleat Group), conformably beneath the Torridon Group, are not seen in contact with the Stoer Group, but are nevertheless believed to be younger. They may have accumulated in a sub-graben.

The age of the three groups is bracketed by a metamorphic event at 1100 Ma in the basement gneisses which they unconformably overlie (Moorbath *et al.*, 1967), and the Lower Cambrian fossils in beds which they unconformably underlie (Cowie and McNamara, 1978). Algal remains in grey shales of the Stoer Group are consistent with a Middle Riphean age according to Downie (pers. comm.), while those in the grey shales of the Torridon Group are Upper Riphean. These attributions fit in quite well with whole-rock Rb–Sr isochron ages obtained, respectively, from siltstones in the Stoer Group (968 ± 24 Ma) and the Torridon Group (777 ± 24 Ma). However, these isochron ages date diagenesis and may, therefore, be as much as 100 Ma too young, as suggested by Smith *et al.* (1983).

8.2 The Stoer Group

The Group was named from the peninsula of Stoer (Stewart, 1969), because the rocks there are sedimentologically representative, structurally simple, and superbly exposed. The main features of the succession at Stoer are shown in Fig. 8.4. The key stratigraphic element is the Stac Fada Member—a unique volcanic sandstone which identifies outcrops of the Group along the coast south of Stoer, as far as Poolewe. The seven red-bed facies which build the Stoer succession are described briefly below.

8.2.1 *Breccia facies*

This facies immediately overlies a hilly landscape of Lewisian gneiss, which at Stoer has over 300 m of relief. The clasts come exclusively from the Lewisian, and it is significant that the proportion of basic (and even

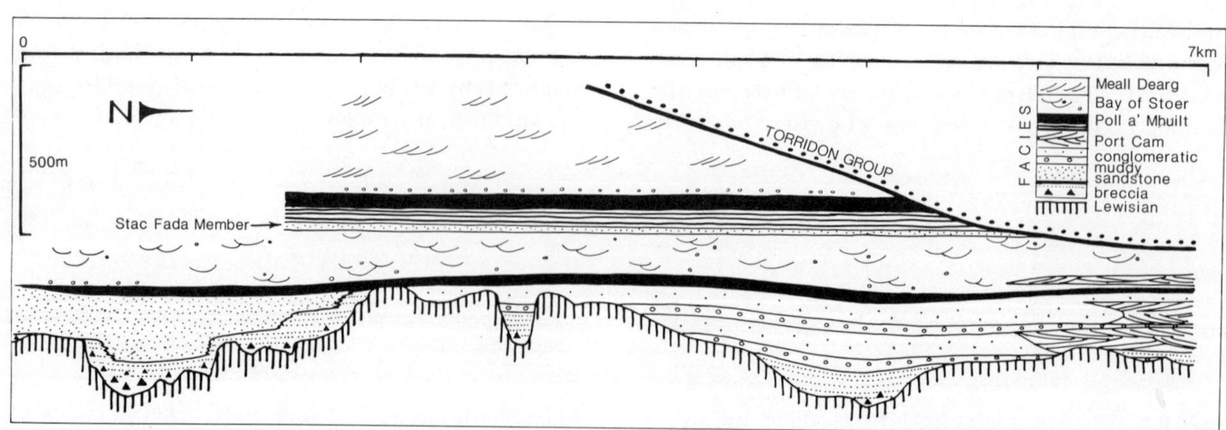

Figure 8.4 Stratigraphic profile of the Stoer Group at Stoer. This is basically a down-dip view of the strata exposed on the peninsula. The Stoer Group is truncated unconformably by the Torridon Group.

ultrabasic) rock types in the breccia is much the same as in the basement nearby. The degree of rounding suggests transport distances of generally less than a kilometre, except near the unconformity where clasts have obviously moved only a few metres. The breccia in contact with the Lewisian is usually massive, with clasts up to about half a metre in size. Stratigraphically upwards a crude stratification appears and the breccias pass into pebbly red sandstone, and sometimes red shale.

There can be no doubt that the facies represents a series of fanglomerates. Maximum fan radius seems to have been about 300 m, the upward fining resulting from upstream retreat of the fan heads.

8.2.2 *Muddy sandstone facies*

This facies includes the Stac Fada Member and consists of reddish-brown rocks, texturally greywackes, with about 40% matrix. They are always lateral equivalents of the finer, distal part of the breccia facies, which geochemically they closely resemble. Sedimentologically, however, there is no resemblance at all. The lowest 130 m of the muddy sandstone facies at Clachtoll are completely devoid of bedding or lamination. This massive development is succeeded upwards by a bedded subfacies in which the beds are about half a metre thick, defined by desiccated sheets of red siltstone or carbonate. The muddy sandstone also shows desiccation patterns (Fig. 8.5) suggesting that the structureless nature of the facies is due to repeated wetting and drying of a sediment originally rich in smectitic clay.

The ponded water mud flats of Hardie *et al.* (1978) would be a suitable setting for the facies, the sediments recording periodic flushes of weathered material from a source area with abundant basic rocks. Percolating groundwater rich in Mg and Ca may also have contributed to smectite production. Very similar massive reddish-brown siltstones are developed stratigraphically close to lacustrine shales in the Lower Jurassic East Berlin Formation of Connecticut. They are probably those described by Demicco and Kordesch (1986) as 'disrupted mudstones' and attributed by them to repeated wetting and drying of lake-marginal clay-rich sediment.

Figure 8.5 Desiccation patterns in the upper part of the muddy sandstone facies at Clachtoll [NC 037 272]. The ruler is 20 cm long.

8.2.3 *Conglomeratic facies*

This consists of upward-fining sequences of coarse, trough cross-bedded red sandstone, and occasionally multistorey conglomerate. Such sequences are typically tens of metres thick. The bases are erosional (Fig. 8.6) and the tops marked by a metre or so of red siltstone (rarely exposed). The conglomerates were derived from local basement, but, unlike the breccia facies, only acid Lewisian detritus is present. This, together with the rotund shape of the pebbles, suggests 5–10 km of transport.

The facies was probably deposited in shallow braided channels. The fining-upward sequences closely resemble those of the Donjek River (Miall, 1977, Fig. 12). However, one of the mapped conglomerates occupies the full width of the palaeovalley (see Fig. 8.4), so that conglomerate deposition must have been essentially synchronous over the whole of the flood plain. There is no evidence of lateral accretion in the conglomerate units, so that an origin by episodic source rejuvenation is preferred to avulsion.

Figure 8.6 Erosive base of a multistorey conglomerate unit (conglomeratic facies) at Stoer [NC 047 329]. The underlying beds belong to the Port Cam facies. The ruler is 20 cm long.

Figure 8.7 Cross-bedding in the Port Cam facies at Stoer [NC 048 328]. The ruler is 20 cm long.

Figure 8.8 Contorted bedding in the Bay of Stoer facies at Clachtoll [NC 035 272]. The 20 cm ruler marks the top of a cusp.

8.2.4 *Port Cam facies*

This facies is striking because of its cross-bedding (Fig. 8.7). Set thickness is generally a few decimetres, but sometimes reaches as much as 10 m. The thinner sets persist laterally for tens of metres, and the thicker ones for much further. The cross-beds are only a few millimetres thick, but can be followed for many metres as they asymptotically approach the base of the set. The cross-beds originally dipped eastwards at angles usually less than 20°, and only rarely more than 25°. The grains forming the rock are well sorted, subangular in shape, and average 0.2 mm in diameter. The maximum is about 2 mm.

In contrast to these well-laminated sandstones, there are also decimetre or metre-thick intercalations of relatively massive sandstone with irregular. erosional, bases. Occasionally these sandstones incorporate gneiss fragments and even lumps of the Port Cam facies, which must have been already partially lithified. The intercalations are quite common where the Port Cam facies is in contact with the breccia and conglomeratic facies (see Fig. 8.4). The base of the Bay of Stoer facies (described below) is also highly erosive where it overlies the Port Cam facies.

The cross-bedding described above is identical to that formed by migrating barchan dunes, while the massive sands evidently record periodic invasions of the dune field by torrential flood water.

8.2.5 *Poll a'Mhuilt facies*

This facies basically consists of thinly bedded red siltstone and fine sandstone. Wave ripples and desiccation cracks are characteristic. These fine-grained sediments form several intercalations, each only a few metres thick, within the Bay of Stoer facies (see below). Though thin, the intercalations can be traced right across the peninsula for about 6 km, with only slight change in thickness. Deposition of the facies was preceded in every case by a decimetre-thick bed of muddy sandstone. The most spectacular example of this association is afforded by the Stac Fada Member and the overlying sequence of red siltstone and sandstone. As mentioned earlier, the Stac Fada Member belongs to the muddy sandstone facies, differing only from the rest of the facies in containing about 30% of devitrified volcanic glass. Petrographic details have been published by Lawson (1972). A new analysis of the glass shows that it was olivine normative and probably undersaturated—a typical feature of rift volcanics.

The silty sediments following the Stac Fada Member, which belong to the Poll a'Mhuilt facies, are about 100 m thick. In addition to the usual red beds, there are also limestones (Upfold, 1984), laminated black shales containing poorly preserved organic-walled microfossils (Cloud and Germs, 1971), and abundant gypsum pseudomorphs (Stewart and Parker, 1979).

The interpretation of the Poll a'Mhuilt facies by Stewart and Parker (1979), based on sedimentology and boron-in-illite data, is that it formed in temporary lakes. The lake associated with the Stac Fada Member must have covered hundreds of square kilometres, with a maximum depth of at least 100 m, and perhaps twice as much. For the first half of its life it must have been perennial and stratified.

8.2.6 *Bay of Stoer facies*

The Bay of Stoer facies simply consists of trough cross-bedded sandstones. Soft-sediment contortions and overturned cross-bedding are common (Fig. 8.8). Well-rounded, centrimetre-sized pebbles of gneiss and orthoquartzite, in roughly equal proportions, are sporadically present throughout the facies.

8.2.7 *Meall Dearg facies*

This facies is entirely built of sandstones petrographically indistinguishable from those in the Bay of Stoer facies. Both planar cross-bedding, and planar bedding with extensive wave-rippled surfaces, are equally common. Pebbles are absent except at the very base at Stoer.

Both the Bay of Stoer and Meall Dearg facies are believed to have been deposited by braided rivers, the latter perhaps deposited in wider channels, on gentler palaeoslopes than the former. A modern analogue of the Bay of Stoer sandstones might be the predominantly trough cross-bedded sands deposited by powerful floods in central Australia (Williams, 1971). In contrast, the Meall Dearg facies resembles the predominantly planar cross-bedded sands deposited in the transverse bars of the relatively sluggish Platte River (Smith, 1970).

The three facies found laterally adjacent to basement hills, namely the breccia, conglomeratic and Port Cam facies, find close modern analogues in areas such as South Yemen (Moseley, 1971). There, gravel fans fringe basement hills of Precambrian gneiss 300–600 m high. The gravels interfinger with the fluvial deposits of an ephemeral river system, over which drift barchanoid dunes. South Yemen lies at latitude 15°, like that deduced from the palaeomagnetism of the Stoer Group (Stewart and Irving, 1974), and has a semi-arid climate.

Palaeocurrents within the Stoer Group at Stoer reverse through 180° at the base of the Bay of Stoer facies and again at the base of the Stac Fada Member (Stewart, 1982), suggesting fault-controlled deposition in a rift valley (Fig. 8.9). The repetition of lacustrine interludes (Poll a' Mhuilt facies) within the fluvial Bay of Stoer facies can also be attributed to episodic fault tilting and disruption of the fluvial drainage net.

All the sandstones in the Stoer Group are arkosic, with oligoclase the dominant feldspar, as would be expected if Scourian granulitic gneisses formed the source area. However, the geochemistry of the sandstones forming the breccia and muddy sandstone facies, which might have been expected to most closely resemble the parent gneisses, tells a different story. The average geochemistry of sandstone from these facies is compared with that for Scourian gneiss in Table 8.1. This shows that the sandstones are much richer in Mg and Rb. The excess Mg is easily accounted for by the observation that the Scourian east of Stoer is much more basic than the norm. The high Rb and correspondingly low K/Rb ratio, however, can only be explained by a substantial presence of undepleted amphibolite-facies gneisses in the source rocks. There is no trace of these gneisses at the present level of erosion, but they must once have existed a kilometre or two structurally higher. The sandstones show only slight depletion in Ca and none at all in Na, suggesting poor drainage together with aridity. The aridity is in line with the palaeolatitude of 15° deduced from palaeomagnetism (Stewart and Irving, 1974).

The sandstones forming the Bay of Stoer and Meall Dearg facies have K/Rb ratios like the others, so that the feldspars, at least, probably came from the same source. The Bay of Stoer and Meall Dearg facies, however, have a much higher ratio of silica to alumina (7.3 as against 4.7), that is, they are much more mature. This is not just due to the breakdown of original mafic minerals to clays and their loss from the depositional system, but to an

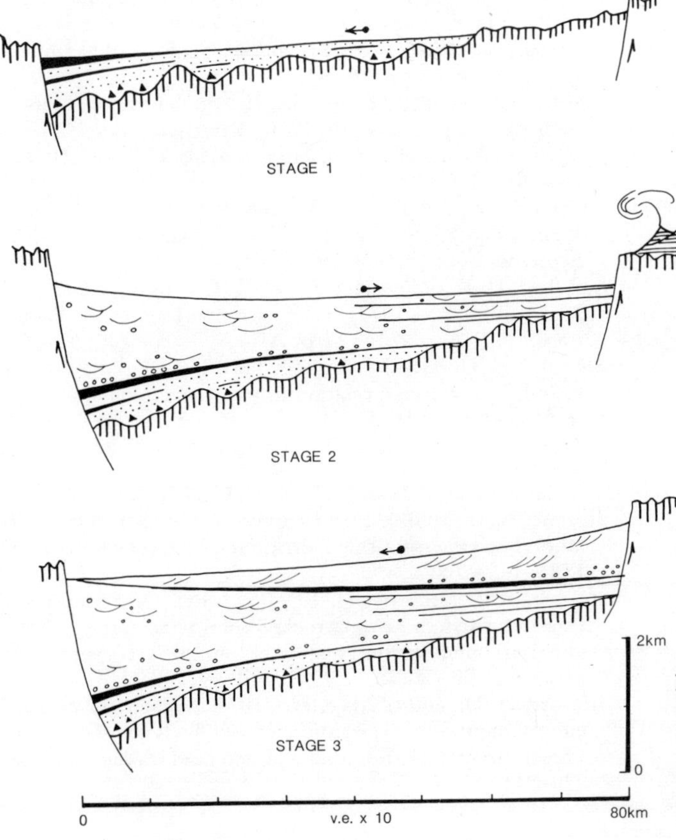

Figure 8.9 Three stages in the development of the Stoer Rift. Stage 1 shows locally derived sediments such as the breccia and conglomeratic facies. Stage 2 shows the Bay of Stoer and stage 3 the Meall Dearg facies. Stoer is located roughly in the middle of the rift. Arrows show palaeocurrent directions. Key as in Fig. 8.4.

Table 8.1 Comparative geochemistry of Scourian basement and adjacent Stoer Group sandstones.

	A	B	C
SiO_2	61.5	60.3	58.1
TiO_2	0.6	0.4	0.7
Al_2O_3	15.5	13.9	13.0
Fe_2O_3(tot.)	6.2	6.0	7.5
MnO	0.1	0.1	0.1
MgO	3.5	7.1	9.0
CaO	5.9	4.6	3.3
Na_2O	4.0	3.8	4.0
K_2O	1.0	1.1	1.5
P_2O_5	0.2	0.1	0.1
Volatiles	1.7	1.5	1.5
Total	100.2	99.2	98.8
Rb (ppm)	9	41	41
K/Rb	922	277	304

Column A: average Scourian
Holland and Lambert (1975) Table 1, analysis 6.

Column B: model Scourian
30% average Scourian + 50% amphibolite gneiss (Sheraton *et al.*, 1973, Table 4C) + 20% ultrabasic rock (Sheraton *et al.*, 1973) Table 3I–J.

Column C: average sandstone
based on representative analyses 81SO80 and 83SO69, Analyst Franz Street, Univ. Reading, Geol. Dept.

additional source of silica, namely siliceous sediments. To produce the observed dilution, these siliceous sediments must have formed about a quarter of the catchment. They also provided the quartzite pebbles so characteristic of the Bay of Stoer facies. The petrography of these pebbles suggests that the parent rocks were fine-grained red beds, quite unlike any exposed in western Scotland today.

References

Blundell, D. J., Hurich, C. A. and Smithson, S. B. (1985) A model for the MOIST seismic reflection profile, N. Scotland, *J. geol. Soc. London* **142**, 245–258.

Chesher, J. A., Smythe, D. K. and Bishop, P. (1983) The geology of the Minches, Inner Sound and Sound of Raasay. *Rep. Inst. geol. Sci. London* **83/6**, 1–29.

Cloud, P. and Germs, A. (1971) New Pre-Paleozoic nannofossils from the Stoer Formation (Torridonian), Northwest Scotland. *Bull. geol. Soc. Am.* **82**, 3469–3474.

Cowie, J. and McNamara, K. J. (1978) *Olenellus* (Trilobita) from the Lower Cambrian Strata of north-west Scotland. *Palaeontol. London* **21**, 615–634.

Demicco, R. V. and Kordesch, E. G. (1986) Facies sequences of a semi-arid closed basin: the Lower Jurassic East Berlin Formation of the Hartford Basin, New England, U.S.A. *Sedimentology* **33**, 107–118.

Evans, D., Chesher, J. A., Deegan, C. E. and Fannin, N. G. T. (1982) The offshore geology of Scotland in relation to the IGS shallow drilling program, 1970–1978. *Rep. Inst. geol. Sci. London* **81/12**, 1–36.

Geikie, A. (1895) *Rep. geol. Surv. Mus. London* for 1894.

Hardie, L. A., Smoot, J. P. and Eugster, H. P. (1978) Saline lakes and their deposits: a sedimentological approach. *Spec. Publ. Intern. Ass. Sedim.* **2**, 7–41.

Harker, A. (1908) The geology of the small isles of Inverness-shire. *Mem. geol. Surv. Scotland* 1–210.

Holland, J. G. and Lambert, R. StJ. (1975) The chemistry and origin of the Lewisian gneisses of the Scottish mainland: the Scourie and Inver assemblages and sub-crustal accretion. *Precambr. Res.* **2**, 161–188.

Kennedy, W. Q. (1951) Sedimentary differentiation as a factor in the Moine–Torridonian correlation. *Geol. Mag.* **88**, 257–266.

Lawson, D. E. (1972) Torridonian volcanic sediments. *Scott. J. Geol.* **8**, 345–362.

MacCulloch, J. (1819) *A Description of the Western Islands of Scotland, Including the Isle of Man, Comprising an Account of their Geological Structure, with Remarks on their Agriculture, Scenery, and Antiques.* 3 vols, London.

MacKie, W. (1923) The source of purple zircons in the sedimentary rocks of Scotland, *Trans. geol. Soc. Edinburgh: Earth Sci.* **11**, 200–213.

McQuillin, R. and Binns, P. E. (1973) Geological structure in the Sea of the Hebrides. *Nature* **241**, 2–4.

Miall, A. D. (1977) A review of the braided-river depositional environment. *Earth Sci. Rev.* **13**, 1–62.

Moorbath, S. (1969) Evidence for the age of deposition of the Torridonian sediments of north-west Scotland. *Scott. J. Geol.* **5**, 154–170.

Moorbath, S., Stewart, A. D., Lawson, D. E. and Williams, G. E. (1967) Geochronological studies on the Torridonian sediments of north-west Scotland. *Scott. J. Geol.* **3**, 389–412.

Moseley, F. (1971) A reconnaissance of the Wadi Beihan, South Yemen. *Proc. geol. Ass. London* **82**, 61–69.

Nicol, J. (1866) *The geology and Scenery of the North of Scotland: being two Lectures given at the Philosophical Institution, Edinburgh, with Notes and an Appendix.* Edinburgh, 1–96.

Plant, J. A. (1984) Regional geochemical maps of the United Kingdom. *NERC News J.* **3(4)**, 1 and 5–7.

Rast, N., Diggens, J. N. and Rast, D. E. (1968) Triassic rocks of the Isle of Mull; their sedimentation, facies, structure and relationship to the Great Glen Fault and the Mull caldera. *Proc. geol. Soc. London* **1645**, 299–304.

Sheraton, J. W., Skinner, A. C. and Tarney, J. (1973) The geochemistry of the Scourian gneisses of the Assynt district. In Park, R. G. and Tarney, J. (eds.) *The Early Precambrian of Scotland and Related Rocks of Greenland.* Geology Department, University of Keele, 13–30.

Smith, N. D. (1970) The braided stream depositional environment; comparison of the Platte River with some Silurian clastic rocks, north-central Appalachians. *Bull. Am. Ass. Pet. Geol.* **81**, 2993–3014.

Smith, R. L., Stearn, J. E. F. and Piper, J. D. A. (1983) Palaeomagnetic studies of the Torridonian sediments, NW Scotland. *Scott. J. Geol.* **19**, 29–45.

Smythe, D. K., Sowerbutts, W. T. C., Bacon, M. and McQuillin, R. (1972) Deep sedimentary basin below northern Skye and the Little Minch. *Nature* **236**, 87–89.

Stewart, A. D. (1962) On the Torridonian sediments of Colonsay and their relationship to the main outcrop in north-west Scotland. *Liverpool Manchester geol. J.* **3**, 121–156.

Stewart, A. D. (1969) Torridonian rocks of Scotland reviewed. *Mem. An. Ass. Pet. Geol.* **12**, 595–608.

Stewart, A. D. (1982) Late Proterozoic rifting in NW Scotland: The genesis of the 'Torridonian'. *J. geol. Soc. London* **139**, 413–420.

Stewart, A. D. and Irving, E. (1974) Palaeomagnetism of Precambrian sedimentary rocks from NW Scotland and the apparent polar wandering path of Laurentia. *Geophys. J. R. astron. Soc.* **37**, 51–72.

Stewart, A. D. and Parker, A. (1979) Palaeosalinity and environmental interpretation of red beds from the late Precambrian ('Torridonian') of Scotland. *Sedim. Geol.* **22**, 229–241.

Upfold, R. L. (1984) Tufted microbial (cyanobacterial) mats from the Proterozoic Stoer Group, Scotland. *Geol. Mag.* **121**, 351–55.

Williams, G. E. (1971) Flood deposits of the sand-bed ephemeral streams of central Australia. *Sedimentology* **17**, 1–40.

9
The Sleat and Torridon Groups

A. D. STEWART

9.1 Introduction

These two groups, as explained in the Introduction to Chapter 8, form a conformable sequence roughly 200 Ma younger than the Stoer Group. They seem to have originated during a later episode of extension across the same rift system which had earlier received the Stoer Group. The petrography and geochemistry of the younger sediments show, however, that the source area was quite different and the climate much wetter. This difference in climate fits the palaeolatitudes deduced from palaemagnetism—15° for the Stoer Group but 30–50° for the Torridon Group (Smith *et al.*, 1983).

Although the Stoer Group is nowhere seen to be overlain by the Sleat Group, there are several localities where it is overlain by the Torridon Group. At all of these the two Groups are separated by an angular unconformity of about 25°. The Stoer Group sandstones were already well lithified by Torridon Group times, for the unconformity has a rugged topography with the valleys choked by pebbles and boulders of Stoer sandstone.

Individual boulders in the basal conglomerate of the Torridon Group have directions of magnetization which are random (Stewart and Irving 1974, Fig. 7). The Stoer Group rocks were evidently not remagnetized when they were deeply buried and warmed beneath the Torridon Group. This suggests that the maximum temperature reached was about 180 °C (Stewart and Irving, 1974). Such a figure is comparable with the results of illite-crystallinity measurements by Rodd (1983, p. 376) and Johnson *et al.*, (1985), which indicate maximum temperatures at the base of the Torridon Group in the range 160–250 °C. These temperatures were probably reached by the end of the Cambrian, prior to emplacement of the Moine Nappe.

9.2 The Sleat Group

The Group consists of 3500 m of coarse grey fluviatile sandstones, with some subordinate grey shales, best exposed between Loch na Dal and Kylerhea in the Sleat of Skye. Although the beds are confined to the Kishorn Nappe, their stratigraphic position is secured by a conformable relationship with the overlying Torridon Group.

The absence of any sequence which resembles the Sleat Group, outside the Kishorn Nappe, may mean that the sediments were deposited in an independent rift, the western edge being a listric normal fault which during Palaozoic compression was transformed into the Kishorn–Suardal Thrust (Fig. 8.2).

The Sleat Group is nowhere seen in contact with the Stoer Group but, as mentioned above, it does conformably underlie the Torridon Group in eastern Skye. In view of the long hiatus between the Stoer and Torridon Groups, it seems almost certain that the Sleat Group must be younger than the Stoer. No Rb–Sr isotopic dating of the Group has ever been attempted, because of the lower greenschist-facies metamorphism, which affected the rocks during the Palaeozoic and probably reset the isotopic system. This low-grade metamorphism also changed the colour of the rocks from red to grey. The colour change is partly due to growth of chlorite, but may also stem from the transformation of some detrital hematite into magnetite (Bailey, 1955, pp. 97 and 134). Coward and Whalley (1979) report significant amounts of magnetite, as well as hematite, in both the Sleat Group and the Applecross Formation of the overlying Torridon Group.

Stratigraphic names, palaeocurrents and framework mineralogy of the sandstones are given in Fig. 9.1. Brief sedimentological details follow.

9.2.1 *Rubha Guail Formation*

The Rubha Guail Formation consists almost entirely of coarse sandstone, coloured green by its content of chlorite and epidote. The sandstone is underlain by gneiss breccia close to the unconformity with the Lewisian north of Loch Alsh (Peach *et al.*, 1907, p. 343), but, unfortunately, neither unconformity nor breccia is exposed in the type section. Trough cross-bedding is typical of the coarser beds, the palaeocurrents coming consistently from the west. Fine-grained, banded sediments, which become more abundant towards the top of the Formation, contain wave ripples and desiccation cracks (Sutton and Watson, 1960; Stewart, 1962, Fig. 11). They are followed by laminated dark grey siltstones and sandstones of the Loch na Dal Formation.

The tendency for the Formation to fine upwards into grey shales suggests that we are looking at a large alluvial fan, building out from the flank of a basement hill or fault scarp into a lake.

9.2.2 *Loch na Dal Formation*

The lower 200 m of the Loch na Dal Formation is composed of laminated, dark-grey siltstones, often phosphatic and frequently punctuated by coarse, or very coarse, sandstone laminae. This unusual juxtaposition of fine and very coarse grain sizes has been noted by several workers (Clough, in Peach *et al.*, 1907, p. 354; Sutton and Watson, 1964) and is identical to

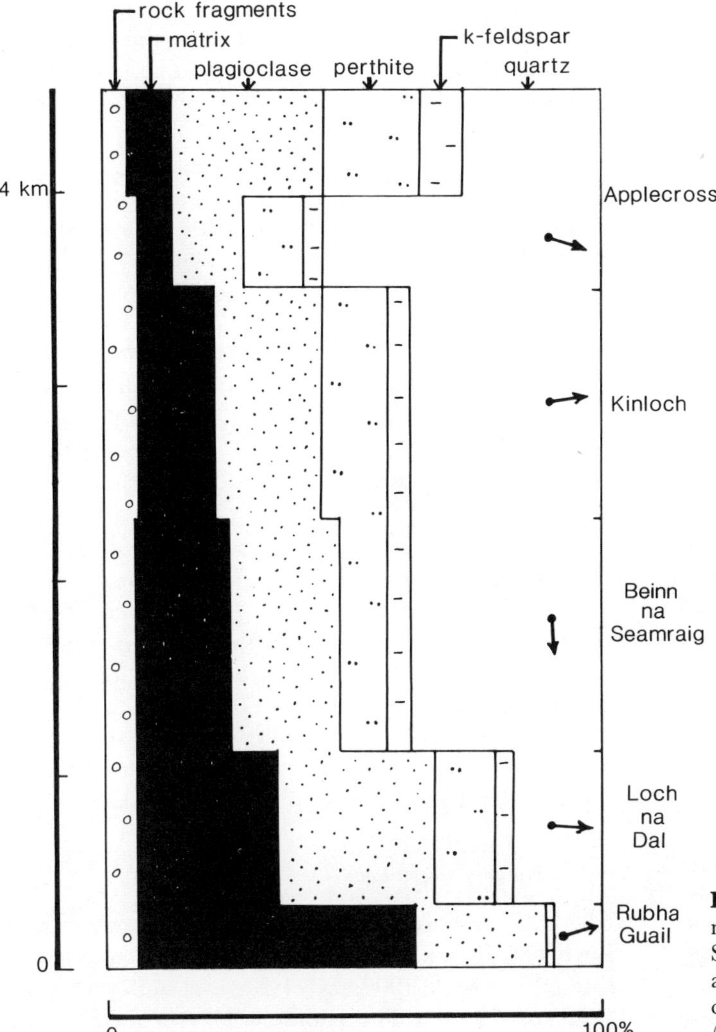

Figure 9.1 Sleat Group and lower Torridon Group mineralogy and vector mean palaeocurrent directions in Skye (arrows). The mineralogy is based on 50 modal analyses by Byers (1972), and the palaeocurrents on 229 cross-bedding directions measured by Sutton and Watson (1960, 1964).

that seen in the Diabaig Formation at Camas a Chlarsair on the south side of Upper Loch Torridon. The upper part of the Loch na Dal Formation is dominated by trough cross-bedded sandstones, still showing palaeocurrents from the west.

The shales probably mark the maximum expansion of a lacustrine or shallow-marine phase, terminated by outward building deltas. The interbanded coarse and fine sediments at Upper Loch Torridon, mentioned above, result from fan toes reaching out only 100 m from the side of a palaeovalley into a lake. Perhaps the Loch na Dal Formation was as close as this to its source.

9.2.3 *Beinn na Seamraig and Kinloch Formations*

The Beinn na Seamraig and Kinloch Formations can conveniently be considered together, for they are much alike. A substantial proportion of both is made up of strongly contorted, cross-bedded sandstones like those in Torridon Group above. Ripple lamination forms metre-thick sequences, especially in the Kinloch Formation. Less commonly there are grey, shaly intercalations resembling those found in the lower part of the Loch na Dal Formation. In the upper part of the Kinloch Formation, these shales form the upper parts of cycles roughly 10 m thick (Stewart 1966*a*). The palaeocurrents measured by Sutton and Watson (1964) show directions in the Kinloch Formation from the west, but in the Beinn na Seamraig Formation they come from the north.

These two formations are thought to be braided river deposits, like the Applecross Formation of the overlying Torridon Group. In Beinn na Seamraig times, the channels were apparently constrained to follow the rift margin.

9.2.4 *Comparison of Sleat and Torridon Groups*

Despite the similarity between the upper formations of the Sleat Group and the overlying Torridon Group, there are two significant differences. Firstly, the pebbly suite in the Sleat contains none of the metasedimentary pebbles which are so common in the Torridon Group of Skye and elsewhere. The vast majority of Sleat pebbles are porphyry, of rhyolitic or rhyodacitic composition. The remainder are acid gneiss, typically quartz – plagioclase ± microcline ± biotite. This suggests different source areas for the groups. The second important difference lies in the composition of the plagioclase, which is variably calcic in the Sleat Group but always albite in the Torridon Group.

The Sleat Group clearly derives from an upper

crustal source, for the sediments have an average K/Rb ratio of 285. The Group as a whole shows a marked upward increase in the proportion of quartz at the expense of plagioclase (Fig. 9.1). Moreover, plagioclase composition becomes progressively less calcic up into the lower part of the Applecross Formation, probably due to more effective weathering in the source area. Source area rejuvenation in Applecross times, responsible for the sudden increase in feldspar content halfway through the Formation, was not, however, accompanied by the reappearance of calcic plagioclase. The most immature sediments in the Group are found, as might be expected, at the base (Rubha Guail Formation). This Formation differs from the rest in having Fe, Ni, Ti, Ca and Mg enriched twofold, the result of its proximity to unusually basic source rocks.

The Tarskavaig Moines, though possessing petrographic and geochemical similarities to the Sleat Group, have K/Rb ~ 500, indicating a completely different source, probably granulitic. This rather undermines the lithostratigraphic correlation formerly advanced (Cheeney, in discussion of Sutton and Watson, 1964; Stewart, 1982), and suggests that they may have been deposited in a different trough.

9.3 The Torridon Group

The Group rests on an old land surface which has a relief of 600 m around Loch Maree, declining to almost nothing in the Cape Wrath area (Geikie, 1888, pp. 400–401; Stewart, 1972). An example of the relief on the unconformity, exhumed in geologically recent time, is shown in Fig. 9.2. There is now no trace of the weathering which generated this ancient topography; the rotten gneiss beneath the Torridon Group near Cape Wrath, attributed by Williams (1968) to Precambrian weathering, probably formed in the Cainozoic (cf. Hall 1985). Cainozoic weathering also affects the Torridon Group itself at some localities (Stewart, in Barber *et al.*, 1978, pp. 35 and 80).

Figure 9.2 Precambrian topography exhumed from beneath the gently dipping beds of the Torridon Group (Stewart, 1972, Figs 4 and 5). The observer is looking north from Slioch across Loch Garbhaig. Torridon Group peaks of An Teallach are visible in the distance at right.

The unconformity generally cuts Lewisian gneiss, but near the mainland coast it truncates the westward-dipping beds of the Stoer Group. There are good exposures at Stoer (Williams, 1966; Stewart in Barber *et al.*, 1978), Achiltibuie (Stewart in Barber *et al.*, 1978), Stattic Point (Lawson, 1976), Bac an Leth-Choin (Stewart, 1966*b*) and Rubha Reidh (Lawson, 1976). There is no doubt, however, that the key locality is Enard Bay (Gracie and Stewart, 1967). The superb coastal section here shows the Stoer Group, including the unique Stac Fada Member, overlain by the two lowest formations of the Torridon Group, one of them containing its diagnostic suite of exotic pebbles. Here, as well as at Achiltibuie and Rubha Reidh, the direction of magnetization of the beds changes abruptly across the unconformity (Stewart and Irving, 1974; Smith *et al.*, 1983).

The Torridon Group can be divided into the Diabaig, Applecross, Aultbea and Cailleach Head Formations, as shown in the restored stratigraphic profiles, Figs 9.3 and 9.4. From these profiles it will be noticed that the Diabaig Formation is confined to the lower half of the palaeovalleys at the base of the Group. The lack of physical continuity with the type area means that this formation generally has only facies status. Brief descriptions of the formations follow.

9.3.1 *Diabaig Formation*

The Diabaig Formation is excellently exposed around Loch Torridon, and especially in the eponymous township (Peach *et al.*, 1907, p. 324; Stewart in Barber *et al.*, 1978, pp. 78–81). The sedimentology of the Diabaig facies in Raasay has been described by Selley (1965*a*, 1965*b*). There are four component subfacies.

9.3.1.1 *Red breccias* mantle the gneiss landscape and choke the lower parts of the palaeovalleys (Fig. 9.5). They are quite similar to those at the base of the Stoer Group, except that the clasts are more angular. Clasts are of sandstone where the facies overlies the Stoer Group, otherwise they are made of local gneiss. Transport distances never exceed 3 km eastward from the source rock (e.g. Peach *et al.*, 1907, p. 315) and are usually negligible. The breccias pass upwards, and also laterally away from the palaeovalley walls, into tabular red sandstone.

9.3.1.2 *The tabular sandstones* usually a few decimetres thick and separated by films of red silt, often show trough and planar cross-bedding, horizontal lamination and extensive wave-rippled surfaces (Fig. 9.6). Shallow channels are locally quite common. However, the sandstones forming many of the beds are well sorted and internally featureless at first glance. Stratigraphically upward and away from the palaeovalley walls this facies interfingers with grey shales.

9.3.1.3 *The grey shales* comprise both millimetre-thick graded units, possibly seasonal and usually desiccated,

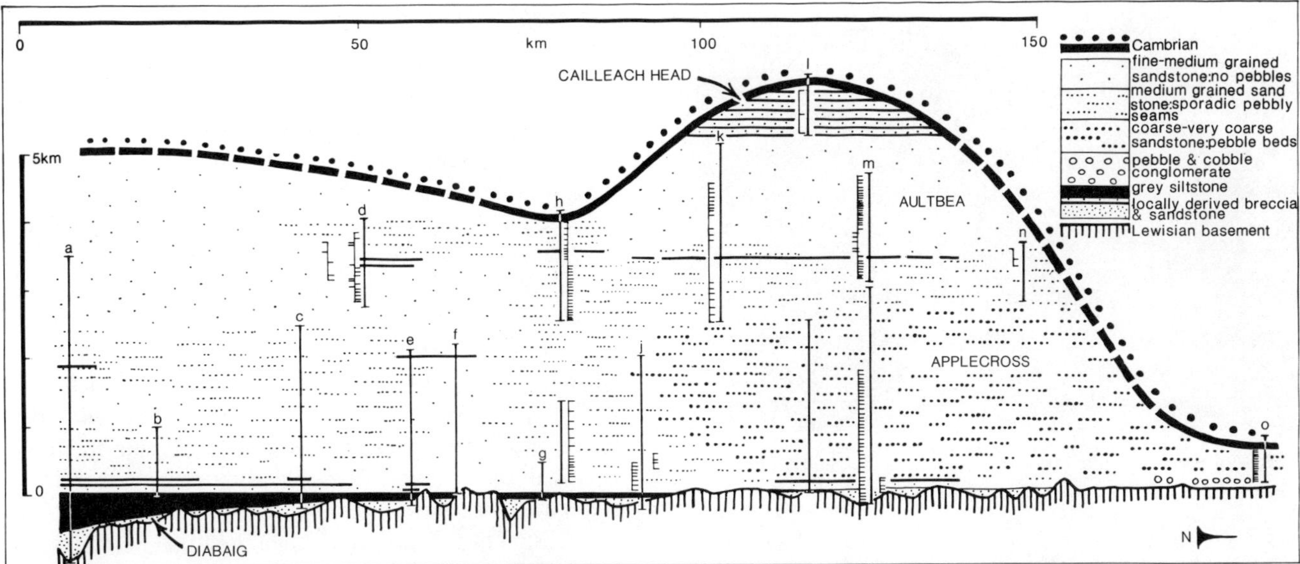

Figure 9.3 Longitudinal profile of the Torridon Group between Rhum and Cape Wrath, perpendicular to the palaeocurrent direction. Key sections are: *a*, Rhum; *b*, Soay; *c*, Scalpay; *d*, Toscaig; *e*, Raasay; *f*, Shieldaig to Applecross; *g*, Diabaig; *h*, Torridon, west and east of the Fasag Fault; *j*,

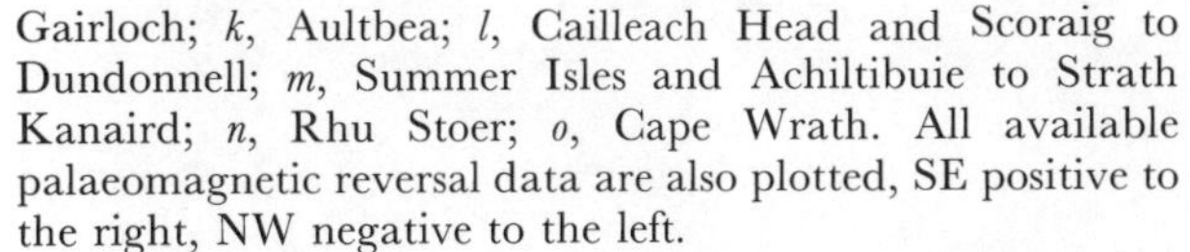
Gairloch; *k*, Aultbea; *l*, Cailleach Head and Scoraig to Dundonnell; *m*, Summer Isles and Achiltibuie to Strath Kanaird; *n*, Rhu Stoer; *o*, Cape Wrath. All available palaeomagnetic reversal data are also plotted, SE positive to the right, NW negative to the left.

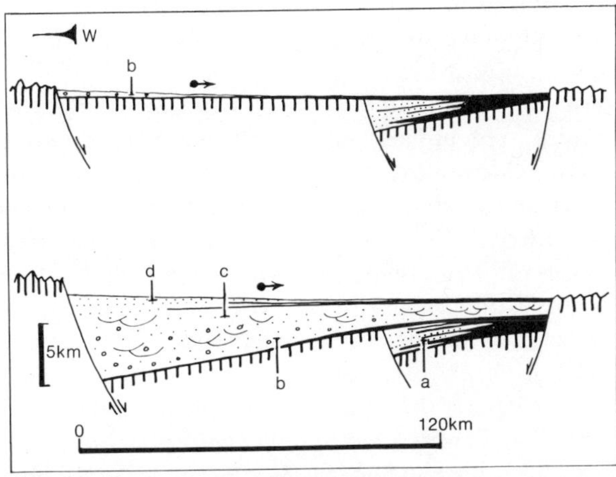

Figure 9.4 Transverse profile of the Sleat and Torridon Groups in the latitude of Skye, restored to show their condition prior Palaeozoic thrusting. Unit *a* belongs to the Sleat Group. Units *b–d* are the Applecross, Aultbea and Cailleach Head Formations of the Torridon Group. Arrows show palaeocurrent directions.

together with fine sandstone bands, millimetres to centimetres thick, showing wave ripples (Fig. 9.7). These fine sandstones fill the desiccation cracks. Phosphatic laminae and pods are common. Cryptarchs are abundant in the shale and are particularly well preserved in the phosphate (Naumova and Pavlovski, 1961; Peat and Diver, 1982; Peat, 1984). Grey shale sequences can be over 100 m thick. Grey sandstone beds appear in the upper part of the facies, increasing in frequency and thickness towards the top.

9.3.1.4 *The grey sandstones* contain about 15% of clayey matrix and are, therefore, petrographically subgreywackes. They are usually several decimetres thick. The beds are typically massive, with sharp bottoms. Ripple-drift lamination, due to currents flowing from the west, is common near the tops of the beds (Fig. 9.8). It is significant that the shales interbedded with the grey sandstones are still desiccated.

The simplest interpretation of the Diabaig facies is that the breccias and tabular sandstones are fan deposits which accumulated in the palaeovalleys, with the grey shales recording ephemeral lakes in the valley bottoms. The arrival of Applecross rivers (see below), before the end of Diabaig times, is evidenced by turbidites (the grey sandstones) in the shales. Ultimately these rivers completely filled the lakes and buried the remaining Lewisian hill tops. Boron-in-illite studies suggest that the shales are non-marine (Stewart and Parker, 1979), as does the lack of primary carbonate and complete absence of evaporites. However, temporary marine

Figure 9.5 Roughly stratified basal breccia of the Diabaig Formation on Beinn Dearg Bheag [NH 020 825]. The photograph shows a bedding plane, about 30 m laterally from a Lewisian hill slope. Some of the loose pebbles strewn across the bedding plane have been liberated from the breccia by recent weathering.

Figure 9.6 Tabular sandstones of the Diabaig Formation showing rippled surfaces, from Balgy Bay [NG 852 547].

Figure 9.7 Desiccated grey shales of the Diabaig Formation on Diabaig shore [NG 796 601].

Figure 9.8 Grey sandstones near the top of the Diabaig Formation on Diabaig Shore [NG 79 276 027].

influences cannot be entirely excluded. The significance of the southward thickening shown by the formation as a whole, and the grey shale in particular (Fig. 9.3), is not clear. It could simply result from increasing palaeo-relief. There seems to be no evidence for trunk streams during Diabaig times, except in the sector Inverpolly Forest–Cam Loch, where gneiss-cobble conglomerate fills some palaeovalleys (Stewart 1972, Fig. 8).

9.3.2 *Applecross Formation*

The Applecross Formation consists of red sandstones in which trough cross-bedding is slightly more abundant than the planar type (but cf. Selley, 1969, Table 1 and Fig. 2). Average width-to-depth ratio for the troughs is 10. Some troughs are 30 m wide, but most are only 1–2 m. The sandstones usually contain pebbles, including highly distinctive types such as porphyry and jasper. Pebble abundance provides a ready index to the Applecross subfacies recognized at Cape Wrath by Williams (1969*b*), and is also used in Fig. 9.3. Red siltstone beds are virtually absent from all but the lowest 100 m of the formation. One or two grey siltstone beds mark the top of the formation and appear to be of regional significance, because they correlate with a sequence of rapid palaeomagnetic reversals (see Fig. 9.3). The formation north of the Loch Maree Fault is geochemically remarkable in showing a monotonic upward decline in the ratio Na_2O/K_2O from unity at the base to almost zero at the top, due to the progressive disappearance of plagioclase.

About half of the beds show soft-sediment contortions which usually take the form of open synclines 0.5–2 m wide, linked by sharp cusps (Fig. 9.9). The cusps frequently have structureless cores, suggesting fluidization by upward-moving porewater (type B pillars of Lowe, 1975). Isolated cusps also occur, along with complex recumbent folds and overturned cross-bedding. Most cusps tend to lie perpendicular to the palaeocurrent direction, slightly overturned in the down-current (or downslope) direction. The great majority of the beds, though not all, have been mobilized only once, for the contortions are usually truncated by the base of the next bed.

The origin of these contortions, which are far more abundant here than in any comparable clastic sequence, is still obscure despite a substantial amount of research (Selley and Shearman, 1962; Selley *et al.*, 1963; Steward, 1963; Selley, 1969). Earthquake shocks are certainly capable of creating structures like these, but it would require an incredible degree of seismic activity to ensure that the deposition of every second bed coincided

Figure 9.9 Contorted Applecross sandstone near Diabaig [NG 78 676 037]. Note that several cross-bedded sets are deformed together and truncated by an erosion surface.

with an earthquake. More likely, perhaps, is the idea of liquefaction resulting from a flood-related process.

9.3.3 *Aultbea Formation*

The Aultbea Formation consists of red sandstones, which differ from the Applecross Formation in being generally finer. Average grain size is slightly less than 0.5 mm and pebbles are generally absent, except for a lens south of Applecross village. Virtually all the beds are contorted. A few grey shale units, individually no more than a metre or two thick, are notable for their sphaeromorphic acritarchs and filamentous sheaths (Zhang Zhongying *et al.*, 1981; Zhang Zhongying, 1982), perhaps the remains of a lake flora.

The interpretation of the Applecross Formation in terms of braided-river deposition is well established from the studies of Selley (1965*a*, 1969) and Williams (1969*b*). Similar reasoning can be applied to the Aultbea Formation. According to Miall (1977, Table V) Applecross sedimentation is Platte type, i.e. deposition was by linguoid and transverse sand bars in very shallow river channels. However, the lowest 500 m of the Applecross south of the Loch Maree Fault shows Donjek type fining-upward cycles with red siltstone tops and erosive bases. These are well exposed on the northern coast of Loch Gairloch, immediately west of Big Sand fishing station.

North of the Loch Maree Fault, the lowest 500 m of the Applecross Formation is built of fining-upward alluvial fan cycles, of the order of 100 m thick, which Williams (1969*b*) showed had their apexes along the Minch Fault, roughly 40–50 km from the present outcrop.

This difference in Applecross stratigraphy north and south of the Loch Maree Fault suggests that it was active in Torridon Group times. It is significant that the line formed by the intersection of the base of the Torridon Group and the base of the Stoer Group is dextrally displaced by the fault at least 17 km, only 5 km of which is post-Cambrian (Peach *et al.*, 1907, pp. 192 and 548).

9.3.4 *Cailleach Head Formation*

The Cailleach Head Beds are only exposed on the cliffs of Cailleach Head. The base of the formation, concealed by sea at this point, may be seen in the north-eastern part of Gruinard Island. The formation consists of cyclothems, averaging 22 m in thickness (Fig. 9.10). Each begins with laminated dark-grey shales, which pass up into tabular red sandstones, internally containing planar cross-bedding and often covered with wave ripples. This tabular facies gives way upwards to trough cross-bedded sandstones, red or green in colour and often very micaceous. Deep desiccation cracks commonly occur near the top of the grey shale, but just as in the Diabaig Formation there are no evaporites or carbonates. Teall (in Peach *et al.*, 1907, p. 287) described and figured microfossils from phosphatic laminae and pods in the shales—the first Precambrian fossils described in Britain.

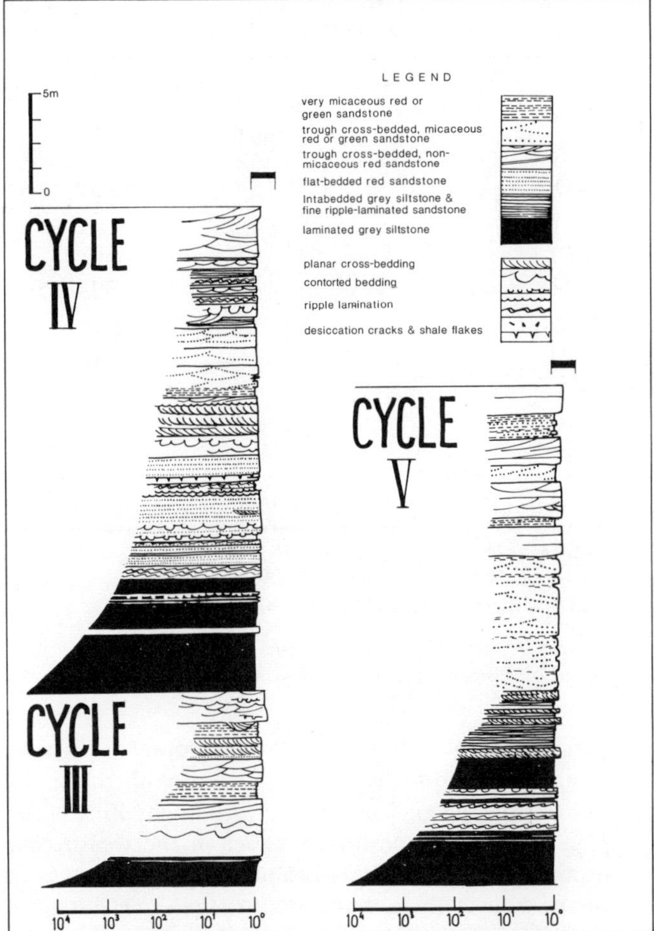

Figure 9.10 Graphic logs of three cyclothems in the Cailleach Head Formation, outcropping on the cliffs immediately south-west of Cailleach Head lighthouse. The base of the measured section is at NG 98 49 9848 and the top at NG 98 46 9835. The horizontal scale gives the lateral persistency of the beds, defined as lateral extent divided by maximum thickness. The grain-size bar represents 0–4 ϕ units.

The lack of evaporites and carbonates from the cyclothems suggests that they represent repeated delta advance into freshwater lakes. The thickness of the tabular facies suggests maximum water depths around 5–6 m.

9.4 The provenance of the Sleat and Torridon Groups

The Diabaig Formation was clearly derived from gneisses which can still be seen outcropping close by, but the rest of the sediments come from areas now inaccessible, the nature of which can only be inferred from the geochemistry and mineralogy of the sediments themselves. For this purpose, 300 whole-rock chemical analyses and 220 model analyses are now available, sampling the whole of the two groups.

A comparison of the geochemistry of Laxfordian gneisses around Diabaig with the average composition of the adjacent Diabaig Formation is shown in Table 9.1. It is clear from this that 80% of the calcium

Table 9.1 Comparative geochemistry of Laxfordian basement and the adjacent Diabaig Formation.

	A	B
SiO_2	68.8	70.7
TiO_2	0.4	0.5
Al_2O_3	15.4	13.5
Fe_2O_3 (tot.)	3.7	4.6
MnO	0.1	0.0
MgO	1.8	3.0
CaO	3.4	0.7
Na_2O	4.3	3.3
K_2O	2.3	3.1
P_2O_5	—	0.0
Volatiles	—	0.9
Total	100.2	100.3

Column A: Laxfordian basement:
based on 90% biotite gneiss + 10% metadolerite (Holland and Lambert 1973, Tables 4G and 6K).

Column B: average Diabaig sediment:
based on 29 analyses by Rodd (1983).

and 25% of the sodium present in the source rocks have disappeared from the system. Virtually all the mafic minerals in the gneiss, together with about 40% of the plagioclase, were destroyed either in the weathering profile or by interstratal dissolution from precursors of the present Diabaig sediments. At the same time, the magnetite originally present in the gneisses must have been transformed to the hematite now found in the sediments. The Si and Al, released as a result of plagioclase breakdown, were removed as clay and redeposited downstream in the shales, along with chlorite and iron minerals from the mafics. The Ca and Na were probably removed in solution in a hydraulically open system. Loss of Ca and Na from the formation during the diagenetic transformation of smectite to illite can be shown to have been relatively slight.

This loss of soluble components during Diabaig times is in strong contrast to their conservation in the lowest sediments of the Stoer Group, suggesting a climate in Diabaig times much wetter than that during the early part of the Stoer Group.

The source areas for the arkoses of the rest of the Torridon Group and the Sleat Group have to be compatible with the following facts:

(i) Gross mineralogy of the sand-sized material and schist pebbles; quartz, orthoclase, microcline, perthite, albite (in the Torridon Group), oligoclase (in the Sleat Group)
(ii) Pebble suite; this comprises muscovite schist, porphyritic rhyolite and dacite, chert and jasper (sometimes oolitic and possibly containing greenalite), banded iron formation, and tourmalinized quartzite (Williams, 1969*a*; Anderton, 1980). In the Sleat Group, however, metasedimentary pebbles are completely lacking
(iii) Laxfordian ages for the volcanic, muscovite schist and microline pebbles (Moorbath *et al.*, 1967)
(iv) Scourian ages for some of the tourmaline-bearing quartzite pebbles (Allen *et al.*, 1974)
(v) K/Rb ratio for the sediments of about 300, similar to that in the upper crust.

Bearing in mind that the source area for most of the Torridon Group was over the present position of the Outer Hebrides, it is not surprising to find that the Laxfordian granite gneiss and pegmatites of Harris and Lewis could have provided geochemically and mineralogically suitable sand-grade material. Details of these rocks have been published by Jehu and Craig (1927, 1934), Dearnley (1963), Myers (1971) and van Breeman *et al.* (1971). The granite gneisses contain about 35% of albite–oligoclase. The absence of oligoclase (and Ca) from the Applecross and Aultbea Formations is not due to albitization, but to its destruction during weathering in the mountainous source regions. The Ca released was removed from the system in solution even more effectively than from the Diabaig, doubtless the result of the higher rainfall in the mountains. The formidable amounts of clay which must have been produced have also vanished.

The upward decline in albite through the Applecross Formation, mentioned earlier, is not matched by any significant reduction in grain size, and probably results from gradual reduction of source area relief after rapid initial uplift. This would mean that as time progressed there would be more effective weathering of the source rocks. First the oligoclase would go, then the albite, in accord with Goldich's stability series. A rather similar trend of increasing maturity with time is seen in the Sleat Group (Fig. 9.1). However, the source area for the Sleat Group, as already explained, seems to have been somewhat different in composition, with more calcic plagioclase than the gneisses in the Outer Hebrides today.

The source of the 'exotic' volcanic and metasedimentary pebbles is less easily specified. According to Moorbath *et al.* (1967) they represent Laxfordian supracrustals. This explains the Rb–Sr ages, the well-preserved igneous textures in the volcanic pebbles, and the fine grain-size of the greenalite-bearing chert pebbles, which suggests $T \sim 200\,°C$, and $P < 2$ kbar (Klein, 1983, pp. 434–440 and Figs 11–16). On this basis, the igneous pebbles could be giving unmodified eruptive ages. The Laxfordian mineral assemblage of the Outer Hebrides today is in the amphibolite facies, corresponding with $P \sim 5$ kbar. So the total uplift of the Outer Hebrides block since Sleat Group times would need to be about 10 km. Considering that 4 km of uplift occurred during the Permo-Triassic alone (Steel and Wilson, 1975), this is perfectly feasible. But the hypothesis fails to account for the Scourian ages for tourmaline in quartzite pebbles. Also, it does not explain why the pebble suite is so persistent through over 3 km of Applecross and Aultbea sediment.

An alternative hypothesis is that the metasediments

and volcanics are *Scourian* supracrustals. The association of porphyritic rhyolite and dacite with cherts, banded iron formation and quartzites is like that found in Archaean greenstone belts and certain early Proterozoic basins. Any basic volcanics or greywackes originally present would not be expected to survive as pebbles, though Black and Welsh (1961) record greywacke pebbles in the Applecross formation of Rhum. The hypothesis requires the Rb–Sr ages to have been reset during Laxfordian metamorphism by temperatures not much more than the 200 °C mentioned above, while at least some of the $^{40}Ar/^{39}Ar$ tourmaline ages remained unaltered. If these Scourian sediments and volcanics were deeply infolded, or thrust into the basement as a result of Laxfordian deformation, then they would be able to contribute clasts to the sediments for a long time. The early Proterozoic sediments around Gairloch (Williams *et al.*, 1985) and in the Outer Hebrides, may represent deeper infolds of similar supracrustals.

The proposition that the Sleat and Torridon Groups fill ancient rifts (Fig. 9.4) satisfies all the provenance data given above, and also the relationship to Palaeozoic thrusting. This relationship involves thrusts lying perpendicular to the palaeocurrents in the Torridon Group, and thrust location just where rift-bounding faults would be expected. If the thrusts are reactivated normal faults then perhaps the eastern boundary of the Torridon Group was not a major fault (cf. Stewart 1982, Fig. 4 with Fig. 9.4 herein).

References

Allen, P., Sutton, J. and Watson, J. V. (1974) Torridonian tourmaline-quartz pebbles and the Precambrian crust northwest of Britain. *J. geol. Soc. London* **130**, 85–91.

Anderton, R. (1980) Distinctive pebbles as indicators of Dalradian provenance. *Scott. J. Geol.* **16**, 143–152.

Bailey, E. B. (1955) Moine tectonics and metamorphism in Skye. *Trans. geol. Soc. Edinburgh: Earth Sci.* **16**, 93–166.

Barber, A. J., Beach, A., Park, R. G., Tarney, J. and Stewart, A. D. (1978) The Lewisian and Torridonian rocks of northwest Scotland. *Geol. Assoc. Guide* **21**, 1–99.

Black, G. P. and Welsh, W. (1961) The Torridonian succession of the Isle of Rhum. *Geol. Mag.* **98**, 265–276.

Byers, P. N. (1972) Correlation and Provenance of the Precambrian Moine and Torridonian rocks of Morar, Raasay, Rhum, and Skye, Northwest Scotland. Ph.D. Thesis, University of Reading, 2 vols.

Coward, M. P. and Whalley, J. S. (1979) Texture and fabric studies across the Kishorn Nappe, near Kyle of Lochalsh, western Scotland. *J. Struct. Geol.* **1**, 259–273.

Dearnley, R. (1963) The Lewisian complex of South Harris with some observations on the metamorphosed basic intrusions of the Outer Hebrides, Scotland. *Q. J. geol. Soc. London* **119**, 243–312.

Geikie, A. (1888) Report on the recent work of the Geological Survey in the north-west Highlands of Scotland based on the field notes and maps of Messrs. B. N. Peach, J. Horne, W. Gunn, C. T. Clough, L. Hinxman and H. M. Cadell. *Q. J. geol. Soc. London* **44**, 378–439.

Gracie, A. J. and Stewart, A. D. (1967) Torridonian sediments at Enard Bay, Ross-shire. *Scott. J. Geol.* **3**, 181–194.

Hall, A. M. (1985) Cenozoic weathering covers in Buchan, Scotland and their significance. *Nature* **315**, 392–395.

Holland, J. G. and Lambert, R. StJ. (1973) Comparative major element geochemistry of the Lewisian of the mainland of Scotland. In Park, R. G. and Tarney, J. (eds.), *The Early Precambrian of Scotland and Related Rocks of Greenland.* Department of Geology, University of Keele, 51–62.

Jehu, T. J. and Craig, R. M. (1927) Geology of the Outer Hebrides. Part IV—South Harris. *Trans. R. Soc. Edinburgh: Earth Sci.* **55**, 457–488.

Jehu, T. J. and Craig, R. M. (1934) Geology of the Outer Hebrides. Part V—North Harris and Lewis. *Trans. R. Soc. Edinburgh: Earth Sci.* **57**, 839–874.

Johnson, M. R. W., Kelley, S. P., Oliver, G. J. H. and Winter, D. A. (1985) Thermal effects and timing of thrusting in the Moine Thrust zone. *J. geol. Soc. London* **142**, 863–873.

Klein, C. (1983) Diagenesis and metamorphism of Precambrian banded iron-formations. In Trendall, A. F. and Morris, R. C. *Iron-Formation; Facts and Problems.* Elsevier, Amsterdam, 417–469.

Lawson, D. E. (1976) Sandstone-boulder conglomerates and a Torridonian cliffed shoreline between Gairloch and Stoer, northwest Scotland. *Scott. J. Geol.* **12**, 67–88.

Lowe, D. R. (1975) Water escape structures in coarse-grained sediment. *Sedimentology* **22**, 157–204.

Miall, A. D. (1977) A review of the braided river depositional environment. *Earth Sci. Rev.* **13**, 1–62.

Moorbath, S. (1969) Evidence for the age of deposition of the Torridonian sediments of north-west Scotland. *Scott. J. Geol.* **5**, 154–170.

Myers, J. S. (1971) The late Laxfordian granite–migmatite complex of western Harris, Outer Hebrides. *Scott. J. Geol.* **7**, 234–284.

Naumova, S. N. and Pavlovsky, E. V. (1961) The discovery of plant remains (spores) in the Torridonian shales of Scotland. *Dokl. Acad. Sci. USSR* **141**, 181–182.

Peach, B. N., Horne, J., Gunn, W., Clough, C. T., Hinxman, L. W. and Teall, J. J. H. (1907) The geological structure of the north-west Highlands of Scotland. *Mem. geol. Surv. G. B.*, 1–668.

Peat, C. J. (1984) Comments on some of Britain's oldest microfossils. *J. Micropalaeont.* **3**, 65–71.

Peat, C. J. and Diver, W. (1982) First signs of life on Earth. *New Scientist* **95**, 776–80.

Rodd, J. A. (1983) The Sedimentology and Geochemistry of the Type Diabaig Formation in the Upper Proterozoic Torridon Group of Scotland. Ph.D. Thesis, University of Reading, 1–596.

Selley, R. C. (1965*a*) Diagnostic characters of fluviatile sediments of the Torridonian Formation (Precambrian) of northwest Scotland. *J. Sedim. Pet.* **35**, 366–380.

Selley, R. C. (1965*b*) The Torridonian succession on the islands of Fladday, Raasay, and Scalpay, Inverness-shire, *Geol. Mag.* **102**, 361–369.

Selley, R. C. (1969) Torridonian alluvium and quicksands. *Scott. J. Geol.* **5**, 328–346.

Selley, R. C. and Shearman, D. J. (1962) Experimental production of sedimentary structures in quicksands. *Proc. geol. Soc. London* **1599**, 101–102.

Selley, R. C., Shearman, D. J., Sutton, J. and Watson, J. (1963) Some underwater disturbances in the Torridonian of Skye and Rhum. *Geol. Mag.* **100**, 224–243.

Smith, R. L., Stearn, J. E. F. and Piper, J. D. A. (1983) Palaeomagnetic studies of the Torridonian sediments, NW Scotland. *Scott. J. Geol.* **19**, 29–45.

Steel, R. J. and Wilson, A. C. (1975) Sedimentation and tectonism (?Permo-Triassic) on the margin of the North Minch Basin, Lewis. *J. geol. Soc. London* **131**, 183–202.

Stewart, A. D. (1962) On the Torridonian sediments of Colonsay and their relationship to the main outcrop in northwest Scotland. *Liverpool Manchester geol. J.* **3**, 121–156.

Stewart, A. D. (1963) On certain slump structures in the Torridonian sandstones of Applecross. *Geol. Mag.* **100**, 205–218.

Stewart, A. D. (1966*a*) On the correlation of the Torridonian between Rhum and Skye. *Geol. Mag.* **103**, 432–439.

Stewart, A. D. (1972) Precambrian landscapes in northwest Scotland. *Geol. J.* **8**, 111–124.

Stewart, A. D. (1966*b*) An unconformity in the Torridonian. *Geol. Mag.* **103**, 462–465.

Stewart, A. D. (1982) Late Proterozoic rifting in NW Scotland: the genesis of the 'Torridonian'. *J. geol. Soc. London* **139**, 413–420.

Stewart, A. D. and Irving, E. (1974) Palaeomagnetism of Precambrian sedimentary rocks from NW Scotland and the apparent polar wandering path of Laurentia. *Geophys. J. R. astron. Soc.* **37**, 51–72.

Stewart, A. D. and Parker, A. (1979) Palaeosalinity and environmental interpretation of red beds from the late Precambrian ('Torridonian') of Scotland. *Sediment. Geol.* **22**, 229–241.

Sutton, J. and Watson, J. (1960) Sedimentary structures in the Epidotic Grits of Skye. *Geol. Mag.* **97**, 106–122.

Sutton, J. and Watson, J. (1964) Some aspects of Torridonian stratigraphy in Skye. *Proc. Geol. Ass. London* **75**, 251–289.

van Breemen, O., Aftalion, M. and Pidgeon, R. T. (1971) The age of the granite injection complex of Harris, Outer Hebrides. *Scott. J. Geol.* **7**, 139–152.

Williams, G. E. (1966) Palaeogeography of the Torridonian Applecross Group. *Nature* **209**, 1303–1306.

Williams, G. E. (1968) Torridonian weathering, and its bearing on Torridonian palaeoclimate and source. *Scott. J. Geol.* **4**, 164–184.

Williams, G. E. (1969*a*) Petrography and origin of pebbles from Torridonian strata (late Precambrian), Northwest Scotland. *Mem. Am. Ass. Pet. Geol.* **12**, 609–629.

Williams, G. E. (1969*b*) Characteristics and origin of a Precambrian pediment. *J. Geol. Chicago* **77**, 183–207.

Williams, P. J., Tomkinson, M. J. and Cattell, A. (1985) Petrology and deformation of metamorphosed volcanic–exhalative sediments in the Gairloch Schist Belt, N. W. Scotland. *Mineralium Deposita* **20**, 302–308.

Zhang Zhongying (1982) Upper Proterozoic microfossils from the Summer Isles, N. W. Scotland. *Palaeontology, London* **25**, 443–460.

Zhang Zhongying, Diver, W. L. and Grant, P. R. (1981) Microfossils from the Aultbea Formation, Torridon Group, Tanera Beg, Summer Isles. *Scott. J. Geol.* **17**, 149–454.

10
The Double Mer Formation

C. F. GOWER

10.1 Introduction

The Double Mer Formation is a sequence of red-bed deposits, confined to grabens within the Grenville Province in eastern Labrador, Canada. An outline of the regional distribution and structural setting of the Double Mer Formation is presented here, followed by a description of the red-beds and comments on their probable environment of deposition. Finally, the age of the sediments is discussed, using both stratigraphic and structural information.

10.2 Regional distribution

The Double Mer Formation is mostly contained in a series of grabens extending through Groswater Bay and Lake Melville in east-central Labrador (Fig. 10.1). These grabens, recently termed the Lake Melville rift system (Gower *et al.*, 1986), have been identified over a distance of some 300 km, extending inland from the coast of Labrador. A smaller basin, the Sandwich Bay graben, is located parallel to, and 100 km south-east of the Lake Melville rift system. It has been speculated (Kumarapeli and Saull, 1966; Kumarapeli, 1985; Gower *et al.*, 1986) that the Lake Melville rift system may link up with parts of the St Lawrence graben system (Fig. 10.1*b*).

Although elements of the Lake Melville rift system have been known for some time (Kindle, 1924; Kranck, 1947, 1953), it is only recently that the rift system has been comprehensively defined. Revised interpretation is based on recent 1:100 000 geological mapping, greatly assisted by bathymetric and seismic information, aeromagnetic data and LANDSAT imagery (see Gower *et al.*, 1986 for reference sources). Using this multiple approach it has been possible to delineate the major boundary faults, one of which is now recognized to extend for over 270 km. In places, the faults can be identified on the ground by broad zones of brecciation and low-grade alteration, and commonly can be traced for considerable distances by sharp breaks in topography.

Major outcrop areas of the Double Mer Formation are restricted to the Double Mer half-graben and the north side of the Lake Melville graben. No outcrops of Double Mer Formation are known from the south-western half of the Lake Melville graben, and only by comparing the physiographic similarity of the region with areas known to be underlain by Double Mer Formation, and then coupling this observation with the regional structural interpretation, is it inferred that similar sediments may underlie the surficial deposits. Outside the Double Mer half-graben and the Lake Melville graben, the Double Mer Formation occurs only as isolated coastal or riverside exposures. Bedrock extrapolation in these areas is based on physiographic and/or geophysical criteria.

10.3 Characteristics of the Double Mer Formation

The Double Mer Formation in both the Double Mer half-graben and the Lake Melville graben (Fig. 10.2) consists of reddish-brown to maroon weathering conglomerate, subarkosic to arkosic sandstone, siltstone and shale. Bedding attitudes in the Double Mer half-graben suggest that the strata have been warped into broad, open folds. In contrast, in the Lake Melville graben the bedding is tilted westward, to outline a gently dipping homoclinal sequence. Only in the most easterly dipping outcrops of the Lake Melville graben do dips differ, being north to north-east, either because of open folding, or post-depositional faulting. The total depositional thickness of strata is inferred to exceed 5 km in the Lake Melville graben (Erdmer, 1984), to be approximately 2 km in the Double Mer half-graben and, from seismic data, to be 3–4 km underlying parts of Lake Melville (Grant, 1975).

In the Double Mer half-graben, data suggest that there is a systematic variation in rock types. Conglomerate is most abundant near the east end of the half-graben, and found intermittently adjacent to the north side of the half-graben. Sandstone occurs throughout the central part of the half-graben, and interdigitates with siltstone and shale in the central and western parts of the basin. No such variation has been recognized in the Lake Melville graben.

In both areas the conglomerate is clast-supported, and consists of rounded to angular quartzofeldspathic and mafic intrusive and high-grade metamorphic rocks comparable with those in the surrounding Grenvillian basement, from which they are almost certainly derived. The inter-clast space is infilled with a quartz- and feldspar-rich sandy to pebbly matrix.

The subarkosic to arkosic rocks exhibit medium-scale cross-stratification in sets up to 50 cm. Cross-bedding is mostly planar (10–40 cm sets), but trough cross-bedding (10–50 cm sets) is dominant in the north-east part of the Double Mer half-graben. Other features include pebbly layers, scour surfaces and beds of planar lamination and thin stratification (Fig. 10.3). In detail, the sandstones consist of grain-supported quartz, plagioclase, microcline and perthitic K-feldspar particles, bound together by a pervasive hematite cement. In addition, heavy-mineral laminations, defined by

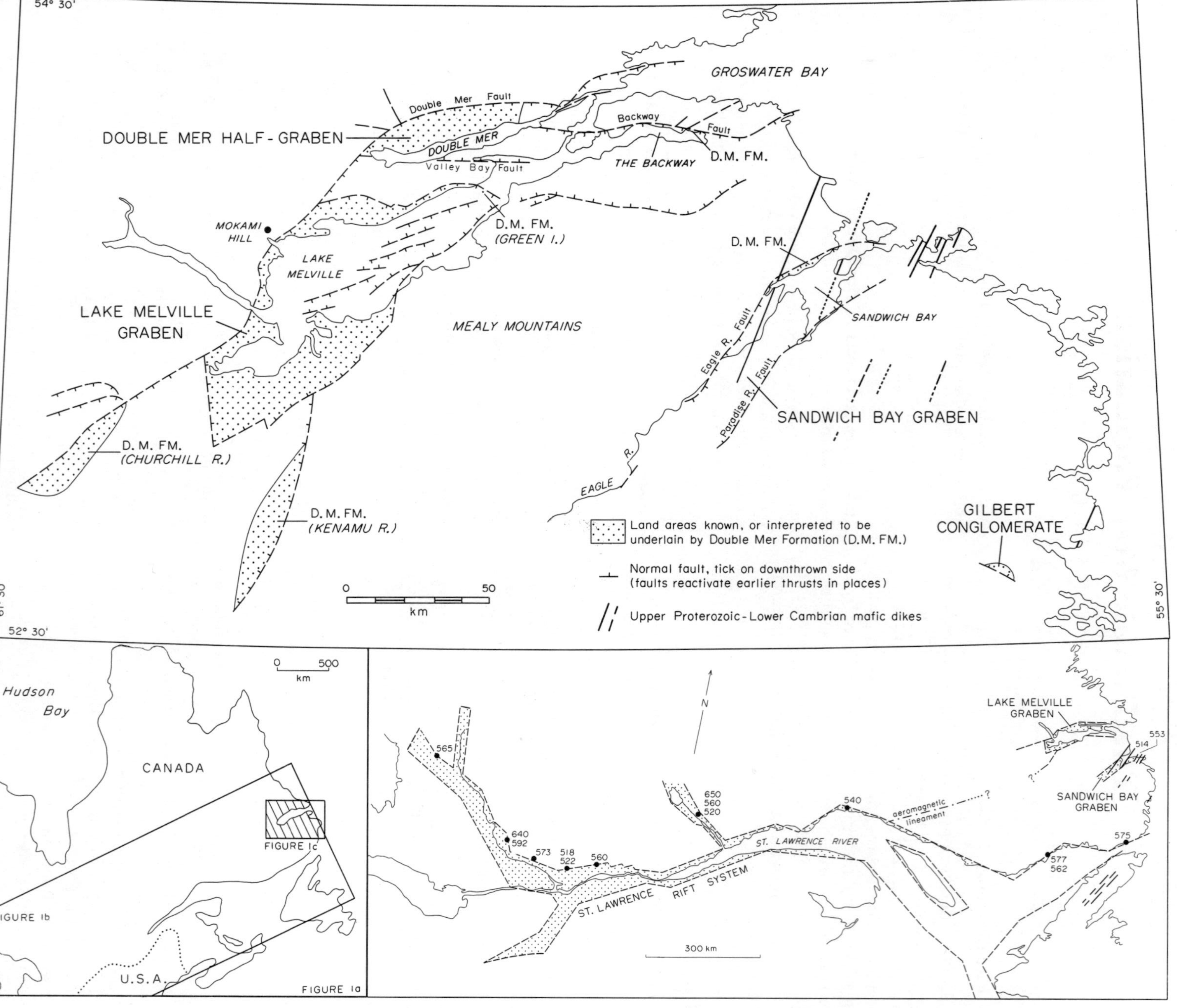

Figure 10.1 (*a*) Location of the area in eastern Canada. (*b*) Spatial relationships between the Lake Melville rift system and the St Lawrence graben system. Age determinations for graben-related intrusions are indicated. The outline of the St Lawrence graben system is taken from Kamarapeli and Saull (1966). (*c*) Regional interpretation of the distribution of the Double Mer Formation, assumed correlative strata, and graben-related faults in eastern Labrador.

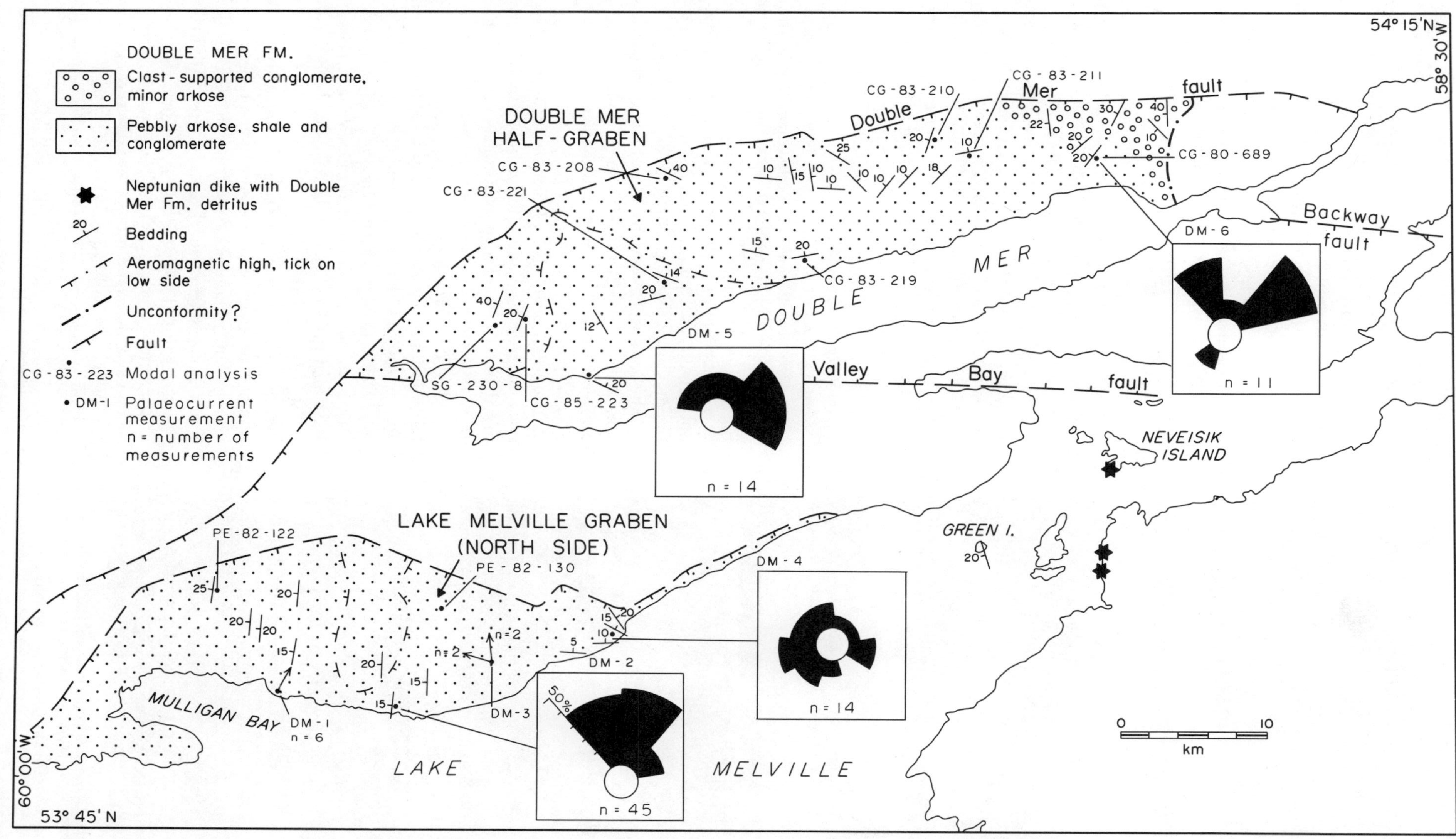

Figure 10.2 The Double Mer Formation in the Double Mer half-graben and on the north side of the Lake Melvelle graben. Modal analyses for indicated samples are given in Table 10.1.

E

Figure 10.3 (*a*) Cross-bedded arkose in the Double Mer half-graben. (*b*) Conglomerate at The Backway. (*c*) Detail of conglomerate at The Backway outcrop, showing clasts of gabbro, minor gneiss and vein quartz. (*d*) Pebbly arkose at Gilbert River, in contact with mylonitized host rock.

opaque minerals, garnet and zircon, occur in places. Other minerals include muscovite flakes, chlorite (after mafic silicates), oxidized biotite and apatite. Moderately sorted, rounded to angular rock fragments, consisting mostly of composite quartz-feldspar grains, derived from the surrounding Grenvillian basement, are found in some rocks. Modal data of sandstone from both the Double Mer half-graben and the Lake Melville graben are given in Table 10.1. Unfortunately, lack of stratigraphic constraints prevent evaluation of the significance of the mineralogical variability that is evident in these rocks.

The red-brown and maroon siltstones and shales, that make up a minor component of many outcrops, and are most common at the western end of the Double Mer half-graben, are generally poorly exposed and have yet to be studied in any detail.

Strata presumed to be correlative with the Double Mer Formation are exposed on the Churchill River, the Kenamu River, The Backway, Sandwich Bay and on the Gilbert River (Gower *et al.*, 1986). Red-beds on the Churchill and Kenamu Rivers were described by Stevenson (1967*a*, *b*) as thick-bedded red conglomerate and arkose, gradational into maroon shale. Both outcrops are located within the southward extension of the Lake Melville rift system.

The Backway outcrop clearly is also related to the Lake Melville rift system. The locality consists entirely of polymictic conglomerate that has been preserved in a pre-existing basement hollow. The clasts include boulders of metagabbro, anorthosite, augen granodiorite, amphibolite and quartzite, similar to the bedrock in the surrounding area (Fig. 10.3*b*, 10.3*c*).

The Sandwich Bay outcrop is also a conglomerate. It is made up of rounded to subrounded boulders of gneiss, granite and gabbro up to 1 m diameter in a compositionally similar, pebbly matrix.

The outcrop on Gilbert River was termed the Gilbert conglomerate by Eade (1962), but is more correctly described as a maroon, arkosic sandstone to pebbly grit (Fig. 10.3*d*). The outcrop is confined to a 5 m wide, parallel-sided vertical zone that is parallel to the Gilbert River Fault. Although Eade considered that the arkose was unconformable on mylonite, it has also been interpreted to be infolded with the basement (Piloski, 1955) and to be a clastic dyke. Gower *et al.* (1987) re-examined the locality in 1986, and favoured Bradley's (1966) interpretation that the outcrop is a fault-bounded clastic dyke. The surrounding area has been mapped as part of a broad region of right-lateral, Grenvillian, ductile shearing, and the Gilbert River Fault is considered to mark a plane of post-Grenvillian, brittle deformation, producing further right-lateral displacement of about 2.5 km (Gower *et al.*, 1987). It seems likely that deposition of the Gilbert conglomerate was associated with this reactivation.

Only reconnaissance sedimentological studies of the Double Mer Formation have been made. In the Double Mer half-graben, the conglomerate is interpreted as fanglomerate deposited adjacent to the northern and

Table 10.1 Model analyses of sandstones from the Double Mer half-graben and Lake Melville graben*.

Mineral	Double Mer half-graben								Lake Melville graben	
	SG-230 8	CG-83 223	CG-83 221	CG-83 208	CG-83 219	CG-83 210	CG-80 211	CG-83 689	PE-82 122	PE-82 130
Quartz	25.8	43.8	32.8	33.6	24.5	21.1	24.6	28.2	40.6	35.1
K-feldspar	24.3	28.5	32.0	29.8	27.5	23.3	30.0	40.0	23.1	26.8
Plagioclase	39.8	16.5	18.6	24.0	40.3	19.3	18.0	21.2	18.6	16.8
Opaques	4.0	2.3	1.6	2.1	2.0	10.0[1]	5.6	3.0	2.1	2.3
White mica	1.1	—	1.1	0.6	0.8	0.3	0.3	1.3	0.1	0.1
Chlorite[2]	1.6	—	2.6	2.5	1.0	7.6	3.3	4.0	1.6	2.1
Other minerals	0.1(gnt)	0.6(gnt)	—	—	—	—	0.3(zir, apt)	—	—	—
Rock fragments	—	1.1	1.5	0.3	0.6	—	2.1	2.0	1.1	1.0
Matrix[3]	3.3	7.2	9.8	7.1	3.3	18.4	15.8	0.3	12.8	15.8

* Modal analyses expressed as volume %, based on 600 points per thin-section. Location of samples shown in Fig. 10.2.
[1] High opaque-mineral percentage due to heavy-mineral laminations.
[2] Chlorite present as an alteration product of mafic silicates (with some opaque minerals), and as a matrix mineral.
[3] Matrix is hematite–impregnated siltstone or fine sandstone.
gnt—garnet, zir—zircon, apt—apatite.

eastern margins of the basin, with the arkosic sandstone and siltstone laid down in successively more distal areas to the south-west. Palaeocurrent directions have been reported by Stevenson (1970), Erdmer (1984) and Gower *et al.* (1986), from which northerly dispersal has been inferred. Gower *et al.* (1986), concurring with Stevenson's earlier suggestion, noted that the data are consistent with a braided fluvial environment, but emphasized that available information does not exclude other environments. They reconciled the north-west current flow (as indicated from cross-bedding orientations) with the south-westward fining of conglomerate and sandstone (away from the northern margin of the basin) by proposing that the Double Mer half-graben had a fault-bounded northern palaeo-edge and a north-sloping floor.

10.4 Interpretation of the age of the Double Mer Formation

Lack of fossils or radiometrically datable material precludes direct determination of the age of the Double Mer Formation. Samples of sandstone have been processed for palymorphs, but no recognizable microfossils were recovered (Erdmer, 1984). Furthermore, the red-beds are not known to have been intruded by younger rocks. The Double Mer Formation must postdate Grenvillian uplift (*c.*, 960 Ma), as the sediments are unmetamorphosed and contain clasts of Grenvillian basement.

Previous authors (Low, 1968; Kindle, 1924; Kranch, 1947; Stevenson, 1970) have emphasized the lithological similarity with the Bradore Formation, which is believed to be early Cambrian. This formation is exposed on the south-east coast of Labrador and in north-west Newfoundland. Other potentially correlatable strata in Labrador are the Simarnekh Formation (Wheeler, 1964) and similar rocks (Stevenson, 1969) in the vicinity of Michikamau Lake, but the age of these strata is also poorly known, and therefore offers few constraints toward inferring an age for the Double Mer Formation. On the Geological Map of Labrador, the Double Mer Formation and the Gilbert conglomerate are assumed to be Upper Proterozoic (Greene, 1972).

In a wider context, some potentially time-equivalent strata are the Torridonian sediments of north-west Scotland, the Sparagmite succession of Scandinavia and (admittedly somewhat improbably) the Keeweenawan rocks of central North America. Ages have been determined within the Torridonian sediments on the Stoer Group, and the younger Sleat and Torridonian Groups (as 968 ± 24 Ma and 777 ± 24 Ma respectively; Moorbath, 1969), although Smith *et al.* (1983) claim that these dates relate to diagenesis and may be up to 100 Ma too young. As several K–Ar dates in east-central Labrador indicate that the Grenvillian basement was still buried at about 960 Ma, the earlier date is too old to have much relevance to the Double Mer Formation. The younger age is compatible with the time span presently permitted for the Double Mer Formation. It is noteworthy that, on a pre-Iapetus reconstruction (Allen *et al.*, 1974; Anderton, 1982; Gower and Owen, 1984), the Double Mer Formation and the Torridonian may originally have been no more than about 800 km apart, increasing the probability that they were subjected to a common tectonic regime. Stewart (1982) has suggested that the Torridonian sediments were deposited in a rift environment, which is consistent with the tectonic setting envisaged for the Double Mer Formation.

Another potentially time-equivalent series is the Sparagmite succession of Scandinavia, which is believed to be late Precambrian in age (Bjørlykke *et al.*, 1976). This sequence is also considered to be rift-related, but it includes carbonate and diamictite, neither of which has been recognized in the Double Mer Formation.

The Keeweenawan volcanic and sedimentary rocks,

although formed during late Proterozoic rifting, are too early (*c.* 1150 Ma), and were formed in a somewhat different overall tectonic setting, for there to be much basis for comparison with the Double Mer Formation. The Keeweenawan rift system is thought to have formed during extension in the foreland, synchronously with Grenvillian orogenesis farther south-east (cf. Gordon and Hempton, 1986), whereas the Double Mer Formation is within the Grenville orogen and is unequivocally post-Grenvillian.

As an alternative method of inferring the age of the Double Mer Formation, Gower *et al.* (1986) proposed a structural model linking the Lake Melville rift system with north-north-east-trending late Proterozoic–early Palaeozoic mafic dykes. In this model the Lake Melville rift system was generated as a result of a left-lateral shear couple, such that the direction of principal extension was normal to the mafic dykes. The dykes have been dated in Labrador at 514 Ma (K–Ar, whole-rock, Grasty *et al.*, 1969) and 553 Ma (K–Ar, biotite; Wanless *et al.*, 1970). Similar dykes in north-west Newfoundland have yielded an age of 605 Ma (^{40}Ar/^{39}Ar; Stukas and Reynolds, 1974).

Acknowledgement

The *Canadian Journal of Earth Sciences* is thanked for permission to use Fig. 10.1, Fig. 10.2 and Fig. 10.3*a* and Table 10.1.

References

Allen, P., Sutton, J. and Watson, J. V. (1974) Torridonian tourmaline-quartz pebbles and the Precambrian crust northwest of Britain. *J. geol. Soc. London* **130**, 85–91.

Anderton, R. (1982) Dalradian deposition and the late Precambrian–Cambrian history of the N. Atlantic region. A review of the early evolution of the Iapetus Ocean. *J. geol. Soc. London* **139**, 423–434.

Bradley, D. A. (1966) Report on exploration in Sandwich Bay—Square Island, Labrador. BRINCO Ltd, Unpubl. Rept. Nfld. Dept. Mines and Energy, Open File 13A(1), 19 pp.

Bjørlykke, K., Elvsborg, A. and Hoy, T. (1976) Late Precambrian sedimentation in the central Sparagmite basin of south Norway. *Norsk. Geol. Tidsskr.* **56**, 233–290.

Eade, K. E. (1962) Geology, Battle Harbour—Cartwright, Labrador. *Geol. Surv. Can., Map 22-1962.*

Erdmer, P. (1984) Precambrian geology of the Double Mer–Lake Melville region, Labrador. *Geol. Surv. Can. Pap.* **84–18**, 37 pp.

Gordon, M. B. and Hempton, M. R. (1986) Collision-induced rifting: the Grenville orogeny and the Keeweenawan rift of North America. *Tectonophysics* **127**, 1–25.

Gower, C. F. and Owen, V. 1984. Pre-Grenvillian and Grenvillian lithotectonic regions in eastern Labrador—correlations with the Sveconorwegian Orogenic Belt in Sweden. *Can. J. Earth. Sci.* **21**, 678–693.

Gower, C. F., Erdmer, P. and Wardle, R. J. (1986) The Double Mer Formation and the Lake Melville rift system, eastern Labrador. *Can. J. Earth Sci.* **23**, 359–368.

Gower, C. F., Neuland, S., Newman, M. and Smyth, J. (1987) Geology of the Port Hope Simpson map region, Grenville Province, eastern Labrador. In *Current Research for 1986*, Mineral Development Divn., *Newfoundland Dept. Mines & Energy, Rep.* **87-1**, 183–199.

Grant, A. C. (1975) Seismic reconnaissance of Lake Melville, Labrador. *Can. J. Earth Sci.* **12**, 2103–2110.

Grasty, R. L., Rucklidge, J. C. and Elders, W. A. (1969) New K–Ar age determinations on rocks from the east coast of Labrador. *Can. J. Earth Sci.* **6**, 340–344.

Greene, B. A. (1972) *Geological map of Labrador.* Mineral Resources Divn., Nfld. Dept. Mines, Agriculture and Resources, Province of Newfoundland and Labrador, scale 1:1 000 000.

Kindle, E. M. (1924) Geography and geology of Lake Melville district, Labrador Peninsula. *Geol. Surv. Can. Mem.* **141**, 71 pp.

Kranck, E. H. (1947) Indications of movements of the Earth's crust along the coast of Newfoundland—Labrador. *Geol. Surv. Can. Bull.* **140**, 89–96.

Kranck, E. H. (1953) Bedrock geology of the seaboard of Labrador between Domino Run and Hopedale, Newfoundland. *Geol. Surv. Can. Bull.* **26**, 45 pp.

Kumarapeli, P. S. 1985. Vestiges of Iapetan rifting in the craton west of the northern Appalachians. *Geosci. Can.* **12**, 54–59.

Kumarapeli, P. S. and Saull, V. A. (1966) The St Lawrence Valley system: a North American equivalent of the East African rift valley system. *Can. J. Earth Sci.* **3**, 639–658.

Low, A. P. (1896) Report on explorations in the Labrador Peninsula, along the Eastmain, Hamilton, Manicouagan and portions of other rivers in 1892–93–94–95. *Geol. Surv. Can., Ann. Rep.* **VIII**, Part L, 387 pp.

Moorbath, S. (1969) Evidence for the age of deposition of the Torridonian sediments of north-west Scotland. *Scott. J. Geol.* **5**, 154–170.

Piloski, M. J. (1955) Geological report on area 'E' Labrador concession. BRINCO Ltd. Unpubl. Rept., Newfoundland Dept. Mines & Energy, Open File, Labrador **186**, 29 pp.

Smith, R. L., Stearn, J. E. F., and Piper, J. D. A. (1983) Palaeomagnetic studies of the Torridonian sediments, N.W. Scotland. *Scott. J. Geol.* **19**, 29–45.

Stevenson, I. M. (1967*a*) Goose Bay map-area, Labrador (13F). *Geol. Surv. Can. Pap.* **67–33**, 12 pp.

Stevenson, I. M. (1967*b*) Minipi Lake. *Geol. Surv. Can. Map 6-1967.*

Stevenson, I. M. (1969) Lac Brûlé and Winokapau Lake map-areas, Newfoundland and Quebec. *Geol. Surv. Can. Paper* **67–69**, 16 pp.

Stevenson, I. M. (1970) Rigolet and Groswater Bay map-areas, Newfoundland (Labrador) (13J, 13I). *Geol. Surv. Can. Pap.* **69–48**, 24 pp.

Stewart, A. D. (1982) Late Proterozoic rifting in NW Scotland: the genesis of the 'Torridonian'. *J. geol. Soc. London* **139**, 415–422.

Stukas, V. and Reynolds, P. H. (1974) ^{40}Ar/^{39}Ar dating of the Long Range dykes, Newfoundland. *Earth Planet. Sci. Lett.* **22**, 256–266.

Wanless, R. K., Stevens, R. D., Lachance, G. R. and Delabio, R. N. (1970) Age determinations and geological studies. *Geol. Surv. Can. Pap.* **69–2A**, 71–78.

Wheeler, E. P. (1964) Unmetamorphosed sandstone in northern Labrador. *Geol. Soc. Am. Bull.* **75**, 339–346.

11
The Colonsay Group

M. BENTLEY

11.1 Introduction

The Colonsay Group is exposed on Colonsay and Islay, two islands in the southern Inner Hebrides of Scotland, which lie at the seaward end of the Firth of Lorne (Fig. 11.1).

The group comprises a suite of deformed, low-grade, metasedimentary rocks which lie with a sheared unconformity upon gneissic basement. The metasediments and the underlying basement are tectonically isolated from the surrounding area by two major faults: the Colonsay Fault, part of the submarine extension of the Great Glen Fault system, and the Loch Gruinart Fault, itself a splay of the Great Glen system (Fig. 11.2). The two faults merge to the north of Colonsay, but remain separate to the south, in which direction the maximum extent of the Colonsay Group is unclear. Therefore, although the group is only exposed over an area of *c.* 150 km², the submarine extent of the group covers a somewhat larger area (> 500 km²).

The stratigraphic age of the group and the regional significance of the structures developed within it are enigmatic. Previous attempts to date the group have rested on correlations with the known Proterozoic sequences of the Scottish mainland, particularly the Torridonian, and more recently the Dalradian. Correlations with mainland units have not proved conclusive. raising the possibility that the Group may be stratigraphically unique in the Scottish Caledonides, or even exotic.

The structure of the Group has recently been reinvestigated (Fitches and Maltman, 1984; Saha, 1984; Bentley, 1986), and isotope studies have been attempted in order to constrain the age of the Group and the timing of the structural history (Bentley, 1986).

11.2 The Colonsay Group Sequence

11.2.1 *Lithostratigraphy*

The Colonsay Group is a 5.5–6 km thick suite of low-grade metasediments. The lowest 1.5 km are exposed on the Rhinns of Islay, the upper 3.5 km on Colonsay itself,

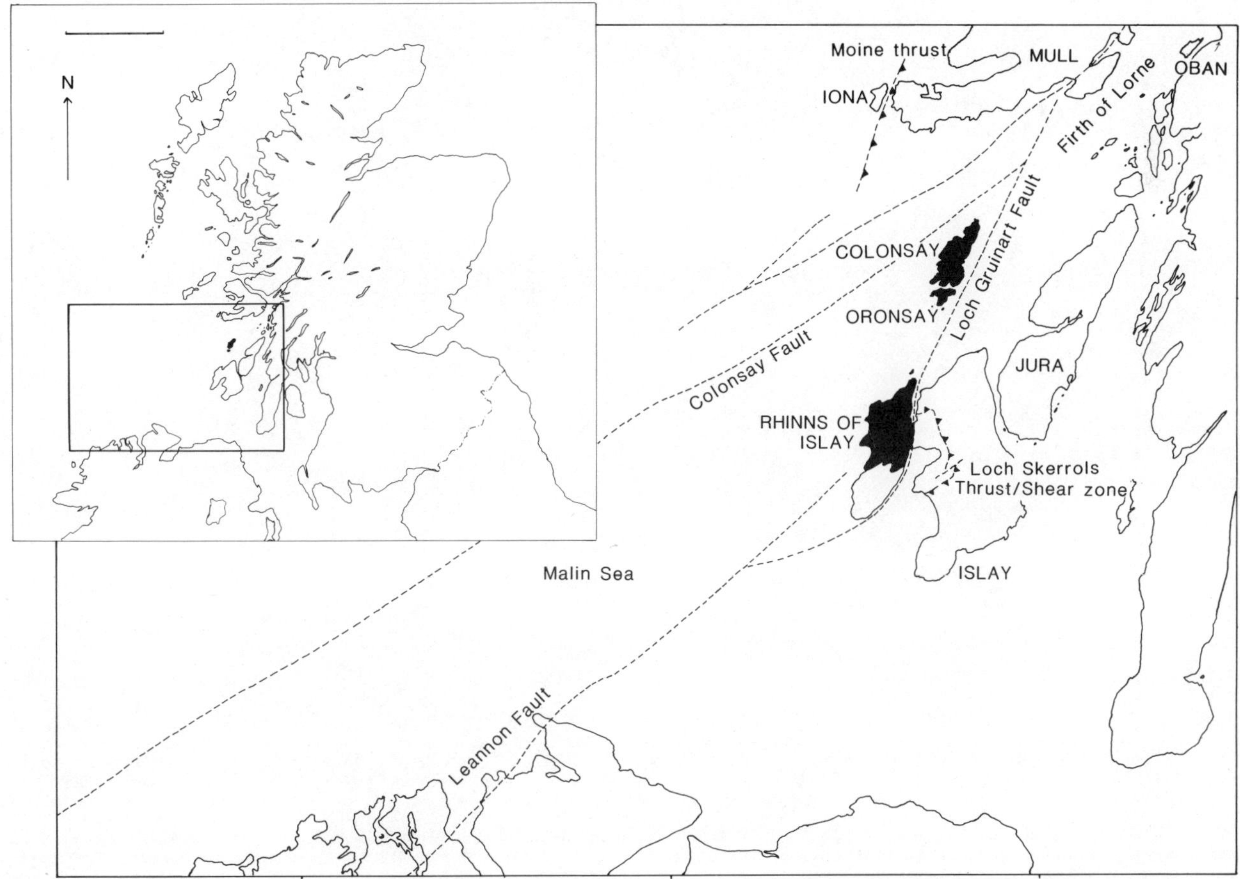

Figure 11.1 The southern Inner Hebridean area and the Malin Sea. The blacked-out area indicates the extent of Colonsay Group exposures on land. Offshore locations of faults are taken from Evans *et al.* (1982).

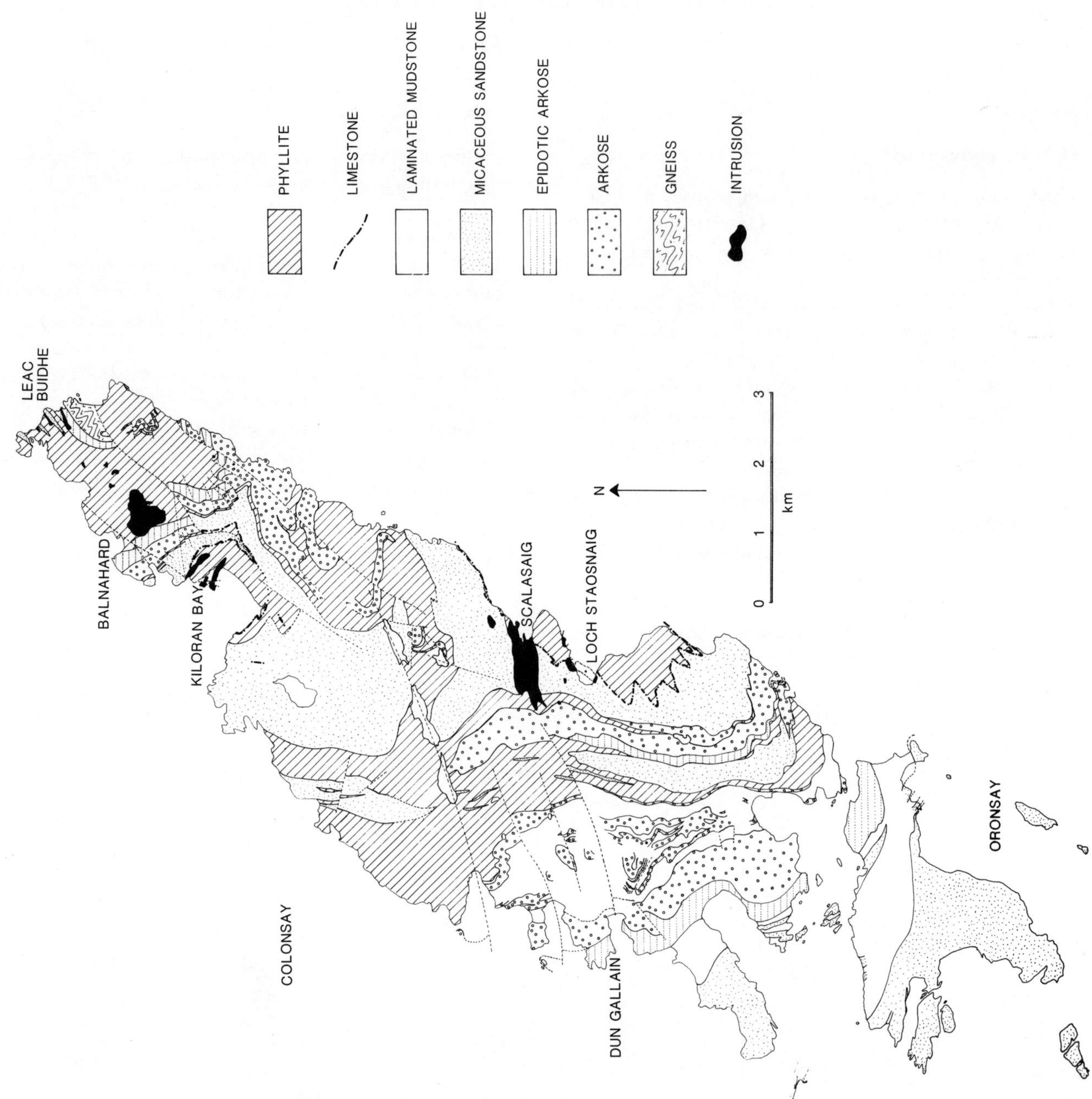
PHYLLITE
LIMESTONE
LAMINATED MUDSTONE
MICACEOUS SANDSTONE
EPIDOTIC ARKOSE
ARKOSE
GNEISS
INTRUSION
LEAC BUIDHE
BALNAHARD
KILORAN BAY
SCALASAIG
LOCH STAOSNAIG
N
0
1
2
3
km
COLONSAY
DUN GALLAIN
ORONSAY

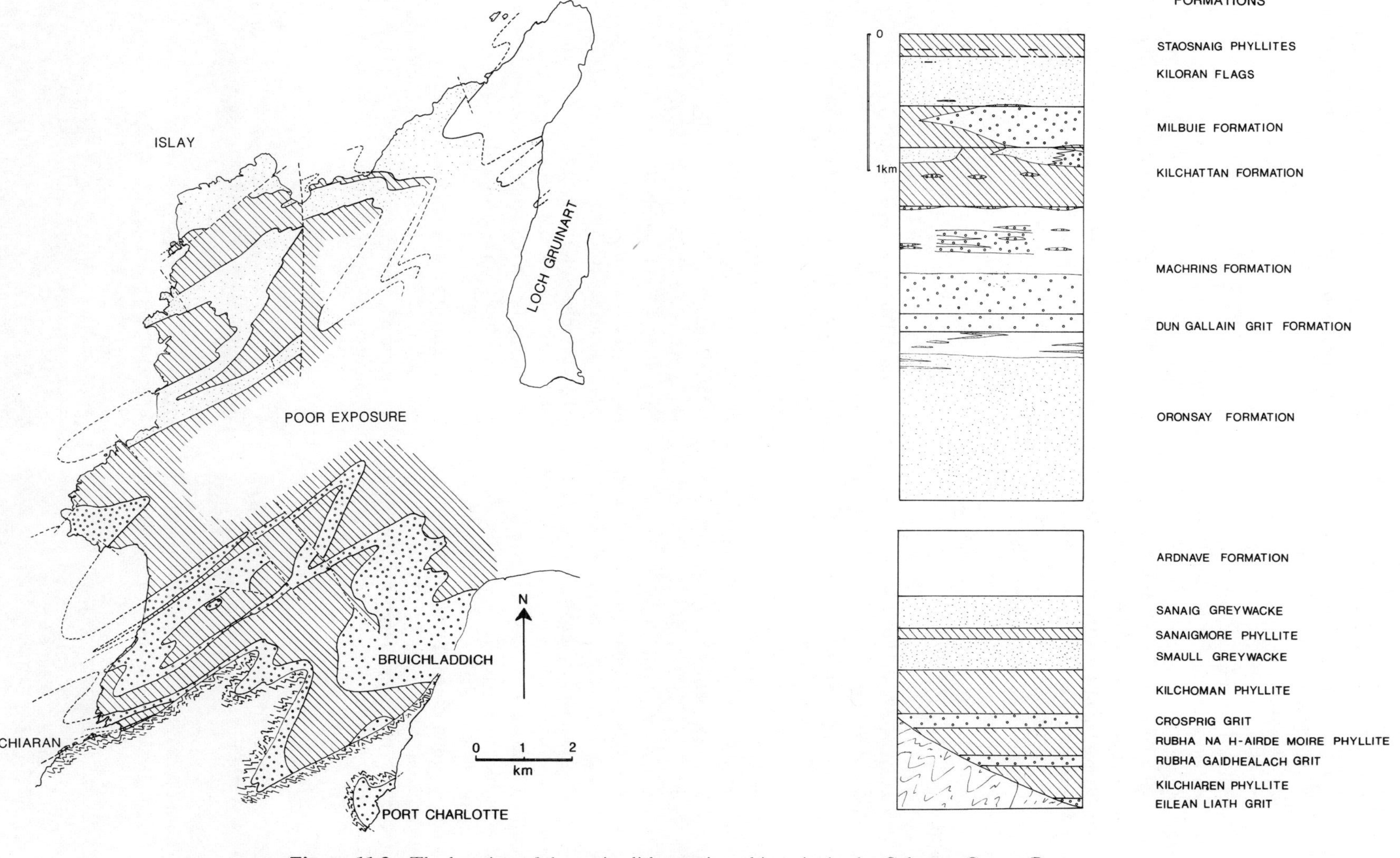

Figure 11.2 The location of the major lithostratigraphic units in the Colonsay Group. Data from the Islay area are taken from Stewart and Hackman (1973). Data for Colonsay are modified from Stewart (1962*a*). Note that the Islay map, for convenience, is drawn at a smaller scale than the Colonsay map.

and a small portion is unexposed in the strait between the two islands. The sequence is divided into eighteen lithostratigraphic subunits: ten on Islay (Stewart and Hackman, 1973) and eight on Colonsay (Cunningham-Craig *et al.*, 1911). The sequence was named the Colonsay Group by Stewart (1975), whereafter the eighteen subunits assumed the status of formations. The lithology of the sequence is described in detail in Stewart (1962*a*, for Colonsay) and Stewart and Hackman (1973, for Islay) and is summarized in Fig. 11.2.

The sedimentary detail is generally well preserved, and primary structures and clast shapes may be identified. The group includes a wide variety of clastic sediments, ranging from coarse, laterally variable sandstones to relatively homogeneous argillites. The sandstones are locally very coarse-grained, and are commonly arkosic, especially in the coarser members. Graded- and cross-bedding are common, so that stratigraphic 'way-up' is generally well defined. Although much of the argillite sequence is now metamorphosed to phyllite, mudstones and siltstones are still widely preserved, particularly in SW Colonsay and Oronsay, commonly in the form of fine-scale interlaminated sequences (Fig. 11.3). Calcareous beds are infrequent, except at the Kiloran Flags–Staosanig Phyllite boundary, where one or more horizons of dolomite or calcareous phyllite occur. This zone, termed the 'Colonsay Limestone', is typically 1–5 m thick, highly deformed, and thoroughly recrystallized.

The exact thickness of the Colonsay Group sequence is unclear, owing to the 7.5-km wide geographical break in exposure between Islay and Colonsay. Given the gentle, predominantly north-westerly dip of beds in the Colonsay Group on Islay and S Colonsay, it is likely that the physical break between the islands represents a stratigraphic gap of no more than a kilometre. When the stratigraphic repetition caused by large-scale folding is taken into account, it is apparent that none of the sequence is necessarily missing. However, given the structural dissimilarity between the sediments on Oronsay and NW Islay (see next section), it is suggested that the Colonsay Group sequences on Colonsay and Islay do not overlap stratigraphically. In this case, the thickness of the sequence is probably in the range 5.5 to 6 km.

11.2.2 *Cover–basement contacts*

The sequence is in contact with gneissic basement both on Islay and in N Colonsay. The contact on Islay occurs at the base of the sequence, where the five lowest Colonsay Group formations are truncated against the basement. The contact zone may be traced across the Rhinns of Islay, from Kilchiaren in the west (GR 1892 5950) to Bruichladdich in the east (GR 2650 6120, Fig. 11.2), although the actual line of contact is rarely exposed. The zone is highly deformed, and was described by Stewart and Hackman (1973) as a tectonic break—'the Bruichladdich Slide'. However, the units close to the contact are generally coarser than elsewhere on Islay (Fig. 11.4), and locally a basal conglomerate has been observed (Wilkinson, 1907, plate III; Saha, 1984). Therefore, despite the deformed nature of the contact, and evidence of local cover–basement interleaving in some areas, it is suggested that the contact represents an original unconformity, albeit sheared, with cover parautochthonous with respect to basement.

The contact in N Colonsay is less extensive, but well exposed on the hillside at Leac Buidhe (GR 42 00). As in Islay, the contact is highly sheared, but a conglomeratic cover sequence is preserved (Fig. 11.5). The conglomerate is unique within the Colonsay Group, containing large, rounded quartzite clasts up to 5 cm in diameter. The conglomerate also contains smaller fragments, some reminiscent of the local basement gneisses. As in Islay, the contact is interpreted as a sheared unconformity, with parautochthonous cover overlying basement. The sediments at Leac Buidhe belong to the Kilchattan Formation, *c.* 4 km stratigraphically above the base of the sequence in Islay. A highly uneven basement topography is therefore implied.

Figure 11.3 Laminated, fine-grained sandstones, siltstones and mudstones of the Dun Gallain Grit Formation. Dun Gallain, Colonsay (GR 3499 9331).

Figure 11.4 Basal Colonsay Group sediments: the Eilean Liath Grit Formation close to the cover–basement contact at Claddach Min, near Kilchiaren, W Islay (GR 1890 5954). Note the prolate shape of the strongly deformed quartz clasts.

Figure 11.5 The quartzite conglomerate close to the cover–basement contact on Colonsay. Leac Buidhe, N Colonsay (GR 4272 0047).

11.2.3 *Sedimentology*

The sedimentology of the group has been discussed by Stewart (1960, 1962*a*, *b*) and Stewart and Hackman (1973). The lower part of the sequence, on Islay, may be considered in two parts: a lower sequence dominated by arkoses and phyllites, and an upper sequence comprising more quartz-rich sandstones and mudstones, described as 'greywackes'. The lower, arkose-phyllite sequence was considered by Stewart and Hackman to represent a deltaic facies, in which the arkosic sheets were deposited on a delta front. The greywackes were interpreted as turbidites, becoming progressively more distal at higher stratigraphic levels.

The sequence on Colonsay describes a change from deep-water turbidites to relatively shallow-marine siliciclastics. The uppermost units in the sequence (the Kiloran Flags and Staosnaig Phyllite Formations) may herald a return to a deeper, possibly pro-deltaic environment, but on a broad scale the higher stratigraphic levels tend to reflect shallower depositional environments. In a general way, therefore, the Colonsay Group sequence on Colonsay is the inverse of that on Islay, portraying progressively shallower-water depositional environments at higher stratigraphic levels.

Stewart (1960, 1962*a*) and Saha (1984, 1985) analysed the clast composition of the Colonsay Group sediments on Colonsay, and envisaged a source area containing both high- and low-grade rocks. For the high-grade rocks, Stewart drew comparisons with basement gneisses on Iona (Fig. 11.1), and Scourian gneisses from the Lewisian Complex of NW Scotland. Saha concluded that the sediments were derived from Scourian granulites, primarily because of the identification of blue quartz among the Colonsay Group sediments. The low-grade terrain was thought to consist of deformed metasediments. Form the consistent pattern of heavy-mineral distribution across the sequence in Colonsay, Stewart (1960) suggested that the sediments had a single, although lithologically diverse source area.

Palaeocurrent directions were determined from the lower part of the sequence on Colonsay by Stewart (1960), and from Islay by Stewart and Hackman (1973). These indicated deposition from predominantly north- or north-easterly flowing currents. In higher units on Colonsay (the Kilchattan Formation) Saha (1984) additionally reported local westerly flowing currents.

From clast composition and palaeocurrent indicators, therefore, the Colonsay Group sediments appear to be derived from an area containing deformed, high-grade gneisses with sedimentary cover, lying broadly to the south of the Colonsay–Islay region. No evidence has been found to link clearly this gneissic source with the known gneisses of NW Scotland or to suggest that the source area contained high-grade metasediments.

11.2.4 *Regional depositional environment*

The sedimentary sequence developed above the gneissic inlier in N Colonsay is consistent with the interpretation of the inlier as a parautochthonous basement block, rather than a tectonic slice detached from deeper-level (3–4 km) basement. The 3–4 km basement relief between Islay and northern Colonsay may be interpreted in terms of normal block faulting, in which the inlier in N Colonsay is part of a major basement horst. Despite this evidence for high-level basement towards the north, the primary source area for much of the sequence lay to the south, and included gneissic basement. Positive basement areas therefore lay to the south and north, and possibly also to the east of the depositional area, implying an intracratonic setting. It is therefore suggested that the Colonsay Group sediments were deposited across the northern margin of a fault-bounded, intracratonic basin. The extent of the basin is unknown, particularly since the exposed northern margin is overstepped by Kilchattan Formation sediments.

Given the predominant NE–SW strike and the general northward younging of units across Islay, Oronsay and S Colonsay, the Colonsay Group sequence effectively represents an oblique, rather than true vertical, stratigraphic section. It is likely that the transition from shallow to deep, then back to relatively shallow-water sedimentation, as indicated in the Colonsay Group sequence, does not simply represent a variation from basin initiation to infilling but is partly a reflection of a lateral facies variation within the basin. Thus, whereas the progressively deepening depositional environments represented on Islay may indicate basin initiation and subsidence, higher stratigraphic levels represent an oblique section towards the basin margin, revealing progressively more basin-marginal deposits and ultimately the overstep of the margin itself in N Colonsay.

11.3 Structure

11.3.1 *Deformation history*

Polyphase deformation in the Colonsay Group was noted by Wright (1908) who described two main 'Earth

Figure 11.6 Near-isoclinal early folds in epidotic arkoses of the Milbuie Formation. Beinn Bhreac, N Colonsay (GR 4148 9862).

Figure 11.7 Large-scale F2 close to the Smaull Greywacke–Kilchoman Phyllite Formation boundary, W Islay (GR 2015 6596).

movements' on Colonsay. Subsequent workers have further subdivided the deformation history into three (Stewart, 1960; Bentley, 1986) or four main episodes (Saha, 1984; Fitches and Maltman, 1984), yet the clearest subdivision remains that into the two major events noted by Wright (1908), broadly supported by all later workers. The structure of the group is described briefly below in terms of these two major events.

The early events are characterized by heterogeneous, subhorizontal, northerly shear (Bentley, 1986). The events are manifest primarily in the form of a widespread L-S fabric, associated with the development of small-scale, asymmetrical, similar folds, in places open to close, but elsewhere tight to isoclinal (Fig. 11.6) and in places sheath-like. Although early folds with wavelengths up to *c.* 100 m have been observed, the folds are typically small scale, and no evidence has been found to support major stratigraphic inversion during the early events.

The character of the small-scale structures varies markedly across the group. Towards the top of the Colonsay Group sequence, the fabric is strong, parallel or subparallel to bedding, and with the linear component predominant. In the middle of the sequence (Oronsay, S Colonsay) the fabric is generally weaker and dominated by the planar component, making a conspicuous angle with bedding. This change is associated with a general tightening of the early folds towards the top of the sequence, and is qualitatively interpreted in terms of a vertical strain gradient, with highest strains in the uppermost stratigraphic levels. Towards the bottom of the sequence (Islay), early folds are less common and the early fabric is weak to inconspicuous, although close to bedding where observed. This heterogeneous pattern is interpreted in terms of shear zone geometry, the northward vergence of the folds and fabrics indicating northward, subhorizontal shear. The structural event is associated with the ubiquitous growth of chlorite and the very local development of chloritoid and biotite, indicating burial to depths of 10–15 km at the top of the sequence during deformation.

The later structural events are typified by upright, class 1C folding on all scales, accompanied by the development of spaced, planar fabrics (Figs 11.7 and 11.8). The later fold pattern results from the interference of two major fold sets (F2 and F3) which are close to coaxial in Islay and Oronsay, but become progressively more non-coaxial across Colonsay to the north. As a result, the map pattern is characterized by 'type 0' (Ramsay, 1967) interference patterns in Islay, and 'type 1', basin and dome interference patterns in N

Figure 11.8 Small-scale F2 in the Kiloran Flags Formation at Cnoc Inebri, Kiloran Bay, Colonsay (GR 4105 9798).

Figure 11.9 Explosion breccia marginal to the Lower Kiloran Bay Intrusion, Colonsay (GR 4041 9835). Note the brittle form of the late fold (right side of frame) compared with the more typical form of late folding illustrated in Fig. 11.8.

Colonsay, with intermediate forms occurring in between. Late structural events were accompanied by the local recrystallization of chlorite, and late growth of stilpnomelane.

11.3.2 *Structural relationships of the main igneous intrusions*

Intrusive centres occur on Colonsay at Kiloran Bay, Scalasaig and Balnahard (Fig. 11.4). All centres are sites of one or more small alkaline or subalkaline intrusions. The intrusions are typically basic or ultrabasic at the margins, with cores of syenite, diorite or monzonite at Kiloran Bay, Scalasaig and Balnahard respectively.

At Kiloran Bay and Scalasaig, a marginal explosion breccia is exposed, containing angular, closely packed, disorganized blocks of metasediment, at least some of which are clearly derived from the immediate country rock (Fig. 11.9). The blocks contain disoriented and truncated examples of folds and penetrative fabrics typical of the early structures, indicating that the intrusive event post-dated the development of the early structures. Late structures only occur in the breccia in the form of weak, microscopic crenulations or rare, meso-scale, fractured folds, indicating a pre-F2/F3 structural age for the intrusions. An intra-deformational age is also indicated at the intrusive margins, which are not protected by the relatively rigid explosion breccia. Here, a strong fabric is locally developed in the margin of the intrusions parallel to that in the adjacent country rock (examples occur at Loch Staosnaig, GR 3951 9361, and Kiloran Bay, GR 4066 9855). The fabric is associated with recrystallization and breakdown of the original igneous mineralogy, and is thus interpreted as a post-intrusive, tectonic feature. In addition, kinked or curved biotite phenocrysts and highly strained quartz grains are commonly observed in thin-section.

The Kiloran Bay and Scalasaig intrusions were thus emplaced between the early and late events, or, at the latest, during the initial development of the late structures. The structural age of the poorly exposed Balnahard intrusion remains unclear.

11.4 Geochronological framework

11.4.1 *Isotope studies*

The three major intrusive centres were sampled for isotopic analysis in an attempt to place a minimum age on the early, flat-lying structures, and a maximum age on the later, upright fold structures.

All samples were analysed by the Rb/Sr whole-rock method, and hornblende and biotite samples, where possible, were separated for K/Ar, Rb/Sr and Ar/Ar mineral analyses. Although some analyses are still in progress, tentative conclusions may be drawn from the results presently available, and these are described below. Details of these analyses may be found in Bentley (1986), and will be published in full on completion of the study.

11.4.1.1 *Rb/Sr analysis.* The intrusions have clearly not remained closed to Sr diffusion since initial crystallization. The most coherent whole-rock data came from the Kiloran Bay samples, which yielded an apparent age of 549 ± 120 Ma (MSWD 10.0). Rb/Sr mineral analysis gave an approximate biotite age of 420 ± 15 Ma (MSWD 11.0) from the Scalasaig intrusion.

11.4.1.2 *K/Ar analysis.* K/Ar mineral ages vary considerably. Biotite ages are clustered in the range 417–456 Ma, whereas hornblende ages are relatively dispersed, all > 600 Ma.

11.4.1.3 *Ar/Ar analysis by stepwise heating.* Ar/Ar analysis on hornblende separates revealed both the preferential retention of radiogenic ^{40}Ar ('excess Ar'), and also a late-stage loss of radiogenic ^{40}Ar. The excess argon effect is common in intrusions emplaced under initially high gas pressures, and leads to anomalously high apparent 'ages' from K/Ar analysis (Dalrymple and Lanphere, 1969). The loss of radiogenic ^{40}Ar, prompted by either a protracted cooling history or a later thermal overprint, tends to produce anomalously young ages—to a small degree in hornblende, but more

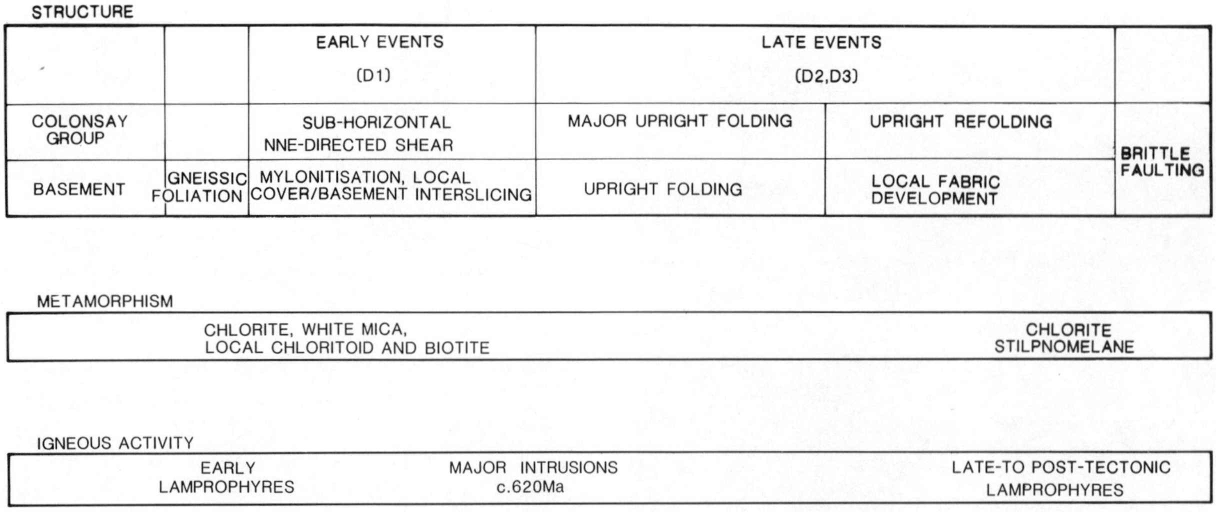

STRUCTURE

		EARLY EVENTS (D1)	LATE EVENTS (D2,D3)		
COLONSAY GROUP		SUB-HORIZONTAL NNE-DIRECTED SHEAR	MAJOR UPRIGHT FOLDING	UPRIGHT REFOLDING	BRITTLE FAULTING
BASEMENT	GNEISSIC FOLIATION	MYLONITISATION, LOCAL COVER/BASEMENT INTERSLICING	UPRIGHT FOLDING	LOCAL FABRIC DEVELOPMENT	

METAMORPHISM

CHLORITE, WHITE MICA, LOCAL CHLORITOID AND BIOTITE	CHLORITE STILPNOMELANE

IGNEOUS ACTIVITY

EARLY LAMPROPHYRES	MAJOR INTRUSIONS c.620Ma	LATE-TO POST-TECTONIC LAMPROPHYRES

Figure 11.10 The relative timing of structural, metamorphic and igneous events within the Colonsay Group.

commonly in biotite (Harrison and McDougall, 1981). Although both these features tend to obscure the initial crystallization age of the mineral separate, the true age may be identified if the effects are sufficiently mild. On a standard Ar/Ar age spectrum, the true age is most closely indicated by a distinct plateau at one age. In the case of the less-distributed hornblende samples from Colonsay, age spectra show plateaux or partial plateaux close to *c.* 620 Ma.

From these results, a tentative intrusive age of *c.* 620 Ma is placed on the Kiloran Bay and Scalasaig intrusions. Although most of the K/Ar and Ar/Ar results were obtained from the Kiloran Bay samples, the Scalasaig intrusion is probably of a similar age, given the general mineralogical similarity and common structural field relationships of the intrusions at the two localities. The age of the Balnahard intrusion remains relatively enigmatic, although one hornblende sample from the intrusion gave a partial plateau close to 600 Ma. The younger K/Ar and Rb/Sr mica ages are interpreted as products of a late metamorphism, manifest in thin section as the partial replacement of biotite.

Tentative minimum and maximum ages of *c.* 620 Ma are therefore given to the early and late structures, respectively.

11.4.2 *The relative timing of events within the Colonsay Group*

The relative timing of the major structural, igneous and metamorphic events is summarized in Fig. 11.10. The main intrusive event partitions the history of the Group into an early, pre-*c.* 620 Ma period, involving substantial deformation and peak metamorphism, and a later, post-*c.* 620 Ma history involving further polyphase deformation and slightly lower-grade metamorphism. Information from Colonsay and Islay alone is insufficient to determine whether or not these two histories are separate or part of a single progressive event.

The age of the Group must exceed 620 Ma, but sufficient time must have elapsed before 620 Ma to allow for deposition and burial of the sediments, the development of the early structures and the achievement of peak metamorphism, since these events clearly predate the period of main igneous activity. The geological events recorded in the Colonsay Group are also recorded in the older gneissic basement, the age of which is unknown.

11.5 Inter-regional correlation

The regional status of the Colonsay Group and its underlying basement is poorly constrained. The gneissic basement is generally assumed to be Lewisian, but has yet to be studied in detail. Early Geological Survey workers considered that the Colonsay Group correlated most clearly with the Torridonian. This correlation was based upon a comparison of the Colonsay Group on Islay with the Torridonian on Skye (Peach, in Wilkinson, 1907, and Clough, in Cunningham-Craig *et al.*, 1911). This opinion prevailed for the ensuing fifty years (Peach and Horne, 1930; Stewart, 1960; Johnstone, 1966), and the Colonsay Group is still described as 'Torridonian' on many recently published maps. However, suggestions by Stewart (1969), that the age of the group is equivocal, and Stewart and Hackman (1973), that the group may be Dalradian, have brought the Torridonian 'age' of the group into question.

The tentative 620 Ma age of intrusion for the Kiloran Bay Syenite allows some constraints to be placed on regional correlations of both the Colonsay Group sediments and the structures contained within them.

11.5.1 *Basement correlatives*

The basement exposures on N Colonsay and Islay constitute the most southerly examples of gneissic basement in the Scottish Highlands, and the only exposures of their type south of the Great Glen Fault in Scotland. By analogy with the gneisses of the NW

Highlands, most workers have assumed a Lewisian age for the Colonsay–Islay basement. However, this assumption is open to question. The basement is strongly retrogressed, especially in Colonsay, where the remnants of an amphibolite-facies metamorphism are now barely recognizable. Also, the dyke swarms which characterize much of the Lewisian Complex appear to be absent on Colonsay and Islay. The geochronology of the gneisses has yet to be tackled, and there are no available geochemical data with which to attempt an inter-regional comparison. Indeed, the most recent published works on the basement are the original Survey Memoirs (Wilkinson, 1907; Cunningham-Craig *et al.*, 1911), in which only brief descriptions of basement are given. At present there are, therefore, no sure grounds for establishing a stratigraphic link between the Lewisian gneisses of NW Scotland and the Colonsay–Islay gneisses. A comparison may equally well be drawn between younger sequences such as the Annagh Gneiss (Aftalion and Max, 1987) or even parts of the Moine. The only age constraint which may be placed on the gneisses is that they must be sufficiently old to have allowed for metamorphism to at least amphibolite-facies conditions before uplift, erosion and the subsequent deposition and deformation of the Colonsay Group (before 620 Ma).

11.5.2 *Colonsay Group correlatives*

The suggested age of the major intrusions indicates that the deposition, burial, initial deformation and metamorphism of the group occurred before *c.* 620 Ma. This affects attempts to correlate the Colonsay Group with the Appin Group of the Lower Dalradian, a correlation advanced by some authors (Stewart and Hackman, 1973; Rock, 1985). The lowest Dalradian sequences have an estimated depositional age in the region of *c.* 700 Ma (Powell and Phillips, 1985). If a Dalradian correlation is pursued, the early deformation on Colonsay must therefore have occurred during the deposition of higher levels of Dalradian sediments between 700 Ma and 620 Ma—a period of relative tectonic stability or extension in the Highland area (Anderton, 1982). The deformation must have occurred without affecting any of the known Dalradian rocks, since the latter are not thought to have undergone deformation until the late Cambrian (Powell and Phillips, 1985). A similar situation has been suggested for the Central Highland area, in which the lowest units of the Grampian Group are considered to have undergone deformation at *c.* 750 Ma, while sedimentation was continuing at a higher level in the same sequence (Piasecki and van Breemen, 1983). However, in the Colonsay Group, the tectonic setting envisaged for the early events involves a substantial overburden on the Colonsay Group during deformation (>10 km), much of the overburden probably also being involved in the deformation event. This effectively rules out an Appin Group, and possibly also a Grampian Group correlation.

Current constraints on the age of the Colonsay Group do not rule out equivalence with either the Torridon or Stoer Groups, or parts of the Moine Assemblage. However, it is suggested that the application of 'Moine' and 'Torridonian' labels to the Colonsay Group is currently inappropriate. In the case of a 'Torridonian' association, it will be argued below that the Colonsay Group lies within the orthotectonic zone of the Caledonides, the Caledonian Front lying to the west of Colonsay and Islay. Since the term 'Torridonian' is generally applied to Precambrian rocks lying either on the foreland or in the marginal nappe zone of the Caledonides (Stewart, 1969), the term is not directly applicable to the Colonsay Group, its use by earlier workers on Colonsay probably stemming from the interpretation of the Loch Skerrols Thrust on Islay, lying to the east of the Colonsay Group, as the southern continuation of the Caledonian Front (Fitches and Maltman, 1984). 'Torridonian' could perhaps be applied to the Colonsay Group if it could be shown that the group lay in a foreland thrust slice marginal to the Caledonides (analogous to the Kishorn Nappe of the NW Highlands), but this is not the case. A Moine correlation is suggested by the presence of a Precambrian deformation in the Colonsay Group, but there are no geochronological or regional geochemical data available to support a stratigraphic correlation. Moreover, the *c.* 1000 Ma ages determined for early Moine structures have not been demonstrated for the early structures in the Colonsay Group.

Among the other, relatively restricted Precambrian sequences in the region, one unit, the Iona Group, may be distinguished by its similarity to the Colonsay Group. The Iona Group is exposed on the eastern coast of Iona, 25 km NNW of Colonsay (Fig. 11.1). The islands on which the two groups are exposed are geographically close. Both groups rest unconformably upon gneissic basement, with contacts which are generally extensively sheared. Both groups have also undergone a mild penetrative deformation, involving substantial shearing subparallel to bedding, associated with the development of a conspicuous L-S fabric, and local small-scale folding. Furthermore, the lithological character of the Iona Group readily matches that observed in areas on Colonsay, and palaeocurrents derived from the two Groups share a common N or NE azimuth (Stewart, 1960). The most significant difference between the two groups is the nature of the polyphase deformation, the sequence on Iona being analogous to the early deformation on Colonsay, without the later deformations superimposed.

11.5.3 *Regional correlation of the Colonsay Group deformation*

The presence of the Loch Gruinart and Great Glen Faults prohibits the direct extrapolation of Colonsay Group structures outside the Colonsay–W Islay area. However, a comparison of deformation sequences in the Colonsay Group and the Dalradian of E Islay indicates that the Colonsay Group has a relatively longer deformation history. Both areas contain one set of major upright folds, and one or more additional sets of smaller-scale upright folds and fabrics. Only in the Colonsay Group, however, was this upright fold event preceded by an additional development of widespread, penetrative, flat-lying structures. The geochronological results outlined in the previous section allow some correlation to be made between these two areas, as the late, upright fold structures in the Colonsay Group post-date the *c.* 620 Ma intrusive event and either precede or partly accompany the later cooling event (reflected in the *c.* 420 Ma K/Ar and Rb/Sr mineral ages). This allows a broad correlation between the late structures on Colonsay and the fold structures in E Islay, which are associated with the regional Grampian orogenic events. However, the early structures in the Colonsay Group, with no clear analogue in the local Dalradian sequences, are provisionally constrained to a pre-Grampian orogenic phase by the > 620 Ma age. The early structures could be placed in a new 'early Caledonian' event—a precursor to the main Caledonian events—but since the suggestion effectively doubles the life of the Caledonian orogeny, and since the > 620 Ma constraint is only a minimum estimate for the age of the early structures, the assertion seems unreasonable at present. A temporal separation between the early and late histories of the Colonsay Group is therefore implied, and a pre-Caledonian correlative for the early events on Colonsay must be sought. Unfortunately, no evidence is currently available to test correlations of the early structures with regional Grenville or Morarian/Knoydartian events.

11.5.4 *Displacement histories on the Great Glen and Loch Gruinart Faults and their bearing on Colonsay Group correlations*

A feature which significantly affects the validity of any regional Colonsay Group correlations, is the nature of the displacements on the two major faults which bound the Colonsay Group 'block': the Loch Gruinart and Great Glen Faults (Fig. 11.1).

The path of the two faults has been a subject of debate, with Bailey (1917), and later Stewart and Hackman (1973), suggesting that the Great Glen Fault displacement may be taken up along the line of the Loch Gruinart Fault, to the east of Colonsay. This interpretation facilitates the correlation of events on Colonsay with those in the Dalradian of the Appin district, assuming a *c.* 105 km sinistral displacement (Kennedy, 1946). Objections have been raised to this interpretation by Westbrook and Borradaile (1978), based on the presence of a magnetic anomaly straddling the fault, which shows no sign of a *c.* 105 km sinistral dismemberment. Westbrook and Borradaile also pointed out the geometric abnormality created by rerouteing the Great Glen Fault through Loch Gruinart, relative to its otherwise typically smooth curvilinear path from Loch Linnhe to the Shetlands. The routeing of the Great Glen Fault to the east of Colonsay is not necessary, given the *c.* 620 Ma age for the main period of intrusion on Colonsay, which precludes direct Colonsay Group–Dalradian and Colonsay Intrusive Suite–Appinite Suite correlations. Furthermore, the continuation of the Great Glen Fault to the west of Colonsay is consistent with geophysical constraints from that area (Evans *et al.*, 1982). Hence, the Loch Gruinart Fault is taken to be a splay of the Great Glen Fault, the main displacement on which runs to the west of Colonsay, approximately along the line of the Colonsay Fault.

The amount of displacement on the two fault is less easy to resolve. Westbrook and Borradaile (1978) suggested a normal displacement for the Loch Gruinart Fault, with an easterly downthrow for the hidden basement surface beneath Colonsay and the Dalradian to the east. However, it was also concluded that considerable reverse faulting had occurred to produce the observed magnetic anomalies, possibly in the form of NW-directed thrusting along the Loch Skerrols Thrust. Since Fitches and Maltman (1984) have cast doubt on the status of the Loch Skerrols structure as a major thrust, and since deep seismic work (Brewer *et al.*, 1983) suggests that the Loch Gruinart Fault may have a significant down-dip extension as a low-angle fault, it is possible that a substantial, presumably relatively early, thrust history may be attached to the structure (as suggested by Fitches and Maltman, 1984). The more local observations of Westbrook and Borradaile, attempting to balance the slight apparent offset on the magnetic anomaly cut by the fault, suggest rather more subdued net displacements of either 9 km (sinistral) or 6 km (dextral). An early sinistral net displacement, followed by a more recent normal downthrow to the south-east, is most consistent with the history of the Leannan Fault—the probable extension of the Loch Gruinart Fault in NW Ireland (Pitcher *et al.*, 1964).

The nature of the displacement on the Great Glen Fault is also a contentious issue. The suggestion of Kennedy (1946) of a net *c.* 105 km sinistral displacement was favoured for a long time and is still followed by many workers (e.g. Watson, 1984; Soper, 1986). However, alternative suggestions have been made of sinistral displacements up to 250 km (Storetvedt, 1974), and dextral displacements up to 110 km (Garson and Plant, 1972). A more extreme estimation, of 2000 km, based on large-scale palaeomagnetic considerations, has been made by Van der Voo and Scotese (1981), although the suggestion has received considerable criticism both on geological (Smith and Watson, 1983) and geophysical grounds (Briden *et al.*, 1984). The available data are clearly insufficient to constrain the displacement history closely, especially if the possibility of pre- or early-Caledonian displacements are considered.

The role of substantial strike-slip movements towards the closing stages of continental collision is gaining wide recognition both in general, and also in specific relation to the Scottish and Irish Caledonides (Phillips *et al.*,

1979; Watson, 1984; Needham and Knipe, 1986). In particular, the 'suspect terrane' hypothesis, in which crustal blocks recording discrete geological histories are tectonically juxtaposed, usually during the closing stages of collision, is being usefully applied to many orogenic belts, including the Scottish Caledonides (Soper, 1986). The Colonsay Group block, largely isolated by tectonic junctions, for which the displacement history is poorly constrained, may be regarded as a suspect terrane relative to the adjacent Caledonian orthotectonic and foreland 'terranes'. This possibility is not prohibited either by the deformation history of the group, or the *c.* 620 Ma age around which its history is divided, since in both respects the Colonsay Group is enigmatic within its region.

A suspect terrane interpretation for the Colonsay Group has been discussed by Bentley *et al.*, (subm.), and would require an early- or pre-Caledonian displacement on the Great Glen Fault system, or its precursor. This would allow docking of the Colonsay Group block prior to the Grampian orogenic events, in which the Colonsay Group was deformed. Although independent evidence is required to test this model, the possibility remains that attempts to correlate regional events across to the Colonsay Group have been problematic because the correlations have been regionally misdirected.

11.6 Conclusions

The Colonsay Group sediments describe sedimentation across the northern margin of a fault-bound, probably intracratonic basin.

The general chracter of the Colonsay Group, in terms of lithology, tectonometamorphic history and cover–basement relationships, is not directly comparable with any of the established Precambrian sequences in Scotland, other than the Iona Group, to which a tentative correlation is extended. Further inter-regional correlations are prohibited by the tectonic isolation of the Colonsay Group, and therefore rest on the available isotope data. Initial findings from one intrusion on Colonsay (the Kiloran Bay Syenite) indicate a tentative *c.* 620 Ma age of intrusion. Data consistent with a *c.* 620 Ma age of intrusion have also been obtained from igneous bodies at Scalasaig and Balnahard. A late Proterozoic age is therefore implied for the major intrusions on Colonsay, and also for the early tectonic history—a feature which makes a direct Colonsay Group–Dalradian (Appin Group) correlation unlikely. The later events on Colonsay, however, may be broadly correlated with Grampian orogenic events. The major upright folds in the Colonsay Group are therefore local contemporaries of the major fold structures in the SW Highlands, such as the Islay Anticline, the Caledonian Front lying to the west of Colonsay and Islay. The age of the Colonsay Group within the Proterozoic, and the regional significance of the early structures remains enigmatic.

Previous Colonsay Group interpretations have rested primarily on attempts to integrate the group with defined sequences on the Scottish mainland. However, the application of 'Moine' or 'Torridonian' labels to the Colonsay Group is not supported here, partly because, currently, these correlations cannot be substantiated, and also because the early history of the Group may be unrelated to its present tectonic setting.

Acknowledgements

The Rb/Sr and K/Ar analysis involved in this work was carried out at the British Geological Survey, London, with the assistance of M. Brook, Dr R. Pankhurst and I. Millar. The Ar/Ar analyses were carried out by Dr D. Rex at the University of Leeds. The constructive comments of Drs A. J. Maltman and W. R. Fitches on earlier drafts of this chapter are gratefully acknowledged.

References

Aftalion, M. and Max, M. D. (1987) U–Pb zircon geochronology from the Precambrian Annagh Division gneisses and the Termon Granite, NW County Mayo, Ireland. *J. geol. Soc. London* **144**, 401–407.

Anderton, R. (1982) Dalradian deposition and the late Precambrian–Cambrian evolution of the Iapetus Ocean. *J. geol. Soc. London* **104**, 99–132.

Bentley, M. R. (1986) The Tectonics of Colonsay, Scotland. Unpublished Ph.D. Thesis, University College of Wales, Aberystwyth.

Bentley, M. R., Maltman A. J. and Fitches W. R. (submitted). Colonsay and Islay: a suspect terrain within the Scottish Caledonides.

Brewer, J. A., Matthews, D. H., Warner, M. R., Hall, J., Smythe, D. K. and Whittington, R. J. (1983) BIRPS deeps reflection studies of the British Caledonides. *Nature* **305**, 206–210.

Briden, J. C., Turnell, H. B. and Watts, D. R. (1984) British palaeomagnetism, Iapetus Ocean, and the Great Glen Fault. *Geology* **12**, 428–431.

Cunningham-Craig, E. H., Wright, W. B. and Bailey, E. B. (1911) The geology of Colonsay and Oronsay with part of the Ross of Mull. *Mem. Geol. Surv. G. B.* **35**, 1–74.

Dalrymple, G. B. and Lanphere, M. A. (1969) *Potassium–Argon Dating*. Freeman, San Francisco.

Evans, D., Kenalty, N., Dobson, M. R. and Whittington, R. J. (1982) Malin sheet (solid rock). *British Geological Survey 1:250 000 series.*

Fitches, W. R. and Maltman, A. J. (1984) Tectonic development and stratigraphy at the western margin of the Caledonides: Islay and Colonsay. *Trans. R. Soc. Edinburgh. Earth Sci.* **75**, 365–382.

Garson, M. S. and Plant, J. (1972) Possible dextral movements on the Great Glen and Minch Faults in Scotland. *Nature* **240**, 31–35.

Harrison, T. M. and McDougall, I. (1980) Investigations of an intrusive contact, northwest Nelson, New Zealand—II. Diffusion of radiogenic and excess ^{40}Ar in hornblende revealed by ^{40}Ar/^{39}Ar age spectrum analysis. *Geochim. Cosmochim. Acta* **44**, 2005–2020.

Johnstone, G. S. (1966) *The Grampian Highlands*. HMSO, Edinburgh.

Kennedy W. Q. (1946) The Great Glen Fault. *Q. J. geol. Soc. London* **102**, 41–76.

Needham D. T. and Knipe, R. J. (1986) Accretion- and collision-related deformation in the Southern Uplands accretionary wedge, southwestern Scotland. *Geology* **13**, 303–306.

Peach, B. N. and Horne, J. (1930) *Chapters on the Geology of Scotland*. Oxford University Press, London.

Phillips, W. E. A., Flegg, A. M. and Anderson, T. B. (1979) Strain adjacent to the Iapetus Suture in Ireland. In Harris, A. L., Holland, C. H. and Leake, B. E. (eds.), The Caledonides of the British Isles—Reviewed. *Spec. Publ. geol. Soc. London* **8**, 257–262.

Piasecki, M. A. J. and Van Breemen, O. (1983) Field and isotopic evidence for a *c.* 750 Ma event in Moine rocks in the Central Highland region of the Scottish Caledonides. *Trans. R. Soc. Edinburgh: Earth Sciences* **73**, 119–134.

Powell, D. and Phillips, W. E. A. (1985) Time of deformation in the Caledonide Orogen of Britain and Ireland. In Harris, A. L. (ed.), The Nature and Timing of Orogenic Activity in the Caledonian Rocks of the British Isles. *Mem. geol. Soc. London* **9**, 17–39.

Ramsay, J. G. (1967) *Folding and Fracturing of Rocks*. McGraw-Hill, New York.

Rock, N. M. S. (1985) Value of chemostratigraphical correlation in metamorphic terranes: an illustration from the Colonsay Limestone, Inner Hebrides, Scotland. *Trans. R. Soc. Edinburgh: Earth Sci.* **76**, 463–465.

Saha, D. (1984) Deformation Phases and Minor intrusions in W. Argyllshire: Regional Implications with respect to Caledonian Tectonics. Unpublished Ph.D. Thesis University of London (Imperial College of Science and Technology).

Saha, D. (1985). Clast composition and provenance of psammites in the Colonsay Group and Bowmore Sandstone, SW Argyllshire. *Scott. J. Geol.* **21**, 1–8.

Smith, D. I. and Watson, J. V. (1983) Scale and timing of movements on the Great Glen Fault, Scotland. *Geology* **11**, 523–526.

Soper, N. J. (1986) The Newer Granite problem: a geotectonic view. *Geol. Mag.* **123**, 227–236.

Stewart, A. D. (1960) On the Sedimentary and Metamorphic Histories of the Torridonian, and the Later Igneous Intrusions of Colonsay and Oronsay. Unpublished Ph.D. Thesis, University of Liverpool.

Stewart, A. D. (1962*a*) On the Torridonian sediments of Colonsay and Oronsay and their relationship to the main outcrop in northwest Scotland. *Geol. J.* **3**, 121–155.

Stewart, A. D. (1962*b*) Greywacke sedimentation in the Torridonian of Colonsay and Oronsay. *Geol. Mag.* **103**, 462–465.

Stewart, A. D. (1969) Torridonian rocks of Scotland—reviewed. In Kay, M. (ed.), North-Atlantic—Geology and Continental Drift, a Symposium. *Mem. Am. Ass. Pet. Geol.* **12**, 595–608.

Stewart A. D. (1975) 'Torridonian' rocks of Western Scotland. In Harris, A. L., Shackleton, R. M., Watson, J., Downie, C., Harland, W. B. and Moorbath, S. (eds.), A Correlation of the Precambrian Rocks in the British Isles. *Spec. Rep. Geol. Soc. London* **6**, 43–51.

Stewart, A. D. (1982) Late Proterozoic rifting in NW Scotland: the genesis of the 'Torridonian'. *J. geol. Soc. London* **139**, 413–420.

Stewart, A. D. and Hackman, B. D. (1973) Precambrian sediments of Western Islay. *Scott. J. Geol.* **9**, 185–201.

Storetvedt, K. M. (1974) A possible large-scale sinistral displacement along the Great Glen Fault in Scotland. *Geol. Mag.* **3**, 23–30.

Van der Voo, R. and Scotese, C. R. (1981) Palaeomagnetic evidence for a large (–2000 km) sinistral offset along the Great Glen Fault during Carboniferous time. *Geology* **9**, 583–589.

Watson, J. V. (1984) The ending of the Caledonian orogeny in Scotland. *J. geol. Soc. London* **141**, 193–214.

Westbrook, G. K. and Borradaile, G. J. (1978) The geological significance of magnetic anomalies in the region of Islay. *Scott. J. Geol.* **14**, 213–224.

Wilkinson, S. B. (1907) The geology of Islay, including Oronsay and portions of Colonsay and Jura. *Mem. Geol. Surv. G. B.*

Wright, W. B. (1908) The two earth-movements of Colonsay. *Q. J. geol. Soc. London* **64**, 287–312.

12
Pre-Dalradian rocks in NW Ireland

J. A. WINCHESTER and M. D. MAX

12.1 Introduction

Rocks currently regarded as pre-Caledonian occur in four areas of NW Ireland:

(i) The Mullet Peninsula and adjacent mainland in NW Co. Mayo
(ii) The NE Ox Mountains and adjacent inliers
(iii) Within the Westport inlier, S of Clew Bay
(iv) Inishtrahull, Tor Rocks and adjacent sea-floor N of Malin Head (Fig. 12.1).

Most of these pre-Caledonian rocks are now believed to have been formed in mid- and late-Proterozoic times before sedimentation of the Dalradian Supergroup was initiated towards the end of the Proterozoic. Since they were first described, more than 100 years ago, controversy has surrounded many of these rocks. Although Hull (1981*a*, *b*) regarded some of the gneisses in the central Mullet Peninsula as 'Laurentian' or 'Precambrian', because of a field similarity to some of the Canadian Shield gneisses, he also regarded the foliated Caledonian granites (now dated at approximately 400 Ma, Long *et al.*, 1984) of Galway, Donegal and the Ox Mountains as Precambrian. Kinahan (1882, 1886) soon showed that these granites were not Precambrian in age, and a discussion ensued, in which the existence of any Precambrian rocks in Ireland was questioned. Coincident with the onset of this lively argument, Geological Survey personnel were transferred from Ireland as the mapping was considered complete.

Kilroe (1907) summarized the evidence for and against the existence of Precambrian rocks in Ireland and concluded that all of Hull's 'Precambrian' rocks were early and synkinematic migmatites, entirely formed during the early Palaeozoic Caledonian orogenesis, using the gneisses of NW Co. Mayo to demonstrate his thesis. Trendall and Elwell (1963) accepted this interpretation of the NW Co. Mayo gneisses even though they recognized that regional Caledonian isograds passed across the margins of the migmatitic gneiss body, which had suffered an earlier retrogression that affected at least the margins of the body. Phillips *et al.* (1969) reiterated Kilroe's view, but Brindley (1969) considered that Hull's original conclusions deserved attention, especially in gneissose complexes, such as the rocks in NW Co. Mayo. In recent years (Sutton and Max, 1969; Max, 1970) detailed structural and petrofabric work, together with geochemistry and isotopic age dating, has confirmed the existence of large areas of Proterozoic basement within the NW Co. Mayo inlier, and other parts of NW Ireland. These Proterozoic basement inliers form parts of separate terranes that were displaced, deformed and metamorphosed in association with younger (but mostly Proterozoic) Dalradian rocks during early Palaeozoic Grampian and Caledonian orogenies.

12.2 Northwest County Mayo inlier

12.2.1 *Erris Complex*

The Erris Complex in NW Co. Mayo (Fig. 12.1) is the westernmost assemblage of Proterozoic rocks exposed in Europe, and this geographical location makes it an important part of any attempt to correlate rocks from Newfoundland with Europe. Sutton and Max (1969)

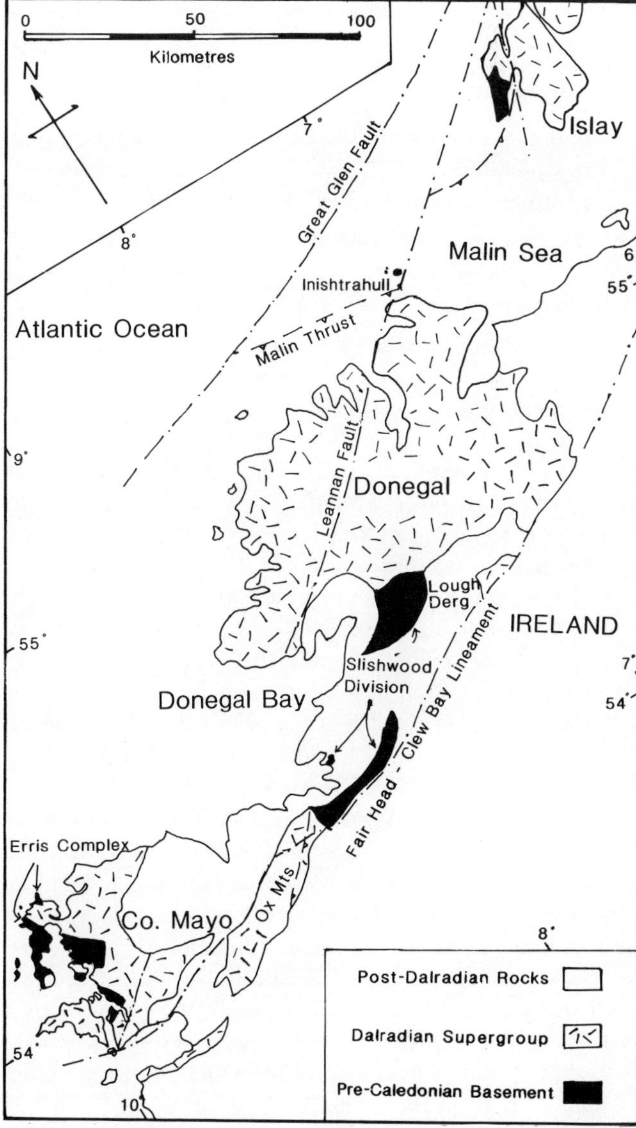

Figure 12.1 Location map showing the distribution of pre-Caledonian basement in NW Ireland.

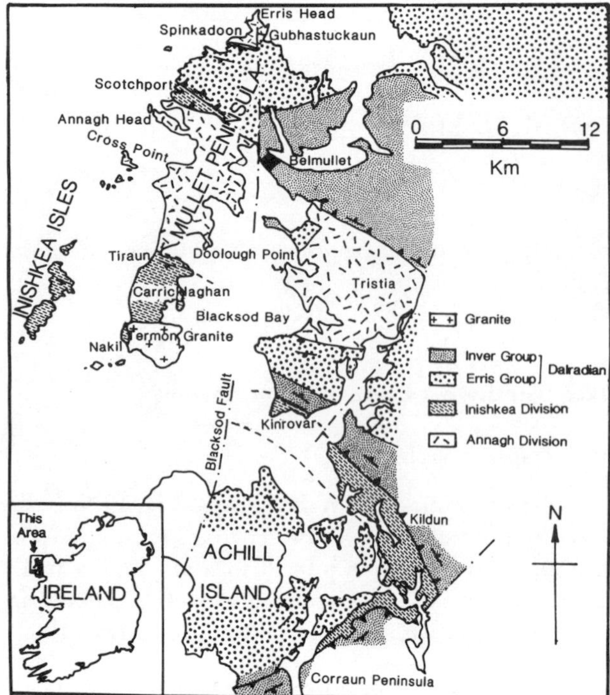

Figure 12.2 Map of part of the NW Co. Mayo inlier, showing the distribution of the Annagh and Inishkea Divisions of the Erris Complex.

originally defined the Erris Complex as comprising all pre-Caledonian rocks that structurally underlie the Dalradian rocks in the NW Co. Mayo inlier. It contains two main divisions:

(i) The Annagh Division, a highly deformed gneissose basement complex
(ii) The Inishkea Division schists, which have a pre-Caledonian metamorphic history and which structurally overlie the Annagh Division (Fig. 12.2).

12.2.1.1 *Annagh Division.* The Annagh Division gneisses (Winchester and Max, 1984) are the only rocks in the Erris Complex which have yielded clear isotopic ages that show that they were affected by Grenville metamorphism. They are best exposed in very clean and accessible beach and cliff exposures on the western shore of the Mullet Peninsula between Annagh Head and Scotch Port, at Cross Point and on the coast SW of Erris Head (Sutton, 1971). They are also well exposed on the Inishglora and Eagle Islands N of Scotch Port, but are not as readily accessible. Inland on the Mullet Peninsula, sufficient patchy exposure occurs to permit the margins of the major units to be traced. East of Blacksod Bay, on the mainland, Annagh Division gneisses are exposed on the headland at Doolough, the northern coast of the Gweesalia headland, and in rare inland exposures, particularly around Tristia (Fig. 12.2). The positive, medium-wavelength magnetic signature associated with the Annagh Division can be followed offshore to the west; hence similar gneisses may be present on, or only thinly buried beneath the sea-floor over considerable areas of the adjacent continental shelf.

Figure 12.3 Annagh Division Grey Gneisses cut by trondhjemitic granodiorite gneiss at Cross Point.

Contacts between the Annagh Division and younger metasedimentary rocks are everywhere tectonic. Immediately S of Belmullet the gneisses are brought into contact with Dalradian schists across the N–S Blacksod Fault, while near Scotch Port Harbour the gneisses are brought into tectonic contact with the structurally overlying Inishkea Division schists, along broad shear zones in which considerable metasomatic change has occurred (Winchester and Max, 1984).

Much of the Annagh Division consists of silicic or feldspathic grey and pinkish migmatitic gneiss, in which all original textures have been obliterated (Fig. 12.3). Some associated amphibolitic dark gneiss has also been migmatized, and two sets of young basic dykes are present throughout. Lensoid, foliated, variously deformed and sometimes agmatized mafic and ultramafic pods, possibly emplaced as dykes, often occur near to relatively undeformed younger basic dykes. Both concordant and discordant pegmatite bodies appear to have developed (Table 12.1). The Annagh Division thus constitutes a gneissose terrain of continental geochemical affinity (Winchester and Max, 1984) which everywhere structurally underlies the Dalradian supracrustal metasediments in the inlier.

Owing to the lack of recognizable metasediments, the sequence of igneous events, and the nature of the constituent gneisses is described below in some detail. The most ancient components of the Annagh Division are an amphibolitic Dark Gneiss and an associated micaceous Grey Gneiss which together underwent a tectonothermal event at about 1300 Ma (Aftalion and Max, 1987), with the development of trondhjemitic granodiorite sheets and localized net vein complexes (Table 12.1). This was followed by the emplacement of ultrabasic dykes, which have since been deformed to produce tectonically isolated pods. Mineral assemblages developed during this early event are not separable from those of the subsequent, and dominant Grenville metamorphism. Immediately before the onset of Grenville metamorphism, dyke-like intrusions of a distinctive peralkaline granite were emplaced at Doolough. The Doolough Granite is chemically similar to penecontemporaneous oversaturated peralkaline in-

Table 12.1 The structural sequence and age of metamorphic events affecting the Annagh Division.

Event	Age
Equilibration of Rb–Sr in amphibole in late metadolerite dyke at Cross Point	*c.* 663 ± 20 Ma
Metadolerite dyke(s?)	
Probable cooling event recorded by Rb–Sr muscovite mineral age from Pink Gneiss specimen	736 Ma
Tectonothermal event recorded by U–Pb lower intercept ages	813 + 166–270 and 869 + 69–118 Ma.
End of major tectonothermal event	
Pegmatitic granitic dykes	1000 ± 30 Ma
Continued deformation	
Basic dykes (pods)?	
Cooling or equilibration event (Pink Gneiss tectosilicate bands) and	1069 ± 72
Pink Gneiss phyllosilicate bands; Rb–Sr whole-rock	995 ± 186
Pink Gneiss	1070 ± 30 Ma
Granite Gneiss	
Onset of major tectonothermal event	
Doolough Granite	1093 + 7–8 Ma
Basic dykes (pods)?	
End of first tectonothermal event	
Granodiorite migmatite	1224 + 192–64, 1297 + 164–100 Ma
Onset of first tectonothermal event	
Ultrabasic dykes (pods)	
Dark Gneiss	
Grey Gneiss	

trusions in the Gardar area of SW Greenland (Winchester and Max, 1987*a*) and may have been intruded during a pre-Grenville rifting event at about 1100 Ma. Intrusion of gneissose granitic sheets and accompanying percolation of associated K-rich migmatitic fluids at about 1070 Ma (van Breemen *et al.*, 1978) marks the onset of the main Grenville tectonometamorphic event, which ended with the intrusion of pegmatitic granitoid dykes at about 1000 Ma (van Breeman *et al.*, 1978). The Rb–Sr equilibration age of 1069 ± 72 Ma in feldspathic gneisses and 995 ± 86 Ma in micaceous gneisses indicate slow cooling following the Grenville metamorphic peak (Max and Sonet, 1979).

During the Grenville event, the presence in the gneisses of some pale-green pyroxene and both perthites and antiperthites indicates that at least low granulite-facies conditions were reached locally (Gray, 1981; Long *et al.*, 1983). However, in the older lithologies in the complex, primarily in the micaceous gneisses, pyroxene is rarely seen in the Doolough area and cannot be demonstrated to have developed throughout the complex. Hence, the lack of characteristic granulite-facies assemblages suggests that granulite-facies conditions, if briefly attained, may not have been pervasive throughout the entire Annagh Division. The presence of perthites may be used to argue that low granulite-facies conditions were attained but high fluid pressures (and almost certainly high gas pressures) accompanied the initial migmatization and metasomatism (Max, 1970; Sutton, 1971). The relict perthites and antiperthites, which occur throughout the division, are most commonly seen within the late pegmatitic granitoid dykes and in some of the granitic gneiss sheets which were apparently emplaced during the Grenville metamorphism.

A later metamorphic event, which is almost certainly associated with some of the retrogression of the Grenville upper amphibolite to granulite-facies assemblages, may be recorded by lower intercept U–Pb ages from micaceous gneisses that do not appear to have been greatly affected by Grenville granitic metasomatism (Aftalion and Max, 1987). Finally, the entire area was also affected by a Caledonian metamorphic overprint, which locally reached epidote-amphibolite to mid-amphibolite facies (Max, 1973).

The micaceous and granitic gneiss lithologies form 90% of the Annagh Division, but minor lithologies are important in interpreting the structural sequence, as the composite foliation can be discriminated on a small scale. The dominant fabric is a sequence of banding which is most strongly developed in the micaceous grey gneisses and feldspathic pink gneisses. This banding is parallel to finer foliation defined by quartz, mica and quartzofeldspathic lithons and stringers which formed during each of the major tectonothermal events. Mineralogical and petrological descriptions are extant (Max, 1970; Sutton, 1971), but mineral chemistries have not yet been determined. Each major constituent of the Annagh Division is mineralogically and structurally distinct (Table 12.2).

(i) *Grey Gneiss* The grey micaceous gneiss is dominantly composed of alternating leucocratic and melanocratic foliae, with wider lithons ranging up to 10 cm. in width. The foliae have irregular margins and lie subparallel with the lithons; none are traceable for more than 5 metres along strike (Fig. 12.3). Garnets ranging up to 15 mm in diameter, with kelyphitic rims of plagioclase and micas, are common. Zoisite, with minor epidote, usually occurs in association with bio-

Table 12.2 Mean compositions and standard deviations of Grey Gneisses from the Annagh Division.

Area	Cross Point		Annagh Head		Erris Head		Doolough Pt		Total	
	$\bar{x}$	s	$\bar{x}$	s	$\bar{x}$	s	$\bar{x}$	s	$\bar{x}$	s
$n=$	6		12		4		7		29	
SiO_2	66.85	4.58	63.91	4.07	76.20	7.13	70.18	6.89	67.72	6.63
TiO_2	0.93	0.50	0.73	0.25	0.37	0.11	0.81	0.46	0.69	0.38
Al_2O_3	14.43	0.79	15.78	0.72	12.15	4.00	13.36	1.69	14.42	2.11
Fe_2O_3	3.01	0.79	2.30	0.73	1.10	0.51	1.51	0.98	2.09	0.99
FeO	2.41	2.20	2.58	1.28	0.85	0.96	2.46	2.10	2.28	1.70
MnO	0.12	0.07	0.08	0.03	0.03	0.0	0.06	0.03	0.08	0.05
MgO	0.79	0.60	2.23	0.90	0.34	0.60	1.67	1.72	1.53	1.27
CaO	2.63	1.28	3.29	1.23	0.86	0.62	2.00	0.87	2.50	1.43
Na_2O	3.34	0.67	3.68	0.89	3.17	1.46	4.00	1.12	3.61	0.90
K_2O	4.54	1.80	3.83	0.89	4.12	1.11	2.78	1.75	3.76	1.44
P_2O_5	0.24	0.17	0.26	0.08	0.06	0.07	0.17	0.14	0.21	0.13
H_2O^+	0.67	0.32	1.26	0.30	0.74	0.27	0.95	0.42	0.99	0.40
Ba*	996	134	1001	268	971	70	457	287	864	323
Cr	11	3	39	14	31	11	43	27	33	20
Nb	29	11	14	4	19	7	16	5	14	8
Ni	3	4	19	8	10	6	20	11	15	10
Rb	99	18	127	24	127	4	107	52	117	32
Sr	338	81	604	123	250	145	367	110	443	180
Y	56	18	26	15	7	2	35	27	32	23
Zr	526	101	206	56	160	46	217	64	150	56
K_2O/Na_2O	1.36		1.04		1.30		0.695		1.04	

* Contents in ppm, oxides as wt. %.

tite. A few pale-green relict pyroxenes occur in a few specimens taken from Doolough Point. Augened and deformed grains of early microcline may form up to 10% of the rock. Micas commonly occur in flakes up to 10 mm wide, but generally vary between 1 to 3 mm wide. In melanocratic foliae, biotite and phlogopite are most common, while muscovite is abundant in leucocratic foliae. Scattered subidomorphic to xenomorphic amphibole occurs as sporadic grains ranging up to 4 mm long. No dominant mineral lineation occurs away from zones of retrogression.

(ii) *Dark Gneiss* Dark amphibolitic gneiss is not widespread, but is best seen at Cross Point. It is usually found in discrete bands which normally have gradational boundaries with the Grey Gneiss (Fig. 12.4). It is a highly schistose plagioclase-amphibole rock containing plagioclase-quartz lithons from 1 mm to 5 mm wide, which are also discontinuous along strike. Small garnets may be present, while 10–50-cm thick bands of biotite-epidote-amphibole gneiss reported by Sutton (1971) have gradational contacts with the Grey Gneiss and appear to be a transitional variety of Dark Gneiss.

Both Grey and Dark Gneiss appear to have undergone the same early tectonothermal history. A geochemical study of the Grey Gneiss (Winchester and Max, 1984), concluded on the evidence of less mobile trace elements that it consisted largely of metamorphosed plutonic rocks of tonalitic to granitic composition, which showed some regional variation (Table 12.2, Fig. 12.5). Only at Doolough was there a suggestion from the chemistry that paragneisses may also be present. The Dark Gneiss likewise has a geochemical signature suggesting an igneous origin: most samples have a modified tholeiitic chemistry, which suggests that it could form a highly deformed suite of tholeiitic dykes within the Grey Gneiss.

(iii) *Ultrabasic pods* These pods are usually small, ranging up to 20 cm across. They occur in clusters often associated with trondhjemitic agmatite. They contain little internal foliation and consist of felted mats of coarse-grained actinolitic amphibole up to 7 mm long. Chlorite and bronze-coloured biotite occur as late alteration products. Geochemical analysis has shown that these rocks characteristically have a Mafic Index reading around 40, MgO exceeding 12%, and unusually high Ni and Cr contents (Winchester and Max, 1984).

Figure 12.4 Associated Dark and Grey Gneiss at Cross Point.

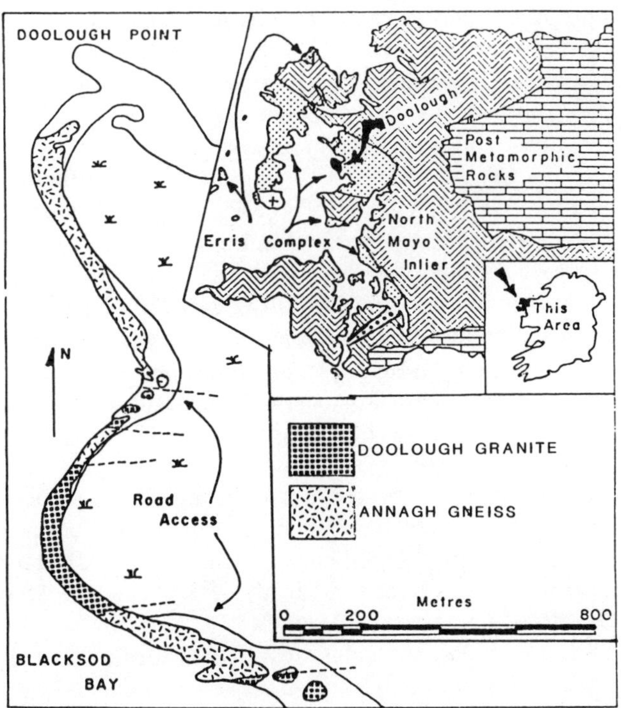

Figure 12.5 Map showing the shoreline outcrops of the Doolough Granite.

(iv) *Basic pods* Basic pods occur commonly throughout the gneisses. Usually ovoid in shape, they may occasionally be highly elongate. Largely composed of amphibole, garnet, biotite, plagioclase (An 26–35), epidote and sphene, they also contain apatite, phlogopite, allanite, opaque ores, quartz and zircon as accessory phases (Table 12. 3). The grain size is often coarse; high percentages of subidiomorphic to xenomorphic amphibole sheaves and large grains of biotite and subidiomorphic epidote commonly over 4 mm long are characteristic. Idiomorphic to subidiomorphic garnets range up to 2.5 mm in diameter, but are not present in all pods, and are commonest in epidote-poor pods. Away from the margins, where quartzofeldspathic foliae commonly parallel the curved margin and the foliation in the immediately adjacent country rock, the other mineral grains range up to 2 mm across and form a subpolygonal aggregate with only weak foliation.

The largest basic pods can be seen in the cliff sections at Annagh Head, where they are associated both with ultrabasic pods and most of the other gneiss types. Occasionally the pods occur in trains and hence they are regarded as representing dismembered basic dykes. Geochemical analysis showed, by the use of relatively immobile trace elements, that they were originally tholeiitic in composition (Table 12.3).

(v) *Granodiorite Gneiss* Granodioritic gneisses,

Table 12.3 Mean compositions and standard deviations of mafic rocks from the Erris Complex.

	Basic pods		Ultramafic pods		Annagh Gneiss metadolerites		Scotch Port metadolerites	
	$\bar{x}$	s	$\bar{x}$	s	$\bar{x}$	s	$\bar{x}$	s
SiO_2	48.66	3.17	51.88	2.73	49.26	1.28	49.74	1.15
TiO_2	1.50	0.20	0.50	0.13	2.45	0.34	3.15	0.58
Al_2O_3	16.66	1.23	8.93	2.55	12.62	0.39	13.09	0.47
Fe_2O_3	4.58	1.00	2.75	0.61	4.12	0.68	3.79	0.60
FeO	6.90	1.86	7.15	1.06	11.01	0.66	12.03	0.96
MnO	0.17	0.04	0.25	0.08	0.22	0.03	0.21	0.04
MgO	5.87	1.13	12.27	1.40	5.74	0.62	5.08	0.44
CaO	7.77	1.12	10.88	1.65	9.66	0.33	8.63	0.96
Na_2O	1.75	1.15	0.37	0.29	1.41	0.49	1.27	0.64
K_2O	3.57	1.08	2.62	0.82	1.08	0.34	0.76	0.16
P_2O_5	0.37	0.16	0.09	0.05	0.23	0.04	0.33	0.10
H_2O^+	2.22	0.24	2.18	1.43	2.15	0.42	2.11	0.60
Ba*	689	210	347	202	161	77	105	71
Ce	55	32	38	16	33	8	55	14
Cr	95	42	1425	293	111	44	107	23
La	24	19	18	10	11	3	21	8
Nb	6	2	9	1	9	3	13	2
Nd	26	9	19	6	28	30	29	6
Ni	69	41	348	62	51	18	46	9
Rb	166	73	126	53	27	17	21	10
Sr	635	277	31	61	188	47	159	70
Y	30	7	27	13	49	9	60	15
Zr	116	45	53	29	161	37	249	62
n =	7		6		13		13	

* Contents in ppm, oxides as wt. %.

which are frequently trondhjemitic, occur in two main forms: as bands ranging from 1 to 10 cm in width with a rather uniform internal texture, and as net-veined zones which are often highly distorted and contain both Grey and Dark Gneiss. Transitional forms are common. Locally, especially in association with Dark Gneiss—where the veining is most abundant, a quasi-agmatitic fabric is developed. Ultrabasic pods are also often found within a trondhjemitic agmatitic matrix, and hence they are interpreted here as entirely predating the tectonothermal event associated with the development of the Granodiorite Gneiss. Within mica-rich bands, a fine foliation similar to that in the Grey Gneiss often occurs in patches, usually inclined at a small angle to the margins of the bands and the coarse foliation. Foliation in the host Grey Gneiss is commonly parallel to the margin of Granodiorite Gneiss bodies near their contacts, but away from the bands foliation is divergent, suggesting that some of the foliation in the surrounding Grey Gneiss may be older than the development of the Granodiorite Gneiss and that it was rotated into parallelism during later deformation. Max (1970) interpreted these Granodiorite Gneiss bands as a product of anatexis of both the Grey and Dark Gneiss, but considered that some of the bands with sharper margins probably reflected bulk mobilization of locally derived magmatic fluids and synkinematic intrusion.

(vi) *Granite Gneiss* Numerous discontinuous sheets of Granite Gneiss surrounded by an envelope of Pink Gneiss range from 0.5 mm to 20 metres in thickness. Their contacts are always subparallel with the foliation in the adjacent Pink Gneiss, although away from the contacts the foliation in the Pink Gneiss often diverges from the orientation of the Granite Gneiss sheets. Pink Gneiss fabrics thus appear to have been rotated into parallelism with the Granite Gneiss sheets, which are variably foliated on a 1 to 5 mm scale, the most massive varieties occurring in the widest bodies. Along strike in the better-defined bands, the margins become indistinct until the banded appearance is lost in a gradational contact with the Pink Gneiss. The Granite Gneiss contains from 20% to 40% microcline occurring as idiomorphic to subidiomorphic single crystals ranging up to 8 mm in diameter. Some 55% to

Table 12.4 Mean analyses of Proterozoic oversaturated alkaline complexes from Ireland, Labrador and S Greenland.

	Doolough Granite	Doolough S. Granite	Arc Lake Granite	Letitia Lake alkali-granite	Flowers Bay peralkaline granite	Lac Brisson fresh granite	Tugtutoq rhyolite
SiO_2	75.37	74.68	74.8	73.64	73.06	71.27	72.32
TiO_2	0.28	0.37	0.3	0.41	0.30	0.29	0.35
Al_2O_3	10.71	12.03	10.6	11.12	11.75	11.26	11.32
Fe_2O_3	4.08	2.36	2.7	2.10	1.73	2.64	5.19
FeO	0.71	0.73	1.7	2.16	2.03	2.31	—
MnO	0.04	0.04	0.05	0.07	0.05	0.11	0.10
MgO	0.16	0.72	0.1	0.20	0.12	0.10	0.10
CaO	0.27	1.35	0.4	0.63	0.57	0.56	0.50
Na_2O	3.20	6.95	4.3	3.32	4.35	5.00	4.80
K_2O	4.88	0.34	5.1	5.04	4.49	4.85	3.97
P_2O_5	0.02	0.015	0.0	0.04	0.00	0.03	0.03
H_2O^+	0.29	0.33	0.57	0.91	0.95	0.42	—
Ba	156	70	66	212	158	75	75
Ce	608	205	135	238	—	820	518
Cr	16	13	13	9	6	—	4
Cu	55	72	7	10	11	14	12
Ga	29	28	31	27	27	—	—
La	331	101	109	165	—	420	245
Nb	255	183	80	77	49	—	199
Nd	179	71	—	—	—	—	192
Ni	2	13	25	38	5	—	2
Pb	45	5	43	39	19	—	64
Rb	162	9	261	217	257	640	205
Sn	16	19	—	—	—	—	—
Sr	78	115	18	39	20	65	28
Th	28	19	20	22	13	—	55
U	9	3	5	5	2	—	—
V	18	—	—	1	5	20	8
Y	238	120	120	167	73	520	154
Zn	146	51	154	172	143	485	285
Zr	1787	1005	1140	1270	1046	3790	1768
A.I.	1.13	0.87	1.18	0.98	1.03	1.20	1.14
$n =$	11	2	9	9	13	7	52

* Contents in ppm, oxides as wt. %.

70% of the rock consists of feldspar. Oriented flakes of mica, dominantly muscovite, ranging up to 5 mm long, and quartz are the other principal phases present. Accessory minerals occur in subpolygonal aggregates and do not include grains exceeding 2.8 mm in diameter.

These Granite Gneiss sheets appear to be synkinematic igneous rocks which intruded the Grey and Dark Gneiss host rocks, and permeated them with metasomatizing granitic fluids producing areas of Pink Gneiss.

(vii) *Doolough Granite* The Doolough Granite occurs in one large body about 400 metres across containing xenoliths of both Grey and Dark Gneiss, which is flanked by two narrower bodies at least 100 metres wide in the south-western beach exposures at Doolough Point. These three bodies are dyke-like, as they roughly conform to the composite foliation orientation and trend eastwards away from the shore exposures. They are concealed by thick peat bogs away from the shore and are not seen inland; they are also not seen on the Mullet Peninsula to the west (Fig. 12.5). Originally thought to be part of the Granite Gneiss suite (Max, 1970), geochemical studies have shown that these bodies are oversaturated peralkaline intrusives, with high concentrations of REEs, Zr, Y, Ta, Hf, Sn and Nb (Winchester and Max, 1983; 1987*a*).

Lacking the large microcline porphyroblasts or profuse micas seen in the Granite Gneiss, which it superficially resembles, the Doolough Granite consists of a characteristically pinkish, medium-grained, non-porphyroblastic, weakly foliated leucocratic rock, containing 20–40% microcline, 20–45% plagioclase and 15–50% quartz. Rare mica, either muscovite or greenish biotite, and greenish amphibole, occur sporadically. No alkaline amphiboles or pyroxenes have been observed: if formerly present, they have presumably been removed during metamorphism. Discrete grains and small aggregates of magnetite are ubiquitous, while zircon, monazite and sphene may occur as accessory phases. Large grains of yellow allanite were observed in some samples, while a few specimens contain relict pink to red almandine garnet, extensively replaced by chlorite.

Geochemical studies of the Doolough Granite (Winchester and Max, 1987*a*) revealed a close compositional similarity with oversaturated peralkaline intrusives in both the Gardar Province of SW Greenland (Emeleus and Upton, 1976; Blaxland *et al.*, 1978; Martin, 1985) and in Labrador (Emslie, 1978*a*; 1978*b*; Thomas, 1981; Hill, 1981, 1982; Collerson, 1982; Hill and Thomas, 1983; Currie, 1985) (Table 12.4). Furthermore, the chemical distinctions noted in the southern body at Doolough (Winchester and Max, 1978*a*) are similar to rocks forming part of the peralkaline suites in the Flowers Bay Complex of Labrador (Collerson, 1982) and the Tugtutoq dyke swarm in SW Greenland (Martin, 1985), suggesting further parallels with those igneous suites. Further links are also suggested by isotopic dating, which has revealed a probable emplacement age of 1093 + 7–8 Ma for the Doolough Granite (Aftalion and Max, 1987). This age is only slightly younger than the mid-Proterozoic age of emplacement for the chemically similar Tugtutoq intrusive complex in SW Greenland. Hence both the chemistry and the age of the Doolough Granite, which is unique in the Proterozoic rocks of the British Isles, suggest links with the Proterozoic rocks of Greenland and Canada.

(viii) *Granitic dykes* Discordant granitic dykes, clearly truncating the gneissose foliation and banding, and possessing sharp contacts with the Grey Gneiss, are very common on the Mullet Peninsula, especially N of Cross Point (Fig. 12.6), where they contain large garnets up to 25 mm in diameter. Elsewhere, these dykes are either less common, or they have been less commonly recognized as distinct from the Granitic and Pink Gneisses, which they may superficially resemble. They consist largely of plagioclase (An 26–35), microcline and quartz. Perthites, but not antiperthites, have been recognized, and biotite and muscovite are common (Table 12.2). Away from zones of later deformation, bifurcating dyke relationships are preserved, but even in the zones of subsequent shearing, the cross-cutting relationship may be preserved, despite extreme rotation towards a common composite foliation (Kelly and Max, 1979). This is particularly well seen on the Mullet Peninsula between Annagh Head and the northern contact with the Inishkea Division N of Port Point (Fig. 12.2). Occasionally preserved tight folds immediately adjacent suggest that these granitic dykes were emplaced in a narrow shear zone, or that marginal deformation affected the banding of the dykes during emplacement. An internal foliation parallel to the margins of the dykes is usually present.

Large, subidiomorphic grains of microcline range up to 70 mm long, but are related to elongated zones of these large crystals within the dykes: these are late 'pegmatitic' zones. None of these internal pegmatitic zones crosses a dyke margin into the country rock. Usually microcline grains are no more than 3 mm across. Mica flakes range up to 5 mm across and some quartz is up to 3 mm across. All other phases form a

Figure 12.6 Discordant granitic dyke cutting Grey Gneiss and earlier concordant granite gneiss at Cross Point.

Figure 12.7 Metadolerite dyke intruding Grey Gneiss SW of Erris Head. Cliff height is approximately 50 m.

porphyroblastic and subpolygonal texture and are no more than 2.5 mm across.

These granitic dykes are clearly late-kinematic intrusives which were rich in fluids.

(ix) *Metadolerite dykes* Throughout the Annagh Division, metadolerite dykes between 0.3 and 10 m wide cut across all the gneissose features, although their general orientation is usually subparallel to the composite foliation in the division (Fig. 12.7). They are weakly foliated, and contain a primary metamorphic mineral assemblage consisting of medium-grained amphibole, plagioclase and quartz, with secondary epidote, biotite and chlorite. Only near the margins and occasional internal shears is schistosity more developed. However, even in the less-deformed areas, no original igneous minerals have been identified, and this distinguishes them from metadolerites of similar appearance which intrude the structurally higher Dalradian rocks (Max, 1970). These metadolerites have a tholeiitic chemistry and plot as 'continental' basalts (Table 12.3; Fig. 12.11).

(x) *Rare lithologies* An unusual elongate pod of garnet-quartz gneiss, 350 cm long and 40 cm wide, occurs on the shore N of Doolough, enclosed by Grey Gneiss. It is foliated on 2–20 mm scale and the garnetiferous foliae also contain biotite, epidote and magnetite. The magnetite grains occur as anhedral blebs less than 2.5 mm long, while garnet grains are less than 1 mm across when euhedral, and less than 0.8 mm across when subhedral. The mineral association in this unusual lithology resembles that developed by Mn-rich cherty ironstones elsewhere and thus suggests that some metasedimentary gneisses may be present within the Annagh Division.

12.2.1.2 *Relationships of the Annagh Division.* The Annagh Division may thus be interpreted as a mid-Proterozoic basement, consisting largely of intrusive rocks. The nature of the rocks that these calc-alkaline plutons intruded is not known, but there is little evidence to link the Annagh Division gneisses with the Archaean Lewisian gneisses of Scotland, which were largely reworked in the early Proterozoic. No evidence for the existence of an Archaean basement has emerged from the Annagh Division, and indeed the evidence of widespread mid-Proterozoic plutonism, the existence of the 1100 Ma peralkaline Doolough Granite, and the high-grade metamorphism of Grenville age all serve to link the Annagh Division with mid-Proterozoic gneissose basements in southern Labrador and perhaps Newfoundland, rather than Scotland.

12.2.1.3 *Inishkea Division.* Inishkea Division rocks crop out in seven main areas: at Kinrovar Point, at Kildun and on the Corraun Peninsula; on the Mullet Peninsula south of Tiraun Point; around Scotch Port and at Spinkadoon and Gubhastackaun near Erris Head; and on the Inishkea Islands, west of the Mullet Peninsula (Fig. 12.2). Shore sections on the Inishkea Islands provide the best exposures, with the greatest variety of rock types, but good, accessible exposures also occur on the west coast of the Mullet Peninsula, south of Tiraun Point, south of Scotch Port Harbour and on the mainland to the SE (Crow *et al.*, 1971). Other, structurally important exposures of the Inishkea Division occur on the mainland in the Kinrovar 'wedge' (Crow and Max, 1976), where has been thrust over the former Erris Group metasediments. The Inishkea Division has now been traced across the Corraun Peninsula (Fig. 12.2) by the re-evaluation of existing work (Crow, 1974; Winchester and Max, 1987*b*), remapping and detailed geochemical characterization of the rocks in the whole NW Mayo inlier, as the tapering end of the Kinrovar 'wedge', which rests on a major D2 Caledonian thrust. Other minor exposures of Inishkea Division rocks include rocks and shoals both north and south of the Inishkea Islands, which also consist of characteristically porphyroblastic schists and gneisses. Several small islets to the southwest of the Nakil area of the Mullet Peninsula consist of Inishkea Division rocks which are hornfelsed by the Caledonian Termon Granite. The Black Rock–Fish Rock group of shoals, forming the westernmost exposures of the division are probably part of a separate wedge that lies to the south of the Duvillaun Islands and the Kinrovar wedge. This wedge may link eastwards with some massive porphyroblastic schists associated with the major Slievemore Slide on Achill Island recognized by Kennedy (1969), which may contain evidence of pre-Caledonian petrofabrics similar to those seen elsewhere in the Inishkea Division (Max, 1972).

Trendall and Elwell (1963) first distinguished the Inishkea Division schists from both the Annagh Division and the overlying Erris Group rocks, on the basis of their lithological contrast with the Dalradian schists. These authors also recognized that the Inishkea Division schists occupied a structural position at the base of the Dalradian succession in the NW Mayo inlier and suggested that they might be pre-Dalradian in age. However, like Kilroe (1907), they considered that the gneisses of the Annagh Division were produced by Caledonian metamorphism, and hence they did not consider the possibility of basement–cover relationships in the NW Mayo inlier.

Inishkea Division rocks are almost entirely metasedimentary, in contrast to orthogneisses of the Annagh Division. Characteristically they consist of rather monotonous flaggy or fissile grey micaceous psammitic and semi-pelitic schists, which frequently contain prominent oligoclase poikiloblasts (Fig. 12.8). No systematic internal stratigraphy has been established: stratigraphic sequences established in certain localities always appear to be cut out tectonically. Except around Scotch Port, where semi-pelitic lithologies are dominant, alternating psammitic and subordinate semi-pelitic lithologies are characteristic of the division. Both muscovite and biotite are abundant, occurring as broad sheaves of aligned laths which are frequently associated with small, often euhedral garnets in more semi-pelitic schists. Quartz and plagioclase are abundant; microcline is usually scarce or absent. In the Inishkea Islands and the southern Mullet Peninsula, grey heavy-mineral seams occur as dense bands concordant with the schistosity, containing abundant sphene, epidote, magnetite and zircon, but only rare apatite and no garnet. These heavy-mineral concentrations, which rarely exceed 2 cm in thickness, are thought to be of sedimentary origin and show some resemblance to the heavy-mineral seams recorded by Richey and Kennedy (1939) in the Moine rocks of western Scotland. Also present in the Inishkea Division rocks from Scotch Port, Gubhastuckaun, the Inishkea Islands and the southern Mullet Penninsula are scarce, thin whitish calc-silicate bands ranging up to 3 cm in thickness. No similar rocks were observed in the Corraun, Kildun or Kinrovar districts, so their distribution appears to be restricted to the northern part of the area. Two types of calc-silicate appear to be present. Those from the Inishkea Islands and the southern Mullet Peninsula contain epidote rather than clinozoisite and are garnet-free, thus resembling the calc-silicates of Arnipol type, now known to be widely distributed in both the Moine rocks and the Grampian Group rocks of Scotland (MacGregor, 1948; Winchester, 1975). Calc-silicates obtained from the Scotch Port Schist, by contrast, contain both clinozoisite and garnet, and resemble the commoner 'whitish' calcsilicates which are also widely distributed in Moine and Grampian Group rocks of Scotland (Fig. 12.9).

In addition to the metasedimentary rocks, small lenticular bodies of amphibolite are also present in Inishkea Division rocks at Scotch Port, Kildun, at Kinrovar Point, and on the Inishkea Islands. These mainly comprise hornblendic amphibole schists, which often contain garnet porphyroblasts and large, randomly orientated porphyroblastic laths of bronze-coloured biotite. These bodies, which are usually conformable with the intense foliation, are probably remnants of dismembered basic dykes.

More than 130 Inishkea Division schists have been analysed for major and selected trace elements (Winchester and Max, 1978*b*). The chemical information obtained has confirmed that chemical differences exist between the Inishkea Division and the Annagh Division, showing also that the Inishkea Division rocks were not formed by shearing Annagh Division gneisses (Fig. 12.10*a*), and revealing the metasedimentary nature of all but the amphibolitic rocks in the Division. However, it does suggest that much of the original sediment had an igneous provenance (Fig. 12.10*b*). It has also revealed the existence of compositional variations between the different isolated areas from which Inishkea Division rocks have been obtained. For example, while all the psammitic rocks plot within the greywacke field defined on a log Na_2O/K_2O–log

Figure 12.8 Graded grey micaceous psammitic schists of the Inishkea Division. SW end of Inishkea North Island.

Figure 12.9 Isoclinally folded 'white' calc-silicate in Inishkea Division Scotch Port Schist, seen on the W side of Scotch Port Harbour.

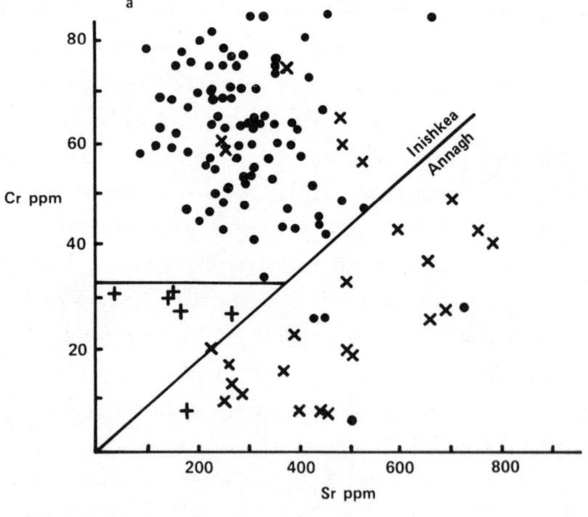

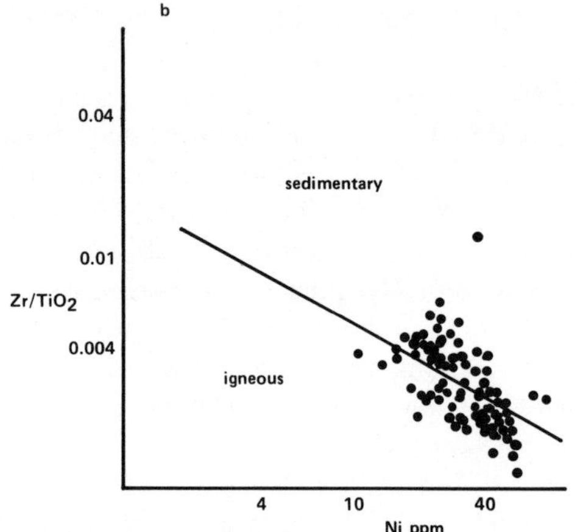

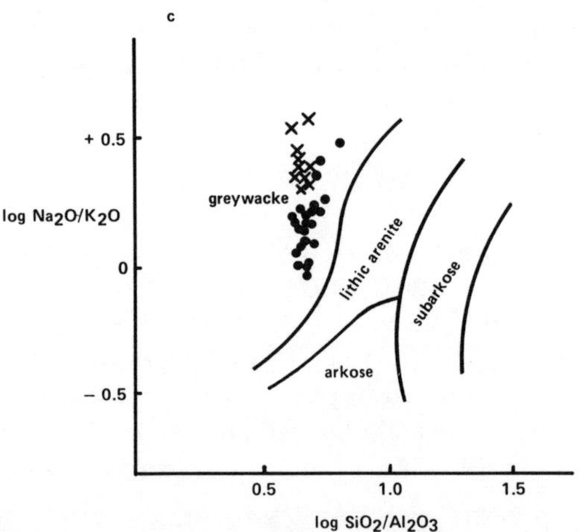

Figure 12.10 Discrimination diagrams illustrating aspects of the geochemistry of the Inishkea Division. (*a*) Cr–Sr diagram; ornament: dots—Inishkea Division schists; diagonal crosses—Annagh Division gneisses; upright crosses—Annagh Division gneisses from late high-strain zones. (*b*) Zr/TiO_2–Ni diagram. (*c*) log Na_2O/K_2O–log SiO_2/Al_2O_3 diagram (after Pettijohn *et al.*, 1973). Ornament: crosses—psammites from Kinrovar, Kildun and Corraun; dots—other Inishkea Division psammites.

SiO_2/Al_2O_3 diagram by Pettijohn *et al.* (1973), the samples from Kinrovar, Kildun and Corraun have a markedly higher Na_2O/K_2O ratio than those from other Inishkea Division rocks (Fig. 12.10*c*). In a similar manner there is a general diminution of Zr/Y ratios from SW to NE, so that highest mean Zr/Y ratios occur in Inishkea Division schists in the Inishkea Islands and the southwesternmost part of the Mullet Peninsula, a change which may be linked with the distribution of heavy-mineral seams. These vague chemical indicators can give only an approximate impression of the nature of the original Inishkea Division sediments, but they suggest that the Inishkea Division consisted of a sequence of greywackes and associated sediments which were probably derived from the south-west, where the greatest concentration of heavy minerals occurs.

Chemical comparisons with rock successions in Scotland of potentially similar age and similar lithological type, such as the Moine and Grampian Group rocks, has revealed no exact parallel, although some chemical similarity with the Moine is apparent. However, the dominance of greywackes and the higher Na_2O and slightly lower K/Rb ratios serve to distinguish the Inishkea Division rocks from the Moine, in which lithic arenites and arkoses are more dominant (Table 12.5). We conclude, therefore, that the Inishkea Division rocks appear to be unrelated to other late Proterozoic metasedimentary suites which crop out in other parts of the British Isles.

A chemical study of the amphibolites (Winchester and Max, 1987*b*) suggests that they were continental tholeiitic intrusives, displaying a wide compositional variation from relatively primitive to highly evolved basaltic types; the latter having been almost exclusively obtained from the Scotch Port area (Fig. 12.11). No clear chemical discrimination between these amphibolites and the late metadolerite dykes in the Annagh Division can be made: it is possible that they are related (Table 12.3).

Isotopic dating of the Inishkea Division has been attempted, using whole-rock specimens obtained from a 250 m coastal exposure on Inishkea South, where field evidence has suggested that the Caledonian metamorphic overprint is relatively light. However, because of the difficulties inherent in dating lithologies containing detrital minerals which have undergone multiple metamorphic events at similar metamorphic grades, a firm pre-Caledonian age has been elusive. Instead a poorly defined regression line of about 800 ± 146 Ma (total error) (Winchester and Max, 1987*b*) has been obtained. While not a precise isochron age, the slope of the line obtained cannot reflect a Caledonian event. This indication of a date around 800 Ma, which may reflect a disturbed Rb–Sr equilibration, coincides approximately with the possible metamorphic event indicated by zircon ages of about 850 Ma from U–Pb dating in Annagh Division gneisses (Aftalion and Max, 1987). Although further isotopic work is needed in the NW Mayo inlier, preliminary results suggest that the first tectonothermal event in the Inishkea Division probably took place around 800–850 Ma, and this event may be marked in the Annagh Division by late mineral growth, and possibly localized retrogression and formation of shear zones. Hence, although it remains possible that the Inishkea Division was related to similar metasedi-

Table 12.5 Comparative compositions of semi-pelitic rocks from the Inishkea Division and the Moine Assemblage of Scotland.

	Inishkea Division	Morar Division	Glenfinnan Division	Loch Eil Division
SiO_2	62.78	62.80	61.30	61.24
TiO_2	0.85	0.93	1.05	1.25
Al_2O_3	16.77	16.90	18.70	17.39
Fe_2O_3	1.71	6.40	7.50	1.75
FeO	4.07	—	—	5.20
MnO	0.10	0.10	0.11	0.09
MgO	2.02	2.00	2.20	3.43
CaO	2.33	2.30	1.80	1.87
Na_2O	3.68	2.80	2.60	2.40
K_2O	3.09	3.80	4.30	5.14
P_2O_5	0.24	0.29	0.25	0.22
H_2O^+	1.86	—	—	—
Ba*	870	990	950	998
Cr	59	—	—	84
Nb	14	14	22	22
Ni	35	22	33	26
Rb	123	109	162	170
Sr	335	328	264	299
Y	33	37	39	53
Zr	223	298	260	537
$n=$	103	39	43	45
Na_2O/K_2O	1.09	0.74	0.60	0.47
K/Rb	208	289	220	251

* Contents in ppm, oxides as wt. %.

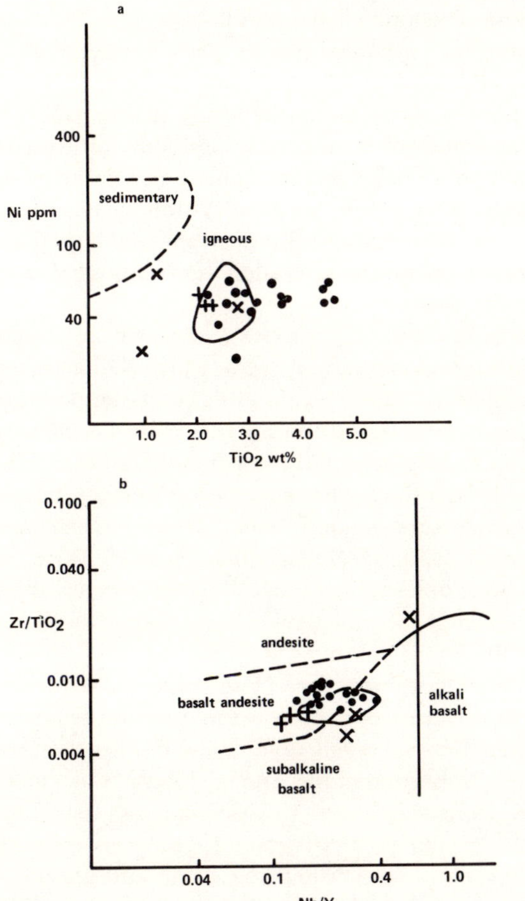

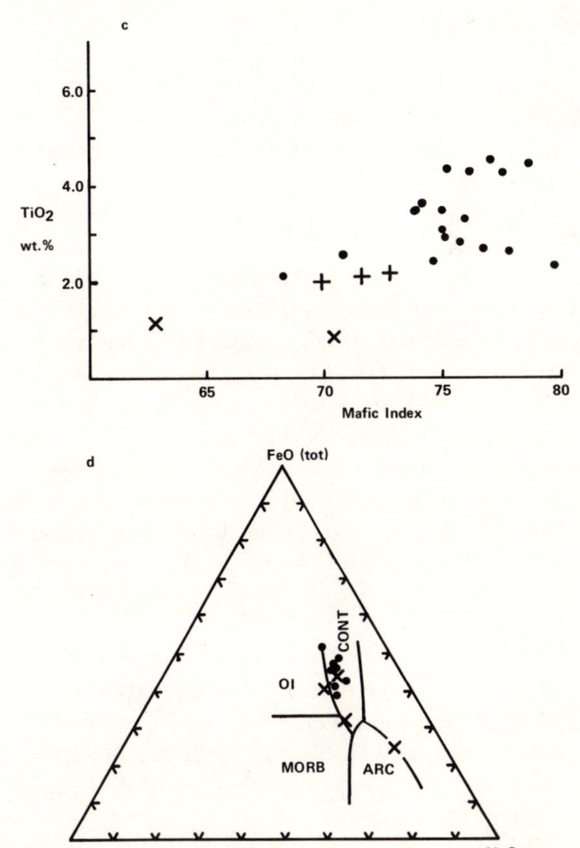

Figure 12.11 Variation diagrams illustrating aspects of the geochemistry of amphibolites in the Inishkea Division. (*a*) Ni–TiO_2 diagram; (*b*) Zr/TiO_2–Nb/Y diagram; (*c*) TiO_2–Mafic Index diagram; (*d*) MgO–FeO (tot.)–Al_2O_3 diagram. ARC—island arc basalts; CONT—continental basalts; MORB—mid Ocean Ridge basalts; OI—ocean island basalts. Ornament: dots—Scotch Port amphibolites; diagonal crosses—Inishkea Island amphibolites; upright crosses—Kinrovar amphibolites. On (*a*) and (*b*), the field of Annagh Division metadolerites is outlined by a continuous line.

ments in the Moine of Scotland, existing isotopic data suggest that the Inishkea Division rocks were not affected by a Grenville metamorphic event and were deposited during the interval between 1000 Ma and 800 Ma. They were then first metamorphosed by a late Precambrian event, which may be linked either with the 'Morarian' ages obtained from Scotland (Lambert, 1969; van Breemen *et al.*, 1974; Piasecki, 1980; Piasecki *et al.*, 1981), or a somewhat earlier event indicated by the ages obtained from the Annagh, Inishkea and Slishwood Divisions.

The amphibolites locally appear to cut across foliation in the field, and are chemically inseparable from the late metadolerites in the Annagh Division. Hence, both sets of amphibolite may form part of a single late dyke swarm, as suggested by Sutton (1971). A mineral age of 663 ± 20 Ma has been obtained from a metamorphic amphibole in one of these dykes occurring within retrogressed Annagh Division rocks near and within contact zones with the Inishkea Division. This indication of pre-Caledonian metamorphism affecting the amphibolite dykes indicates that many of the retrogressive features in the Annagh Division which were originally considered to be the result of the Caledonian (Grampian) event (Sutton and Max, 1969), may have formed during a late Precambrian tectonothermal event, while the effects of the Caledonian event appear to have had only a slight effect on these rocks.

If the Inishkea Division was deposited between 1000 Ma and 800 Ma ago, it is possible that no contemporaneous sedimentary succession is preserved within the Caledonides of the Scottish Highlands.

12.3 Slishwood Division

Rocks assigned to the Slishwood Division (Max and Long, 1985) crop out in three separate inliers, which can be linked by sharing a common lithological character and tectonothermal history (Fig. 12.1). The north-eastern part of the Ox Mountains lying east of the Ladies Brae Fault, the north-eastern part of the Rosses Point Inlier, and the Lough Derg inlier in the southernmost part of the Donegal metamorphic inlier, form windows which expose part of the same magnetic and gravity-defined block (Young, 1974; Riddihough and Max, 1975). These rocks were formerly regarded as Moine in the north-east Ox Mountains and the north-east part of the Rosses Point Inlier (Lemon, 1952, 1971) and in the Lough Derg Inlier (Anderson, 1948; Church, 1969). More recent work, however, has queried the Moine affinities of all these rocks: hence the introduction of the name Slishwood Division (Max *et al.*, 1984; Max and Long, 1985).

The rocks of the Slishwood Division dominantly consist of psammitic paragneisses associated with less abundant pelites, semi-pelites, calc-silicates and rare marbles. Minor intrusive metabasite lenses, containing a garnet-clinopyroxene-plagioclase assemblage, are thought to represent tectonically disrupted tholeiitic dykes (Lemon, 1971; Sanders, 1979). Some tonalitic and younger granitic pegmatite bodies are also present (Molloy and Sanders, 1983). Serpentinites associated with less-disrupted inlier rocks may in part be Precambrian in age (Sanders *et al.*, 1987), but serpentinites in major Palaeozoic fault zones, which probably root in deep-seated structural lines associated with Caledonian terrane boundaries (Max and Long, 1985) may be younger.

Granulite-facies mineral assemblages occur throughout and comprise quartz, perthite and antiperthite and orthoclase, with minor garnet and kyanite, with associated secondary plagioclase, biotite and muscovite: the latter commonly occurring in large idiomorphic crystals. Rare fibrolite may predate kyanite (Sanders, pers. comm.). In the least retrogressed area around Slishwood, the rocks display a coarse-grained granulitic texture. In many localities, textures are blastomylonitic (Lemon, 1971; Sanders, 1979). Coexisting K-feldspar and kyanite led Phillips *et al.* (1975) to deduce that the minimum P–T conditions of the first and highest-grade metamorphism were at least 10 kilobars (representing a depth of about 30 km) and 800 °C. More recently, Sanders (1979) suggested that temperatures may have approached 900 °C on the basis of coexisting garnet and clinopyroxene compositions. Sanders *et al.* (1987) now suggest that temperatures reached nearly 900 °C, while maximum P approached 14.5 kilobars.

Cooling was accompanied by perthitic segregation, and was followed at a lower temperature by deformation, possibly associated with the upward movement of the basement towards its present position. Subsequently, amphibolite-facies metamorphism and lower-grade retrogression were superimposed (Yardley *et al.*, 1979). This retrogression may be distinct from, and later than, a higher-temperature alteration that amphibolitized metabasite lenses.

The Slishwood Division was subjected to a late Proterozoic tectonothermal event, which is interpreted as the period during which the high-temperature granulite-facies metamorphism occurred. Ten Rb–Sr whole-rock samples from the north-east Ox Mountains and the Lough Derg inlier have yielded a 895 ± 60 Ma (total error) regression line, which is effectively defined by six semi-pelitic specimens obtained solely from the Lough Derg inlier (Max *et al.*, 1984). However, these ages can best be interpreted as cooling dates for metasediments which were subjected to the Grenville tectonometamorphic event, and were subsequently raised in the crust through the equilibration temperature of the Rb–Sr system, about 100 Ma later than the Rb–Sr equilibration in the Annagh Division basement already described. It is also possible, though less likely, that the granulite-facies metamorphism was the product of a separate late Proterozoic event. Late growth of zircon in the Annagh Division at about 850 Ma, and the poor *c.* 800 Ma Rb–Sr regression line from the earliest metamorphic event recorded in the Inishkea Division, suggest that such an event may have occurred. However, mineral textures in the Slishwood Division indicate that prolonged metamorphism, rather than separate events, occurred. Sanders *et al.* (1987) suggest that an event dated at 605 ± 37 Ma probably post-dated the high-pressure granulite-facies event, which itself may have taken place after the intrusion of basic bodies at about 1317 Ma. This earliest event might also be reflected in the *c.*1300 Ma earliest event recorded in the Annagh Division; and the younger event may be the same as that which metamorphosed the amphibole in the metadolerite dykes intruding the Annagh Division.

Evidence of the Palaeozoic Caledonide overprinting in the area is also widespread. A ^{39}Ar–^{40}Ar cooling age of about 440 Ma, from hornblende forming a retrogressive replacement of clinopyroxene, was obtained from the west of the Slishwood area (Phillips *et al.*, 1975), and this may be interpreted as dating the last time that the area cooled through 500 °C. It corresponds well with ages of the end of Caledonide metamorphism obtained from the Highlands of Scotland.

Until geochemical data become available, attempts to relate Slishwood Division metasediments to other Proterozoic rocks can only be based upon lithological resemblances and sparse isotopic age dates. However, if both the Annagh and Slishwood Divisions suffered the same early metamorphic event, the Slishwood Division metasediments are unlikely to be related to those of the Inishkea Division. Present information suggests that Slishwood Division rocks may be unique in Ireland as well as elsewhere in the British Isles. If they prove to be a part of the local pre-Caledonian basement, metamorphosed by a 'Grenville' event, equivalent rocks are most likely to be exposed in Labrador and NW Newfoundland. They could even be speculatively interpreted as displaced Ketilidian rocks, modified by later 'Grenville' metamorphism. Conclusions must clearly await the results of extensive further research.

12.4 Northernmost Ireland

Although the Inishtrahull gneisses off northernmost Co. Donegal have been dated as being at least 1750 Ma old (Fig. 12.1), structural and metamorphic evidence suggests that they are substantially older and related to the Lewisian rocks of Scotland (Roddick and Max, 1983). MacIntyre *et al.* (1975) considered that these gneisses and their associated metamorphosed basic dykes could be older than 2000 Ma. Therefore, because these rocks probably form part of the Archaean and early Proterozoic Lewisian craton, they are not discussed here.

However, a triangular area to the south-east of the gneisses of the Inishtrahull Platform (Evans *et al.*, 1983) (Fig. 12.12) consists entirely of low amphibolite-facies garnet-muscovite-biotite metasedimentary schist. This schist is nowhere exposed on land, but the few samples recovered suggest a resemblance to either Moine or Erris Group rocks. They are tectonically separated from the Gneiss of the Inishtrahull platform by the seaward continuation of the Leannan Fault, and from the greenschist-facies Dalradian rocks seen immediately to the south on the Inishowen Peninsula, by an eastward

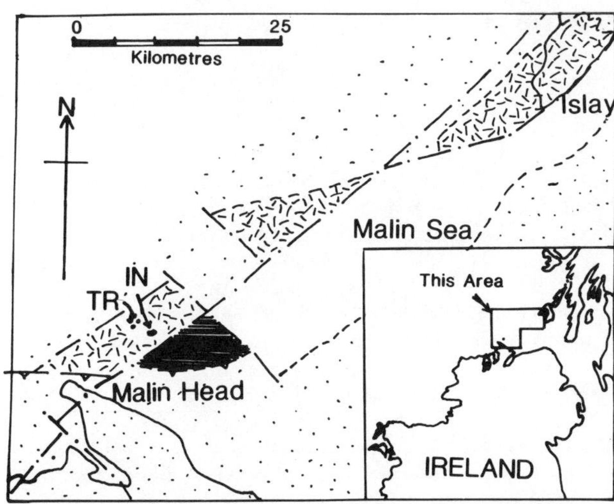

Figure 12.12 Map illustrating the distribution of later Proterozoic rocks NE of Malin Head. IN—Inishtrahull; TR—Tor Rocks. Ornament: randomly oriented dashes—Lewisian; stipple—Dalradian; horizontal lines—garnet-mica schist.

continuation of the Malin Thrust. The Malin thrust may also be the south-western continuation of the Loch Skerrols Thrust on Islay, which carries Dalradian rocks over the late Proterozoic Bowmore Sandstone on Islay. The eastern boundary of the schists is a NW–SE-trending fault which now marks the western margin of a late Palaeozoic sedimentary basin.

These schists are only exposed on the sea floor at depths exceeding 30 metres, and therefore only a few specimens, obtained by divers in a traverse, have been recovered (Evans, 1983); these are currently being geochemically studied. They are distinct from the Inishtrahull gneisses as they do not display a variety of igneous lithologies or the complex geological history of the gneisses. Their metamorphic grade is higher than, and their lithology and tectonothermal history is equally distinct from the Dalradian rocks to the south: hence it seems likely that they are pre-Caledonian in age. As they bear a superficial resemblance to the schists and gneisses of the Inishkea Division of NW Co. Mayo and the late to mid-Proterozoic Moine Assemblage of Scotland, they are thought likely to be late Proterozoic schists, unless new data suggest otherwise.

12.5 Other Proterozoic inliers in NW Ireland

The small bodies of schist lying within the Westport Inlier, associated with a major fault zone bounding the south side of Clew Bay, and the schist septum within the Tyrone Inlier are presently of unknown age and affinity (Max and Long, 1985). Although it is most likely that they are schists formed during the Caledonian orogeny, it is also possible that they represent older Proterozoic rocks. Little is known about them and they are not discussed here.

References

Aftalion, M. and Max, M. D. (1987) U–Pb zircon geochronology from the Precambrian Annagh Division gneisses and the Termon Granite, NW Co. Mayo, Ireland. *J. geol. Soc. London* **144**, 401–406.

Anderson, J. G. C. (1948) The occurrence of Moinian rocks in Ireland. *Q. J. geol. Soc. London* **103**, 171–188.

Blaxland, A. B., van Breemen, O., Emeleus, C. H. and Anderson, J. G. (1978) Age and origin of the major syenite centres in the Gardar Province of South Greenland: Rb–Sr studies. *Bull. Geol. Soc. Am.* **89**, 231–244.

Brindley, J. C. (1969) Caledonian and Pre-Caledonian intrusive rocks of Ireland. In Kay, M. (ed.), North Atlantic—Geology and Continental Drift. *Am. Ass. Pet. Geol. Mem.* **12**, 336–353.

Church, W. R. (1969) Metamorphic rocks of Burlington Peninsula and adjoining areas of Newfoundland, and their bearing on continental drift in the North Atlantic. In Kay, M. (ed.) North Atlantic Geology and Continental Drift. *Am. Ass. Pet. Geol. Mem.* **12**, 212–233.

Collerson, K. D. (1982) Geochemistry and Rb–Sr geochronology of associated Proterozoic peralkaline and subalkaline anorogenic granites from Labrador. *Contrib. Miner. Pet.* **81**, 126–147.

Crow, M. J. (1974) Geology of Metamorphic Rocks in Part of NW Co. Mayo, Ireland. Unpublished University of Dublin Ph.D. Thesis.

Crow, M. J. and Max, M. D. (1976) The Kinrovar Schist. *Sci. Proc. R. Dublin Soc.* **5A**, 429–441.

Crow, M. J., Max, M. D. and Sutton, J. S. (1971) Structure and stratigraphy of the metamorphic rocks in part of NW Co. Mayo, Ireland. *J. geol. Soc. London* **127**, 579–585.

Currie, K. L. (1985) An unusual peralkaline granite near Lac Brisson, Quebec–Labrador. In Current Research Part A. *Geol. Surv. Can., Paper* **85-1A**, 73–80.

Emeleus, C. H. and Upton, B. G. J. (1976) The Gardar Period in southern Greenland. In Escher, A. and Watt, W. S. (eds.), *Geology of Greenland.* Grφnlands Geologiske Undersφgelse, Copenhagen, 153–181.

Emslie, R. F. (1978*a*) Anorthosite massifs, rapakivi granites and late Proterozoic rifting of North America. *Precamb. Res.* **7**, 61–98.

Emslie, R. F. (1978*b*) Elsonian magmatism in Labrador: age, characteristics and tectonic setting. *Can. J. Earth Sci.* **15**, 438–453.

Evans, D. (1983) Malin Sheet 55°N–08°NW 1:250 000 Series. *Inst. Geol. Sci. Map.*

Evans, D., Kenalty, N., Dobson, M. R. and Whittington, R. J. (1979) The geology of the Malin Sea. *Inst. Geol. Sci. Rep.* **79/15.**

Gray, J. R. (1981) Regional Metamorphism in NW Mayo, Eire, and its Bearing on the Regional Geology. Unpublished Ph.D. Thesis, University of East Anglia.

Hill, J. D. (1981) Geology of the Flowers River area, Labrador. *Miner. Devel. Div., Newfoundland Dept. Mines & Energy, Rep.* **81–6**, 40 pp.

Hill, J. D. (1982) Geology of the Flowers River–Notakwanon River area, Labrador. *Miner. Devel. Div., Newfoundland Dept. Mines & Energy Rep.*, **82–6**, 140 pp.

Hill, J. D. and Thomas, A. (1983) Correlation of two Helikian peralkaline volcanic centres in central Labrador. *Can. J. Earth Sci.* **20**, 753–763.

Hull, E. (1881*a*). On the Laurentian beds of Donegal and other parts of Ireland. *Geol. Mag.* **8**, 506–507.

Hull, E. (1881*b*) On the Laurentian rocks of Donegal and of other parts of Ireland. *Trans. R. Dublin Soc.* **1**, 244–256.

Kelly, T. J. and Max, M. D. (1979) The geology of the northern part of the Murrisk Trough. *Proc. R. Ir. Acad.* **79B**, 191–206.

Kennedy, M. J. (1969) The structure and stratigraphy of the Dalradian rocks of north Achill Island, Co. Mayo, Ireland. *Q. J. geol. Soc. London* **125**, 47–81.

Kilroe, J. R. (1907) The Silurian and metamorphic rocks of Mayo and north Galway. *Proc. R. Ir. Acad.* **10B**, 129–160.

Kinahan, G. H. (1882) Jukes and the supposed Laurentian rocks in Donegal, Ireland. *Geol. Mag.* **9**, 46–47.

Kinahan, G. H. (1886) On metamorphic rocks. *Geol. Mag.* **12**, 7–10.

Lambert, R. StJ. (1969) Isotope studies relating to the Precambrian history of the Moinian of Scotland. *Proc. geol. Soc. London* **1652**, 243–245.

Lemon, G. G. (1952) The Moinian inlier of Rosses Point, near Sligo, Eire. *Geol. Mag.* **89**, 124–128.

Lemon, G. G. (1971) The Precambrian rocks of the NE Ox Mountains, Eire. *Geol. Mag.* **108**, 193–200.

Long, C. B., Max, M. D. and Yardley, B. W. D. (1983) Compilation metamorphic map of Ireland. In Shenck, P. E. (ed.), Regional Trends in the Geology of the Appalachian–Caledonian–Hercynian–Mauretanide Orogen. *NATO ASI Series C* **116**, 221–233.

Long, C. B., Max, M. D. and O'Connor, P. J. (1984) Age of the Lough Talt and Easky Adamellite in the central Ox Mountains, NW Ireland, and their structural significance. *Geol. J.* **19**, 389–397.

MacIntyre, R. M., van Breemen, O., Bowes, D. R. and Hopgood, A. M. (1975) Isotopic study of the gneiss complex, Inishtrahull, Co. Donegal. *Sci. Proc. R. Dublin Soc.* **5A**, 301–309.

Martin, A. R. (1985) The Evolution of the Tugtutoq–Illimausaq Dyke Swarm, SW Greenland. Unpublished Ph.D. Thesis, University of Edinburgh.

Max, M. D. (1970) Mainland gneisses southwest of Bangor in Erris, Co. Mayo, Ireland. *Sci. Proc. R. Dublin Soc.* **3A**, 275–291.

Max, M. D. (1972) A note on the stratigraphy and structure of Achill Island. *Geol. Surv. Ire. Bull.* **1**, 223–230.

Max, M. D. (1973) Caledonian metamorphism in part of NW Co. Mayo, Ireland. *Geol. J.* **8**, 375–386.

Max, M. D. and Long, C. B. (1985) Pre-Caledonian basement in Ireland and its cover relationships. *Geol. J.* **20**, 341–366.

Max, M. D. and Sonet, J. (1979) A Grenville age for pre-Caledonian rocks in NW Co. Mayo, Ireland. *J. geol. Soc. London* **136**, 379–382.

Max, M. D., O'Connor, P. J. and Long, C. B. (1984) New age data from the Pre-Caledonian basement of the NE Ox Mountains and Lough Derg inliers, Ireland. *Geol. Surv. Ire. Bull.* **3**, 203–382.

Molloy, M. A. and Sanders, I. S. (1983) The NE Ox Mountains inlier. In Archer, J. B. and Ryan P. D. (eds.), Geological Guide to the Caledonides of Western Ireland. *Geol. Surv. Ire., Guide Ser.* **4**, 52–55.

Pettijohn, F. J., Potter, P. E. and Siever, P. (1973) *Sand and Sandstone.* Springer-Verlag, New York.

Phillips, W. E. A., Kennedy, M. J. and Dunlop, G. M. (1969) Geologic comparison of western Ireland and northeastern Newfoundland. In Kay, M. (ed.), North Atlantic–Geology and Continental Drift. *Mem. Am. Ass. Pet. Geol.* **12**, 194–211.

Phillips, W. E. A., Taylor, W. E. G. and Sanders, I. S. (1975) An analysis of the geological history of the Ox Mountains Inlier. *Sci. Proc. R. Dublin Soc.* **5B**, 311–329.

Richey, J. E. and Kennedy, W. Q. (1939) The Moine and Sub-Moine Series of Morar, Inverness-shire. *Bull. Geol. Surv. Gt. Br.* **2**, 26–45.

Riddihough, R. P. and Max, M. D. (1976) A geological framework for the continental shelf to the west of Ireland. *Geol. J.* **11**, 109–120.

Roddick, C. and Max, M. D. (1983) A Laxfordian age from the Inishtrahull Platform, Co. Donegal, Ireland. *Scott. J. Geol.* **19**, 97–102.

Sanders, I. S. (1979) Observations on eclogite and granulite facies rocks in the basement of the Caledonides. In Harris, A. L., Holland, C. H. and Leake, B. E. (eds.), The Caledonides of the British Isles—Reviewed. *Geol. Soc. London, Spec. Publ.* **8**, 96–100.

Sanders, I. S., Daly, J. S. and Davies, G. R. (1987) Late Proterozoic high-pressure granulite facies metamorphism in the north-east Ox Mountains inlier, north-west Ireland. *J. metam. Geol.* **5**, 69–85.

Sutton, J. S. (1971) The Pre-Caledonian rocks of the Mullet Peninsula, Co. Mayo, Ireland. *Sci. Proc. R. Dublin Soc.* **4A**, 121–136.

Sutton, J. S. and Max, M. D. (1969) Gneisses in the north-west part of Co. Mayo, Ireland. *Geol. Mag.* **106**, 284–290.

Thomas, A. (1981) Geology along the southwestern margin of the Central Mineral Belt, Labrador. *Mine. Devel. Div. Newfoundland Dept. Mines & Energy, Rep.* **81–4**, 40 pp.

Trendall, A. F. and Elwell, R. W. D. (1963) The metamorphic rocks of northwest Mayo. *Proc. R. Ir. Acad.* **62B**, 217–247.

van Breemen, O., Halliday, A. N., Johnson, M. R. W. and Bowes, D. R. (1978) Crustal additions in late Precambrian times. In Bowes, D. R. and Leake, B. E. (eds.), Crustal Evolution in Northwestern Britain and Adjacent Regions. *Geol. J. Spec. Issue* **10**, 81–106.

Winchester, J. A. and Max, M. D. (1983) A note on the occurrence of stanniferous granitic gneiss in Co. Mayo. *Bull. geol. Surv. Ire.* **3**, 113–119.

Winchester, J. A. and Max, M. D. (1984) Geochemistry and origin of the Annagh Division of the Precambrian Erris Complex, NW Co. Mayo, Ireland. *Precambr. Res.* **25**, 397–414.

Winchester, J. A. and Max, M. D. (1987*a*) A displaced and metamorphosed peralkaline granite related to the late Proterozoic Labrador and Gardar suites: the Doolough Granite of Co. Mayo, NW Ireland. *Can. J. Earth. Sci.* **24**, 631–642.

Winchester, J. A. and Max, M. D. (1987*b*) The Pre-Caledonian Inishkea Division of NW Co. Mayo, Ireland: its geochemistry and probable stratigraphic position. *Geol. J.* **22** (in press)

Yardley, B. W. D., Long, C. B. and Max, M. D. (1979) Patterns of metamorphism in the Ox Mountains and adjacent parts of western Ireland. In Harris, A. L., Holland, C. H. and Leake, B. E. (eds.). The Caledonides of the British Isles—Reviewed. *Geol. Soc. London Spec. Publ.* **8**, 369–374.

Young, D. G. G. (1974) The Donegal Granite—a gravity analysis. *Proc. R. Ir. Acad.* **74B**, 63–73.

13
The Grampian Group, Scotland

J. A. WINCHESTER and B. W. GLOVER

13.1 Introduction

Rocks now classified as part of the Grampian Group (Harris *et al.*, 1978) crop out extensively in the Central Highlands of Scotland (Fig. 13.1). Formerly referred to collectively as the 'Central Highland Granulites' before 1978, these rocks were assigned to the Moine Assemblage, which they closely resemble both in lithology and geochemistry. However, recent mapping and isotopic work has revealed that a gneissose assemblage of rocks occurring near Laggan, Kincraig, and north of Carrbridge (Fig. 13.1) appears to be older than, and has been reported to be unconformably overlain by dominantly psammitic rocks currently assigned the Grampian Group (Piasecki and van Breemen, 1979*a*, 1979*b*; Piasecki, 1980). If so, the areas in which these gneissose rocks occur may be interpreted as inliers. In some recent references the term 'Grampian Division' has been applied to the Grampian Group, while yet other papers have used the term 'Younger Moine'. The earlier assemblage, termed by Piasecki (1980) the 'Central Highland Division', has yielded isotopic ages similar to those obtained from Moine Assemblage rocks which crop out west of the Great Glen Fault (Brewer *et al.*, 1976; Brook *et al.*, 1977; Piasecki and van Breemen, 1979*a*). In addition the Central Highland Division contains paragneiss lithologies and minor amphibolite pods, which texturally, mineralogically and geochemically mirror those of the Glenfinnan Division of the Moine Assemblage. Hence it is the Central Highland Division migmatitic gneisses which appear to represent the Moine Assemblage east of the Great Glen and the Grampian Group rocks thus appear to form a cover sequence to the Moine Assemblage in the Grampian Highlands.

The contact between the Grampian Group rocks and the Central Highland Division is usually obscured by a persistent high-strain zone termed by Piasecki (1980) the Grampian Slide. However, in a few isolated locations an unconformable relationship is claimed (Piasecki, 1984), although persistent high strain makes it hard to prove. No evidence has been produced to suggest that the Grampian Group rocks were subjected to the *c.* 1000 Ma tectonometamorphic event which

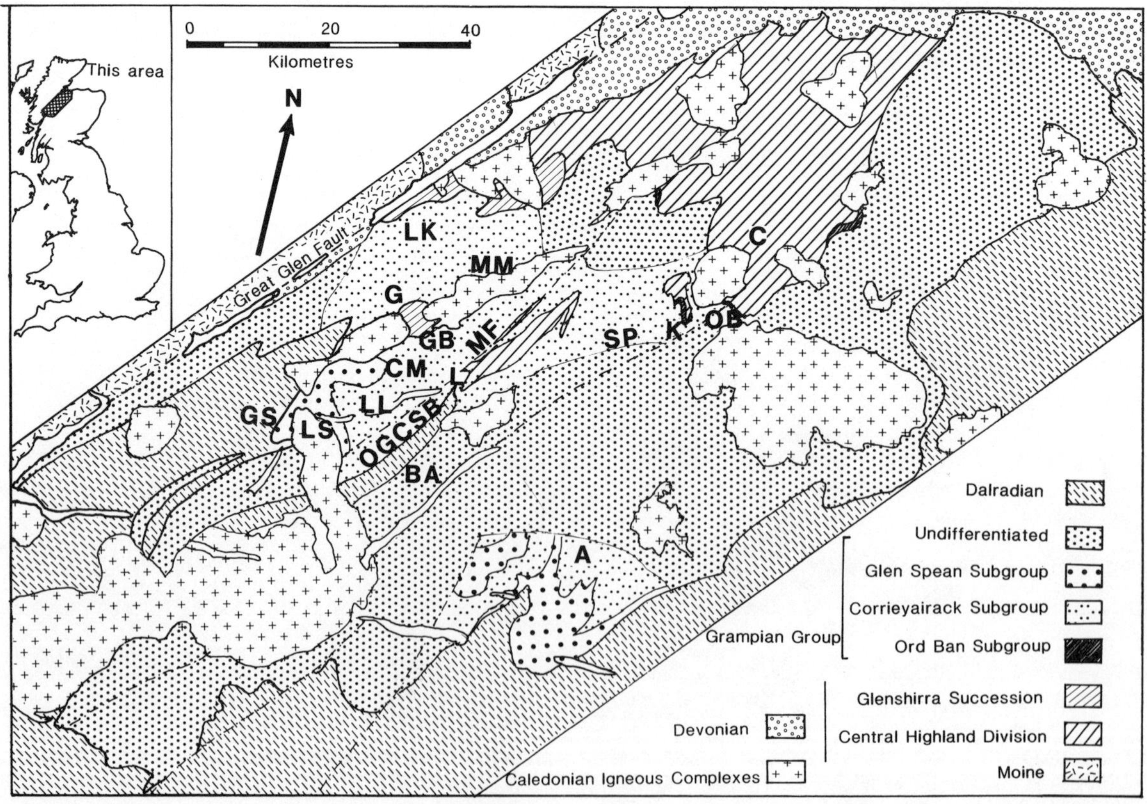

Figure 13.1 Location map showing the extent of the Grampian Group in Scotland. *A*—Atholl; *BA*—Ben Alder; *C*—Carrbridge; *CM*—Creag Meagaidh; *G*—Gairbeinn; *GB*—Garva Bridge; *GS*—Glen Spean; *K*—Kincraig; *LK*—Loch Killin; *LL*—Loch Laggan; *LS*—Loch Spean; *MF*—Markie Fault; *MM*—Monadhliath Mountains; *OB*—Ord Ban; *OGCSB*—Ossian–Geal Charn Steep Belt; *SP*—Speyside.

affected the Moine Assemblage, and, if deposition of the Grampian Group post-dated this event, its basal formations were apparently desposited on an already metamorphosed Moine basement. Because of this relationship, and because the Grampian Group rocks pass up with apparent continuity into the Appin Group of the Dalradian Supergroup, Harris *et al.* (1978) proposed the name 'Grampian Group' and suggested that the rocks so named should be considered as the lowest group of the Dalradian Supergroup, notwithstanding their chemical and lithological differences from the rest of the Supergroup. Whether the Grampian Group rocks should be included as a part of the Dalradian Supergroup is currently being debated. While the Grampian Group appears to share a common structural and metamorphic history with overlying Dalradian rocks (favouring its inclusion within the Dalradian Supergroup, as preferred by Harris *et al.*, 1978), its lithological and geochemical differences from Dalradian rocks argue that it should be classified separately. These differences have been clearly recorded by authors noting the 'Moine-like' nature of the Grampian Group (Lambert *et al.*, 1982; Plant *et al.*, 1984) and a recent review of the British Caledonides (Kelling *et al.*, 1986) has omitted the Grampian Group from the Dalradian Supergroup. It therefore seems most fitting to divide the Caledonides of the Scottish Highlands into three major rock sequences: the Moine Assemblage, incorporating all metasediments affected by a *c.* 1000 Ma metamorphic event; the Grampian Group, embracing all the metasediments lithologically similar to the Moine, but deposited after the 1000 Ma event; and the Dalradian Supergroup incorporating all the varied and very different metasediments laid down in the last part of the Proterozoic.

The outcrop area of the Grampian Group is bisected by the NE-trending Ossian–Geal Charn Steep Belt (Thomas, 1979, 1980), which was formerly interpreted as the root zone of the major Caledonian nappes in the Scottish Highlands. However, more recent mapping on Speyside has not confirmed this interpretation, and has suggested instead that the major lithological formations may be traced through this steep zone from west to east. For this reason a tentative correlation of all the recently mapped Grampian Group rocks is now possible.

In this study the lithotype, sedimentary environment, distribution and thickness of the main formations so far mapped within the Grampian Group will be discussed in turn. In many areas the detailed stratigraphy remains to be worked out, especially in the east and north-east, but, on the basis of recent work in the Monadhliath Mountains, the Creag Meagaidh Range and upper Glen Spean, a stratigraphic model has been established which may be applicable elsewhere in the Grampian Group. It is this stratigraphic framework which is described here.

13.2 Stratigraphic Framework

Detailed stratigraphic work in the Grampian Group has only been undertaken recently. The main areas studied have been separted either by major faults or by unmapped ground, and therefore correlation between them has proved difficult. However, enough is now known to set up an outline stratigraphic model.

In the west, mapping in the Monadhliath Mountains and the Creag Meagaidh Range enabled the recognition of two distinct stratigraphic successions (Haselock, 1982; Haselock *et al.*, 1982; Okonkwo, 1985). Neither of these successions displayed the pervasive migmatization so characteristic of the Central Highland Division and appeared to share some lithological, sedimentological and geochemical characteristics, and so they were initially classified as separate successions within the Grampian Group.

However, recent work by Haselock (pers. comm.) has suggested that the lowest structural rocks in the western area of the Grampian Group, termed the Glenshirra succession, (Haselock *et al.*, 1982), contain a variety of lithologies, which have since been mapped south of

Table 13.1 A simplified stratigraphic correlation chart for the Grampian Group.

Subgroup	West	Speyside	East
Glen Spean	Inverlair Formation Clachaig Formation		Strath Tummel succession
	Creag Meagaidh Formation		Drumochter succession
	Monadhliath Schist		
Corrieyairack	Glen Doe Psammite	Glen Markie Psammite	
	Coire nan Laogh Semi-pelite ~~~~ Slide ~~~~	Pattack Schist	
Ord Ban		Kincraig Limestone and Quartzite succession (thin) ~~~~ Slide ~~~~	

Inverness as a part of the Central Highland Division which was subjected to less pervasive migmatization. In the central Monadhliath Mountains, rocks assigned to this succession are restricted in distribution to parts of the area between the Great Glen and Markie Faults, and they may thus constitute separate inliers of Central Highland Division rocks (Fig. 13.1). Recent mapping, both on Speyside (Piasecki, Temperley, pers. comm.) and in the Glen Spean–Loch Laggan area, suggests that the Grampian Group is divisible into three unequal subgroups, each characterized by different sedimentary environments (Table 13.1).

Sporadically developed in Speyside at the base of the Grampian Group is a thin series of quartzites, impure marble and pelitic schist which is so distinct from the overlying rocks that it is here distinguished as a separate subgroup. The type location is at Ord Ban, 4 km S of Aviemore.

The overlying Corrieyairack Subgroup, formerly referred to as the Corrieyairack succession (Haselock *et al.*, 1982), usually rests directly upon the Central Highland Division, although it also overlies the Ord Ban Subgroup where the latter is present. The Corrieyairack Subgroup comprises the bulk of the Grampian Group, and consists dominantly of enormously thick psammitic formations which were probably turbiditic and which display geochemical compositions typical of greywackes.

The topmost part of the Grampian Group consists of a series of psammitic and semi-pelitic rocks which preserve sedimentary structures indicative of a shallowing environment, and which may consist largely of fluviodeltaic deposits. Preliminary geochemical studies indicate a change in composition consistent with the change in environment. The entire series is well exposed in Glen Spean, and is therefore termed here the Glen Spean Subgroup. Rocks which display similar sedimentary structures and lithologies also occur in the Ben Alder area and in Atholl (Thomas, 1979, 1980), where they also form the local top of the Grampian Group.

13.2.1 *Glenshirra succession*

The stratigraphic affinities of the Glenshirra succession are not currently resolved, and therefore a brief description of its constituent lithotypes is included here. The Glenshirra succession crops out in two principal areas: in the Glenshirra Dome, and between Fort Augustus and Dumnaglass on the SE side of the Great Glen (Fig. 13.1). Poor exposure and the location of the Foyers intrusion prevent the full thickness of this succession from being exposed in either of these areas, while the correlation between them is based on lithological similarity and their shared structural level within the local stratigraphic succession.

The Glenshirra succession is distinctive in containing a high proportion of coarse clastic deposits for which an alluvial origin has been postulated (Haselock, 1982; Okonkwo, 1985). Rapid lateral facies changes are characteristic of this succession, which was divided into four formations within the Glenshirra Dome north of the River Spey (Haselock *et al.*, 1982). However, this division is not readily applicable to the SE near Garva Bridge, where the succession remains divisible on lithological and geochemical grounds into an upper and lower portion only. The lower portion, comprising the Creag Mhor and Carn Dearg psammites (Haselock *et al.*, 1982) and the Chathalain Formation (Okonkwo, 1985) consists in part of rhythmic units, fining up on a scale of several metres, dominated by psammite. Trough and tabular cross-bedding are occasionally preserved in these psammites, which are often overlain by thin-bedded, more micaceous psammites, exhibiting parallel bedding and occasional laterally impersistent semi-pelitic units in repeated fining-upwards sequences, considered by Okonkwo (1985) to be characteristic of deposition in possible fluviodeltaic channels. However, Haselock (1982) recorded dominantly parallel-bedded psammites from the northern part of the Glenshirra Dome at a comparable stratigraphic level, and suggested a shallow-marine origin for them, while admitting the scarcity of positive lithological evidence.

In the upper part of the Glenshirra Subgroup, pebbly beds become increasingly abundant, especially towards the NW. East of Garva Bridge, pebbly beds are scarce and tend to show a preserved 'gritty' texture where they are present, while their thickness is so reduced that they no longer form a distinctive formation. On Gairbeinn (Fig. 13.1) 700 metres of pebbly psammites were recorded, and were interpreted as braided stream deposits (Haselock, 1982). Within this formation, clasts range up to 6 cm in diameter (where relatively undeformed) and occur within graded units ranging up to 20 cm in thickness. These graded units often have sharp, possibly erosional bases and may grade up into semi-pelitic schists representing the finer silty portion of the unit. Essentially similar relationships can also be observed NE of Fort Augustus, where Parson (1982) recorded a thickness of metaconglomerates exceeding 2000 metres. These were also interpreted as braided stream deposits. In this area, clasts may often exceed 6 cm in diameter, suggesting greater proximity to the sediment source than the Glenshirra Dome area.

Rocks akin to the Glenshirra succession are only recorded in the NW part of the Central Highlands. At Dumnaglass, migmatized gneisses of the Central Highland Division have been tectonically emplaced along a ductile thrust above psammitic and pebbly rocks akin to the Glenshirra succession. If both groups of rock are in future to be incorporated within the Central Highland Division, the ductile thrust at Dumnaglass forms a major tectonic break within the Central Highland Division similar to the Sgurr Beag Thrust in the Northern Highlands. As in the Northern Highlands, the lower nappe, exposed to the west, is characterized by less migmatized rocks which appear to have been subjected to lower metamorphic grade before the Caledonian metamorphic overprinting during the early Palaeozoic. If this analogy is correct, the Glenshirra succession may occur in a structural and stratigraphic position similar to that of the Morar Division in the Northern Highlands. Until more information is available the implications of such observation remain speculative.

13.2.2 *Ord Ban Subgroup*

In scattered localities on Speyside, a thin sequence of distinctive rocks is present below the lowest formation of the Corrieyairack Subgroup. Since it incorporates such distinctive lithologies, this sequence has been termed the Ord Ban Subgroup; it consists of thin quartzite, pods of impure marble and pelitic schists which are often highly aluminous and locally contain kyanite. The entire sequence rarely exceeds 200 metres in thickness and is frequently absent from the base of the Grampian Group succession. It has been mapped near Grantown on Spey, Kyllachy, Ord Ban and Kincraig on Speyside (Piasecki, 1980) (Fig. 13.1). Piasecki also states that at Kincraig a gradational contact between the Ord Ban Subgroup and the overlying striped semi-pelite is visible. At the base of the succession fine-grained whitish quartzites occur. They are overlain by, and partly interfinger with, lenticular and laterally impersistent bodies of impure diopsidic marble, which are in turn overlain by a coarse pelitic schist, which on Ord Ban reaches 300 metres in thickness. At this locality it consists of graphitic kyanite-tourmaline schist, within which thin psammitic units may be present. On Speyside a thick, foliated garnet-amphibolite sill, of tholeiitic composition, is frequently associated with this succession. The Kinlochlaggan Limestone (Treagus, 1969) shows some similarities with the Ord Ban Subgroup, but is also associated with a boulder bed, which has been interpreted as a tillite, and used as evidence that the associated rocks should be equated with the Appin Group. Its age, therefore, and whether it belongs to the Ord Ban Subgroup, remains under discussion.

However, because the succession associated with the Kinlochlaggan Limestone is everywhere in tectonic contact with Grampian Group rocks, its true stratigraphic position will be hard to resolve. Because the Ord Ban Subgroup appears to be overlain conformably by Corrieyairack Subgroup rocks, it can only be interpreted on existing evidence as a locally developed basal sequence to that subgroup. However, it may be significant that the Ord Ban Subgroup is nowhere seen west of the Markie Fault. If (as the basal sequence within the Grampian Group) the Ord Ban Subgroup records the earliest development of the depositional basin, it may simply have been laid down in the deepest part only. If so, the wider distribution of the overlying Corrieyairack Subgroup rocks may in turn record a considerable extension of the basin towards the west and south as it developed.

13.2.3 *Corrieyairack Subgroup*

This subgroup encompasses the formations which overlie the Glenshirra Succession west of the Markie Fault, or the Central Highland Division basement or the Ord Ban Subgroup east of the fault. The subgroup incorporates the Monadhliath Schist of Killin (Haselock *et al.*, 1982; Haselock and Winchester, 1981) and the underlying psammitic and semi-pelitic formations. Within the psammites which overlie the Monadhliath Schist there occurs both a change in lithology and geochemistry, and for this reason the upper psammites are assigned to a separate subgroup. The formations which comprise the Corrieyairack Subgroup have already been described from the Corrieyairack area (Haselock *et al.*, 1982). However, additional mapping in the Creag Meagaidh Range (Okonkwo, 1985) has clarified some of the lateral facies changes that occur within the subgroup, while a tentative correlation with Grampian Group rocks mapped on Speyside may also be made. The Corrieyairack Subgroup may be divided into four formations.

13.2.3.1 *Coire nan Laogh semi-pelite* West of the Markie Fault this formation is traceable in two areas: it describes a ring around the Glenshirra Dome, cut by Caledonian granitic rocks. It is exposed on the NW slope of Gairbeinn (NN 460985) and on the south side of Upper Strath Spey between Dirc Beannain Beaga (NN 460930) and Meall an Domhnaich (NN 540960). North of Loch Killin this formation also appears east of the River Fechlin (Fig. 13.1). In the exposures in the SE this formation may be divided into two principal lithofacies: a lower semi-pelitic member and an upper rhythmic member, both described below.

(i) The semi-pelite locally displays small-scale quartzofeldspathic segregations, preserves no original sedimentary structures, and is lithologically uniform in the lower part. At higher levels the appearance of thin psammitic beds represents a transition into (ii).

(ii) Rhythmically interbedded semi-pelite and micaceous psammite, which consists of units not exceeding 20 cm in thickness. Within these units gradational increase in biotite content may reflect original sedimentary grading. Some of the psammitic units are calcareous and may grade laterally into thin calc-silicate bands of probable diagenetic origin. This lithofacies, referred to by Okonkwo (1985) as the 'Transitional Semi-Psammite' (Fig. 13.2) closely resembles that of the 'basal rhythmite' formation of the Grampian Group, recorded on Speyside east of the Markie Fault. Correlation between these formations seems likely, although firm evidence for the validity of this correlation has not yet been obtained. Towards the top of the formation, the micaceous psammite units become dominant in a lithofacies which was interpreted by Okonkwo (1985) as a transition up into the Glen Doe Formation.

13.2.3.2 *Glen Doe Formation* By far the greatest portion of the Corrieyairack Subgroup consists of the Glen Doe and Knockchoilum Psammites, now considered to be lateral equivalents, which were originally distinguished according to the type of calc-silicate pod present (Whittles, 1981). In the Loch Killin area, the Glen Doe

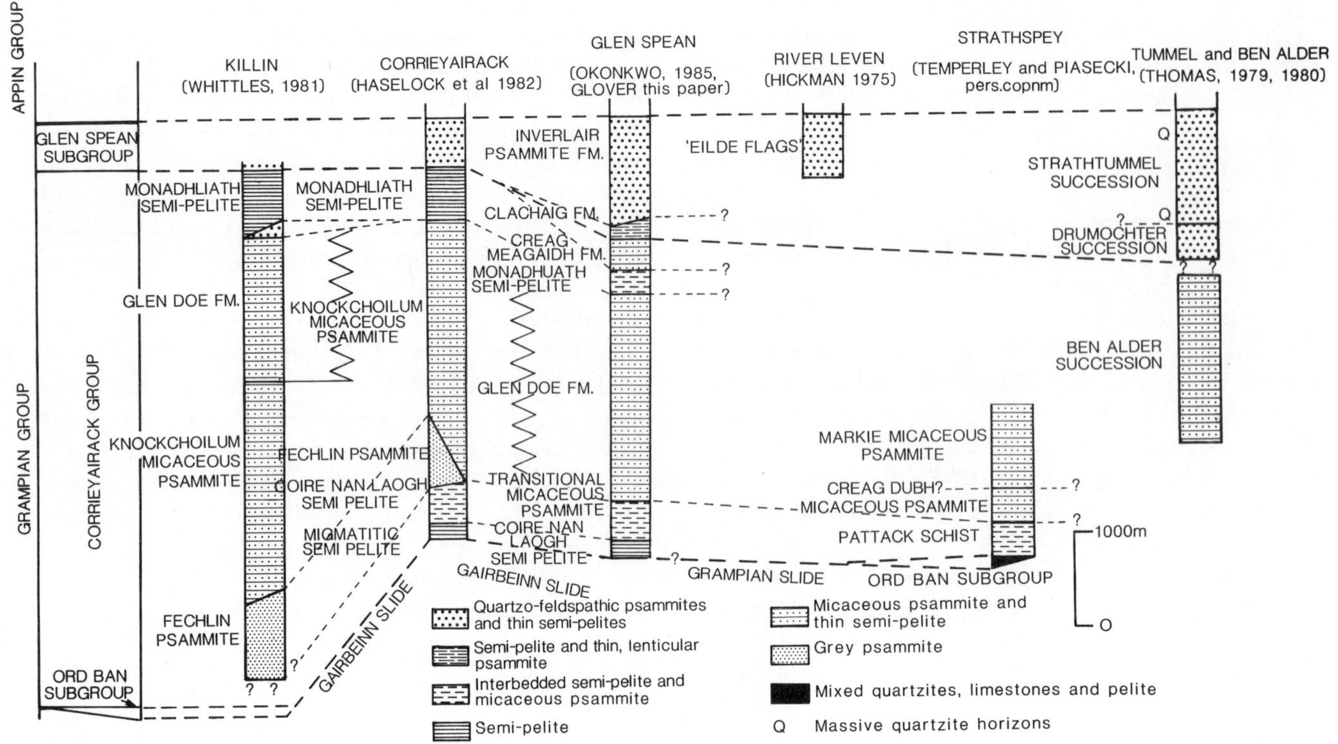

Figure 13.2 Chart of likely stratigraphic correlations across the Grampian Group.

Psammite, which contains whitish clinozoisite-bearing calc-silicates (typical of those widely developed in both the Grampian Group and the Moine) overlies the Knockchoilum Psammite which is characterized by the sporadic occurrence of greenish actinolitic calc-silicate pods. Further west, around the Corrieyairack Pass, the Knockchoilum Psammite with greenish calc-silicates is dominant (Haselock, 1982), while to the south and east, in the Creag Meagaidh Range, no greenish calc-silicates have been recorded, and the presence of white clinozoisite-bearing calc-silicates classifies the entire psammite succession as Glen Doe Psammite.

Locally developed at the base of the Glen Doe Formation is the Fechlin Psammitic Member, which was previously described from the River Fechlin section (Whittles, 1981; Haselock *et al.*, 1982). It is restricted to the north-western part of the area, and has not been recorded in the Creag Meagaidh Range. Within the Glen Doe Formation it includes the only sequence of quartzose psammitic rocks, which form graded sequences with sharp, possibly erosive bases. Well-developed loading is preserved in the River Fechlin section [NH 495 144] (Fig. 13.3).

Towards the SE there is an increase of thickness, from 1200 metres near Corrieyairack, to a thickness possibly exceeding 3000 metres near Loch Laggan (Fig. 13.2). Within the Glen Doe Formation, in the latter area, four principal lithofacies exist:

(i) Graded micaceous psammites, which occur in laterally persistent beds ranging 10–100 cm in

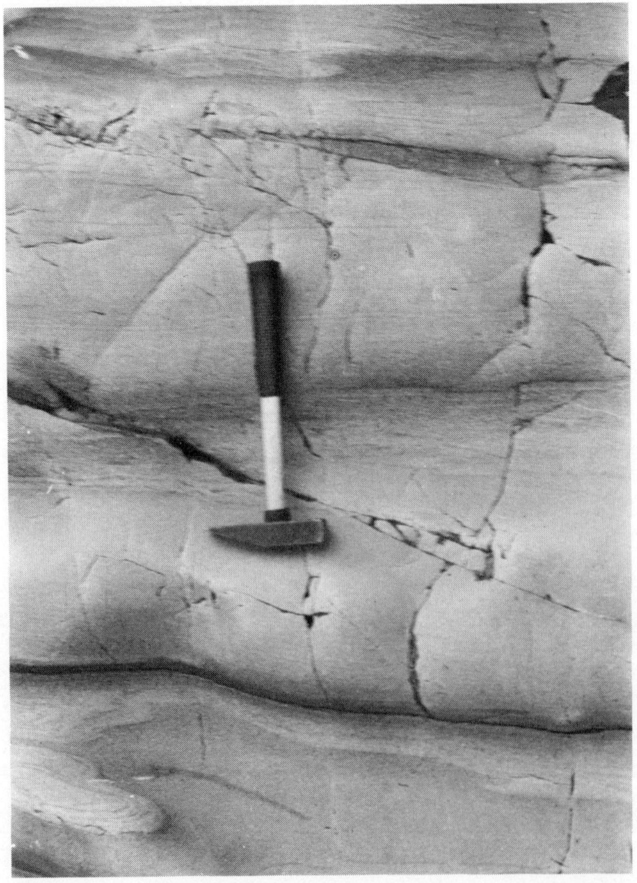

Fig. 13.3 Graded units and load casting in the Fechlin Psammite, River Fechlin.

Figure 13.4 Turbidites within the Glen Doe Formation at Rubha na Magach, showing Bouma a, c, and dewatering structure.

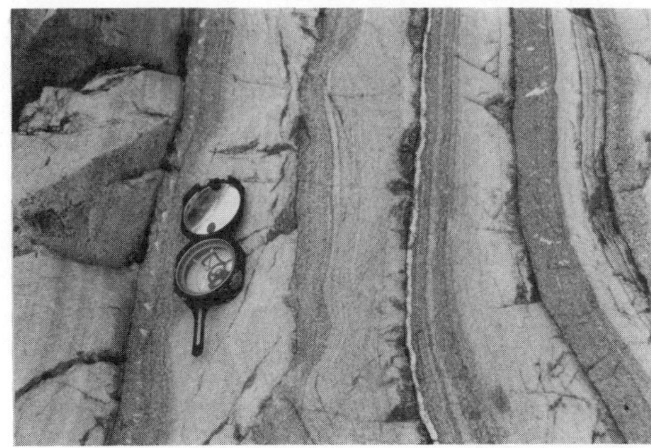

Figure 13.5 Graded units within the uppermost part of the Glen Doe Formation, Allt nam Beith.

thickness. Many of these graded units exhibit the classical Bouma sequence, which is particularly well displayed at Rubha na Magach on the shore of Loch Laggan [NN 460 849] (Fig. 13.4) where the units are interbedded with thinner dark micaceous psammites (1–15 cm thick) and semi-pelite couplets. In association with these are numerous examples of loading, convolute lamination and, more rarely, mud-flake breccias.

(ii) Cross-laminated micaceous psammite occurs in beds associated with the graded micaceous psammites. Ripple-drift cross-lamination is occasionally developed, and the foreset dips are unidirectional, indicating sediment transport from the SW. The current-rippled units may also show convolute lamination and water-escape structures, and climbing-ripple laminations may also be present (Okonkwo, 1985).

(iii) Massive micaceous psammites occur as units with sharp bases, ranging in thickness from 30 to 50 cm. Commonly associated with graded micaceous psammites, these units show little obvious internal structure, although X-radiography has shown a variety of sedimentary structures to be present.

(iv) Wave-rippled micaceous psammites, which were only observed in the Allt Crunachdan, 3 km N of the E end of Loch Laggan, in association with massive micaceous psammites.

Occurring locally at the top of the Glen Doe Formation is a series of striped micaceous psammites and semi-pelites, with thin lenticular orthoquartzites sporadically developed. Recorded in the River Tarff (Haselock, 1982) and NE of Creag Meagaidh (Okonkwo, 1985), this impersistent lithofacies association may contain graded beds ranging up to 1 metre in thickness (Fig. 13.5). Planar lamination and rarer trough cross-bedding are the dominant sedimentary structures found in the dark micaceous psammites, while the quartzites may either be massive or display planar bedding. Whittles (1981) recorded a pinkish tinge to some of the quartzites, resulting from the presence of disseminated hematite. North of the River Spey, thinner graded units are also present (Fig. 13.5).

13.2.3.3 *Monadhliath Formation* With an increase in the proportion of semi-pelite present, the Glen Doe Formation is succeeded by the Monadhliath Formation, previously referred to (together with numerous semi-pelitic rocks which detailed mapping has shown to be unrelated) as the Monadhliath Schist (Anderson, 1948, 1956). This formation shows only slight variations in thickness, and is traceable with only minor discontinuities of outcrop (Haselock *et al.*, 1982) from E of Loch Killin to the ground S of Loch Spean, a distance along strike of over 40 km. It is therefore a useful stratigraphic marker formation. Dominantly semi-pelitic in the north and north-east, where it contains numerous well-developed whitish calc-silicate bands, it becomes significantly less semi-pelitic towards the south. Calc-silicate bands are abundant within the formation in the north-east and south-east, but they become progressively scarcer towards the north-west, and are entirely missing in the area west of the Corrieyairack Pass.

In the Loch Spean area, the Monadhliath Formation displays four main lithofacies:

(i) Semi-pelite and pelite units, occurring in laterally persistent beds, ranging in thickness from fine laminae to beds several metres thick

(ii) Thin-bedded semi-pelite and micaceous psammite beds, occurring as couplets. These couplets rarely exceed 10 cm in thickness and are commonly normally graded and flat bedded

(iii) Massive micaceous psammites, which are laterally persistent and which consist of a number of amalgamated turbidites. Their bases are locally erosive and may occupy channels up to 2 metres deep

(iv) Massive quartzites, up to several metres thick, but laterally impersistent.

In the Loch Killin area the semi-pelitic lithofacies is dominant, although quartzites occur near the base of the Formation. South of the Corrieyairack Granodiorite the semi-pelitic lithofacies becomes interspersed with massive micaceous psammite and thin-bedded semi-pelites, and the relative scarcity of the semi-pelitic lithofacies makes the Monadhliath Formation more difficult to recognize in the south.

This lithofacies distribution suggests that slower rates of sedimentation prevailed in the N and NE where the semi-pelite is most abundant, while a greater clastic input was occurring to the S and SW. In association with this enhanced clastic input is a thinning of the Monadhliath Formation and the appearance of an overlying turbiditic formation in the SW (Creag Meagaidh Formation). It is highly probable that the contact between the two Formations is diachronous and that the overlying Creag Meagaidh Formation may be, in part, a lateral time-equivalent of part of the Monadhliath Formation.

13.2.3.4 *Creag Meagaidh Formation* The Creag Meagaidh Formation is traceable for 10 km along strike, from the SE margin of the Corrieyairack Granodiorite southwards to the shores of Loch Spean. To the NE the Formation progressively loses its distinctiveness and merges laterally into the Monadhliath Formation. The lithofacies recognized include:

(i) Thin-bedded micaceous psammites. The beds range in thickness from 20 cm to a few mm, and are often separated by persistent thin pelitic laminae which impart a flaggy appearance to the rock that is very characteristic of the Creag Meagaidh Formation. Normal grading is occasionally seen, but sharp tops to the micaceous psammites are more common. The thinner beds tend to be lenticular, with base profiles being concave upwards.

(ii) Interbedded semi-pelites and micaceous psammites resemble the previous lithofacies, and differ only in including thicker semi-pelite beds. The micaceous psammite units may exhibit planar lamination, normal grading and rare low-angle scouring. Semi-pelitic beds are laterally persistent over several metres, although rarely more than a few cm thick. They usually display sharp contacts.

(iii) 'Gritty' turbidites. These rocks exhibit a superficially 'coarse'-grained appearance, which is produced by the presence of quartz-muscovite aggregates, which do not reflect original grain size (Fig. 13.6). The turbiditic psammites characteristically exhibit Bouma a, b, d, e units which occur in stacked units up to 5 metres thick, sometimes exhibiting erosive bases. Several stacked units exhibit thinning and fining upwards, possibly reflecting their deposition within a submarine channel during a phase of abandonment.

(iv) Thick-bedded turbidites are represented as stacked, parallel-sided micaceous psammites. The stacked units are up to 8 metres thick and consist of numerous parallel-sided turbiditic beds exhibiting Bouma a, b divisions.

These lithofacies are distributed between three members, within the Creag Meagaidh Formation, and the lithofacies associations distinguish these members. On the eastern slopes of Beinn a Chaoruinn, stacked gritty turbidite units are separated by several metres of interbedded semi-pelite and micaceous psammite and together typify the Uamha Member. By contrast, to the south the gritty turbidite facies disappears and the thin-bedded, flaggy micaceous psammite lithofacies becomes dominant, constituting the Quarry Member.

Figure 13.6 Thin turbiditic units within the Creag Meagaidh Formation.

The spatial relationships between these lithofacies are clearly significant clues to the depositional environment of the Creag Meagaidh Formation. The lateral variation suggests that the more coarsely clastic deposition was mainly confined to submarine channel system, while the flaggy micaceous psammites reflect more 'distal' sedimentation. As sediment supply dominantly seems to be derived from the W and S, the relative locations of the different lithofacies are consistent with deposition in a submarine fan system, which was characterized by frequent channel abandonment and by avulsion occurring throughout sedimentation.

13.2.4 *Glen Spean Subgroup*

Rocks assigned to this subgroup everywhere overlie the Corrieyairack Subgroup conformably. Originally termed collectively the Carn Leac Psammite (Haselock *et al.*, 1982), they are distinguished from the underlying rocks by the presence of lithofacies associations, sedimentary structures and, where checked, geochemistry consistent with their deposition in a shallowing environment. Two formations are recognizable within this subgroup in the Glen Spean area: a thin lower formation, and a thicker, dominantly psammitic formation above, which is overlain by the Appin Group.

13.2.4.1 *Clachaig Formation* This formation does not exceed 200 metres in total thickness in the type area north of the Ossian Granite. It comprises three main lithofacies:

(i) Thin-bedded psammites and semi-pelites form the bulk of the formation (Fig. 13.7). The psammite beds, which are less than 10 cm thick, are separated by thinner semi-pelitic units. Planar laminations and, less commonly, ripple laminations may be developed within the psammitic beds.

(ii) Lenticular psammites, consisting of cross-

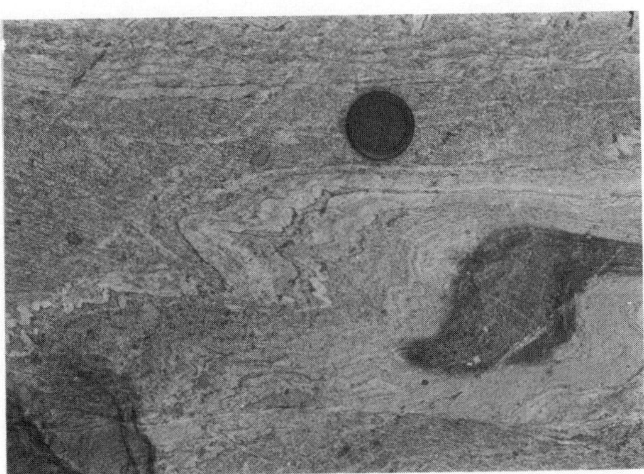

Figure 13.7 Thin semi-pelitic and psammitic beds within the Clachaig Formation, deformed by D2 folds.

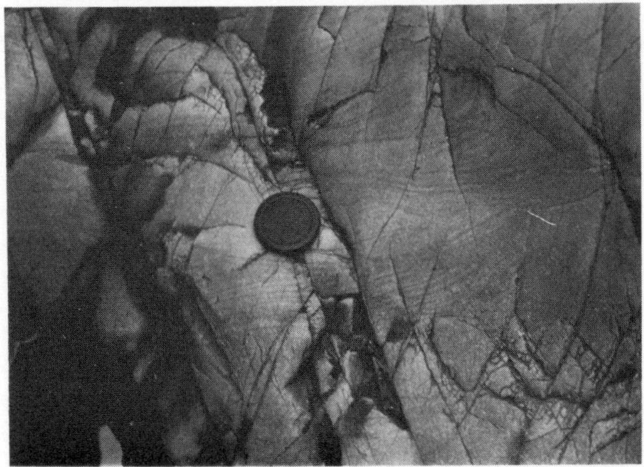

Figure 13.8 Thick cross-bedded psammitic units within the Inverlair Formation, exposed below the road bridge at Inverlair Falls.

bedded and planar bedded units several metres thick, form up to 10% of the formation. They tend to become thicker and more abundant towards the top of the formation.

(ii) Semi-pelite units up to 2 metres thick comprise the remainder of the Clachaig Formation. Microscope study has shown that some semi-pelites may contain apatite-rich bands 2–3 mm thick, suggesting that the original argillaceous sediment contained phosphatic bands. No other stratigraphic formation within the Grampian Group has been reported to contain similar phosphate concentrations.

The significant proportion of semi-pelite within this formation suggests that slow rates of deposition in a low-energy environment usually prevailed. The interbedded psammitic horizons may thus represent the sporadic influx of mature sands, possibly brought into the area by the agency of storms. From this limited evidence, a shallow-marine environment is postulated for the entire Clachaig Formation.

13.2.4.2 *Inverlair Psammite Formation* The boundary between the Clachaig and Inverlair Formations is transitional, the base of the Inverlair Formation being marked by the development of thicker and laterally more extensive psammites and the increasing scarcity of semi-pelitic units. In the Glen Spean area, the Inverlair Formation is approximately 900 metres thick and typically consists of three distinctive lithofacies:

(i) Channel psammites, which comprise at least half the total thickness of the formation. The base of each channel psammite is marked by an erosive surface, succeeeded by a psammite fill displaying slumping at the base. The thickness of the channels varies from 3–10 metres, while the widths may exceed 100 metres, although constraints imposed by the limits of outcrop and structural deformation make this figure hard to estimate. Trough cross-bedding and planar bedding, with individual sets up to 20 cm thick are also occasionally preserved within these psammites (Fig. 13.8). The tops of the channel deposits may be sharp, but often they consist of a thin fining- and thinning-upwards sequence interpreted as channel abandonment deposits.

(ii) Parallel-sided psammites and semi-pelites, interpreted as overbank deposits. They consist of sharp-topped planar-bedded psammites up to 40 cm thick, separated by thin beds of semi-pelite. The bases of these units are usually flat, but locally low-angle scouring has been seen.

(iii) Planar and trough cross-bedded psammites occur occasionally in association with parallel-sided psammites and semi-pelites. They differ from the planar-bedded psammites only in thickness (ranging up to 1 metre and in the development of trough cross-bedding at the top of each unit.

The Inverlair Formation is thus devoid of turbiditic psammites, and was laid down in a fluviodeltaic environment. The three lithofacies described may reflect confined flow conditions (channel psammites) alternating with unconfined overbank and crevasse splay deposition, represented by the parallel-sided psammite and trough cross-bedded lithofacies respectively. However, at present, the sedimentological data remain relatively limited, and this interpretation must remain tentative pending the acquisition of further information.

13.2.4.3 *Correlation with the Glen Spean Subgroup* As the Glen Spean Subgroup represents the highest rocks of the Grampian Group in the Glen Spean area, other Grampian Group rocks underlying the base of the Appin Group should be lateral equivalents. At present only the area N of Schiehallion, and well exposed in a series of new road-cuttings along the main A9 trunk road, has been studied in any detail (Thomas, 1980). In this area a local stratigraphic succession within the Grampian Group was established (Thomas, 1980), and recent investigations have shown that a wealth of sedimentary structures are preserved. While no exact correlation with the Glen Spean area is yet possible, the dominantly psammitic Strath Tummel succession

(Thomas, 1980) appears to share many lithofacies with the Glen Spean Subgroup, notably the well-displayed channel psammites and planar and trough cross-bedded psammites (Figs 13.9; 13.10; 13.11). Thomas (1980) estimated the thickness of the Strath Tummel succession as reaching a maximum of 3000 m in this area. Underlying the Strath Tummel succession, the Drumochter succession (Thomas, 1980) contains monotonous flaggy psammites and semi-pelites, lithologically akin to the Creag Meagaidh Formation. On a lithological basis, therefore, there seems to be similarity between the Grampian Group stratigraphic succession of the Glen Spean–Loch Laggan area, and the area N of Schiehallion, and this similarity is also being checked using chemostratigraphic correlation. Preliminary studies of uppermost Grampian Group rocks exposed in the River Spean W of Spean Bridge, and in the 'Eilde Flags' exposed in the River Leven SE of Kinlochleven (Treagus, 1974; Hickman, 1975) have suggested that these rocks display a lithological similarity to the Inverlair Formation in Glen Spean. If such a correlation is proved by further work, various formations within the Appin Group will be shown to rest directly upon the same uppermost Grampian Group formation in different parts of Lochaber. At present no detailed stratigraphic information is available about the Grampian Group rocks N of the Cairngorm Granite, and hence no correlation is possible with the NE area of outcrop.

Figure 13.9 Tabular cross-bedding in the Tummel Psammitic Schist, exposed by the A9 trunk road.

Figure 13.10 Channelling within the Tummel Psammitic Schist, exposed by the A9 trunk road.

Figure 13.11 Channelling exposed within the Tummel Psammitic Schist at Clunes Cut on the A9 trunk road.

13.3 Geochemistry

A brief geochemical study may reveal information otherwise destroyed during metamorphism and recrystallization of the rocks. Such information may relate to the nature of the original sediment and provide clues about the sedimentary environment, tectonic setting of deposition and the provenance of the original clastic material. For this reason chemical variations within the Grampian Group have been noted, and because almost all the recognized formations include both semi-pelitic and psammitic rocks, these rock types have been treated separately. No attempt has been made to calculate a mean overall composition for any of the formations, as the variations in the proportion of psammitic and semi-pelitic rocks may obscure more significant chemical differences.

The 204 semi-pelitic rocks analysed have been divided into six categories, broadly corresponding with the formation from which each was obtained. No subdivision into formations has been made within the Glen Spean Subgroup, as insufficient data is currently available (Table 13.2). Six samples were obtained from semi-pelite on Ord Ban as being representative of the Ord Ban Subgroup, while 32 samples collected from the Strath Tummel area represent the highest stratigraphic formations of the Grampian Group. For reasons already stated, the latter are considered to be lateral equivalents of the Glen Spean Subgroup. The remaining 166 semi-pelitic samples were obtained from each of the four constituent formations of the Corrieyairack Subgroup. The status of the Glenshirra succession as part of the Grampian Group is currently uncertain, and therefore no chemical data obtained from the succession have been included. However, recent work has shown that Glenshirra succession rocks are chemically distinct from those of the Corrieyairack Subgroup (Haselock, 1984).

Compositional variations may be graphically

Table 13.2 Mean compositions of semipelitic rocks from Grampian Group formations.

	Ord Ban	Coire nan Laogh	Glen Doe	Monadh-liath	Creag Meagaidh	Glen Spean
	$\bar{x}$	$\bar{x}$	$\bar{x}$	$\bar{x}$	$\bar{x}$	$\bar{x}$
SiO_2	67.24	59.61	59.26	59.52	62.77	62.63
TiO_2	0.92	0.95	0.92	0.95	1.06	1.09
Al_2O_3	16.68	18.58	17.64	18.83	17.08	16.32
Fe_2O_3	5.50	2.36	7.43	7.47	7.03	1.35
FeO	—	4.76	—	—	—	4.81
MnO	0.08	0.11	0.13	0.14	0.10	0.10
MgO	1.85	2.20	2.53	2.61	2.78	3.30
CaO	1.03	2.03	2.30	2.16	1.85	1.88
Na_2O	1.88	2.74	2.92	2.92	2.56	2.17
K_2O	4.34	3.63	3.94	3.74	4.71	5.72
P_2O_5	0.21	0.30	0.29	0.27	0.23	0.20
H_2O^+	—	1.92	1.63	1.45	1.59	—
S	0.26	—	—	0.08	0.02	0.01
Ba*	795	829	1019	971	994	1007
Ce	63	—	—	—	106	120
Cr	45	62	81	69	51	64
Cu	19	—	—	29	23	32
Ga	—	—	—	—	23	20
La	30	—	—	—	48	54
Nb	20	21	17	20	19	21
Nd	—	—	—	—	41	49
Ni	10	32	38	36	28	22
Pb	—	—	—	—	23	25
Rb	164	153	155	150	173	204
Sr	172	285	301	313	286	249
Th	—	—	—	—	15	15
V	—	—	—	—	4	5
Y	27	36	38	35	42	49
Zn	70	—	—	130	99	88
Zr	230	196	239	222	362	472
$n =$	6	27	37	75	27	32

* Contents in ppm, oxides as wt.%.

illustrated using a 'spidergram', with compositions normalized against averages of post-Archaean non-calcareous terrigenous shale (Taylor and McLennan, 1985). For comparative purposes, and in an attempt to use chemical composition as an indicator of tectonic depositional setting, the mean compositions of semi-pelitic rock formations from western North America (Van de Kamp and Leake, 1985) and Western Australia (Taylor and McLennan, 1985), which had been assigned by geological association and by major-element composition to defined tectonic settings, have been plotted on a spidergram of trace and minor elements arranged in a sequence of incompatibility (Fig. 13.12). The traces of these rocks showed that the

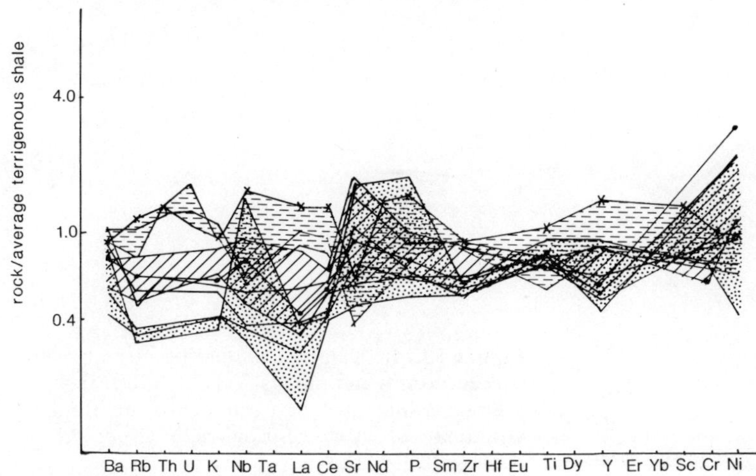

Figure 13.12 A 'spidergram' plot showing the multi-element profiles characteristic of unmetamorphosed shales deposited in different tectonic environments. Values are normalized against average post-Archaean shales (Taylor and McLennan, 1985); sediment type data and mean geochemical analyses are obtained from Taylor and McLennan (1985) and Van de Kamp and Leake (1985). Ornament denotes the tectonic setting of shales: stipple—oceanic island arc; stripes—continental arc; dashes—active continental margin; single unornamented line—continental basin.

large ion lithophile elements (LILE) Ba, Rb, Th, K, and the light rare-earth elements (LREE) La and Ce were most depleted in sediments derived from and associated with oceanic island arcs. Shales associated with continental island arcs and active continental-margin settings showed a less marked depletion in the same elements, while shales deposited in a passive continental setting showed relative enrichment. Differences in concentration of the other elements were less marked, and considerable overlap occurs. Concentration of Ni and Cr may be principally controlled by the proportion of mafic material in the source area, and this may be highly variable in any tectonic setting.

When the semi-pelitic rocks from the different formations of the Grampian Group, and the pelitic schists from the overlying Lochaber Subgroup of the Appin Group were plotted on a similar spidergram, all were found to have relatively high LILE and LREE contents, where information is available (Fig. 13.13). This suggests that all these rocks were deposited in a passive continental margin or intracratonic basin setting, an environment entirely consistent with the geological information. All the formations display depletion in Ni and Cr, suggesting a relatively small proportion of mafic source material, and most formations possess enhanced Zr and Y contents, consistent with their derivation from broadly granitoid material. On the same diagram it is also evident that differences between some of the formations are present. Within the Grampian Group, the Ord Ban Subgroup is distinguished by containing lower Cr, Ni, Sr, CaO, Fe_2O_3(tot.), MnO and MgO (Table 13.2). Indeed, the lower Fe_2O_3(tot.), CaO, Na_2O, and Sr are characteristics which are shared with semi-pelites from the Lochaber Subgroup of the Appin Group, which are included for comparison. However, the Ord Ban Subgroup semi-pelites also differ from those in the Lochaber Subgroup in possessing higher Sr/Y and lower Y, Rb, Rb/Sr and Y/P_2O_5, differences which are highlighted in a second diagram which is designed to emphasize chemical contrasts between semi-pelites in the different formations (Fig. 13.14).

The semi-pelites from the four formations in the Corrieyairack Subgroup have very similar compositions. The chemistry thus supports the inclusion of all these formations within the same subgroup. Characteristically they all contain high Na_2O, CaO and Sr concentrations, together with low Rb and K_2O, typical

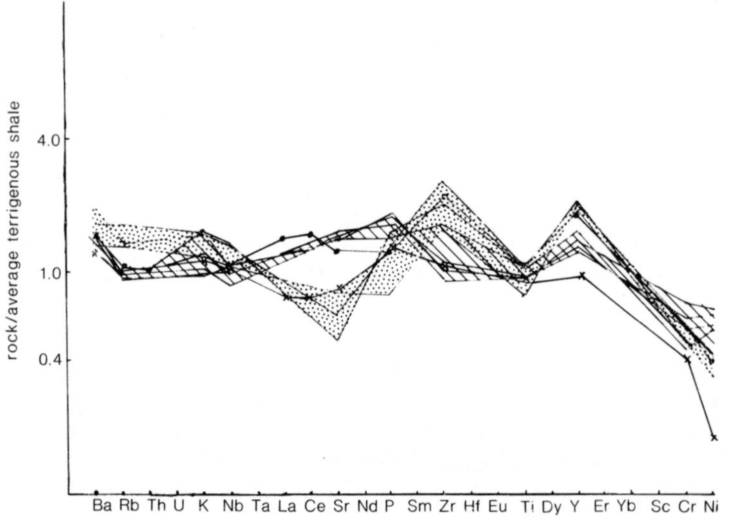

Figure 13.13 A 'spidergram' plot comparing mean semi-pelitic compositions of formations within the Grampian and Appin Groups. Ornament: stipple—Appin Group (Lochaber Subgroup only); line connecting spots—Glen Spean Subgroup; stripes—Corrieyairack Subgroup; line connecting crosses—Ord Ban Subgroup.

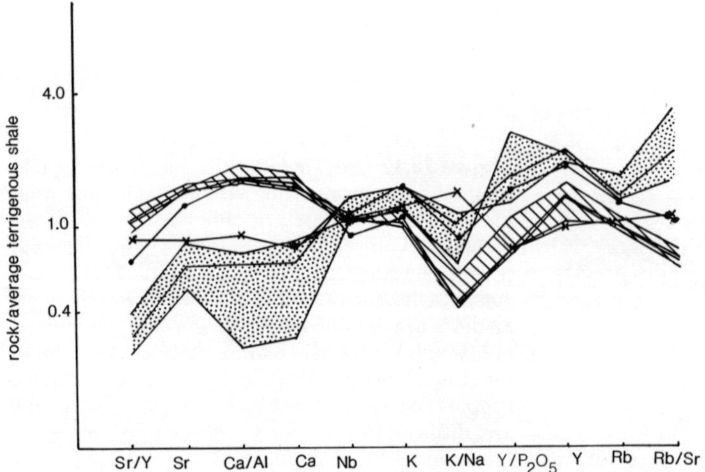

Figure 13.14 Diagram comparing mean element concentrations and ratios selected to illustrate compositional differences between semi-pelites from the Grampian Group and Lochaber Subgroup. Ornament as in Fig. 13.4.

of immature sediments (Lambert *et al.*, 1982). The Sr/Y and CaO/Al_2O_3 ratios are high, and K_2O/Na_2O and Rb/Sr ratios are low, so a declining trend is displayed on the diagram showing chemical contrast (Fig. 13.14), suggesting immaturity of sediment. Previous studies (Lambert *et al.*, 1981) suggested that such sediments contained clastic plagioclase and a relatively low proportion of clay minerals in the original shale. The highest formation of the Corrieyairack Subgroup, the Creag Meagaidh Formation, contains semi-pelites with somewhat higher K_2O, Rb and Y than the others in the subgroup, and thus heralds increasing sediment maturity, a trend which is continued in the Glen Spean Subgroup. The semi-pelites from the Strath Tummel area possess lower mean Sr/Y, Sr and higher K_2O, K_2O/Na_2O, Y, Y/P_2O_5, Rb and Rb/Sr than any in the Corrieyairack Subgroup, suggesting increased sediment maturity. Enhancement of Zr and Y contents is also characteristic of the Glen Spean Subgroup, suggesting derivation of heavy minerals from a granitoid source. The lack of comparable increase in TiO_2 and Fe_2O_3(tot.) contents suggests that any increased heavy-mineral supply was not derived from basic rocks enriched in these elements or Ni and Cr.

The main chemical contrast occurs at the base of the Appin Group, confirming that this contact reflects a marked change in sedimentary environment. The chemistry does not indicate whether or not an unconformity exists at this contact: it merely shows that an environmental change took place. Semi-pelites from the Lochaber Subgroup all possess higher Rb and Y contents, and lower Sr and CaO contents than the Grampian Group semi-pelites; characteristics consistent with deposition of mature clays under quiet conditions. The contrast of chemical composition between these semi-pelites and the Monadhliath Semi-pelitic Formation has already been described (Lambert *et al.*, 1982), and this difference extends to the entire Grampian Group. Of all the semi-pelitic rocks in the Grampian Group, the Ord Ban semi-pelite shows the closest chemical resemblance to the Lochaber Subgroup semi-pelites. These semi-pelites, which may have been deposited in quiet-water conditions in association with thin limestones and well-sorted quartz-rich sandstones, may have been laid down in a sedimentary environment similar to that which prevailed during the deposition of the Lochaber Subgroup rocks, and their chemistry reflects this. However, the Ord Ban semi-pelites do not share the high Y, Rb and Rb/Sr of the

Table 13.3 Mean compositions of psammitic rocks from Grampian Group formations.

	Coire nan Laogh	Glen Doe	Monadhliath	Creag Meagaidh	Inverlair (Eilde Flags)
	$\bar{x}$	$\bar{x}$	$\bar{x}$	$\bar{x}$	$\bar{x}$
SiO_2	71.56	72.80	76.08	72.85	72.03
TiO_2	0.61	0.51	0.42	0.61	0.70
Al_2O_3	13.68	12.15	10.98	12.80	12.77
Fe_2O_3	1.34	0.86	0.82	0.27	3.14
FeO	2.79	2.48	2.22	3.45	
MnO	0.06	0.08	0.11	0.07	0.05
MgO	1.43	1.37	1.23	1.32	1.81
CaO	1.51	1.94	2.13	2.18	2.20
Na_2O	3.05	3.15	3.34	2.95	3.11
K_2O	2.34	2.45	1.16	2.48	3.26
P_2O_5	0.15	0.13	0.13	0.16	0.06
H_2O^+	1.23	1.53	0.82	0.86	—
Ba	548	592	328	581	776
Ce	—	—	—	—	—
Cr	36	39	32	45	—
Cu	—	—	—	—	—
Ga	—	—	—	—	—
La	—	—	—	—	—
Nb	18	19	22	16	15
Nd	—	—	—	—	—
Ni	14	22	23	19	—
Pb	—	—	—	—	—
Rb	102	88	50	91	84
Sr	248	225	344	291	179
Th	—	—	—	—	—
V	—	—	—	—	—
Y	17	19	17	30	26
Zn	—	—	—	—	—
Zr	229	208	216	236	698
n =	4	11	9	4	24

* Content in ppm, oxides as wt.%.

Lochaber Subgroup semi-pelites, and hence the chemistry does not support a correlation between them. Such a correlation might otherwise have been made on lithological grounds, as the succession associated with the Kinlochlaggan Limestone is similar to the Ord Ban Subgroup rocks on Speyside, but has been correlated with the Appin Group (Treagus, 1969). However, while the chemical composition of the Kinlochlaggan Limestone itself has been compared with the Appin Group Lismore Limestone (Hickman and Wright, 1983), no comparison has been made with limestones from the Ord Ban Subgroup, and so the stratigraphic position of the Kinlochlaggan Limestone remains uncertain.

The chemistry of the semi-pelites in the Grampian Group and Lochaber Subgroup thus suggests that an intracratonic basin developed, deriving its sedimentary fill from a passive continental source area. In the early stage of basin development, mature shales were deposited in restricted areas, and were then overlain by immature semi-pelitic sediments associated with turbiditic deposits. Subsequently, an increase in maturity of sediment accompanied the filling and shallowing of the basin, and this was followed in turn by renewed basin development, as illustrated by the Lochaber Subgroup rocks.

However, the bulk of the rocks deposited were psammitic, and hence it is also essential to study the chemical variations within the psammite (Table 13.3). When the mean concentrations of the psammitic rocks

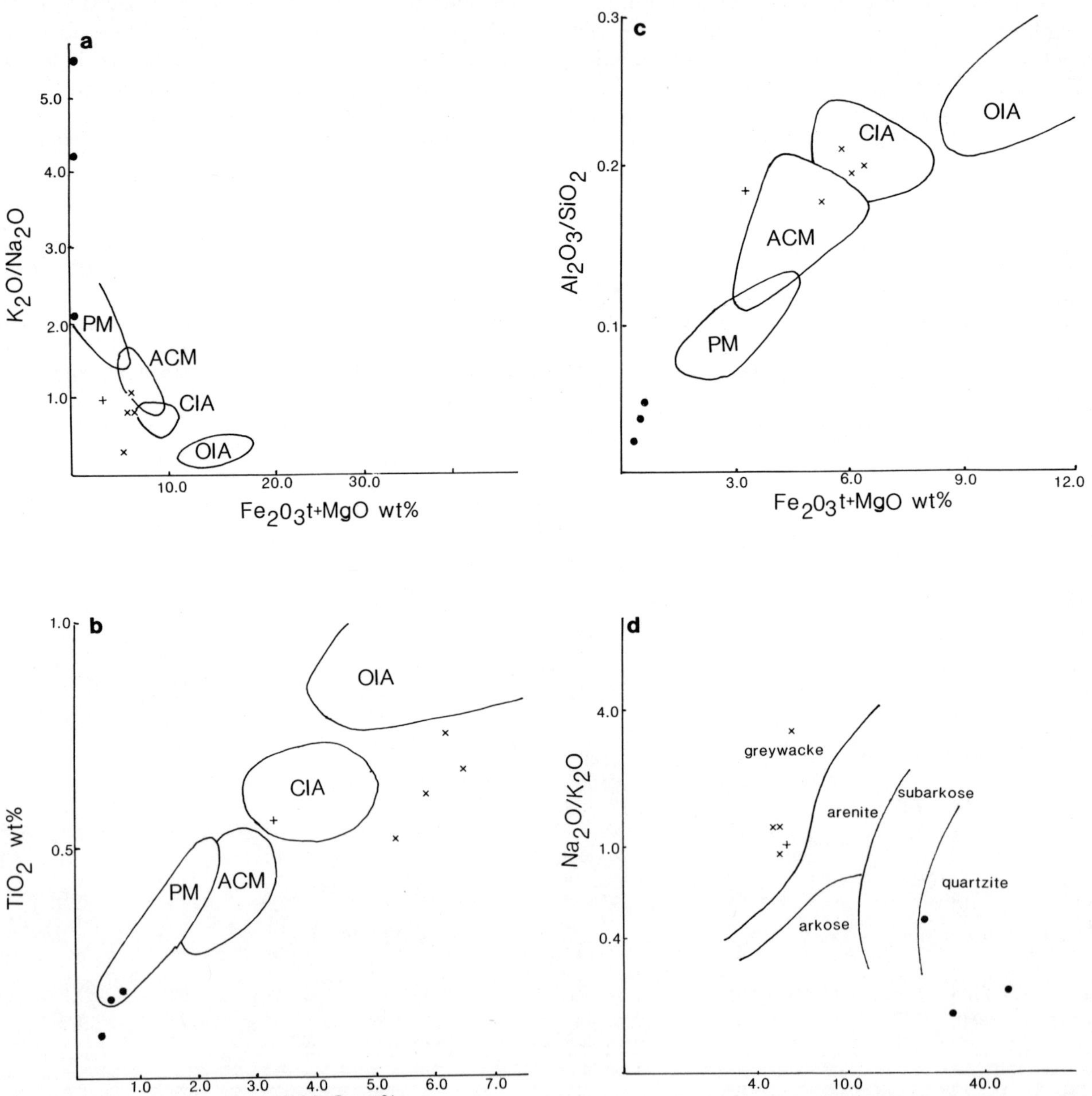

Figure 13.15 (*a–c*) Diagram to illustrate the tectonic settings of sandstones/psammites (after Bhatia, 1983). *OIA*—oceanic island arc; *CIA*—continental island arc; *ACM*—active continental margin; *PM*—passive margin. (*d*) Diagram discriminating between sandstone types (after Pettijohn *et al.*, 1973). Symbols (for all diagrams): spots—Lochaber Subgroup quartzites; upright crosses—Glen Spean Subgroup; diagonal crosses—Corrieyairack Subgroup.

are plotted on tectonic setting discriminant diagrams (Bhatia, 1983) (Fig. 13.15), the psammites from the Grampian Group tend to plot within the continental island arc field. By contrast, the Lochaber Subgroup quartzites plot nearest to the passive margin field. This difference may be more correctly attributed to differences of sandstone type, rather than tectonic setting, as indicated in a log Na_2O/K_2O–log SiO_2/Al_2O_3 discrimination diagram (Pettijohn *et al.*, 1973). On this diagram the contrast is plain between the Grampian Group psammite means, which fall within the greywacke field, and the Lochaber Subgroup quartzites, which plot as quartzites.

Where psammitic rocks deposited in different tectonic settings were plotted on a spidergram, using values normalized according to average post-Archaean non-calcareous terrigenous shale (Taylor and McLennan, 1985), differences of composition were most marked among the LILE. In particular, the sandstones deposited in oceanic island arc and continental island-arc settings displayed relative depletion of LILE, and so are characterized by a rising trend. By contrast, the trends of sandstones deposited on active continental margins and passive margins tend to be level (Fig. 13.16).

When the psammites from the Grampian Group and the quartzites from the Lochaber Subgroup were plotted on a similar diagram (Fig. 13.17), a spurious contrast was produced as a result of the different SiO_2 contents. However, none of the plotted formations displayed a rising trend, suggesting that neither the Grampian Group psammites nor the Lochaber Subgroup quartzites were deposited in association with island arcs. Instead they emerge as sediments deposited on a passive margin or in an intracratonic basin, as suggested by the geological evidence.

There is thus a contradiction between the geochemical indicators suggested by Bhatia (1983), and the trends revealed by spidergrams. On account of the total absence of volcanic deposits associated with either the Grampian Group or the Lochaber Subgroup, the geological evidence suggests that both sedimentary sequences were deposited in basins spatially unrelated to island arcs, and hence the geological indicators used by Bhatia are misleading in this instance.

Equally, however, it might be argued that wholesale metasomatic activity during metamorphism has altered the original premetamorphic concentrations of several mobile LILE, producing deceptive results. Such a suggestion cannot be entirely discounted, but, while there is evidence that localized metasomatism has occurred, the preserved chemical differences between adjacent rocks of different compositions within both the Grampian Group and the Lochaber Subgroup argue

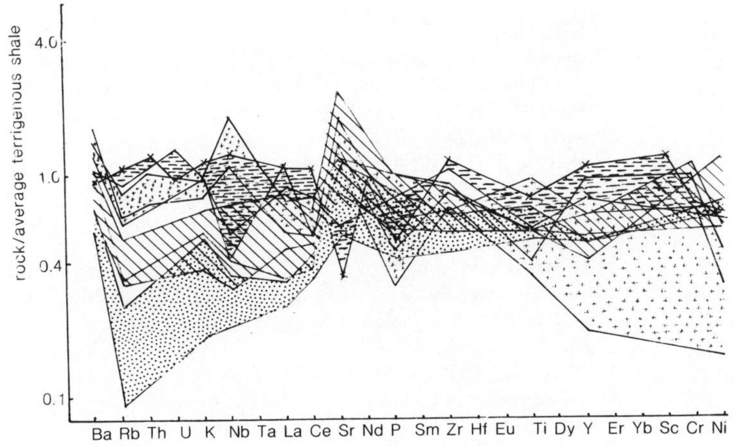

Figure 13.16 A 'spidergram' plot showing the multi-element profiles characteristic of unmetamorphosed sandstones deposited in different tectonic environments. Values are normalized against average post-Archaean shales (Taylor and McLennan, 1985); sedimentary data are obtained from Taylor and McLennan, 1985 and Van de Kamp and Leake, 1985. Ornament denotes tectonic setting of sandstones: stipple—oceanic island arc; stripes—continental island arc; dashes—active continental margin; crosses—continental basins.

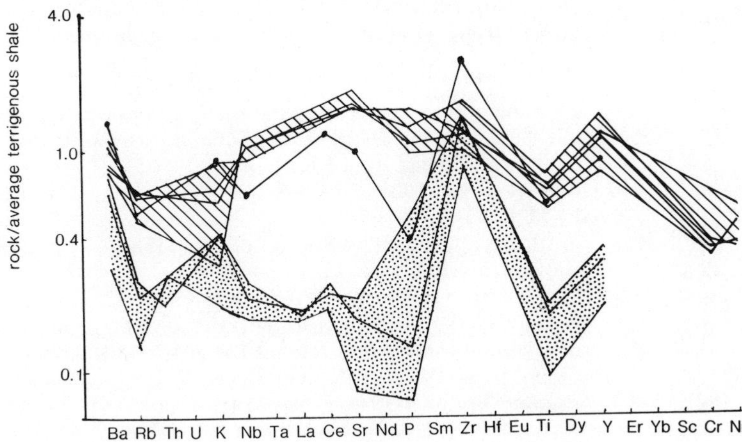

Figure 13.17 A 'spidergram' plot comparing mean psammitic compositions from formations in the Grampian Group and Lochaber Subgroup. Ornament as in Fig. 13.4. Lower values reflect the higher SiO_2 concentration in psammites.

against wholesale chemical alteration of the entire system.

13.4 Conclusions

Both the sedimentological and geochemical information suggest that the deposition of the Grampian Group began in quiet-water conditions in a restricted part of the basin. Only the western margins of these deposits, represented by the Ord Ban Subgroup, are currently exposed, and deposition did not extend further west than Speyside. The eastward extent of these deposits is unknown.

Subsequent accelerated subsidence was accompanied by the formation of a more extensive ensialic basin. Rapid turbidite deposition in deeper water characterizes this stage of basin development, and these deposits collectively form the Corrieyairack Subgroup. Deposited directly upon the pre-Grampian Group basement in the west, these deposits increase in thickness towards the east and south-east. No volcanogenic material is associated with the entire succession, and hence the development of the basin appears to have been unaccompanied by volcanic activity. Even the large sills associated with the Ord Ban Subgroup may have been emplaced after the deposition of the Grampian Group.

The closing stages of Grampian Group sedimentation were marked by more rapid deposition than subsidence, producing a shallowing of the basin and the introduction of possibly fluviodeltaic sands at the top of the succession represented by the Glen Spean Subgroup. The lithological changes at the base of the overlying Lochaber Subgroup represent a change of sedimentary environment, with renewed subsidence initiated in a new, extensive basin.

Whether or not this renewed subsidence occurred without a break in sedimentation is not indicated by geochemical data alone. However, the model, suggested from the sedimentological characteristics of the Grampian Group, does not preclude the possibility that around the margins of the main basin localized emergence and erosion may have occurred at the top of the Grampian Group. It therefore remains possible that locally the Lochaber Subgroup may rest unconformably upon the Grampian Group, while elsewhere the relationship is entirely conformable. Such a model may account for the absence of the lower formations of the Lochaber Subgroup north of the River Spean, and also the thinning of Grampian Group formations towards the NW.

Thus, on account of its lithological and geochemical differences from the overlying Dalradian rocks, and because it was apparently deposited in an earlier sedimentary basin, the Grampian Group emerges as a distinct sedimentary succession from the later Dalradian rocks. Hence, although it shares, with the overlying rocks a common experience of metamorphism and deformation during the Caledonian event, it may be more appropriate to classify the Grampian Group as an earlier and separate rock suite from the Dalradian Supergroup.

References

Anderson, J. G. C. (1948) The Kinlochlaggan Syncline, Southern Inverness-shire. *Trans. geol. Soc. Glasgow* **21**, 97–115.

Anderson, J. G. C. (1956) The Moinian and Dalradian rocks between Glen Roy and the Monadhliath Mountains, Inverness-shire. *Trans. Roy. Soc. Edinburgh: Earth Sci.* **63**, 15–36.

Bhatia, M. R. (1983) Plate tectonics and geochemical composition of sandstones. *J. Geol.* **91**, 611–627.

Brewer, M. S., Brook, M. and Powell, D. (1979) Dating of the tectonometamorphic history of the southwestern Moine, Scotland. In Harris, A. L., Holland, C. H. and Leake, B. E. (eds.), The Caledonides of the British Isles—Reviewed. *Spec. Publ. geol. Soc. London* **8**, 129–137.

Brook, M., Powell, D. and Brewer, M. S. (1976) Grenville age for rocks in the Moine of northwestern Scotland. *Nature* **260**, 515–517.

Harris, A. L., Baldwin, C. T., Bradbury, H. J., Johnson, H. D. and Smith, R. A. (1978) Ensialic basin sedimentation: the Dalradian Supergroup. In Bowes, D. R. and Leake, B. E. (eds.), Crustal Evolution in Northwestern Britain and Adjacent Regions. *Geol. J. Spec. Issue* **10**, 115–138.

Haselock, P. J. (1982) The Geology of the Corrieyairack Pass Area, Inverness-shire. Unpublished Ph.D. Thesis, University of Keele.

Haselock, P. J. (1984) The systematic geochemical variation between two tectonically separate successions in the southern Monadhliaths, Inverness-shire. *Scott. J. Geol.* **20**, 191–205.

Haselock, P. J. and Winchester, J. A. (1981) A note on the stratigraphic relationship of the Leven Schist and the Monadhliath Schist in the Central Highlands of Scotland. *Geol. J.* **16**, 237–241.

Haselock, P. J. and Winchester, J. A. and Whittles, K. H. (1982) The stratigraphy and structure of the southern Monadhliath Mountains between Loch Killin and upper Glen Roy. *Scott. J. Geol.* **18**, 275–290.

Hickman, A. H. and Wright, A. E. (1983) Geochemistry and chemostratigraphic correlation of slates, marbles and quartzites of the Appin Group, Scotland. *Trans. Roy. Soc. Edinburgh: Earth Sci.* **73**, 251–278.

Kelling, G., Phillips, W. E. A., Harris, A. L. and Howells, M. F. (1986) The Caledonides of the British Isles: a review and appraisal. In Gee, D. G. and Sturt, B. A. (eds.), *The Caledonide Orogen—Scandinavia and Related Areas*. Wiley, New York.

Lambert, R. StJ., Holland, J. G. and Winchester, J. A. (1981) Comparative geochemistry of pelites from the Moinian and Appin Group (Dalradian) of Scotland. *Geol. Mag.* **118**, 477–490.

Lambert, R. StJ., Holland, J. G. and Winchester, J. A. (1982) A geochemical comparison of the Dalradian Leven Schists and the Grampian Division Monadhliath Schists of Scotland. *J. geol. Soc. London* **139**, 71–84.

Okonkwo, C. T. (1985) The Geology and Geochemistry of the Metasedimentary Rocks of the Loch Laggan—Upper Strathspey area, Inverness-shire. Unpublished Ph.D. Thesis, University of Keele.

Parson, L. M. (1982) The Precambrian and Caledonian Geology of the Ground near Fort Augustus, Inverness-shire. Unpublished Ph.D. Thesis, University of Liverpool.

Pettijohn, F. S., Potter, P. E. and Siever, P. (1973) *Sand and Sandstone*. Springer-Verlag, New York.

Piasecki, M. A. J. (1980) New light on the Moine rocks of the

Central Highlands of Scotland. *J. geol. Soc. London* **137**, 41–59.

Piasecki, M. A. J. (1984) Ductile thrusts as time markers in orogenic evolution: an example from the Scottish Caledonides. In Galson, D. A. and Mueller, S. T. (eds.), First European Geotraverse Workshop: the Northern Segment. *Eur. Sci. Found. Publ. Strasbourg* 109–114.

Piasecki, M. A. J. and van Breemen, O. (1979*a*) A Morarian age for the 'younger Moines' of central and western Scotland. *Nature,* **278**, 734–736.

Piasecki, M. A. J. and van Breemen, O. (1979*a*) The 'Central Highland Granulites': cover–basement tectonics in the Moine. In Harris, A. L., Holland, C. H. and Leake, B. E. (eds.), The Caledonides of the British Isles—Reviewed. *Spec. Publ. geol. Soc. London* **8**, 139–144.

Plant, J. A., Watson, J. V. and Green, P. M. (1984) Moine–Dalradian relationships and their palaeotectonic significance. *Proc. R. Soc. London* **A395**, 185–202.

Taylor, S. R. and McLennan, S. M. (1985) *The Continental Crust: its Composition and Evolution.* Blackwell Scientific, Oxford.

Thomas, P. R. (1979) New evidence for a Central Highland root zone. In Harris, A. L., Holland, C. H. and Leake, B. E. (eds.), The Caledonides of the British Isles—Reviewed. *Spec. Publ. geol. Soc. London* **8**, 205–212.

Thomas, P. R. (1980) The stratigraphy and structure of the Moine rocks N of the Schiehallion Complex, Scotland. *J. geol. Soc. London* **137**, 469–482.

Treagus, J. E. (1969) The Kinlochlaggan Boulder Bed. *Proc. geol. Soc. London* **1654**, 55–60.

Van de Kamp, P. C. and Leake, B. E. (1985) Petrography and geochemistry of feldspathic and mafic sediments of the northeastern Pacific margin. *Trans. R. Soc. Edinburgh: Earth Sci.* **76**, 411–449.

Whittles, K. H. (1981) The Geology and Geochemistry of the Area West of Loch Killin, Inverness-shire. Unpublished Ph.D. Thesis, University of Keele.

14
The Erris Group, Ireland

J. A. WINCHESTER, M. D. MAX and C. B. LONG

14.1 Introduction

The Erris Group comprises a series of dominantly psammitic schists which crop out within the metamorphic inlier of NW Co. Mayo, Ireland (Fig. 14.1). Max (1970) defined it as the lower part of the Bangor stratigraphic succession that occurs within the Broad Haven Nappe (Max *et al.*, in press; Long and Max, in prep). The Group consists of a related series of monotonous psammites, micaceous psammitic schists and impure quartzites with subordinate semi-pelitic schists and sporadically distributed thin calc-silicate bands and heavy-mineral seams. The Group underlies the lithologically distinct Appin Group of the Dalradian Supergroup, with which it shares a common structural and metamorphic history. It therefore occupies a stratigraphic and structural position analogous to the Grampian Group of the Scottish Highlands, with which

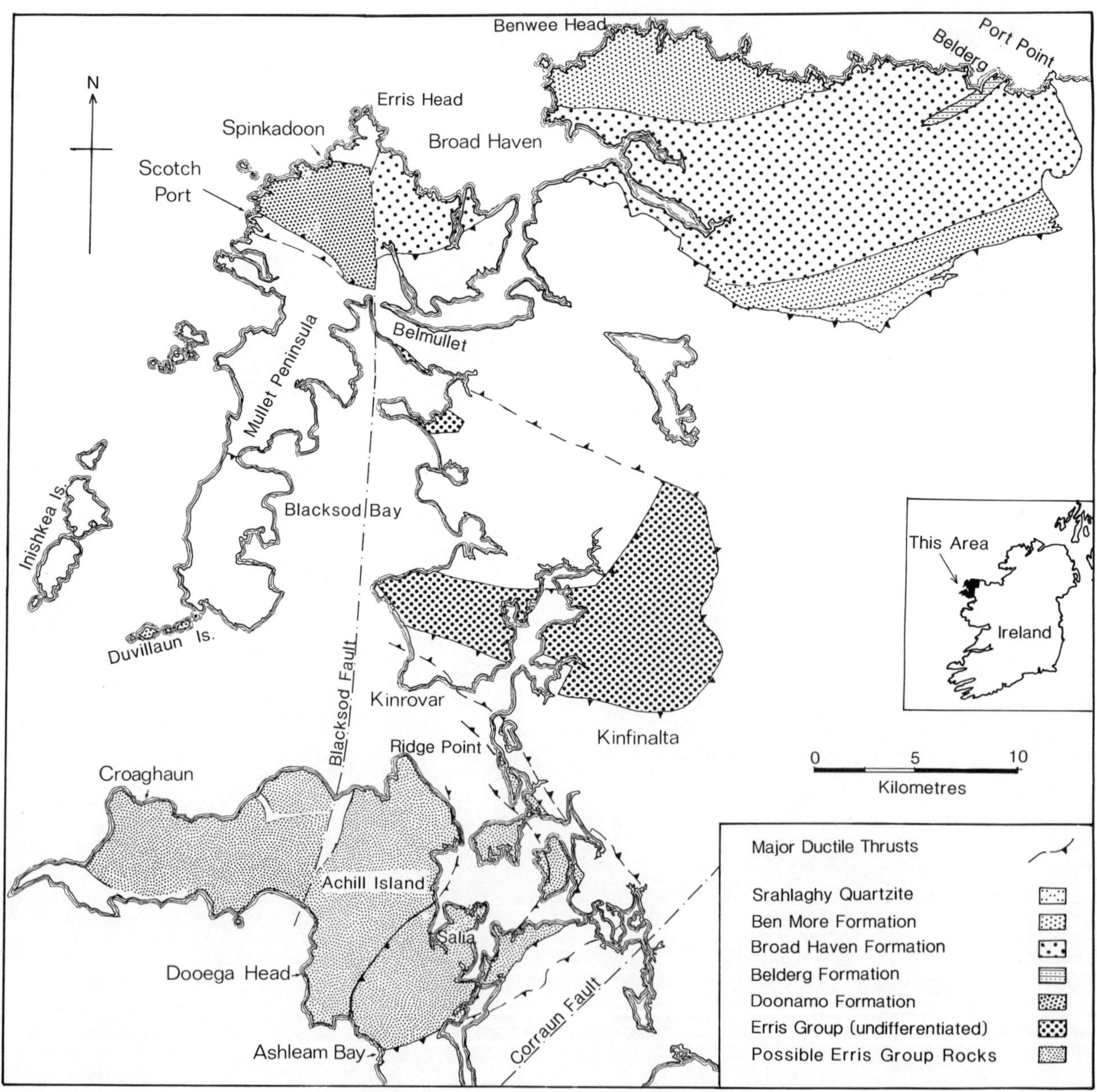

Figure 14.1 Location map showing the distribution of Erris Group rocks and their likely correlatives NW of the Corraun Fault in the NW Mayo Inlier.

it may be generally correlated. It structurally overlies the Annagh and Inishkea Divisions, but the contact can nowhere be described as conformable.

In the northern part of the Mullet Peninsula and north of the tectonically emplaced Kinrovar wedge and the Corraun Fault (Fig. 14.1), structural relationships with the underlying basement are clear: the cross-bedded metasediments of the Erris Group young away from all basement contacts. Although the basement contacts are always localized along high-strain zones, there is no evidence within the Group for imbrication or structural repetition of the stratigraphic sequence. Equally, because the basal contacts are always tectonic, it is not clear that the lowermost Erris Group rocks preserved represent the original base of the succession. Such uncertainty hinders stratigraphic correlation between the main areas in which rocks currently assigned to the Erris crop out. For example, stratigraphic correlation between the Erris Group rocks of the northern Mullet Peninsula (the Doonamo Formation), which have their basal tectonic contact exposed, and the succession in the Belderg area east of the Blacksod Fault—where Erris Group rocks attain their greatest thickness, but the base is not exposed—is not clearly established. Current geochemical studies reveal differences between the rocks in the two areas, although we continue to include the Doonamo Formation as part of the Erris Group on the basis of lithological comparability. It is probable that the Doonamo Formation represents the lowest part of the Erris Group succession, which is not seen in the Belderg area, because the Belderg Anticline, which exposes the lowest stratigraphic levels around Belderg Harbour, exposes in its core no older rocks.

Exposures of the Erris Group are restricted to the NW Mayo inlier. In the Donegal inlier around Lough Derg, where high-grade Slishwood Division gneisses are in contact with Dalradian schists, the contact is tectonic and the Erris Group and lower formations of the Dalradian Supergroup are absent. Hence the lateral extent of the Erris Group beyond the NW Mayo inlier is not known. Also, while there is a stratigraphic and lithological parallel with the Grampian Group of Scotland, it is not clear whether the two groups were deposited in the same or different, contemporary sedimentary basins. The Erris Group is a monotonous sequence of psammites, quartzites, semi-pelites and, rarely, psephites, with a thin discontinuous white quartzite at the top. Dominantly quartzofeldspathic, it often has a banded appearance, with alternating psammitic and grey micaceous psammitic schists. In areas of relatively low strain, sedimentary structures are often preserved: these usually consist of cross-bedded sets with bedding surfaces defined by concentrations of biotite, epidote and magnetite. Although no detailed sedimentological work has yet been done on the Erris Group rocks, a shallow-water depositional environment is inferred from profuse cross-bedding, channelling and compaction structures. Cross-bedding also commonly occurs in psammitic units even where the metasediments are finely banded. Bedded sets vary from 2 cm to 4 metres in thickness and are discontinuous along strike: this appears to be an original sedimentary feature.

14.2 Stratigraphy

Within the Erris Group different stratigraphic formations have been distinguished:

14.2.1 *Doonamo Formation*

In the north-western part of the Mullet Peninsula, west of the Blacksod Fault (Fig. 14.1), Erris Group rocks are represented by the Doonamo Formation, which bears a lithological resemblance to the Erris Group rocks on the mainland. However, because of the large throw of the Blacksod Fault and the absence of good lithological marker horizons, direct correlation with other Erris Group rocks has not so far proved possible. The Doonamo Formation structurally overlies the Inishkea Division rocks at Scotch Port and Spinkadoon, where the contact is obscured along a zone of high strain. However, chemical studies have shown that the movement on this zone of high strain was accompanied by scant metasomatism, in contrast to the extensive metasomatic change, affecting even the usually immobile elements, which characterizes other high-strain zones near Scotch Port (Winchester and Max, 1984) within the Erris Complex. It therefore seems unlikely that major displacement has occurred and, while the exact nature of the contact at the base of the Doonamo Formation is obscured within the high-strain zone, it seems unlikely that much of the original stratigraphic succession has been tectonically removed. Hence the Doonamo Formation probably represents beds near the lowest part of the Erris Group succession.

The Doonamo Formation typically comprises banded creamy-white psammites, alternating with greyer micaceous psammites which often contain prominent biotite porphyroblasts. All the psammites contain abundant quartz, which is associated with microcline and plagioclase, while foliae of muscovite and biotite define the S1 foliation. Accessory epidote, magnetite and sphene are usually present. In semi-pelitic bands, which are more abundant towards the base of the formation, micas comprise a higher proportion of the rock. Also present are sporadically developed, weakly calcareous epidotic calc-silicate bands, which may represent primary concentrations of detrital epidote formed in a manner similar to that proposed for Arnipol-type calc-silicates in the Moine and Grampian Group successions in Scotland (MacGregor, 1948; Winchester, 1975). Narrow heavy-mineral bands up to 1 cm thick are also present. They contain concentrations of magnetite, ilmenite, epidote, sphene and sometimes pyrite, and are likewise thought to represent original sedimentary heavy-mineral concentrates. More siliceous psammites occur at the top of the succession: these possess a composition quite distinct from the Srahlaghy Quartzite (see below) and it is possible that they might be stratigraphically equivalent to the Belderg Formation in Erris.

East of the Blacksod Fault, the Erris Group is divided into four type formations (Crow *et al.*, 1971), which form the basis for correlation in the bulk of the Erris Group succession in NW Mayo. In the main area, however, the base of the Erris Group is not exposed, and the lowest formation recognized—the Belderg Formation—may thus not represent the local base of the Erris Group.

14.2.2 *Belderg Formation*

This formation is only seen in the core of the recumbent, north-facing Belderg Anticline, which is exposed around Belderg Harbour. It is at least 350 metres thick, but is repeated by recumbent F1 folding and the base is not seen (Fig. 14.2). The formation consists of banded grey to creamy psammites with subordinate semi-pelitic beds. Lateral changes in mica contents suggest original lateral facies changes: some of the psammites become quartzitic locally. All the psammites contain abundant quartz, associated with plagioclase: microcline is only locally present. Muscovite and biotite are present as aligned laths defining the S1 foliation, and almandine garnet occurs sporadically. Apatite, magnetite and epidote are present as accessory minerals.

14.2.3 *Broad Haven Formation*

Monotonous banded to massive, white to tan-coloured quartz-rich psammites, assigned to the Broad Haven Formation, overlie the Belderg Formation. The contact is gradational, and the best exposures of the Broad Haven Formation occur in the high sea-cliffs forming the bold north Mayo coastline. Inland exposure is frequently poor, and for this reason, and because of the discontinuous nature of the few distinctive units, such as pebbly horizons, it has not been possible to subdivide the Broad Haven Formation stratigraphically, despite its thickness of at least 3800 metres. Heavy-mineral bands up to 1 cm thick become common in the upper part of the formation, but they are rare elsewhere. Near the top of the cliffs at Kilgalligan (Fig. 14.1) there is a thin (< 4 metres thick) deeply weathered garnet-biotite-muscovite schist, seen nowhere else in the northern inlier. Across Broad Haven, the psammites forming the north-eastern part of the Mullet Peninsula—east of the Blacksod Fault—form a westward continuation of the Broad Haven Formation. In all the psammites abundant quartz is associated with plagioclase, while microcline is often present. Muscovite laths define the S1 foliation; biotite is also frequently present. Magnetite is usually present as an accessory mineral: apatite and epidote may also be present. Other than these minor concentrations of magnetite in preserved bedding foresets, heavy-mineral bands are absent.

14.2.4 *Benmore Formation*

The Benmore Formation has a gradational contact with the Broad Haven Formation, which it resembles lithologically, except that it lacks the characteristic tan colour of the Broad Haven Formation, and tends to contain more quartzitic units. It is well exposed on coastal sections around Benwee Head (Fig. 14.3). Thin orthoquartzite bands up to 10 cm thick are present, but they form less than 3% of the formation, which is about 1500 metres thick. Some pebbly horizons occur. Heavy-mineral bands are also present. These frequently show original bedding, defined by variation in concentration of the heavy minerals magnetite, zircon, sphene and epidote. Marked concentrations of zircon may occur: one heavy-mineral band analysed contained over 2% Zr.

14.2.5 *Srahlaghy Formation*

The exposures of this formation are restricted to the Srahlaghy area, where they occur immediately below a major zone of high strain which forms the local structural base of the Appin Group Dalradian rocks. Its maximum thickness is 45 metres, but it has been subjected to tectonic thinning. The lower contact with

Figure 14.2 Isoclinally folded psammites of the Belderg Formation at Belderg Harbour.

Figure 14.3 Folded psammites of the Benmore Formation at Benwee Head.

the Benmore Formation is relatively sharp, but appears to be a conformable stratigraphic passage rather than a tectonic junction. The formation consists of pale, banded quartzites, which are frequently highly deformed, so that cross-bedding, where preserved, is greatly distorted. Thinning and the local absence of the Srahlaghy Formation elsewhere in the type area is attributed to attenuation and some truncation along the major ductile shear zone marking the base of the Bangor Nappe (Max *et al.*, in press).

All but the Srahlaghy Formation, and its lateral equivalents in other parts of the NW Mayo inlier, can be correlated on stratigraphic grounds with the Grampian Group of Scotland (Harris *et al.*, 1978). Although Max (1970) placed the Srahlaghy Formation within the Erris Group because it was a quartzofeldspathic unit of restricted occurrence, he noted that it might mark the onset of changed sedimentation conditions. At that time, more work was needed to confirm its stratigraphic position as a major stratigraphic break which everywhere overlies the Srahlaghy Formation in the type area. More recent mapping, however, suggests that the Srahlaghy Formation is best correlated with the Eilde Quartzite, which is the basal formation of the Appin Group in Scotland. An exact correlation is impossible to confirm: the greater thickness of the Srahlaghy Formation in parts of NW Mayo, and the development of subordinate muscovite-quartz schist units within it, may imply that the Srahlaghy Formation is a lateral equivalent of much of the Lochaber Subgroup—but with the quartzite members forming a much greater part of the succession. However, as the thin schists within the Srahlaghy Formation do not form continuous, mappable units, it is not possible to subdivide the formation into thinner members. On structural and lithological grounds, therefore, the Erris Group below the Srahlaghy Formation is a likely equivalent of the Grampian Group, and the Srahlaghy Formation is roughly equivalent to the Lochaber Subgroup of the Appin Group in Scotland.

14.2.6 *Rocks correlated with the Erris Group*

Elsewhere in the NW Mayo inlier, psammitic rock successions have been correlated with the Erris Group, on the basis of lithology and their stratigraphic position below the Appin Group Dalradian rocks. In particular, the psammitic rocks around Kinfinalta (Fig. 14.1), SE of Blacksod Bay can be confidently correlated with the Erris Group, and this correlation has been confirmed by a chemical study, which will be summarized later.

Correlation of the psammitic and semi-pelitic rocks of the Duvillaun Islands (Fig. 14.1 and Fig. 14.4) with the Erris Group is less certain because of their geographical isolation. In a detailed study of the geology of the islands, a stratigraphic succession has been established (Max *et al.*, 1970), and a recent chemostratigraphic study has confirmed the likely correlation of the entire Duvillaun succession with the Erris Group, with the closest resemblance to the Doonamo Formation.

Figure 14.4 Slump bedding in psammite from the Duvillaun succession, on the SE point of Duvillaun More Island.

On Achill Island, Erris Group rocks may be more widely distributed than previously thought. Recent mapping (Max *et al.*, in press) suggests that revision of Kennedy's stratigraphic correlation is needed. This is because Kennedy (1969) could make little reference either to correlation with Erris Group rocks in NW Mayo, or to the scale and geometry of the tectonic dislocations within the NW Mayo inlier, as they had yet to be delineated. In NE Achill Island, however, Kennedy did equate the Ridge Point Psammite and Doogort Quartzite with what are now recognized as the Benmore and Srahlaghy Formations on the mainland respectively. In central Achill Island, east of the southern continuation of the Blacksod Fault, the thick sequence of psammites younging south-eastwards towards Ashleam Bay may also represent the upper portion of the Erris Group. Remapping has shown that the probably stratigraphically equivalent quartzites of the Dooega Head Formation and the Portnahally Formation (immediately to the NW of Ashleam Bay), and the psammites, occur in different nappes and are thus not part of an unbroken stratigraphic succession. Both quartzite units are interpreted as representatives

Figure 14.5 Desiccation cracks on a bedding surface within the quartzites of the Portnahally Formation, N side of Ashleam Bay, Achill Island.

Table 14.1 Correlation chart showing the stratigraphic succession in the Broad Haven Nappe in Erris, (NW of the Corraun Fault), and likely stratigraphic equivalents in Achill Island and in Scotland.

Broad Haven Nappe	West Achill succession	NE Achill succession	South Achill succession	Scotland
		Doogort Slide		
Srahlaghy Quartzite Formation	Gubroe Quartzite Member	Doogort Quartzite Formation	Dooega Head Formation	Lochaber
			Portnahally Formation	subgroup
Benmore Formation	Croaghaun Formation	Ridge Point Psammitic Formation	Formations within the Salia Nappe	Grampian group
Broad Haven Formation				
Belderg Formation	?			
	Duvillaun succession			
Doonamo Formation				
~~ slide ~~	~ Kinrovar Slide ~	~ Kinrovar Slide ~		~ Grampian Slide ~
Erris Complex				Central Highland Division

of the Srahlaghy Formation, as the uppermost unit youngs into the Ashleam Bay succession, the lower part of which is correlated with Appin Group.

In SE Achill Island, the whole of the Salia Nappe, which comprises psammitic, semi-pelitic and quartzitic rocks underlying rocks correlated with the Appin Group at Ashleam Bay, may also be tentatively correlated with the Erris Group. At the top of this succession is a quartzitic member, which may be the lateral equivalent of the Srahlaghy Formation. Within this member is a coarse 'rubbly' quartzite, exposed on the north side of Ashleam Bay, which preserves desiccation cracks on bedding surfaces (Fig. 14.5).

In W Achill Island, west of the southern continuation of the Blacksod Fault, the relatively monotonous quartzitic cross-bedded Slievemore Psammitic Formation was correlated by Kennedy (1969) with the Ballachulish Subgroup, and the thick, monotonous, banded and cross-bedded psammites of the Croaghaun Formation were correlated with part of the Argyll Group succession. Remapping (Max *et al.*, in press) has suggested little lithological resemblance between the Croaghaun Formation and the Argyll Group rocks recognized elsewhere in the NW Mayo inlier. There is, however, a great similarity with the Erris Group. Middle and upper Argyll Group rocks are highly bed-differentiated, thin banded quartzites, semi-pelitic schists and thin calcareous semi-pelitic schists, in the Bangor stratigraphic succession on the north Mayo mainland, where correlation of stratigraphic successions is more certain than on Achill Island—in contrast to the beds of the Croaghaun Formation. Within the Croaghaun Formation there is a psammitic lower part: a more siliceous psammite upper portion, overlain by a thin white quartzite uppermost member (the Gubroe Quartzite Member). Above, there is a sedimentary transition to the pelitic Dooagh Schist Formation (Kennedy, 1969). This sequence shows a clear resemblance to the type Erris Group succession, with the psammite equivalent to the Benmore Psammite, the Gubroe Quartzite Member to the Srahlaghy Quartzite, and the overlying Dooagh Schist representing the base of the Inver Group succession (Table 14.1). North of the Slievemore Slide on Achill Island, thick cross-bedded psammites of the Slievemore Psammitic Formation crop out on the summit and northern slopes of Slievemore, and have a close lithological resemblance to both the Croaghaun and Benmore Formations. Geochemical studies, currently in progress, should reveal whether these tentative lithological correlations are valid.

In the southern part of the NW Mayo inlier, SE of the Corraun Fault, occur other rocks which may be correlated with the Erris Group. They include the white, cross-bedded, quartzitic Cullydo and Srahmore Formations, which overlie the monotonous psammites and massive psammitic schist members of the Anaffrin Formation in the Glennamong Nappe (Max *et al.*, in press). As 'undifferentiated sediments' these rocks have been previously correlated with Argyll Group Dalradian rocks (Phillips *et al.*, 1969) but, apart from the upper quartzitic formation, these rocks are neither as quartzitic nor as varied as Argyll Group rocks seen elsewhere in Ireland or Scotland. Crow (1974) observed the likelihood that these rocks are Erris Group equivalents. Their generally monotonous lithological character, the quartzitic top to the succession and the structural position below younger Dalradian rocks suggests that they are more probably correlated with the Erris Group, rather than with the Argyll Group. Geochemical studies are currently in progress to discover whether or not this correlation is valid.

14.3 Geochemical Studies

Major- and trace-element chemical studies of many rocks within the NW Mayo inlier are currently in progress, in an attempt to recognize the geochemical signatures of the major formations, which can then be used carefully in stratigraphic correlation. Within the Erris Group type area 182 metasediments and 45 amphibolites have been collected from the major formations and analysed.

14.3.1 *Psammites*

Preliminary results show that there are major chemical differences between the Doonamo and Broad Haven

Formations, while the Benmore Formation tends to be chemically very similar to the Broad Haven Formation. The Doonamo and Belderg Formations consist of metamorphosed sandstones, with the higher Na/K, Fe_2O_3(tot.), MgO and CaO contents characteristic of a greywacke sequence, and possessing higher Cr and Ni contents indicative of a higher proportion of clastic material with a mafic provenance. By contrast, the Broad Haven and Benmore Formations comprise more feldspathic metamorphosed sandstones, characterized by lower Na/K ratios and lower contents of Fe_2O_3(tot.), MgO, CaO, Ni, Sr and V. The Sr/Rb and Sr/Y ratios are also much lower than in the Doonamo Formation, and these differences are interpreted as evidence that the Broad Haven and Benmore Formations were deposited in a very different sedimentary environment to the greywackes of the Doonamo Formation (Fig. 14.6). While the Belderg Formation tends to plot with the Doonamo Formation, suggesting that the basal part of the Erris Group in the type area consists in part of greywackes similar to the Doonamo Formation, the Benmore Formation is shown to be very similar to the Broad Haven Formation, and the Srahlaghy Formation emerges as a highly siliceous suite of very mature sands carrying exceptionally low Fe_2O_3(tot.), MgO, CaO and Sr contents (Table 14.2). Hence the broad

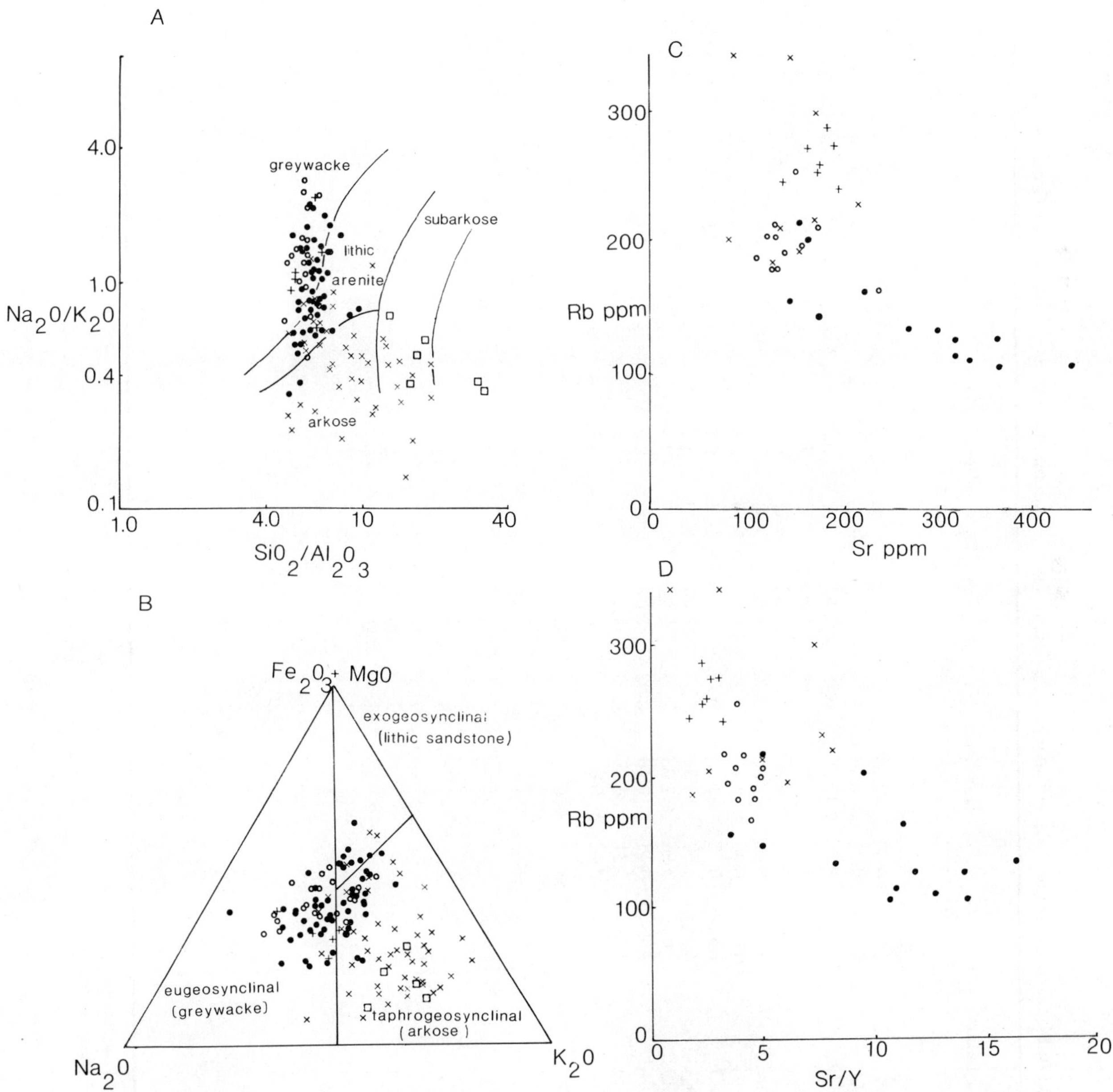

Figure 14.6 (*A*) Diagram discriminating between sandstone types (after Pettijohn *et al.*, 1973)—log Na_2O/K_2O–log SiO_2/Al_2O_3. (*B*) Ternary diagram distinguishing sandstone types using Na_2O–K_2O–(Fe_2O_3 (tot.) + MgO). (*C*) Rb–Sr variation diagram distinguishing semi-pelites in the Doonamo Formation from those higher in the stratigraphic succession. (*D*) Rb–Sr/Y discrimination diagram separating Doonamo Formation semi-pelites from those higher in the succession. Ornament: filled dots—Doonamo Formation; open circles—Duvillaun succession; upright crosses—Belderg Formation; diagonal crosses—Broad Haven and Benmore Formations; squares—Srahlaghy Formation.

Table 14.2 Mean compositions and standard deviations for the psammites from the principal Erris Group Formations.

Formation	Doonamo		Duvillaun		Belderg		Broad Haven		Benmore		Kinfinalta		Srahlaghy Quartzite	
	$\bar{x}$	s	$\bar{x}$	s	$\bar{x}$	s	$\bar{x}$	s	$\bar{x}$	s	$\bar{x}$	s	$\bar{x}$	s
SiO_2	73.99	9.60	73.70	1.90	75.09	2.40	80.98	6.36	80.07	1.64	77.73	4.47	91.55	2.44
TiO_2	0.47	0.28	0.41	0.12	0.47	0.18	0.37	0.21	0.22	0.13	0.38	0.28	0.18	0.06
Al_2O_3	11.57	1.07	12.40	0.82	12.67	1.12	9.21	3.06	10.61	1.13	10.93	1.76	3.93	1.29
Fe_2O_3	2.51	1.34	2.81	0.66	0.51	0.13	1.53	1.18	0.79	0.21	1.55	1.27	0.19	0.23
FeO	0.64	0.59	0.62	0.25	1.62	0.43	0.47	0.44	0.18	0.05	0.79	0.71	0.21	0.04
MnO	0.05	0.02	0.05	0.02	0.05	0.02	0.02	0.02	0.001	—	0.03	0.04	0.002	0.004
MgO	0.69	0.30	0.67	0.23	0.53	0.19	0.46	0.44	0.19	0.09	0.58	0.49	0.09	0.09
CaO	1.66	0.75	1.57	0.77	1.59	0.46	0.44	0.44	0.32	0.19	0.44	0.57	0.03	0.02
Na_2O	2.91	0.62	3.39	0.57	3.06	1.39	1.72	0.88	2.43	1.22	2.02	0.49	0.86	0.44
K_2O	3.01	0.86	2.88	0.56	3.32	0.99	3.45	0.97	4.31	0.98	3.98	0.76	1.76	0.45
P_2O_5	0.09	0.13	0.10	0.02	0.07	0.02	0.07	0.04	0.10	0.05	0.05	0.04	0.05	0.01
H_2O^+	0.78	0.31	1.06	0.23	1.00	0.54	0.84	0.52	0.53	0.23	1.16	0.73	0.25	0.11
S	0.01	0.02	0.01	—	0.01	—	0.02	0.01	0.01	—	0.02	0.02	0.01	—
Cl*	372	353	680	211	457	107	721	1104	712	368	1421	2080	361	87
Ba	950	201	911	193	813	190	896	260	1018	273	1061	178	460	226
Ce	33	27	42	27	59	21	35	23	20	6	31	26	7	7
Cr	38	11	34	5	18	7	31	9	28	6	28	11	19	4
Cu	2	2	1	1	2	2	2	7	0.01	—	6	15	1	1
Ga	14	2	13	3	12	2	12	3	13	1	14	2	3	4
La	19	12	17	9	34	9	16	12	9	7	16	13	6	5
Nb	8	5	6	3	10	3	6	4	3	3	7	5	5	2
Nd	34	11	31	14	41	12	31	12	22	8	41	10	37	11
Ni	13	4	15	3	5	2	8	5	4	2	9	6	2	3
Pb	11	4	12	5	15	2	7	7	7	1	6	2	4	1
Rb	87	32	90	20	100	19	103	39	129	42	121	32	57	13
Sr	302	118	304	132	282	81	142	64	159	30	116	45	52	23
Th	7	4	6	3	9	4	6	4	4	3	7	5	6	1
V	52	24	52	12	44	8	41	22	20	6	49	29	15	6
Y	18	7	18	6	27	7	16	8	16	4	18	12	19	2
Zn	40	13	44	7	31	6	34	16	24	3	40	23	10	6
Zr	162	97	129	40	291	145	159	74	133	83	176	112	196	35

* Contents in ppm, oxides as wt. %

trend of geochemical change within the Erris Group of the type area seems to reveal that the sandstones deposited became increasingly mature with time.

A comparison of the chemistry of Erris Group rocks from the Kinfinalta area, with that of the type Erris Group, suggests a close correlation with the Broad Haven and Benmore Formations. None of the samples analysed resembled the Doonamo Formation chemically, suggesting that the lower part of the local Erris Group succession has been tectonically removed adjacent to the inlier of Annagh Division gneisses (Fig. 14.1).

On the Duvillaun Islands, by contrast, the chemical characteristics of psammites from the lower four mem-

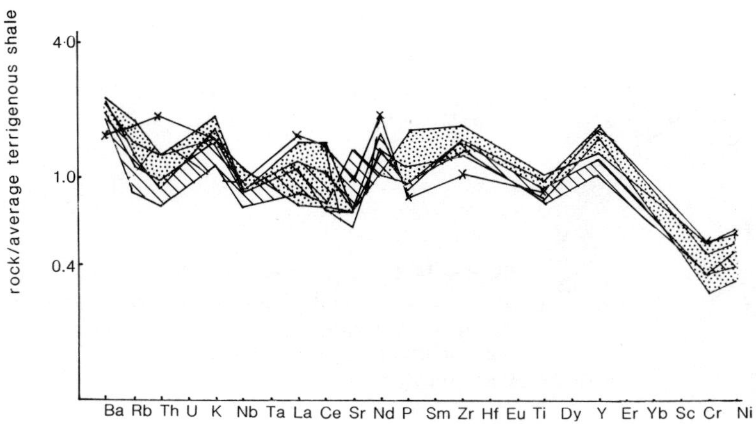

Figure 14.7 A 'spidergram' plot showing comparative mean semi-pelitic compositions of formations in the Erris Group and superjacent rocks. Ornament: stipple—Benmore and Broad Haven Formations, together with Erris Group rocks in the Kinfinalta area; stripes—Doonamo Formation and Duvillaun succession; line with crosses—Belderg Formation.

Table 14.3 Mean compositions and standard deviations for the semi-pelitic rocks from the principal Erris Group formations.

Formation	Doonamo		Duvillaun		Belderg		Broad Haven		Benmore		Kinfinalta	
	$\bar{x}$	s	$\bar{x}$	s	$\bar{x}$	s	$\bar{x}$	s	$\bar{x}$	s	$\bar{x}$	s
SiO_2	66.83	1.97	65.84	3.09	61.14	1.79	61.21	4.92	60.86	4.95	64.25	8.20
TiO_2	0.77	0.24	0.90	0.24	1.03	0.08	1.05	0.19	0.79	0.23	0.96	0.42
Al_2O_3	14.42	1.55	14.56	1.77	17.97	0.65	16.97	2.66	18.59	1.47	15.66	2.29
Fe_2O_3	3.62	1.77	4.92	0.79	1.71	0.37	5.83	0.68	4.82	2.44	3.90	1.10
FeO	1.90	2.21	1.26	0.35	5.37	0.42	1.71	1.10	0.62	0.27	2.47	1.58
MnO	0.07	0.02	0.07	0.03	0.12	0.01	0.08	0.04	0.02	0.01	0.095	—
MgO	1.56	0.46	1.86	0.24	1.72	0.22	1.96	0.33	1.54	0.58	1.85	0.91
CaO	1.82	0.74	1.14	0.35	1.30	0.19	0.71	0.39	0.54	0.14	0.92	0.40
Na_2O	3.02	1.54	0.94	0.66	1.81	0.27	0.76	0.58	0.93	1.07	1.14	0.20
K_2O	4.33	1.00	6.15	0.77	5.30	0.30	5.50	0.82	6.91	0.74	5.21	0.20
P_2O_5	0.15	0.06	0.14	0.03	0.22	0.10	0.26	0.03	0.18	0.05	0.15	0.15
H_2O^+	1.27	0.52	2.06	0.40	2.15	0.67	2.82	0.82	1.25	0.59	2.87	0.86
S	0.01	0.01	0.01	—	0.01	—	0.03	0.02	0.01	0.01	0.02	—
Cl*	576	634	555	199	458	113	457	254	474	267	770	486
Ba	1213	267	1447	368	1020	128	1395	149	1446	88	1291	20
Ce	62	34	87	47	175	18	60	27	113	71	57	26
Cr	41	9	41	7	51	3	56	25	33	12	49	10
Cu	5	4	11	19	10	7	11	14	2	2	19	26
Ga	18	3	22	3	23	2	23	4	25	4	18	1
La	33	16	46	21	83	12	28	12	54	33	42	1
Nb	14	4	17	5	22	1	21	6	16	9	17	8
Nd	34	10	52	16	70	12	27	17	43	28	43	21
Ni	22	5	26	3	28	4	30	4	19	11	27	11
Pb	13	4	12	3	25	4	13	5	17	6	13	8
Rb	138	52	204	21	271	17	235	60	296	65	177	20
Sr	267	87	138	34	172	19	142	42	140	47	120	11
Th	11	4	13	3	27	3	18	8	19	10	14	12
V	84	25	108	28	109	11	121	23	72	27	118	36
Y	28	7	33	7	62	7	35	15	41	35	47	20
Zn	71	18	77	17	116	12	104	18	66	23	91	32
Zr	280	96	303	89	270	37	359	73	266	17	308	132

* Contents in ppm, oxides as wt. %

bers described (Max *et al.*, 1970) show a closer resemblance to the Doonamo Formation. Only the upper members show chemical differences, and they may belong to an intermediate part of the succession, perhaps analogous to the Belderg Formation in the type area. These upper members are separated from the lower members of the Duvillaun succession by a fault of unknown displacement. No clear chemical discrimination between individual members of the lower part of the Duvillaun succession has yet emerged.

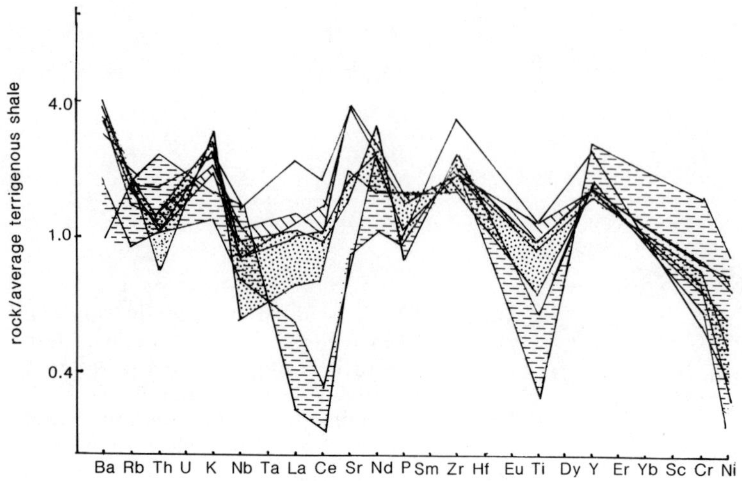

Figure 14.8 A 'spidergram' plot showing comparative mean compositions of psammites in Erris Group Formations. Ornament: as in Figure 14.3, but including dashes—denoting Srahlaghy and Portnahally Formations.

Table 14.4 Mean compositions and standard deviations for the heavy-mineral bands within some Erris Group Formations. Compositions of heavy-mineral bands from the Inishkea Division and the Inver Group Dalradian rocks are also listed for comparison.

Formation	Doonamo		Benmore		Duvillaun (Member 1)		Duvillaun (Member 6)		Inishkea		Inver	
	$\bar{x}$	s	$\bar{x}$	s	$\bar{x}$	s	$\bar{x}$	s	$\bar{x}$	s	$\bar{x}$	s
SiO_2	55.09	15.97	61.47	16.24	49.95	4.03	24.47	0.93	46.27	8.51	74.10	3.99
TiO_2	3.44	1.90	6.94	7.56	3.05	0.00	7.51	0.71	4.87	3.03	3.74	1.98
Al_2O_3	9.70	2.43	10.81	5.70	11.35	4.23	4.03	0.77	13.44	4.16	5.30	3.85
Fe_2O_3	19.13	13.77	8.50	8.52	19.80	15.42	54.42	0.34	9.80	8.11	11.80	7.53
FeO	2.12	1.42	0.94	0.57	2.89	0.84	2.19	0.16	8.98	2.74	1.15	0.43
MnO	0.14	0.09	0.13	0.17	0.19	0.06	0.22	0.10	0.29	0.15	0.02	0.02
MgO	1.88	1.35	0.67	0.24	2.87	0.98	2.00	0.38	2.99	0.89	0.30	0.27
CaO	2.63	1.62	3.89	4.77	3.91	2.49	2.43	0.01	6.23	2.27	0.09	0.16
Na_2O	1.53	0.79	2.12	2.44	0.10	0.11	0.22	0.02	0.58	0.72	1.09	1.58
K_2O	2.80	1.19	3.03	1.44	3.39	1.77	1.38	0.01	4.21	1.50	1.23	0.76
P_2O_5	0.17	0.10	0.37	0.38	0.19	0.06	0.15	0.02	0.39	0.17	0.06	0.06
H_2O^+	1.37	0.33	1.06	0.26	2.35	0.40	1.15	0.19	2.05	0.69	0.75	0.26
S	—	—	—	—	0.01	—	—	—	0	—	—	—
Cl*	444	105	523	77	423	4	900	629	0	—	367	298
Ba	745	311	809	263	928	353	429	147	708	360	316	249
Ce	292	132	350	292	284	64	630	132	0	—	58	26
Cr	136	138	61	47	117	—	0	—	131	43	54	19
Cu	48	47	29	30	25	6	57	4	0	—	8	5
Ga	26	9	28	12	32	4	46	5	0	—	16	5
La	100	29	118	101	91	60	233	135	0	—	24	12
Nb	51	20	150	164	44	28	138	7	50	21	85	50
Nd	86	23	115	70	108	11	155	51	0	—	40	22
Ni	44	41	11	4	55	12	79	19	25	6	10	3
Pb	29	15	35	31	36	10	58	—	0	—	21	11
Rb	121	78	129	58	141	50	65	25	198	45	55	35
Sr	253	91	158	55	414	156	116	42	337	148	59	76
Th	37	21	137	185	42	8	124	37	0	—	33	13
V	284	217	271	279	248	28	657	96	0	—	183	93
Y	60	25	169	145	69	37	129	23	146	56	50	16
Zn	133	89	54	24	158	20	192	62	0	—	24	5
Zr	1284	681	6035	8411	1427	588	4142	125	2544	1637	3137	1520
n =	6		5		2		2		8		4	

* Contents in ppm. Oxides as wt%.

14.3.2 *Semi-pelites*

Semi-pelitic rocks from the Erris Group show a slightly different pattern of change. While Sr content is higher, and Rb and Y contents are lower in the Doonamo Formation than in the Broad Haven and Benmore Formations, Y–Sr and Rb–Sr/Y discriminant plots show that the semi-pelites from the Belderg Formation are more akin to those of the Broad Haven and Benmore Formations. The semi-pelites from the Duvillaun Islands tend to plot in an intermediate field, with considerable overlap into both the Doonamo and Broad Haven fields. The relatively higher Al_2O_3, Rb and Y, and correspondingly low CaO, Na_2O and Sr in the semi-pelites of the Broad Haven and Benmore Formations, indicates that the original argillite contained little detrital plagioclase. This is in contrast to the semi-pelites from the Doonamo Formation, in which detrital plagioclase must have been a major constituent of the original sediment (Fig. 14.6). The Sr/Y ratios show an equally marked contrast, which probably reflects the proportions of such Y-rejecting, Sr-accepting phases as plagioclase (Lambert and Holland, 1974). In the Doonamo Formation semi-pelites, the Sr/Y ratio is 9.5, whereas in semi-pelites from the Broad Haven Formation the mean Sr/Y ratio is 3.8. Alkali concentrations also vary systematically. Na_2O/K_2O varies from 0.70 in the Doonamo Formation to 0.36 in the Broad Haven Formation, reflecting reduced plagioclase/mica ratios in the higher part of the Erris Group succession. Semi-pelites from the Belderg Formation are similar to those in the Broad Haven and Benmore Formations, while

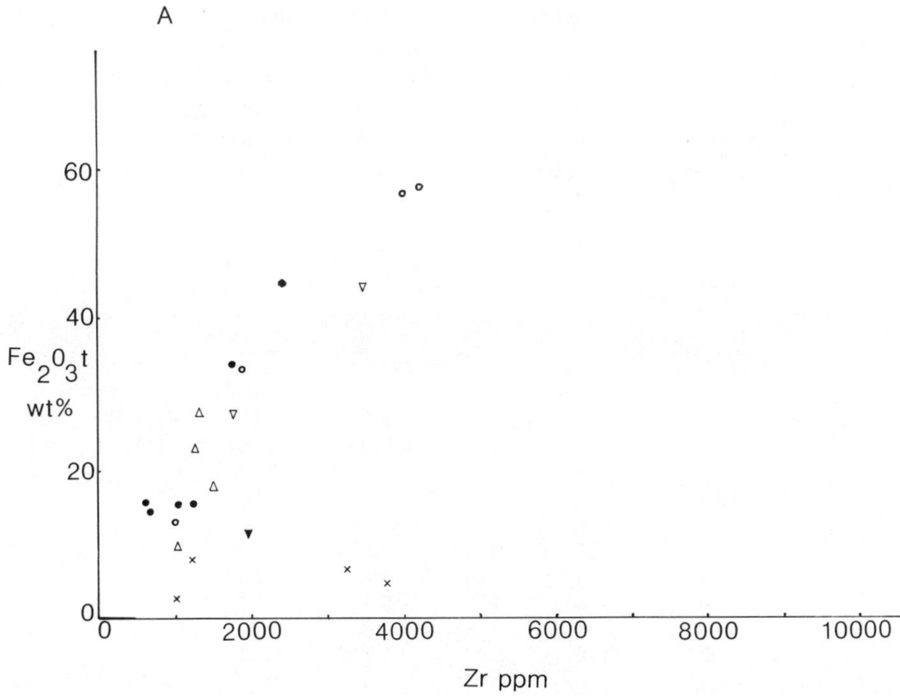

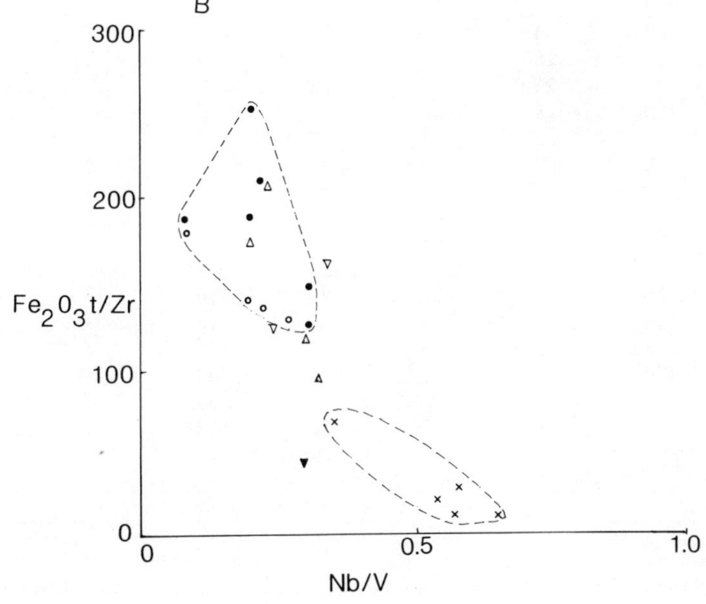

Figure 14.9 Variation diagrams contrasting the compositions of heavy-mineral bands from different Erris Group formations. (*A*) Fe_2O_3 (tot.)–Zr; (*B*) Fe_2O_3(tot.) Zr–Nb/V. Ornament (both diagrams): filled dots—Doonamo Formation; open circles—Duvillaun succession; upright crosses—Belderg Formation; diagonal crosses—Benmore Formation; triangles—South Achill succession; inverted triangles—Ridge Point Psammite; filled inverted triangle—Salia Nappe.

those from the Duvillaun succession occupy an intermediate position between the Broad Haven and Doonamo fields (Fig. 14.6). The chemical evidence, therefore, suggests that there was a progression from immature greywacke sedimentation at the base of the Erris Group, up into mature, possibly shallow-water well-sorted sands at the top.

When the mean concentrations of trace and minor elements in Erris Group semi-pelitic rocks are plotted on a spidergram, with values normalized against average terrigenous shale (Taylor and McLennan, 1985), all the formations show a broadly level trend—distinguished only by depletion in Cr and Ni, and the reduction of the Sr content higher in the stratigraphic succession (Fig. 14.7) (see also Table 14.3). These broadly level trends are consistent with the development of these formations in an ensialic basin or on a passive continental margin—by analogy with the pattern displayed by Phanerozoic sediments, using data from Van de Kamp and Leake (1985) (Fig. 13.16).

In the same way, a spidergram showing variations in psammitic chemistry in the Erris Group likewise shows a broadly level trend similar to those of younger continent-derived sandstones (Fig. 13.16). However, there are marked fluctuations in element concentrations compared with average shale (Fig. 14.8). Lower overall values reflect the greater SiO_2 concentrations in the psammites. Particularly notable is the reduction of La, Ce, Sr and Ti concentrations in formations higher in the stratigraphic succession, which may be stratigraphic equivalents of the Lochaber Subgroup in Scotland. However, this pattern again suggests deposition in an ensialic basin.

Chemical results from most formations sampled on Achill Island and in the area on the mainlands (SE of the Corraun Fault) are not yet complete, and only limited chemical confirmation of the suggested correlations is available.

14.3.3 *Heavy-mineral bands*

The differences between the Doonamo and Benmore Formations are exaggerated by comparisons of the compositions of the thin, heavy-mineral bands which occur in both formations. One distinction between the Broad Haven and Benmore Formations is the scarcity of heavy-mineral concentrations in the former formation. In the Doonamo Formation, magnetite is the dominant

Table 14.5 Mean compositions and standard deviations for amphibolite bodies intruding Erris Group Formations.

Formation	Doonamo		Belderg		Broad Haven		Srahlaghy		Duvillaun	
	$\bar{x}$	s	$\bar{x}$	s	$\bar{x}$	s	$\bar{x}$	s	$\bar{x}$	s
SiO_2	48.80	0.62	49.43	0.79	49.04	0.32	48.93	0.32	47.65	1.31
TiO_2	2.34	0.55	2.62	0.37	3.13	0.58	2.65	0.39	1.99	0.27
Al_2O_3	13.65	0.80	12.87	0.44	12.81	0.38	13.70	0.63	13.57	0.82
Fe_2O_3	4.00	1.62	5.18	1.01	3.63	1.30	1.73	0.36	5.81	0.57
FeO	9.65	1.37	10.69	1.08	11.06	0.96	11.64	1.01	7.73	0.82
MnO	0.20	0.02	0.23	0.01	0.22	0.02	0.20	0.02	0.28	0.09
MgO	6.07	0.71	5.22	0.50	5.64	0.44	5.62	0.52	6.38	0.83
CaO	9.47	1.06	9.30	0.38	8.36	0.78	9.15	0.51	7.44	0.73
Na_2O	1.99	0.78	1.90	0.25	2.41	0.20	2.61	0.32	2.77	1.00
K_2O	1.01	0.60	0.70	0.08	0.75	0.11	1.08	0.38	1.93	1.06
P_2O_5	0.29	0.08	0.35	0.04	0.40	0.08	0.34	0.09	0.17	0.03
H_2O^+	2.09	0.69	1.63	0.14	2.35	0.36	1.49	0.17	3.70	1.93
S	0.02	0.02	0.03	0.02	0.01	0.004	0.01	0.005	0.02	0.01
Cl*	660	592	359	75	672	307	517	131	514	125
Ba	149	122	121	30	189	84	174	31	407	196
Ce	18	13	21	10	32	15	31	8	7	8
Cr	119	55	94	28	127	20	74	33	228	35
Cu	139	62	269	54	280	34	202	93	219	162
Ga	22	2	20	2	24	2	23	2	23	4
La	6	6	15	5	5	2	9	9	4	2
Nb	13	3	14	2	16	4	16	2	10	3
Nd	39	11	41	14	42	10	45	10	26	15
Ni	72	14	65	10	67	4	62	6	77	8
Pb	14	5	19	2	14	3	24	22	14	21
Rb	31	29	16	4	16	3	36	23	76	37
Sr	227	47	179	38	239	29	244	30	519	272
Th	2	2	4	2	0.2	0.3	3	2	3	2
V	388	85	488	35	448	29	373	25	366	75
Y	39	7	47	11	50	5	42	5	35	6
Zn	121	28	130	10	132	10	117	7	169	33
Zr	161	49	170	26	223	49	176	21	112	17
$n=$	24		8		8		5		5	

* Content in ppm, oxides as wt. %

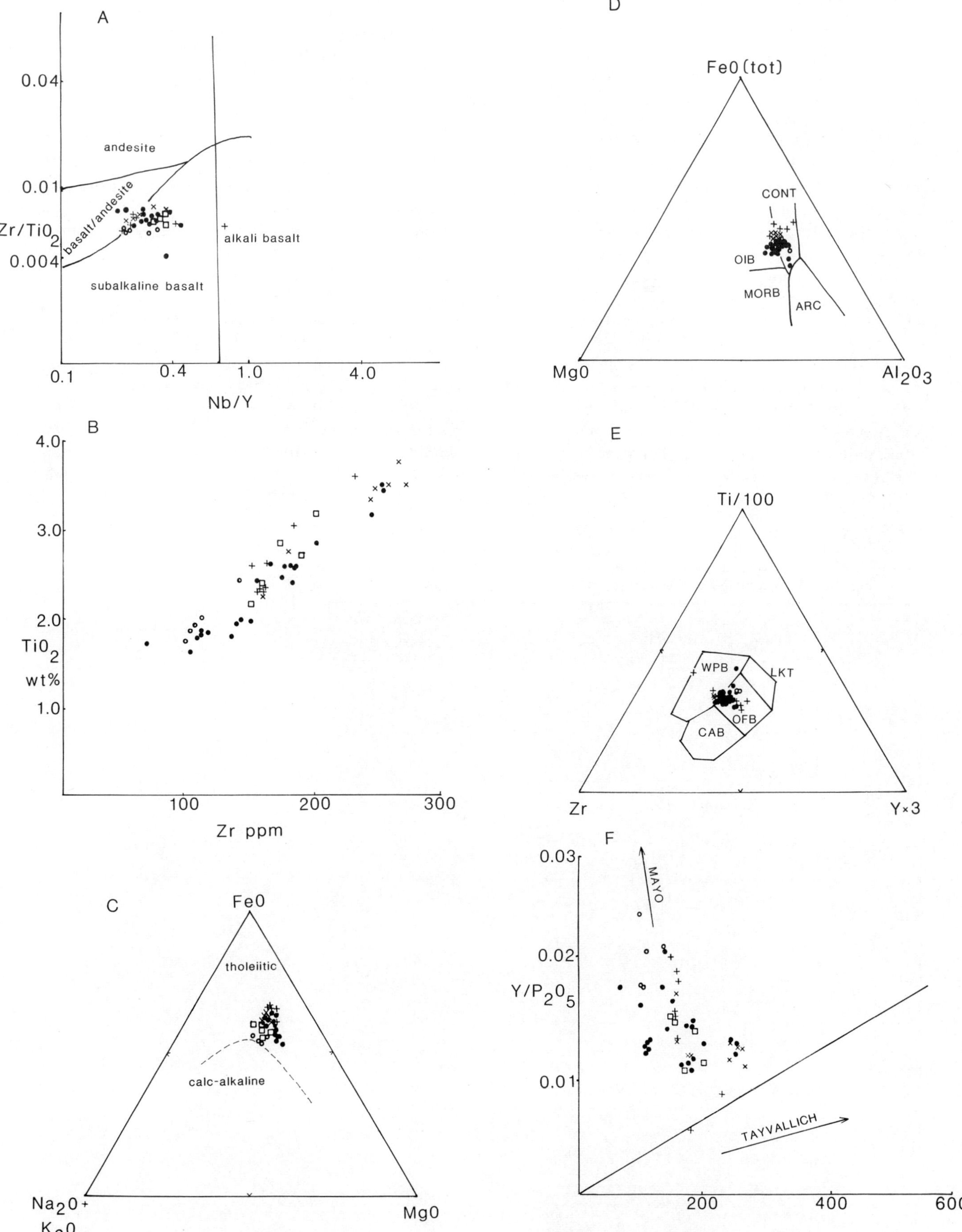

Figure 14.10 Variation diagrams relating to orthoamphibolites in the Erris Group. (*a*) Zr/TiO_2–Nb/Y; (*b*) TiO_2–Zr, showing fractionation; (*c*) AFM diagram, revealing the tholeiitic nature of the original basalts; (*d*) FeO (tot.)–MgO–Al_2O_3 ternary diagram (after Pearce *et al.*, 1977) indicating tectonic setting: CONT—continental, ARC—island arc, MORB—mid-ocean ridge, OIB—oceanic island; (*e*) Ti–Zr–Y ternary diagram (after Pearce and Cann, 1976) indicating tectonic setting CAB—calc-alkali basalt, LKT—low-K tholeiite, OFB—ocean-floor basalt, WPB—Within-plate basalt; (F) Y/P_2O_5–Zr discrimination diagram distinguishing Tayvallich metavolcanics from those in NW Mayo. Discriminant line from Winchester *et al.*, in press.

mineral in heavy-mineral bands, and consequently Fe_2O_3(tot.)/Zr ratios exceed 80, whereas, in the Broad Haven Formation, zircon is much more abundant and many Fe_2O_3(tot.)/Zr ratios are less than 20 (Fig. 14.9) (Table 14.4). Samples of heavy-mineral bands from the Duvillaun Islands also contain high Fe, thus providing confirmation of the correlation between the Duvillaun succession and the Doonamo Formation. However, the heavy-mineral bands sampled from the lowest member of the Duvillaun succession (Member 1) (Table 14.4) are compositionally most similar to those from the Doonamo Formation. Heavy-mineral bands from near the top of the Duvillaun succession (Member 6) have a higher Fe content than any sampled from the Doonamo Formation, although their Fe_2O_3(tot.)/Zr ratio remains similar (Fig. 14.9). Heavy-mineral bands, sampled from the probable Erris Group psammites of South Achill Island and the Ridge Point Psammite of NE Achill Island, also show similar Fe_2O_3(tot.)/Zr ratios, different from those in the Benmore Formation, which they might otherwise be equated with. This difference illustrates that correlation of the possibly Erris Group rocks of Achill Island with those on the mainland is still somewhat speculative.

14.3.4 *Amphibolites*

Dykes and pods of amphibolite occur throughout the Erris Group in the type area. Chemical studies have revealed that all are tholeiitic in composition (Table 14.5, Fig. 14.10A, C). While some compositional variation occurs as a result of fractionation (Fig. 14.10B), no discrimination could be made between the bodies obtained from different formations of the Erris Group, and hence all appear to belong to the same magmatic suite. Tectonic setting diagrams imply that all were emplaced in a continental setting (Fig. 14.10D). Amphibolites collected from Duvillaun Beg Island also share the same chemistry, perhaps confirming once again that the Duvillaun succession should be equated with the Erris Group. The fractionation trends of these rocks are different from those of the

Table 14.6 A comparative table, showing the compositions of Erris Group amphibolites, and those intruding the Scotch Port Schist of the Inishkea Division, younger metadolerites from Annagh Division, and Dalradian Argyll Group basic metavolcanic rocks from the NW Mayo and Ox Mountains inliers.

	Scotch Port Schist	Annagh metadolerites	Erris group	NW Mayo Argyll Group in metavolcanics
SiO_2	49.74	49.26	48.77	48.66
TiO_2	3.15	2.45	2.55	1.73
Al_2O_3	13.09	12.62	13.32	14.88
Fe_2O_3	3.79	4.12	4.07	4.68
FeO	12.03	11.01	10.15	6.62
MnO	0.21	0.22	0.23	0.17
MgO	5.08	5.74	5.79	6.95
CaO	8.63	9.66	8.74	8.47
Na_2O	1.27	1.41	2.34	2.92
K_2O	0.76	1.08	1.09	0.35
P_2O_5	0.33	0.23	0.31	0.22
H_2O^+	2.11	2.15	2.25	3.50
S	—	—	0.02	—
Cl*	—	—	544	—
Ba	105	161	208	57
Ce	55	33	22	29
Cr	107	111	128	281
Cu	—	—	222	44
Ga	—	—	22	19
La	21	11	8	11
Nb	13	9	14	12
Nd	29	28	39	19
Ni	46	51	69	130
Pb	—	—	17	10
Rb	21	27	35	11
Sr	159	188	282	300
Th	—	—	2	2
V	—	—	413	263
Y	60	49	43	30
Zn	—	—	134	102
Zr	249	161	168	134
n =	13	13	50	61

* Content in ppm, oxides as wt. %

Southern Highland Group Tayvallich metavolcanics from the SW Highlands of Scotland (Wilson and Leake, 1972), but comparable with the Argyll Group metavolcanics, which crop out in the southern part of the NW Mayo inlier and the SW Ox Mountains inlier (Winchester *et al.*, in press); hence they may be related to the latter group of metavolcanic rocks (Fig. 14.10F and Table 14.6). We conclude, therefore, that the amphibolite pods scattered throughout the Erris Group may represent feeder dykes, which supplied magma to the Argyll Group Dalradian (late Precambrian) metavolcanics that are a widespread feature of NW Ireland.

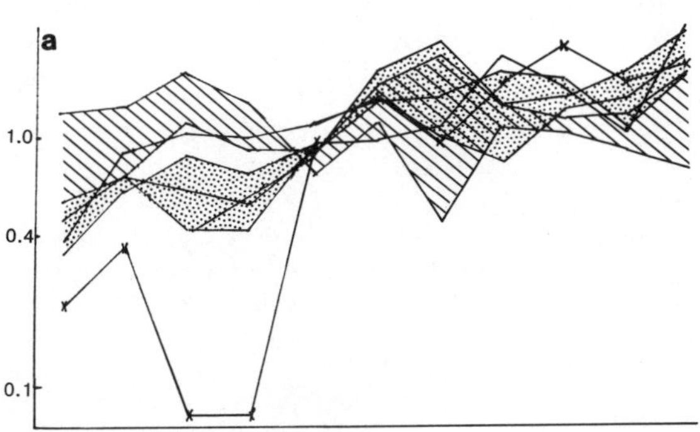

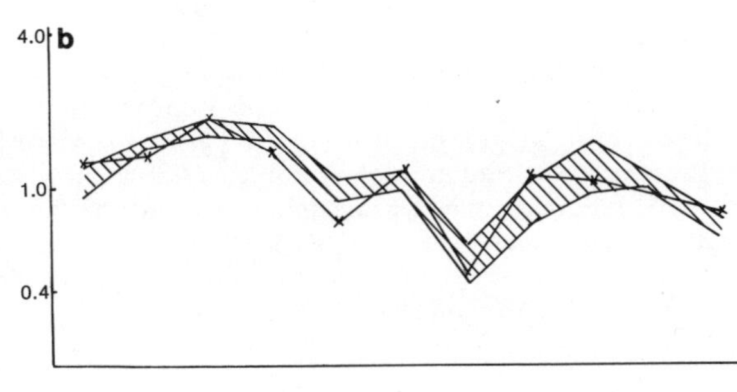

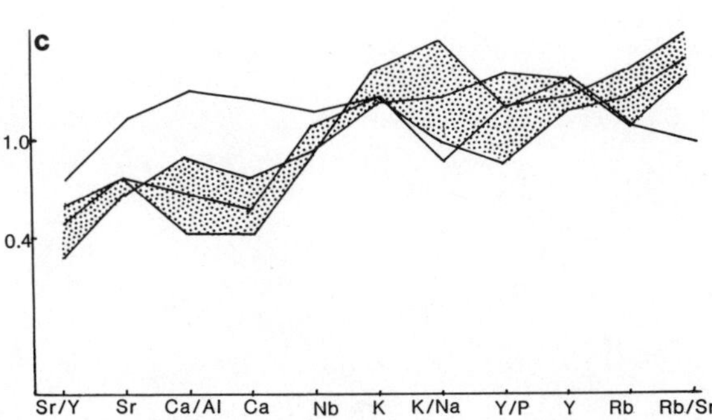

Figure 14.11 Diagrams comparing mean element concentrations and ratios, selected to illustrate differences between semi-pelites from the Grampian Group and Lochaber Subgroup of the Dalradian Appin Group. Erris Group rocks are plotted for comparative purposes. (*a*) Erris Group semi-pelitic rocks; ornament: stipple—Broad Haven and Benmore Formations, together with Erris Group rocks from Kinfinalta; stripes—Doonamo Formation and Duvillaun succession; line—Belderg Formation; line with crosses—Portnahally Formation. (*b*) A comparison between the Doonamo Formation (line with crosses) and the Corrieyairack Subgroup (Grampian Group) (shown in striped ornament). (*c*) A comparison between the Broad Haven Formation, Benmore Formation, and Kinfinalta Erris Group and the Glen Spean Subgroup rocks (Grampian Group), the latter shown as a single line.

14.4 Correlation with the Grampian Group

Both the Erris Group in Ireland and the Grampian Group in Scotland are directly overlain by rocks assigned to the Appin Group of the Dalradian Supergroup, and therefore it is logical to compare the two groups. Both consist dominantly of psammitic rocks, and some correlation might therefore seem feasible between them, despite their 400 km separation. Chemical studies of both groups do not support close correlation of individual formations, but similar broad geochemical changes occur with time, in both groups. When semi-pelitic rocks from Erris Group formations are plotted on a diagram designed to emphasize their chemical differences (Fig. 14.11A), the changes are found to be similar to those shown by Grampian Group rocks (Fig. 13.14). Thus, the Doonamo Formation is typified by the high Sr/Y, Sr, Ca/Al and Ca, and low K/Na, Y, Rb and Rb/Sr which also characterize the Corrieyairack Subgroup of the Grampian Group (Fig. 14.11B). Likewise, the Broad Haven and Benmore Formations, together with the rocks from the Kinfinalta area, possess the lower Sr/Y, Sr, Ca/Al and Ca also shown by the Glen Spean Subgroup in the Grampian Group (Fig. 14.11C). The semi-pelites associated with the quartzites which overlie these rocks in Achill Island, which should perhaps be assigned to the base of the Appin Group, show a very distinctive chemistry, with markedly low Sr/Y, Ca/Al and Ca, and high Rb/Sr which are also characteristic of the lower semi-pelites within the Appin Group in the type area.

14.5 Conclusions

To conclude, therefore, the chemical signatures of Erris Group formations broadly support their correlation with the Grampian Group. Differences between the two groups are not surprising because of their geographical separation. These differences include the lesser thickness of the Doonamo Formation compared with the Corrieyairack Subgroup, and the much thicker development of the higher formations of the Erris Group compared with the modest thickness of the Glen Spean Subgroup in Glen Spean. A greater abundance of heavy-mineral bands is also a feature of the Erris Group. No sedimentological work has yet been undertaken in the Erris Group, and, as a result, insufficient information is currently available to allow a detailed model for an Erris Group–Grampian Group basin to be described. However, the evidence of emergence and subaerial conditions, displayed in the quartzites at the top of the Erris Group in NW Mayo, confirms the shadowy chemical evidence suggesting that Grampian Group/Erris Group deposition occurred in a basin separated from (and earlier than) that of the Appin Group sedimentation. This evidence reinforces the suggestion that neither of these groups are integral part of the Dalradian Supergroup.

References

Crow, M. J. (1974) Geology of Metamorphic Rocks in Part of NW Co. Mayo, Ireland. Unpublished Ph.D. Thesis, University of Dublin.

Crow, M. J., Max, M. D. and Sutton, J. (1971) Structure and stratigraphy of the metamorphic rocks in part of NW Co. Mayo, Ireland. *J. geol. Soc. London* **127**, 579–585.

Harris, A. L., Baldwin, C. T., Bradbury, H. J., Johnson, H. D. and Smith R. A. (1978) Ensialic basin sedimentation: the Dalradian Supergroup. In Bowes, D. R. and Leake, B. E. (eds.), Crustal Evolution in Northwestern Britain and Adjacent Regions. *Geol. J. Spec. Issue* **10**, 115–138.

Kennedy, M. J. (1969) The structure and stratigraphy of the Dalradian rocks of north Achill Island, Co. Mayo, Ireland. *Q. J. geol. Soc. London* **125**, 47–81.

Lambert, R. StJ. and Holland, J. G. (1974) Yttrium geochemistry applied to petrogenesis utilising calcium–yttrium relationships in minerals and rocks. *Geochim. Cosmochim. Acta* **38**, 1393–1414.

Long, C. B. and Max, M. D. (in prep) Contribution to terrane map of the circum-Atlantic region.

MacGregor, A. G. (1948) Resemblances between Moine and 'Sub-Moine' metamorphic sediments in the western Highlands of Scotland. *Geol. Mag.* **85**, 265–275.

Max, M. D. (1970) Mainland gneisses southwest of Bangor in Erris, Co. Mayo, Ireland. *Sci. Proc. R. Dublin Soc.* **3A**, 275–291.

Max, M. D., Phillips, W. E. A. and Bruck, P. M. (1970) Geology of the Duvillaun Islands. *Sci. Proc. R. Dublin Soc.* **3A**, 257–268.

Max, M. D., Long, C. B. and MacDermot, C. V. (in press) The solid geology of Sheet 6, North Mayo. *Geol. Surv. Ire. Map* (1:126 720).

Pearce, J. A. and Cann, J. R. (1973) Tectonic setting of basic volcanic rocks determined using trace element analyses. *Earth Planet. Sci. Letters* **19**, 290–300.

Pearce, T. H., Gorman, B. E. and Birkett, T. C. (1977) The relationship between major element chemistry and tectonic environment of basic and intermediate volcanic rocks. *Earth Planet. Sci. Letters* **36**, 121–132.

Phillips, W. E. A., Kennedy, M. J. and Dunlop, G. M. (1969) Geologic comparison of western Ireland and northeastern Newfoundland. In Kay, M. (ed.), North Atlantic—Geology and Continental Drift. *Mem. Am. Ass. Pet. Geol.* **12**, 194–211.

Taylor, S. R. and McLennan, S. M. (1985) *The Continental Crust: its Composition and Evolution.* Blackwell Scientific, Oxford.

Van de Kamp, P. C. and Leake, B. E. (1985) Petrography and geochemistry of feldspathic and mafic sediments of the northeastern Pacific margin. *Trans. R. Soc. Edinburgh: Earth Sci.* **76**, 411–449.

Wilson, J. R. and Leake, B. E. (1972) The petrochemistry of the epidiorites of the Tayvallich Peninsula, North Knapdale, Argyllshire. *Scott. J. Geol.* **8**, 215–251.

Winchester, J. A. (1975) Epidotic calc-silicates of Arnipol type—a widely distributed minor sedimentary facies of the Moinian Assemblage, Scotland. *Geol. Mag.* **112**, 175–181.

Winchester, J. A. and Max, M. D. (1984) Element mobility associated with syn-metamorphic shear zones near Scotchport, NW Mayo, Ireland. *J. metamorph. Geol.* **2**, 1–11.

Winchester, J. A., Max, M. D. and Long C. B. (1987) Trace element geochemical correlation in the reworked Proterozoic Dalradian metavolcanic suites of the western Ox Mountains and NW Mayo inliers, Ireland. In Pharaoh, T. C., Beckinsale R. D. and Rickard, D. T. (eds.), Geochemistry and Mineralization of Proterozoic Volcanic Suites. *Spec. Publ. geol. Soc. London.* **33**, 489–502.

15
The Appin Group

A. E. WRIGHT

15.1 Introduction

The Appin Group consists of a very varied group of largely shelf sediments, which can be recognized throughout the whole of the Dalradian outcrop, from Connemara in western Ireland to the Banff coast of north-east Scotland. Possible correlatives of this group also occur in Shetland. The type area from which it takes its name is in the north-western part of Argyllshire (now the south-western part of the Highland Region and the northern part of Strathclyde Region), and in this area the rocks are, in places, at a relatively low grade of metamorphism and have suffered moderate to low amounts of strain. It is therefore only in this area that any sedimentological studies are possible, although even here there are very few significant pieces of work (Hickman, 1975; Litherland, 1980; Stedman, 1986). It was also in this area that chemostratigraphy was first shown to be useful in unravelling this tectonically complex Dalradian terrain (Hickman, 1975; Hickman and Wright, 1983), and further geochemical studies have since been undertaken by the present author and by Rock (1985*a*, *b*; 1986).

15.2 Stratigraphy

The Appin area is historically important for the use of sedimentary structures to indicate the way-up of strata. In the lower quartzites of the Appin Group, Vogt (1930) and Bailey (1930) demonstrated that cross-bedding and grading could be used to determine the order of deposition. Such is the structural complexity, with many of the rocks inverted or vertical, that previous interpretations had the sequence inverted from the then newly discovered order (Bailey, 1917, 1930, 1934).

Although the stratigraphy was known in some detail from the early part of this century onwards (largely from the work of Bailey, 1917, 1922), the foundations of modern stratigraphic subdivision, and particularly correlation of these formations over the whole of the Dalradian, are due to the work of the Precambrian Subcommittee of the Geological Society of London (Harris and Pitcher, 1975). This work of compilation, which also involved consultation with all workers studying the Dalradian at that time, produced an invaluable basic stratigraphic framework, on which further sedimentological, structural and geotectonic work could build. The Dalradian Supergroup was divided into three major groups, the Appin, Argyll and Southern Highland—the Appin and Argyll Groups being divided into subgroups. The rocks below the Appin Group had been traditionally regarded as belonging to the Moinian Assemblage, but in 1978, Harris and his co-workers proposed that these rocks should be regarded as part of the Dalradian, forming a fourth major group, the Grampian Group (Harris *et al.*, 1978). This has not won universal acceptance, and Lambert *et al.* (1982) have shown that there is a distinct geochemical difference between the rocks of the Grampian Group and Dalradian semi-pelitic schists, such as the Lochaber Schist Formations, although Rock and his co-workers (Rock *et al.*, 1986) suggest that pelitic rocks in the Grampian Group are closely similar to those of the Dalradian, and distinctly different from Moinian and Lewisian pelites.

The stratigraphic terminology of the Dalradian is confused. Old usage, whereby the same geographical name is used for several formations, e.g. Ballachulish Slates, Ballachulish Limestone, Appin Quartzite, Appin Phyllite, Appin Limestone, has been compounded by the Geological Society Committee in using Appin for a group name and Ballachulish for a subgroup name; and Islay has been used for a subgroup name that does not even include the Islay Limestone Formation. This type of multiple usage still persists in recently erected stratigraphic units equivalent to these older units (e.g. Smith and Harris, 1976; Treagus and King, 1978). The reason why this is bad practice can be illustrated by the case of the Appin Limestone and Appin Phyllite, which actually appear to be an interbedded formation of limestones and phyllites. Due to the incorrect usage it is not possible simply to use Appin Formation instead.

The sequence now recognized (see Figs 15.1–15.3) follows on conformably (in the Appin area) from the semi-pelitic flags of the Grampian Group, starting with the Lochaber Subgroup of semi-pelitic schists and interbedded quartzites. In the type area there are three major quartzites, but these are known to die out stratigraphically northwards (Bailey, 1960; Hickman, 1975), and in most other areas where this subgroup is recognized there is only one major quartzite. The subgroup ends with the Leven Schist, a formation of rather mixed facies, comprising sands and silts with a significant carbonate component towards the top. The Ballachulish Subgroup, which follows, is composed of the following formations in the type area: Ballachulish Limestone, Ballachulish Slate, Appin Quartzite, Appin Phyllite and Limestone. The Ballachulish Limestone and Ballachulish Slate both seem to be fairly widespread, but the Appin Quartzite, with its transitional member (the 'striped transition' of Bailey, 1960), displays quite wide variations within the Appin area itself and further afield is not always recognizable. The Appin Phyllite and Limestone Formation is a

Group	Subgroup	CONNEMARA	S.W. MAYO, ACHILL I.	N.W. MAYO	CENTRAL DONEGAL
		Tanner and Shackleton, 1979			Pitcher and Berger, 1972 Central Donegal Succession
	Islay	Cleggan BB Fmn		Briska BB Fmn	Gaugin BB
APPIN	Blair Atholl	Barnanoraun Fmn Connemara Mbl Fmn Clifden Fmn	Doogort Lst Doogort Sch Doogort Qzt	Srahlaghy Mbl Fmn Inishderg Fmn	Owengarve Fmn
APPIN	Ballachulish		Kildownet Lst Ashleam Head Psammite Ashleam Bay Pel & Lst	Inver Sch Fmn Doon-a-dell Fmn Kinfinalta Fmn	
APPIN	Lochaber		Kildun Sch Sraheens Qzt Cashel Qzt Minaun Qzt Slievemore Qzt	Bellagarvann Fmn Pollacappal Fmn	
Grampian				Srahlaghy Qzt — Erris Group	

Group	Subgroup	GLENCOLUMBKILLE	MAAS-ROSBEG	CREESLOUGH	FANAD
		Pitcher and Berger, 1972 Kilmacrenan Succession	Pitcher and Berger, 1972 Creeslough Succession	Pitcher and Berger, 1972; Pitcher, 1977 Creeslough Succession	Pitcher and Berger, 1972 Kilmacrenan Succession
	Islay				
APPIN	Blair Atholl	Glencolumbkille Dol Fmn 25 Glencolumbkille Sch Fmn 100	Fintown Pel Fmn Loughros Fmn Rosbeg Pel Fmn Portnoo Lst Fmn Maas semipel Fmn	Fintown Pel Fmn 125 Loughros Fmn 155 U. Falcarragh Pel Fmn 250 Falcarragh Lst Fmn 170 200 L. Falcarragh Pel Fmn 500	Fanad Dolomitic Lst Fmn 120 Rosnakill Pel Sch Fmn 120
APPIN	Ballachulish	Sessiagh Clonmass Fmn: Mulamin Calc-silicate Flag Mbr, Mulamin siliceous Flag Mbr Cor Qzt Fmn	(see Maas-Rosbeg column)	Port Lst Mbr Sessiagh Banded Qzt Mbr Marble Hill Lst Mbr 350 450 Clonmass Banded Qzt Mbr Clonmass Lst Mbr Ards Qzt Fmn 310 615 Ards Black Schist Fmn Altan Lst Fmn 255 400	Sessiagh Clonmass Fmn
APPIN	Lochaber			Creeslough Fmn 310 725: Duntally Lst Mbr, Brackagh Lst Mbr, Creevagh Lst Mbr	
Grampian					

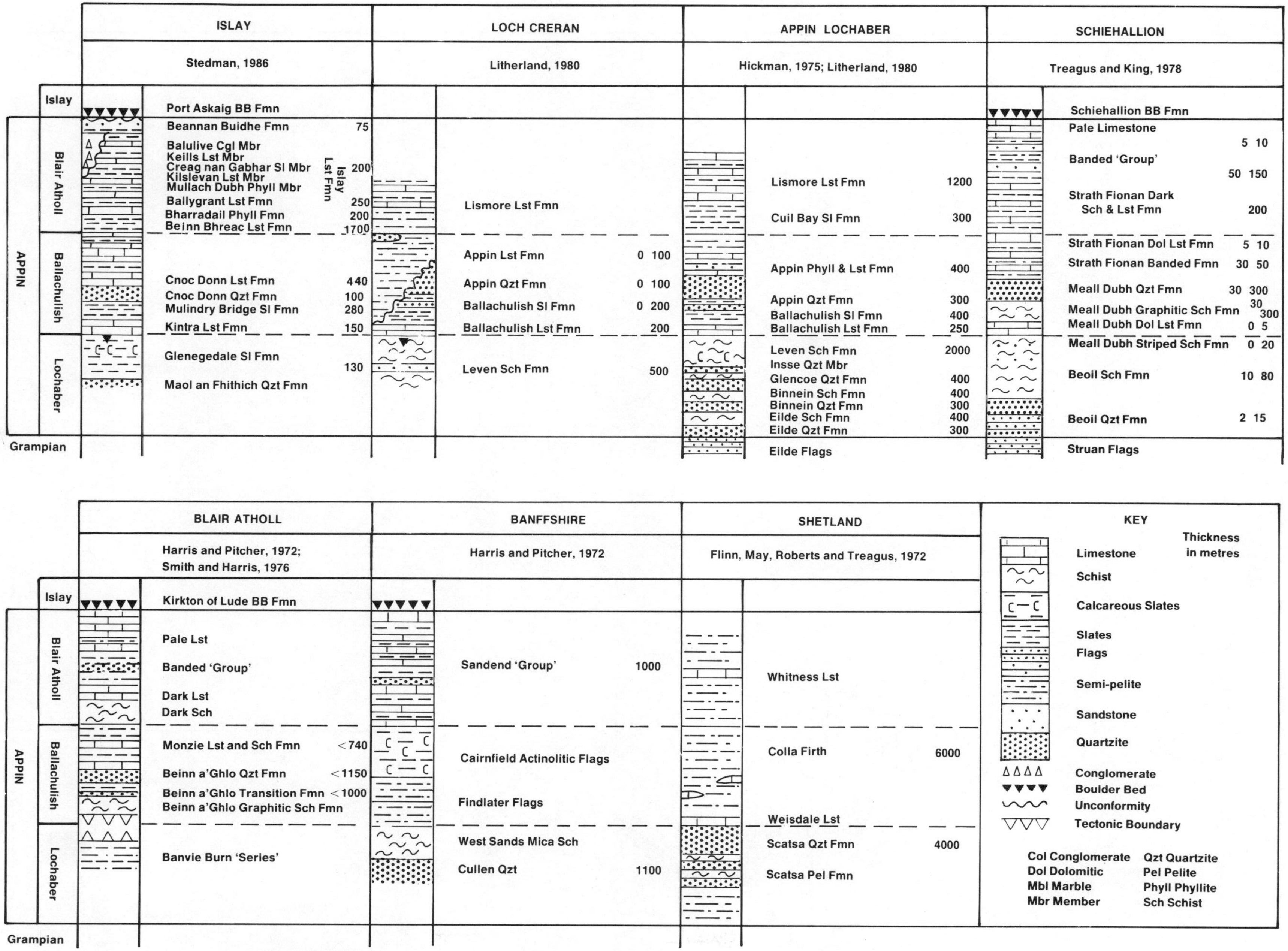

Subgroup	ISLAY (Stedman, 1986)	Thickness
Islay	Port Askaig BB Fmn	
Blair Atholl	Beannan Buidhe Fmn	75
Blair Atholl	Islay Lst Fmn: Balulive Cgl Mbr; Keills Lst Mbr; Creag nan Gabhar Sl Mbr; Kilslevan Lst Mbr; Mullach Dubh Phyll Mbr	200
Blair Atholl	Ballygrant Lst Fmn	250
Blair Atholl	Bharradail Phyll Fmn	200
Blair Atholl	Beinn Bhreac Lst Fmn	1700
Ballachulish	Cnoc Donn Lst Fmn	440
Ballachulish	Cnoc Donn Qzt Fmn	100
Ballachulish	Mulindry Bridge Sl Fmn	280
Ballachulish	Kintra Lst Fmn	150
Lochaber	Glenegedale Sl Fmn	130
Lochaber	Maol an Fhithich Qzt Fmn	
Grampian		

Subgroup	LOCH CRERAN (Litherland, 1980)	Thickness
Blair Atholl	Lismore Lst Fmn	
Ballachulish	Appin Lst Fmn	0 100
Ballachulish	Appin Qzt Fmn	0 100
Ballachulish	Ballachulish Sl Fmn	0 200
Ballachulish	Ballachulish Lst Fmn	200
Lochaber	Leven Sch Fmn	500

Subgroup	APPIN LOCHABER (Hickman, 1975; Litherland, 1980)	Thickness
Blair Atholl	Lismore Lst Fmn	1200
Blair Atholl	Cuil Bay Sl Fmn	300
Ballachulish	Appin Phyll & Lst Fmn	400
Ballachulish	Appin Qzt Fmn	300
Ballachulish	Ballachulish Sl Fmn	400
Ballachulish	Ballachulish Lst Fmn	250
Lochaber	Leven Sch Fmn	2000
Lochaber	Insse Qzt Mbr	
Lochaber	Glencoe Qzt Fmn	400
Lochaber	Binnein Sch Fmn	400
Lochaber	Binnein Qzt Fmn	300
Lochaber	Eilde Sch Fmn	400
Lochaber	Eilde Qzt Fmn	300
Grampian	Eilde Flags	

Subgroup	SCHIEHALLION (Treagus and King, 1978)	Thickness
Islay	Schiehallion BB Fmn	
Blair Atholl	Pale Limestone	5 10
Blair Atholl	Banded 'Group'	50 150
Blair Atholl	Strath Fionan Dark Sch & Lst Fmn	200
Ballachulish	Strath Fionan Dol Lst Fmn	5 10
Ballachulish	Strath Fionan Banded Fmn	30 50
Ballachulish	Meall Dubh Qzt Fmn	30 300
Ballachulish	Meall Dubh Graphitic Sch Fmn	30 300
Ballachulish	Meall Dubh Dol Lst Fmn	0 5
Lochaber	Meall Dubh Striped Sch Fmn	0 20
Lochaber	Beoil Sch Fmn	10 80
Lochaber	Beoil Qzt Fmn	2 15
Grampian	Struan Flags	

Subgroup	BLAIR ATHOLL (Harris and Pitcher, 1972; Smith and Harris, 1976)	Thickness
Islay	Kirkton of Lude BB Fmn	
Blair Atholl	Pale Lst	
Blair Atholl	Banded 'Group'	
Blair Atholl	Dark Lst	
Blair Atholl	Dark Sch	
Ballachulish	Monzie Lst and Sch Fmn	<740
Ballachulish	Beinn a'Ghlo Qzt Fmn	<1150
Ballachulish	Beinn a'Ghlo Transition Fmn	<1000
Ballachulish	Beinn a'Ghlo Graphitic Sch Fmn	
Lochaber	Banvie Burn 'Series'	
Grampian		

Subgroup	BANFFSHIRE (Harris and Pitcher, 1972)	Thickness
Blair Atholl	Sandend 'Group'	1000
Ballachulish	Cairnfield Actinolitic Flags	
Ballachulish	Findlater Flags	
Lochaber	West Sands Mica Sch	
Lochaber	Cullen Qzt	1100

Subgroup	SHETLAND (Flinn, May, Roberts and Treagus, 1972)	Thickness
Blair Atholl	Whitness Lst	
Ballachulish	Colla Firth	6000
Ballachulish	Weisdale Lst	
Lochaber	Scatsa Qzt Fmn	4000
Lochaber	Scatsa Pel Fmn	

KEY

Limestone; Schist; Calcareous Slates; Slates; Flags; Semi-pelite; Sandstone; Quartzite; Conglomerate; Boulder Bed; Unconformity; Tectonic Boundary

Thickness in metres

Col Conglomerate; Dol Dolomitic; Mbl Marble; Mbr Member; Qzt Quartzite; Pel Pelite; Phyll Phyllite; Sch Schist

Figure 15.1 Stratigraphic sequences of Appin Group rocks. Subsequent columns run from SW to NE from locations shown on Figs 15.2 and 15.3. Thicknesses are in metres.

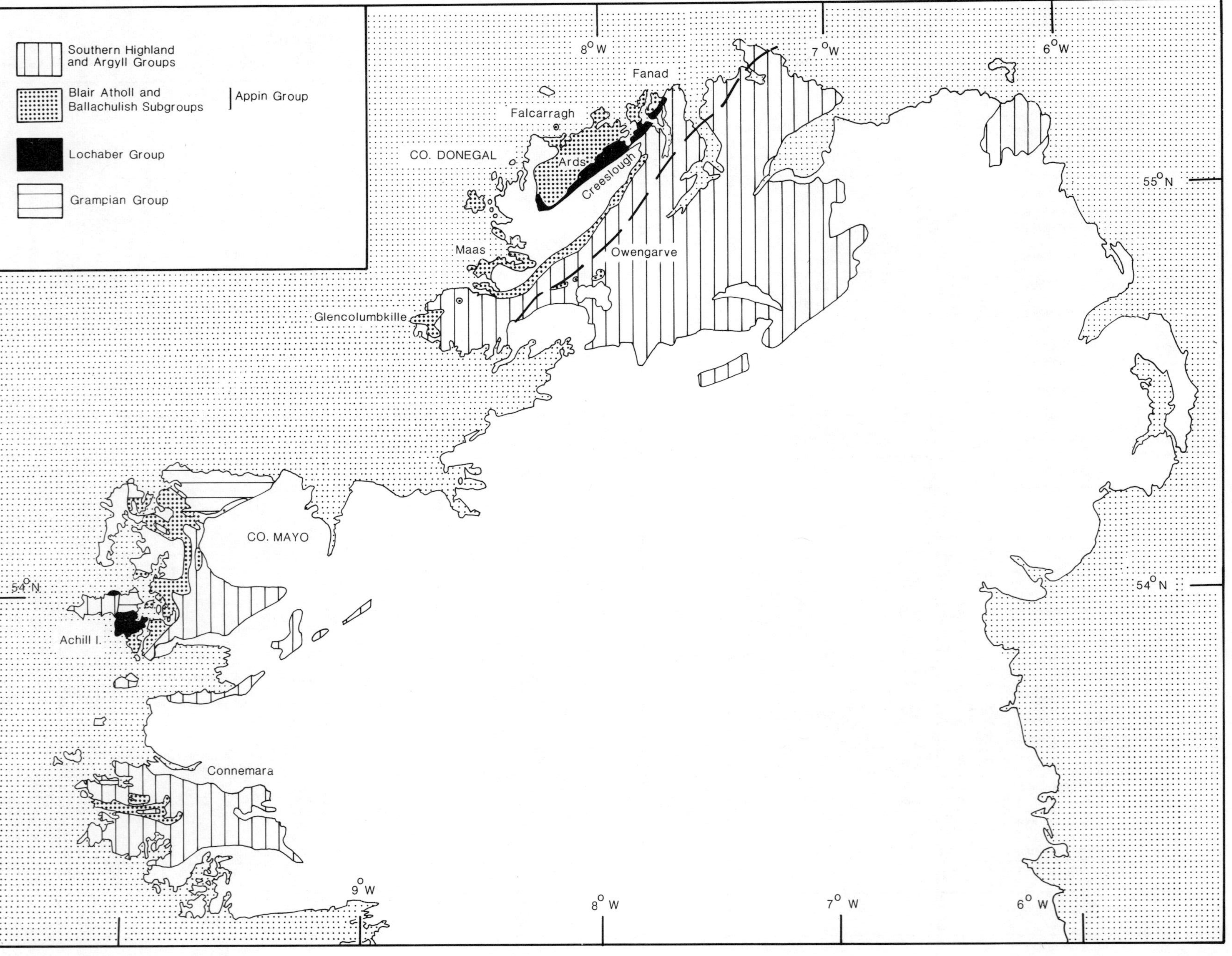

Figure 15.2 The Appin Group in Ireland.

mixed limestone, phyllite and quartzite unit. Originally recognized as the Appin Limestone followed by the Appin Phyllite, it is now regarded as interbedded limestones and phyllites. Traced laterally, there are quite wide variations in rock type within this formation. The Blair Atholl Subgroup above is composed of alternating slates and limestones. The Cuil Bay Slate and Lismore Limestone is followed on Islay by the Mullach Dubh Phyllite and Islay Limestone. An interbanded limestone and slate sequence, which varies in detail but is generally of considerable thickness, occurs at this horizon over much of the Dalradian.

Within the upper two subgroups, Litherland (1980) recognized considerable lateral facies variation across the strike of the orogen (probably also across the sedimentary basin) (see Fig. 15.8), although considerable constancy of sedimentation within the Blair Atholl Subgroup, along the strike from Ireland to Banffshire, is implied by the geochemistry (Hickman and Wright, 1983). Following Rast and Litherland's (1970) recognition of the stratigraphic continuity of the Appin Group and the Argyll Group in Islay, it has become clear that the sequence above the Lismore Limestone, which is missing in the type area, continues with further phyllites and limestones on Islay, and this is probably generally developed elsewhere, although within Islay this formation shows lateral facies changes (Stedman, 1986).

The basal member of the Islay Subgroup, within the succeeding Argyll Group, is the Port Askaig Tillite, a thick sequence of boulder beds in the type area. Boulder beds, usually less well developed, are found throughout the Dalradian area, and form a marker horizon with which it is possible to accurately identify the lower limit of the succeeding Argyll Group although in many cases there is a stratigraphic hiatus or a structural break at about this position in the sequence.

Following the sedimentological studies of Hickman (1975), there has been little published work on the sedimentology of this group, although the stratigraphy is now known in more detail from the Central Highland stretch of the Dalradian, and some further work has been done in Ireland. The mapping does, however, indicate that some formations are laterally very consistent, while others show distinct facies changes, sometimes within quite short distances.

15.2.1 *Lochaber Subgroup*

The sequence in Lochaber is the most varied and thickest sequence developed anywhere in the Dalradian. It cannot therefore be thought of as typical. It has long been recognized in this area (see Bailey, 1960) that the 'Eilde Flags' (uppermost Grampian Group) pass conformably into the quartzites and semi-pelitic schists of this subgroup. There are three major quartzites recognized along the sides of Loch Leven, the Eilde, Binnein and Glen Coe Quartzites, with two intervening schist units: the Eilde and Binnein Schists. Above the Glen Coe Quartzite is a further schist, the Leven Schist, which is much thicker and more variable than the lower schist formations; its lower parts consist of dark psammitic and semi-pelitic schists, which pass upwards into greener more pelitic schists with some calcareous and psammitic horizons towards the top.

Bailey (1960) and Hickman (1975) showed that, traced northeastwards, both the Glen Coe Quartzite and Eilde Quartzite thin and die out stratigraphically. The three schist units thus coalesce, and where only one quartzite occurs this seems to be lithologically more similar to the Binnein Quartzite than to the other two quartzite units. However, on the west side of the Appin Syncline there is a single quartzite between the Ballachulish Limestone and the Eilde Flags in the Spean Bridge, Bohuntine Hill area, and this is not regarded as directly equivalent to any of the quartzites. Hickman also showed that as the Glen Coe Quartzite is traced northwards it changes from an impure arkosic rock to a pure quartzite: very similar in appearance and chemistry to the Binnein Quartzite (Hickman and Wright, 1983).

Kruhl and Voll (1975) have suggested that the sequence along Loch Leven is not repeated by sliding, so that there are four quartzite units. Their interpretation has not been accepted by other workers. Outside the Lochaber district, the Grampian Group is generally succeeded by a succession which can broadly be correlated with the Lochaber Subgroup, composed of quartzites and schists. The calcareous component at the top of the uppermost schist unit is quite widespread, with, below, a rather pure quartzite which some authors correlate with the Binnein Quartzite (Treagus and King, 1978). The sequence at Kinlochlaggan (Treagus, 1969; Roberts and Treagus, 1977*a*) above the Grampian Group contains quartzites, above which is a very thin and persistent boulder bed with granite cobbles, followed by the Kinlochlaggan Limestone. Such extrabasinal granite cobbles and pebbles are found at a similar horizon on Islay (Rast and Litherland, 1970; Stedman, 1986) and Loch Creran (Litherland, 1980), although not associated with a major quartzite unit. In the Schiehallion area, a thick, pure quartzite unit (the Beoil Quartzite) lies between a thin, tectonically attenuated schist above, and the Struan Flags (uppermost Grampian Group) below. Near the top of the Struan Flags is another quartzite which may correlate with the Eilde Quartzite (Treagus and King, 1978). On Islay there is only one quartzite formation, the Maol an Fhithich Quartzite, which occurs at the base of the exposed sequence, with above it a semi-pelitic schist becoming calcareous towards the top. The presence of a few granitic clasts suggests that this quartzite may be correlative with the Kinlochlaggan Quartzite.

In view of the impersistent nature of the three quartzites in the type area, and the change in facies from pebbly and feldspathic quartzite to pure quartzite within the Glen Coe Quartzite itself—when traced from Loch Leven to Glen Spean (Hickman, 1975), it seems unlikely that individual quartzites are laterally very persistent. A similar change, from 'Eilde

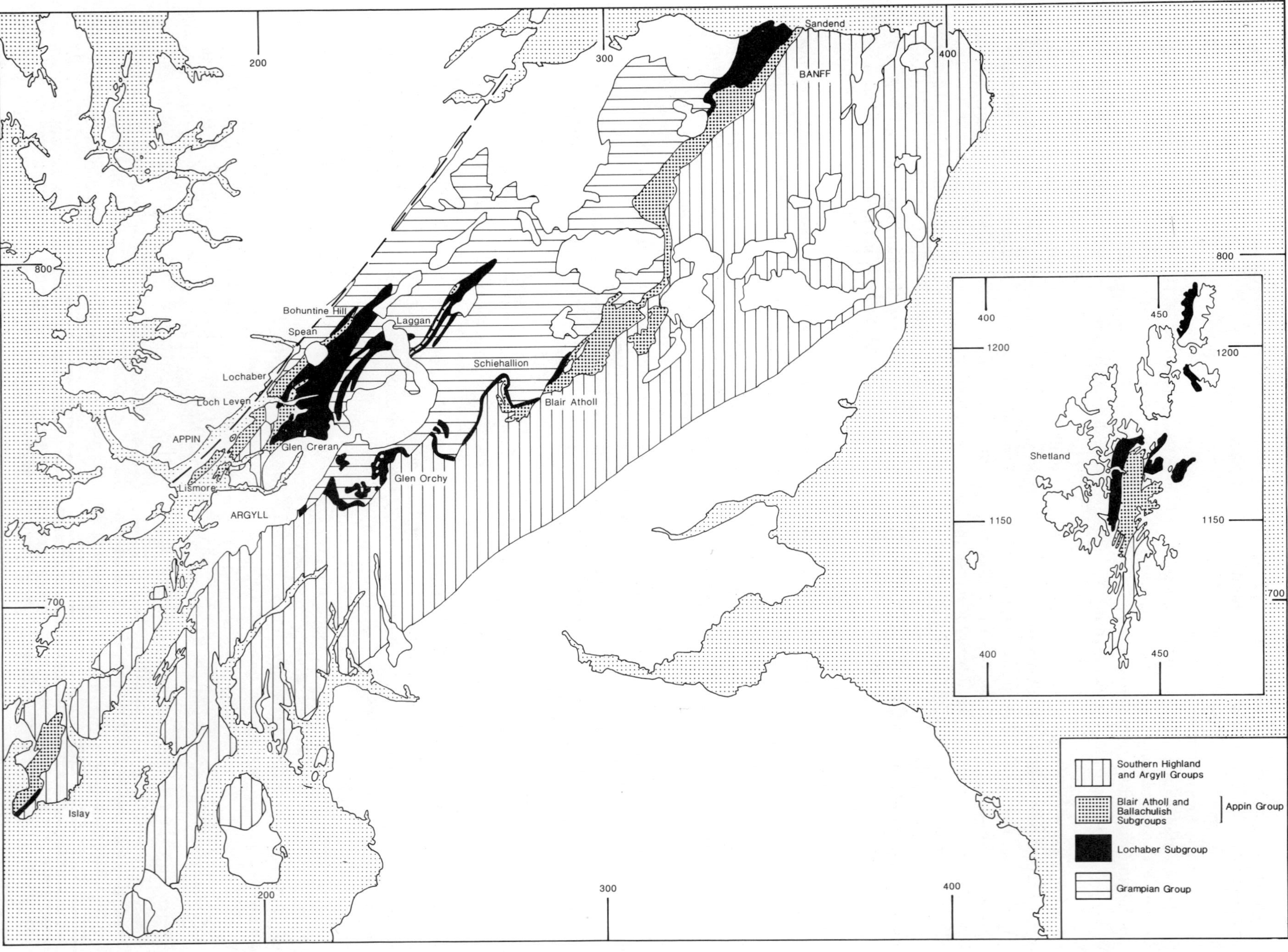

Figure 15.3 The Appin Group in Scotland.

Quartzite-like' on one limb of a fold to a pure 'Binnein Quartzite' aspect on the other, has been noted in the Dalmally area (Roberts and Treagus, 1975). Therefore there is little to be gained by suggesting that where one or two quartzites occur in the sequence they are necessarily correlative of any particular quartzite at Loch Leven.

In Ireland (see Fig. 15.3) there are two areas where rocks of this subgroup occur. In Donegal the Dalradian occurs in three structurally distinct units. One of these, the Creeslough succession, correlates broadly with the Ballachulish and Blair Atholl subgroups and this has, at its base, a quartzite, the Lackagh Quartzite, and a sequence of banded pelites, semi-pelites and calcareous pelites, the Creeslough Formation, above (McCall, 1954; Rickard, 1962; Pitcher and Berger, 1972). This is probably the equivalent of the Leven Schist, although in the Creeslough area, McCall (1954) does recognize the occurrence of three distinct limestone members: which is the lowest recorded occurrence of significant limestone units in the Dalradian. Much of this part of the sequence, however, lies within the aureole of the Donegal granites.

On Achill Island in Co. Mayo, there is a very thick sequence of cross-bedded psammites with intervening semi-pelitic units, which passes up into more pelitic beds, some of which are calcareous, and with local marbles and quartzites similar to the Creeslough Formation of Donegal and to the top of the Leven Schists. There is a very great thickness of psammitic beds in this area, and the base of the Lochaber Subgroup is taken to be the lowest of the quartzite members, the Slievemore Quartzite. Three other quartzites, the Minaun, Cashel and Sraheens Quartzites, occur, and the units between contain minor quartzites and psammites, as well as semi-pelite. This stratigraphy is only given in the review papers (Harris and Pitcher, 1975; Phillips, 1981); no detailed description of it currently appears to be in the literature. No exact equivalence of any of these quartzites with particular formations in Lochaber is envisaged, and it seems that western Ireland was also a local area of thick psammitic sedimentation.

From the Central Highlands of Scotland, the Dalradian extends through the area of maximum Caledonian metamorphism and granite intrusion into north-east Scotland, where the main outcrops are along the Banffshire coast. The sequences there are more difficult to correlate with the main Dalradian area, but at the base of the sequence is a quartzite, the Cullen Quartzite, followed by the West Sands Micaschist, which may be the equivalent of the Lochaber Subgroup (Harris and Pitcher, 1979). Further north-eastwards, Shetland also contains groups of rocks which are regarded as Dalradian. One of these, the Scatsa Division of quartzites and mica schists, has been proposed as the equivalent of the Lochaber Subgroup (Flinn *et al.*, 1972), although recent work (Flinn, this volume) has since raised doubts about this correlation.

It is thus apparent that there is a very widespread phase of alternating quartzite and psammitic silt and shale developed after the Grampian Group lithologies. Although this is a facies correlation, individual formations, when laterally well exposed, are of considerable extent. However, in the best exposed and most extensive area of outcrop, the type Lochaber area, individual quartzites do eventually taper out.

15.2.2 *Ballachulish Subgroup*

The classic stratigraphic sequence of the Lochaber area, largely worked out by the meticulous structural–stratigraphic mapping of Bailey in the early part of this century (Bailey, 1960), was as follows: the lowest unit, the Ballachulish Limestone, is followed by the Ballachulish Slate, with a transitional quartzite and slate passing into the Appin Quartzite, followed by the Appin Limestone and the Appin Phyllite. This stratigraphy has been widely used, and in many other areas such a sequence has been recognized. All the formations show, to a greater or lesser extent, a transitional passage to the formation above, so that although there has been a tendency to recognize a 'striped transition formation' below the Appin Quartzite, it is here included within the Appin Quartzite Formation, as in the original description (Bailey, 1960). The Appin Limestone and Appin Phyllite Formations present a different problem, in that carbonates of characteristic type (the 'Tiger Rock') occur at several stratigraphic levels, interbedded with great thicknesses of phyllite. Litherland (1980) suggests that these are best combined into one formation of alternating carbonates and pelites, termed the Appin Limestone Formation; most subsequent authors have used Appin Phyllite and Limestone Formation, and this is the usage suggested here.

This sequence is well exposed in the type Lochaber area around Ballachulish, Onich and Ardsheal, although nowhere is there a complete sequence from the Leven Schist below to the Cuil Bay Slate (in the subgroup above) without some form of tectonic hiatus. On Islay the sequence does seem to be more or less complete, and for a long time this whole succession was lumped into one unit: the Mull of Oa Phyllite. In 1970 it was shown that a much more varied sequence of rocks formed a substantial part of the sequence, and that a sequence of limestones, slates and quartzites similar to that of the Lochaber area could be recognized (Rast and Litherland, 1970). Recent mapping (Stedman, 1986) has confirmed this in detail and, although in part poorly exposed, it is of very similar apparent thickness. In the Appin area, south of Lochaber, Litherland (1980) again found the same sequence, but recognized that, in the south-easterly direction, the lower part of the sequence was gradually cut out by the Appin Limestone Formation as it changed facies laterally.

Along the strike in the Central Highlands, very similar sequences of a blue dolomitic limestone, followed by a dark phyllite, transitional into a massive quartzite, and then a green phyllite and light carbonate horizon have been recognized (Smith and Harris, 1976; Treagus and King, 1978). To the north-east, the

Garron Point or Cairnfield Flags and the Findlater Flags probably correlate with this subgroup, but if so, there is a marked change in facies. In Shetland the rocks above the Scatsta Division are the Weisdale Limestone and the Colla Firth Group: a sequence of psammites, limestones and a small amount of pelite. The Weisdale Limestone may correlate with the Ballachulish Limestone, but whether the Colla Firth rocks bear any simple relationship to the rest of the Ballachulish and Blair Atholl Subgroups remains unclear (Flinn *et al.*, 1972).

In Ireland, north-west Donegal and north-west Mayo are the only areas where this subgroup crops out. In Donegal the Creeslough succession contains the following sequence: an impersistent limestone at the top of the Creeslough Formation: the Altan Limestone (Rickard, 1962), is generally correlated with the Ballachulish Limestone (Pitcher and Berger, 1972; Harris and Pitcher, 1975; Phillips, 1981). It may be that the three limestones lower in the Creeslough succession mentioned above may also be correlative with the Ballachulish Limestone, and that the episode of calcareous deposition was more intermittent here than in Scotland.

The Ards Black Schist which follows with a transition into the Ards Quartzite is a very similar sequence to the Ballachulish Slate and Appin Quartzite of Scotland. The sequence above, composed of limestone, quartzite and pelite, is more complex and has been termed the Sessiagh–Clonmass Formation (Harris and Pitcher, 1975; Phillips, 1981). Lateral facies changes are visible within quite small areas. Three limestones are very persistent, and there is a greater proportion of quartzitic horizons than in Scotland. However, as the Appin Phyllite and Limestone Formation is variable even in the type area, the correlation of this formation with that in Appin is taken to be reasonable.

In Co. Mayo there is a similar limestone, slate, psammite, limestone/pelite sequence, but the Inver 'Group' (Crow *et al.*, 1971) has not yet been given a detailed description in the literature.

15.2.3 *Blair Atholl Subgroup*

This is composed of dark limestones and black slates as major monolithological units and as interbanded transitional units. The interbanded nature is so predominant in some areas that some authors have suggested just one formational name for the whole subgroup. In the type area, however, there is a very thick slate at the base, the Cuil Bay Slate, which is succeeded by a very thick sequence of pure and banded limestones with only very thin slate members, the Lismore Limestone, and a distinction is thus merited.

There are no higher beds exposed in Appin and Lochaber, but on Islay the sequence continues with what has traditionally been called the Islay Limestone. This has now been shown (Stedman, 1986) to be composed of four members, the Mullach Dubh Phyllite, the Kilslevan Limestone, the Creag nan Gabhar Slate and the Keills Limestone. However, within quite a short distance (on the west limb of the Islay Anticline), these are all replaced by a limestone conglomerate, the Balulive Limestone Conglomerate Member. Above this, Stedman (1986) described the Beannan Buidhe Formation as a sequence of slates, quartzites and carbonates below the lowest tillite on Islay, composed of five members which are cut out laterally by the erosive unconformable tillite immediately above.

Since the resolution of the Ballappel/Islay Divisions of the Dalradian into one succession by Rast and Litherland (1970), it has been recognized that the Blair Atholl Limestone of Perthshire is the equivalent, at least in part, of the Lismore Limestone. This has traditionally been mapped as a Dark Schist, Dark Limestone, Banded Unit, Pale Limestone sequence. In the Schiehallion area (Treagus and King, 1978) the lowest units are termed the Strath Fionan Dark Limestone and Schist, and are composed of three alternations of limestone and black pelite, although the thicknesses in all these Perthshire areas are considerably attenuated from the type area. It is probable that the upper 'Pale' Blair Atholl Limestone is the equivalent of the Islay Limestone. In north-east Scotland the Sandend Formation bears a much closer similarity to the Blair Atholl Subgroup than does the lower part of the sequence, and although metamorphism was at higher grade, it is clear that there was originally a considerable thickness of interbedded limestone and pelite, similar in facies to the Lismore Limestone. The geochemistry seems to indicate that it is most similar to the Lismore Limestone, although this is not by any means as certain as in some other cases, and Rock (1986) favours correlation on geochemical grounds with the Ballachulish Limestone. However, such a correlation would render all the previous suggested correlations between the Buchan area and Argyll invalid.

In Ireland, the upper part of the Creeslough succession contains a widespread major limestone, best developed at Falcarragh. This has a major pelite both below and above, and the Lower Falcarragh Pelites are generally equated with the Cuil Bay Slates, and the Falcarragh Limestone with the Lismore Limestone. Elsewhere in Donegal (Pitcher and Berger, 1972), there are facies variations in the lower part of this sequence, although the limestone is quite persistent. Above these, the upper Falcarragh Pelites and their correlatives are generally more semi-pelitic and pass up into quartzites, flags and semi-pelites of the Loughros Formation, and slates of the Fintown Pelite. Thus, in the Creeslough Division of Donegal, although there is a considerable thickness of strata above the Lismore Limestone, there is no apparent equivalent to the Islay Limestone sequence or to the tillites of the succeeding Argyll Group. However, the tectonically separate Kilmacrenan succession does contain a well-developed tillite horizon throughout northern and western Ireland (Kilburn *et al.*, 1965), below which there is an equally well-marked limestone. This is usually a complex sequence of limestones and pelites, the Glencolumbkille Limestone and Glencolumbkille Pelite, which may be equivalent in places to both the Islay and Lismore Limestones. The sequence outcrops from Donegal, through Mayo to Connemara. There are some facies variations, and much of this sequence is at a fairly high grade of metamorphism, but it is possible that there is very little overlap in stratigraphy between the Kilmacrenan and Creeslough successions, the tectonic break coming mostly at about the middle of the Blair Atholl Subgroup. In Central Donegal–to the south-east of the main part of the Dalradian–the Central Donegal succession is almost completely different to that of the rest of Ireland (Pitcher *et al.*, 1964). The sequence does, however, contain a boulder bed, and this is taken to correlate with the Port Askaig Tillite. The rocks immediately below are again limestones and calc-silicate bands, separated by pelites (the Owengarve Formation). How far down the Appin Group sequence this extends is unknown, but it seems likely that the Owengarve represents part, at least, of the Blair Atholl Subgroup.

It seems, therefore, that south-eastwards within the Blair Atholl Subgroup, both in Ireland and in Scotland (as shown in the Loch Creran area), there is a general change in facies, with the limestones becoming less massive or disappearing, and the generally mature, well-sorted, banded quartzite/pelite/limestone sequences giving way to more monotonous semi-pelites and pelites.

15.3 Geochemistry

Trace-element geochemistry has been used to a greater extent in the Appin Group than in any other comparable sequence, for stratigraphic and palaeoenvironmental studies.

15.3.1 *Lochaber schists*

Lambert and his co-workers (1981, 1982) have shown that the Lochaber Subgroup schists have a different geochemistry to Moine pelites, and that the Leven Schists are so distinctly different to the Monadhliath Schists that the latter should be regarded as part of the Grampian Group, which has closer geochemical affinities with the Moine Assemblage than with the Dalradian Supergroup.

Average analyses of the Eilde Schist, Binnein Schist and Leven Schist are given in Table 15.1. Compared with normal slates (Table 15.1, col. 1), they are high in Al, K, Ba, Zr, Y and Rb, and low in Ca, Na, Sr, Ni and Cu. In all cases the Eilde and Binnein Schists are very similar to each other, and the Leven Schists are not quite so extreme in the differences noted above. This chemistry indicates that these rocks were probably derived from a plagioclase-poor source, with a high proportion of illite, and with a very low proportion of basic igneous rocks in the source region, and thus low in

Table 15.1 Chemical composition of Lochaber Subgroup schists.

	1	2		3		4		5	
	$\bar{x}$	$\bar{x}$	σ	$\bar{x}$	σ	$\bar{x}$	σ	$\bar{x}$	σ
SiO_2	61.90	60.00	5.51	61.8	2.8	62.9	2.7	62.6	1.0
TiO_2	0.82	0.71	0.21	1.06	0.19	1.07	0.15	0.79	0.37
Al_2O_3	16.54	21.38	3.65	20.9	2.5	19.6	1.4	21.2	1.0
FeO(tot.)*	6.07	6.13	1.83	5.7	0.8	5.0	0.3	5.4	0.4
MnO	Tr.	0.02	0.02	0.10	0.02	0.09	0.01	0.08	0.01
MgO	2.45	2.95	1.25	1.9	0.3	2.1	0.3	2.1	0.2
CaO	1.07	0.95	0.63	0.4	0.3	0.9	0.5	1.1	0.4
Na_2O	2.57	1.24	0.34	1.6	0.5	1.8	0.5	2.1	0.3
K_2O	3.15	3.50	0.95	5.6	0.5	5.6	0.8	4.3	0.4
P_2O_5	0.04	0.09	0.06	0.24	0.2	0.21	0.07	0.13	0.02
Ni′	68	20	9	19	3	18	3	24	2
Cu	45	29	15	11	5	8	5	22	8
Zn	95	81	24	103	20	83	18	118	8
Rb	140	141	34	264	30	220	28	213	17
Sr	300	145	106	100	17	132	37	167	18
Y	26	21	9	51	19	58	9	56	4
Zr	160	206	58	442	120	543	86	331	33
Nb	11	14	5	25	5	25	3	20	1
Ba	580	588	144	1070	119	1260	270	848	83
n	8	103		10		10		71	

* FeO (tot) = total iron as FeO; oxides as wt%; trace elements as ppm; Tr = trace; bdl = below detection limit; $\bar{x}$ = mean; σ = standard deviation; n = number of samples

1. Average shale (Clarke, 1924; trace elements—Turekian and Wedepohl, 1961).
2. Dalradian black slates (Hickman and Wright, 1983, revised).
3. Eilde Schist (Lambert *et al.*, 1981).
4. Binnein Schist (Lambert *et al.*, 1981).
5. Leven Schist (Lambert *et al.*, 1981).

chlorite in the sediment. Factor analysis by Lambert *et al.* (1982) shows that the Leven Schist has a much more complex chemistry than Moine pelites, with seven independent functions indicating a complex, rather mature, mixed sedimentary origin. It also shows a distinct change in chemistry from SW to NE, with a relative enrichment of 'muscovite' over 'chlorite' in that direction, indicative of derivation of the sediment from the W (Fig. 15.4). Isotope chemistry is only available for the Leven Schists, and strongly indicates a sedimentary derivation for the Sr and Rb, with a 'sedimentary'-type isochron of 655 ± 25 Ma (Sr_i 0.7232 ± 12), but with four anomalous points lying on a separate line (640 ± 30 Ma, Sr_i 0.7358 ± 7), possibly indicative of a separate source area with more radiogenic micas (Fig. 15.5). The very high initial ratio is noteworthy and again consistent with illite-rich sedimentation (Lambert *et al.*, 1982). It is much higher than the Moine, Laxfordian and Scourian rocks at the time of deposition, and seems to indicate a completely independent source, perhaps derived from the more K-feldspar-rich Torridonian, or, given that this terrane may not have been adjacent to north-west Scotland during its deposition, perhaps from some entirely different area with a more normal granitic basement.

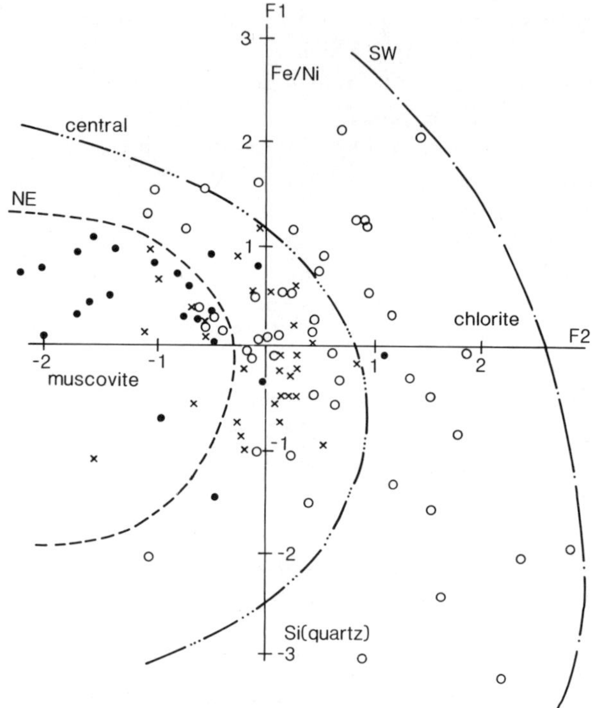

Figure 15.4 Factor scores for Leven Schists; circles; samples from NE of the River Spean; crosses: samples from the River Spean; dots: samples from SW of the River Spean. This demonstrates a progressive change in composition (from Lambert *et al.*, 1982).

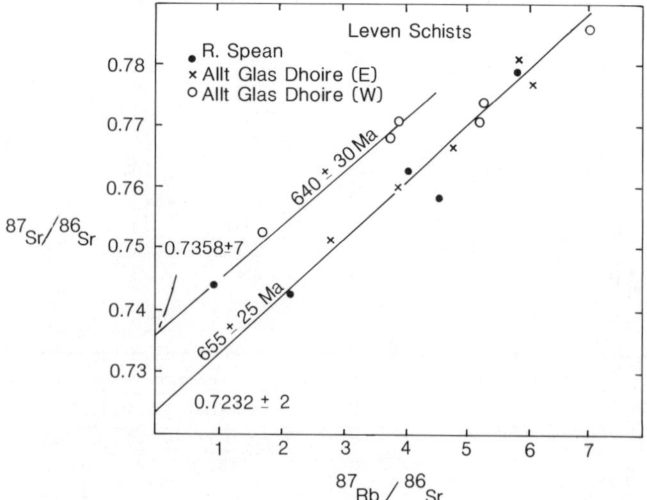

Figure 15.5 Rb–Sr data for Leven Schists (from Lambert *et al.*, 1982).

15.3.2 *Quartzites*

All four major quartzite units have been analysed, and the results described in detail (Hickman and Wright, 1983). The Eilde and Glen Coe Quartzites are indistinguishable chemically (Table 15.2), with La the only element different at the 99% level. The Binnein Quartzite, however, has a completely different chemistry, being much purer and with different tenors of Mn, Rb, Sr, and Ba, probably due to the lower Fe and K contents. However, it also has lower K/Rb ratios and high Fe/Mn ratios, which help to distinguish it. The change in facies of the Glen Coe Quartzite, as it is traced northwards is well illustrated by the chemistry, which becomes much more silica rich, and ultimately bears a close similarity to the Binnein Quartzite (Table 15.2, columns 5 and 6). Further to the north and west, a quartzite lies between the Ballachulish Limestone and the Eilde Flags that are adjacent to the Fort William Slide. This quartzite was regarded by Hickman (1975) as the Glen Coe Quartzite, but is shown as the Eilde Quartzite on the Geological Survey Sheet 63E. The geochemistry (Table 15.2, column 7) indicates that it is an impure quartzite, but as the Glen Coe Quartzite of the main outcrop has become a pure quartzite by this latitude, it is perhaps unlikely to be a direct correlative. It is not possible to distinguish geochemically between the various possibilities, but it is perhaps most likely to be a completely separate sand body, not directly related to either. The Insse Quartzite Member (Table 15.2, column 8), a separate thin quartzite above the Glen Coe Quartzite, within the Leven Schist, can be distinguished geochemically from the three major quartzites (Hickman and Wright, 1983).

Although the schists and semi-pelites of the Lochaber Group have a different chemistry from those of the Grampian Group which immediately precedes it (Lambert *et al.*, 1982), such a difference cannot be readily discerned in the quartzites. Thin, pure quartzite units are widespread in the uppermost Glen Spean Subgroup of the Grampian Group, and the base of the Lochaber Subgroup is taken to be the occurrence of the first thick quartzite unit. The pure quartzites in the Eilde Flags are similar in chemistry to the Eilde and Glen Coe Quartzites (Table 15.2, column 1).

Table 15.2 Chemical compositions of Appin Group quartzites

	1		2		3		4		5		6		7		8		9	
	$\bar{x}$	σ	$\bar{x}$	σ	$\bar{x}$	σ	$\bar{x}$	σ	$\bar{x}$	σ	$\bar{x}$	σ	$\bar{x}$	σ	$\bar{x}$	σ	$\bar{x}$	σ
SiO_2^*	90.17	4.73	92.39	3.91	96.83	2.98	94.93	3.66	90.92	5.98	98.10	1.08	93.14	2.91	98.57	0.99	91.94	6.13
TiO_2	0.11	0.06	0.19	0.14	0.09	0.09	0.15	0.13	0.21	0.14	0.06	0.03	0.07	0.03	0.05	0.03	0.15	(0.01–0.77)
Al_2O_3	5.63	2.70	4.13	2.15	1.73	(0.02–7.40)	2.69	2.13	5.02	3.34	0.87	0.61	3.98	1.71	0.79	0.55	4.39	3.33
FeO(tot.)	0.56	0.33	0.53	0.48	0.34	0.34	0.43	0.43	0.58	0.42	0.36	0.21	0.34	0.16	0.21	(0.06–0.69)	0.51	(0.05–3.04)
MgO	0.13	0.08	0.18	0.16	0.10	(0.01–1.09)	0.14	(0.0–1.1)	0.32	(0.0–1.66)	0.16	(0.01–0.87)	0.14	0.08	0.06	0.06	0.49	(0.0–4.44)
CaO	0.14	0.13	0.29	(0.01–3.71)	0.07	(0.01–1.63)	0.25	(0.01–5.07)	0.17	(0.02–1.10)	0.03	0.02	0.02	0.01	0.03	0.03	0.29	(0.01–5.74)
Na_2O	1.29	0.59	0.76	(0.0–4.47)	0.25	(0.01–1.22)	0.26	0.25	0.41	(0.0–1.52)	0.29	(0.0–0.77)	0.78	0.57	0.09	(0.07–0.60)	1.27	(0.05–6.05)
K_2O	1.97	1.18	1.55	0.77	0.64	(0.01–3.13)	1.19	1.04	2.27	1.43	0.16	0.10	1.54	0.63	0.22	0.18	1.34	1.05
P_2O_5	0.01	0.00	0.02	0.02	0.01	0.01	0.06	0.05	—		—		—		—		0.02	(0.0–0.16)
Mn	45	38	58	(4–272)	22	(3–156)	62	(4–347)	65	39	11	5	32	24	22	20	44	0–428)
Zn	5	2	8	8	5	3	9	9	13	11	3	1	5	2	5	1	4	3
Ga	5	0	4	2	8	4	—	—	6	1	—	—	bdl	bdl	bdl	bdl	7	1
Rb	46	25	41	20	21	(0–97)	34	27	62	36	6	3	37	14	7	4	26	17
Sr	73	57	33	29	16	(0–487)	40	(1–745)	28	17	5	3	23	10	3	2	20	19
Y	7	4	9	5	5	2	8	(2–57)	7	4	5	2	6	0	4	2	5	4
Zr	154	85	278	233	175	147	276	273	321	255	127	43	164	100	136	118	360	(40–2406)
Ba	545	214	425	261	222	149	457	189	676	245	329	26	498	73	237	102	333	301
La	8	5	7	5	6	3	4	3	3	2	3	0	1	1	2	1	6	5
Ce	21	13	21	13	14	6	17	11	20	8	8	4	6	3	6	4	14	13
Pb	8	7	9	2	14	(0–101)	—		—	—	—	—	—	—	—	—	10	7
Th	4	1	3	0	4	3	—		—	—	—	—	—	—	—	—	4	3
K/Rb	309		317		216		264		292		203		337		228		392	
K/Ba	27		31		20		18		25		3		24		9		33	
Fe/Mn	157		129		183		134		89		281		131		89		533	
n	21		48		80		37		22		10		5		7		64	

$SiO_2^* = 100\%$—other oxides. Other abbreviations as Table 15.1.

1. Quartzites in Eilde Flags.
2. Eilde Quartzite Formation.
3. Binnein Quartzite Formation.
4. Glencoe Quartzite Formation.
5. Glencoe Quartzite, south of Loch Leven.
6. Glencoe Quartzite, Stob Coire na Ceannain.
7. Quartzite, Bohuntine Hill.
8. Inssc Quartzite Member, Leven Schist Formation.
9. Appin Quartzite Formation.

(From Hickman and Wright, 1983, revised data.)
NB Ratio data in Hickmann and Wright (1983) are incorrect.

Table 15.3 Chemical compositions of Appin Group black slates.

	1	2		3	4		5	6		7	
	$\bar{x}$	$\bar{x}$	σ	$\bar{x}$	$\bar{x}$	σ	$\bar{x}$	$\bar{x}$	σ	$\bar{x}$	σ
SiO_2	52.46	56.24	4.09	57.56	58.11	2.98	62.85	66.12	2.72	63.71	4.81
TiO_2	0.56	0.61	0.19	0.77	0.63	0.17	0.89	0.81	0.13	0.88	0.21
Al_2O_3	25.08	23.66	3.43	20.58	19.64	1.80	24.01	19.61	3.03	20.14	3.45
FeO (tot.)	7.27	7.08	1.74	6.29	7.01	1.16	4.14	3.97	0.70	5.85	0.75
MnO	0.02	0.02	0.01	0.04	0.02	0.02	bdl	bdl	0.00	0.01	0.00
MgO	2.96	3.63	1.23	5.08	3.23	0.84	2.32	2.14	0.46	1.93	0.34
CaO	1.20	0.80	0.61	0.65	1.44	0.93	0.92	0.92	0.36	0.72	0.11
Na_2O	0.98	1.20	0.24	0.64	1.32	0.43	1.22	1.41	0.35	1.17	0.27
K_2O	4.84	3.99	1.07	4.01	3.43	0.50	3.20	2.86	0.61	3.14	0.73
P_2O_5	0.23	0.04	0.02	0.01	0.07	0.03	0.22	0.17	0.03	0.09	0.03
Total	95.60	96.99		95.64	94.68		99.77	98.01		97.32	
S	1264	1397	(118–5978)	2951	1548	(288–5631)	2037	1316	(101–5479)	996	(53–5165)
Cr	94	93	15	87	110	32	205	183	35	146	35
Ni	22	27	8	15	24	7	24	11	6	15	7
Cu	24	37	18	58	28	9	32	22	4	19	6
Zn	105	97	25	98	84	9	68	60	15	63	11
Ga	50	27	4	23	23	2	25	24	3	23	4
Rb	169	158	36	146	132	20	140	131	30	127	27
Sr	61	93	87	63	107	33	353	279	70	133	55
Y	28	31	6	19	20	4	14	10	4	15	5
Zr	163	191	61	268	182	53	218	257	40	196	32
Nb	14	12	5	12	11	4	16	17	3	16	3
Ba	703	594	179	692	573	116	742	591	99	533	112
La	60	50	13	35	42	10	42	35	33	41	13
Ce	119	97	24	73	77	21	67	50	27	63	23
Pb	21	23	7	20	20	6	14	7	3	16	9
Th	13	15	3	15	14	2	14	16	2	14	2
n	2	36		3	21		3	20		18	

Abbreviations as Table 15.1.

1. Black slates in Leven Schist Formation.
2. Ballachulish Slate Formation.
3. Ards Slate (Ballachulish Slate), Ards Point, Donegal.
4. Cuil Bay Slate Formation.
5. Manse Slate Member of Lismore Limestone Formation.
6. Fiart Slate Member of Lismore Limestone Formation.
7. Kilcheran Slate Member of Lismore Limestone Formation (from Hickman and Wright, 1983, revised).

The only other major quartzite unit in the Appin Group is the Appin Quartzite in the Ballachulish Subgroup, although other thinner and less persistent quartzite horizons occur within the Appin Phyllite and Limestone. The Appin Quartzite is variable in facies, and this is borne out in its geochemistry (Hickman and Wright, 1983). It has a wide variation in composition along its outcrop, with the pebbly and rather more immature quartzites in the Lynn of Lorn area having a sandstone rather than quartzite composition. It is characteristically distinctly lower in Zn, Rb, Sr, Ba, Y and Ce than the Lochaber Quartzites. Even the very pure Binnein Quartzite has higher Zn, Y, La and Ce than the Appin Quartzite. The presence of abundant plagioclase in the Appin Quartzite is the main reason for these differences, although some trace-element differences probably indicate a different source, and it is also a much more immature deposit.

15.3.3 *Black slates*

Dark grey and black slate units occur at several levels within the group. Some of the schists of the Lochaber Subgroup have the occasional black pelite unit (Table 15.3, column 1), but most of the schists are much more immature and, although they are now most frequently of higher metamorphic grade (being mica schist rather than slate), there does seem to be a real original sedimentary difference. Black slates form the Ballachulish Slate, Cuil Bay Slate and Mullach Dubh Phyllite Formations, and there are slate members in the Lismore and Islay Limestones. Of these, only the geochemistry of the three lower units has been investigated, but these prove to be readily distinguishable by geochemistry (Hickman and Wright, 1983). The elements P and Y distinguish the Ballachulish and Cuil Bay Slates, and many trace elements (principally Sr, Cr, Zn) distinguish the Lismore Slates from the others. The slates are rather aluminous with the Ballachulish and Cuil Bay Slates being rather richer in Fe and Mg than the Lismore Slates, suggesting that montmorillonite was a major component of the original clays of the former formations. The Lismore Slates are higher in SiO_2, P and Zr, suggesting a more neritic

Table 15.4 Chemical compositions of Appin Group dolostones.

	1		2		3	
	$\bar{x}$	σ	$\bar{x}$	σ	$\bar{x}$	σ
SiO_2	27.47	6.45	14.87	6.68	14.3	0.9
TiO_2	0.11	0.05	0.07	0.05	0.06	0.06
Al_2O_3	5.25	2.16	2.77	1.41	1.4	0.9
FeO (tot.)	2.95	1.85	2.15	1.46	0.63	0.29
MgO	9.80	1.41	12.11	3.57	19.0	1.9
CaO	21.55	5.55	27.31	6.03	29.7	3.2
Na_2O	bdl		0.60	0.00	0.21	0.05
K_2O	0.95	0.47	0.40	0.34	0.25	(0–1.39)
P_2O_5	0.08	0.03	0.04	0.03	0.04	0.03
S	260	(12–2345)	341	(56–1561)	—	—
Cr	19	10	10	8	24	7
Mn	793	(269–4410)	1667	1079	—	—
Ni	7	3	2	3	6	3
Cu	10	8	7	3	2	(0–14)
Zn	26	14	13	8	28	19
Ga	5	2	2	1	2	1
Rb	32	19	11	7	8	8
Sr	293	112	237	120	318	164
Y	12	5	10	8	5	3
Zr	99	51	58	29	31	18
Nb	3	1	2	1	—	—
Ba	262	(10–1578)	65	(13–340)	116	56
La	12	6	13	6	5	3
Ce	51	19	16	(0–45)	9	7
Pb	6	6	3	2	—	—
Th	4	2	2	1	—	—
Ca/Sr	608		933		976	
n	29		11		22	

Abbreviations as Table 15.1.

1. Ballachulish Limestone Formation dolostones (< 35% SiO_2).
2. Appin Limestone Formation dolostones (< 25% SiO_2).
3. Connemara Marble Formation dolostones (< 25% SiO_2).

(From Rock, 1986, and Hickman and Wright, 1983, revised).
NB Ca/Sr ratios in Hickman and Wright (1983) are incorrect.

environment. Even within the Lismore Limestone Formation, the two slate members investigated have a sufficiently different chemistry to allow them to be distinguished (Hickman and Wright, 1983). Although there are differences in chemistry, the Cuil Bay Slate is much more similar to the Ballachulish Slate than the Lismore Slate Members, so that the geochemistry does not support the amalgamation of the Cuil Bay Slate and Lismore Limestone into one formation of alternating slates and limestones. The chemistry suggests that there is a real change in environment between the Cuil Bay Slate and the slates interbedded with the overlying limestones.

The Ballachulish Slate seems to have had a more granitic source area, judged by the higher proportion of rare earths and none has a particularly important basic rock component. The rather greater proportions of Sr in the Lismore slates may be the result of association with the particularly Sr-rich Lismore Limestone environment (either original or diagenetic).

15.3.4 *Carbonate rocks*

The limestones and dolostones of the Ballachulish and Blair Atholl Subgroups have also been subjected to detailed study (Hickman and Wright, 1983; Rock, 1985*a*, 1986). The four main stratigraphic horizons where carbonate rocks occur are the Ballachulish Limestone, Appin Limestone, Lismore Limestone and Islay Limestone Formations. Both limestones and dolostones are present in these formations and, as noted by Hickman and Wright (1983), the limestones are generally much purer carbonate rocks—although some pure dolostone horizons are present.

The Ballachulish Limestone Formation is composed of limestones and very impure dolostones in almost equal proportions. The limestones are mostly grey, interbedded with shales and with a cream-coloured pure dolostone facies. The impurities indicate some quartzose content as well as clay minerals. The trace-element characteristics of the limestones are relatively

Table 15.5 Chemical composition of Ballachulish Subgroup limestones.

	1	2		3		4		5
	$\bar{x}$	$\bar{x}$	σ	$\bar{x}$	σ	$\bar{x}$	σ	$\bar{x}$
SiO_2	11.5	8.3	7.2	10.2	7.4	5.6	4.1	16.0
TiO_2	0.04	0.04	0.03	0.03	0.02	0.04	0.03	0.25
Al_2O_3	2.4	2.0	1.8	2.1	1.5	1.1	0.8	5.2
FeO (tot.)	0.55	1.2	(0.1–4.7)	0.64	0.51	0.67	0.32	1.1
MgO	3.3	3.6	1.5	2.7	1.7	1.3	0.7	2.3
CaO	45.2	45.5	9.6	45.7	6.8	49.0	2.9	39.5
Na_2O	bdl	bdl		0.14	0.02	0.15	0.12	0.52
K_2O	0.32	0.33	(0.02–1.59)	0.35	0.28	0.25	0.18	1.12
P_2O_5	0.03	0.04	0.03	0.04	0.02	0.07	0.05	0.09
S	735	1185	1066	719	(50–500)	—		—
V	—	—		—		3	2	15
Cr	9	8	5	14	6	13	11	10
Mn	476	490	413	308	193	556	525	
Ni	bdl	1	(0–10)	1	1	1	0	7
Cu	18	15	2	14	5	10	8	8
Zn	6	8	8	26	(3–271)	23	(3–169)	19
Ga	2	12	(0–100)	2	1	2	1	—
Rb	9	9	(1–47)	10	6	12	6	40
Sr	1128	563	213	547	164	1825	314	1777
Y	7	9	9	6	4	3	3	9
Zr	110	57	25	52	29	89	45	75
Nb	2	1	1	1	0	2	1	1
Ba	68	64	(9–309)	85	66	48	37	205
La	9	6	5	6	4	5	3	10
Ce	35	44	20	18	1	31	4	30
Pb	9	4	(0–35)	16	(0–115)	8	4	2
Th	2	2	1	2	1	3	2	2
U	—			1	1	1	1	2
Ca/Sr	349	609		639		200		159
n	4	21		15		18		2

Abbreviations as Table 15.1.

1. Marble Hill Limestone, Donegal.
2. Appin Limestone Formation, Lochaber.
3. Ballachulish Limestone Formation, Lochaber.
4. Kinlochlaggan Limestone, Kinlochlaggan.
5. Weisdale Limestone, Shetland.

(From Rock, 1986, revised.)

Table 15.6 Chemical composition of Blair Atholl Subgroup limestones.

	1		2		3	4		5		6		7		8
	$\bar{x}$	σ	$\bar{x}$	σ	$\bar{x}$	$\bar{x}$	σ	$\bar{x}$	σ	$\bar{x}$	σ	$\bar{x}$	σ	$\bar{x}$
SiO_2	3.9	0.7	16.1	11.7	5.5	1.6	0.66	8.0	5.0	7.5	2.2	9.2	8.2	7.4
TiO_2	0.02	0.01	0.08	0.08	0.01	0.02	0.01	0.03	0.02	0.16	0.13	0.09	0.08	0.06
Al_2O_3	1.3	0.5	4.6	1.7	1.2	0.38	0.18	2.5	1.9	1.6	0.68	2.2	2.1	1.3
FeO (tot.)	0.5	0.1	1.4	0.5	0.61	0.39	0.17	0.68	0.51	0.75	0.16	0.76	0.63	—
MnO	0.03	0.02	0.03	0.02	0.04	0.05	0.04	0.02	0.01	0.02	0.02	0.03	0.01	—
MgO	4.3	1.4	2.5	0.8	1.5	1.5	1.2	1.1	0.7	0.76	0.47	0.86	0.14	—
CaO	47.1	1.2	42.6	5.2	51.4	52.6	0.9	49.1	4.4	48.2	2.4	47.6	6.7	49.2
Na_2O	0.13	0.05	—		—	0.09	0.04	0.21	0.0	0.31	0.15	0.31	0.20	0.17
K_2O	0.29	0.18	0.71	0.65	0.14	0.09	0.03	0.37	0.23	0.30	0.15	0.51	0.51	0.22
P_2O_5	0.03	0.01	0.06	0.03	0.04	0.05	0.05	0.07	0.03	0.05	0.02	0.05	0.04	0.07
S	—		1157	750	460	—		402	357	—		—		—
V	—		—		—	bdl		—		2	(0–11)	22	(5–73)	bdl
Cr	18	2	29	13	9	bdl		13	10	13	7	46	(9–144)	bdl
Ni	4	1	5	(0–40)	1	bdl		1	1	7	(1–31)	3	(0–8)	2
Cu	bdl		18	2	19	21	(0–43)	20	5	12	12	5	3	bdl
Zn	12	5	18	(4–91)	6	8	3	7	6	11	8	14	12	12
Ga	2	1	5	2	2	2	2	2	1	—		7	(0–26)	—
Rb	8	5	24	18	5	2	2	13	8	12	6	42	(5–141)	5
Sr	223	80	1499	330	2095	1627	409	2098	368	1954	453	1351	497	2161
Y	3	1	7	3	6	4	1	3	2	4	2	9	9	5
Zr	16	6	118	40	112	4	3	121	25	20	11	74	(17–216)	17
Nb	—		3	2	—	bdl		1	0	1	1	5	(0–20)	bdl
Ba	88	19	194	118	10	15	6	77	77	93	38	237	(14–952)	60
La	3	1	5	3	1	bdl		5	2	bdl		16	(0–45)	10
Ce	bdl		46	20	—	3	(0–10)	26	17	16	13	27	(0–94)	20
Pb	—		5	(0–39)	—	8	8	2	(0–18)	10	(0–35)	10	5	4
Th	—		4	1	—	bdl		2	1	1	1	5	(1–17)	bdl
Ca/Sr	1754		209		181	241		171		180		315		163
n	5		13		4	5		51		6		6		1

Abbreviations as Table 15.1.

1. Connemara Limestone, Connemara.
2. Falcarragh Limestone, Donegal.
3. Fanad Limestone, Donegal.
4. Islay Limestone, Islay.
5. Lismore Limestone, Lismore and Lochaber.
6. Blair Atholl Limestone, Blair Atholl.
7. Sandend Limestone, Buchan.
8. Whiteness Limestone, Shetland.

(From Rock, 1986, revised.)

low S, Mn and Sr, and high Pb and Y (Table 15.5). The Sr content (*c.* 550 ppm) and Ca/Sr ratio (*c.* 600) are indicative of a deep saline origin. The dolostones are very impure, most having over 25% SiO_2. They are much richer in Mn than the limestones, but otherwise show similar characteristics, although Sr is lower, at *c.* 300 ppm (Table 15.4).

The Appin Limestone Formation has two facies, a lower pure white limestone (the Onich Limestone Member), and a banded brown and cream unit of interbedded carbonates and sandy carbonates, which has been given the name 'Tiger Rock'. This facies includes both limestones and dolostones. The impurities are rather more argillaceous than those of the Ballachulish Limestone and include a considerable proportion of iron. This is matched by the very considerable (though highly variable) amounts of S and Mn, suggesting that much of this horizon has iron oxides or sulphides in greater abundance than the other carbonate units. The Sr proportion and Ca/Sr ratio (*c.* 600 for limestones, 700 for dolostones; Table 15.5) are suggestive of a deep-water origin (Veizer and Demovič, 1974).

The Lismore Limestone Formation is very distinctive in its chemistry (Table 15.6). Almost all the formation is limestone, low in impurities. It is particularly rich in Sr (over 2000 ppm), and low in Mn. Features associated with its insoluble residue are high Ba and Rb, due to the insolube residue being much more clay-rich than the Ballachulish Subgroup limestones, and high Zr, Cr and P, probably indicative of a near shore environment.

Although only a few analyses are available (Rock, in 1986) (Table 15.6, column 4), the Islay Limestone Formation does seem to show somewhat similar characteristics. It is composed of pure limestones relatively high in Sr. It does not, however, have the low Mn, and high Zr and Cr exhibited by the Lismore Limestone. Chemical correlation of these four major carbonate horizons along the strike of the Dalradian basin, has been attempted by Rock (1986).

Carbonate units thought to be stratigraphically low in the Appin Group (Treagus, 1969), have been analysed only for Kinlochlaggan and Shetland (the Weisdale Limestone). Both are fairly similar in chemistry to the Ballachulish Limestone, although the Ballachulish Limestone Formation is much more dolomitic. The major element that differs is Sr, which is very high indeed, very similar to the Lismore and Islay proportions, and hence the chemical results do not conclusively prove a correlation. The Appin Limestone has also not been analysed along the length of the outcrop. In Donegal, the Marble Hill Limestone is not particularly similar (Hickman and Wright, 1983), although Rock (in press) suggests that the correlation is statistically reasonable.

The two Blair Atholl Subgroup limestones have generally not been separated outside the Islay area, although analyses are available from a wide area (Table 15.6). The Falcarragh and Fanad Limestones of Donegal, the Blair Atholl and Sandend Limestones of Central and north-east Scotland and the Whiteness Limestone of Shetland, all show that pure limestones are predominant, and all have the characteristic high Sr and low Mn of the Lismore Limestones. In Connemara, in the far western limit of the outcrop, the Connemara Marble is in the stratigraphically equivalent position to the Islay/Lismore Limestones (Leake *et al.*, 1975) but is predominantly a dolostone (Rock, 1985*b*) although, as it is at such a high metamorphic grade, much has been converted to calc-silicates and even ophicalcites (magnesian rocks with > 25% MgO). It may well be that there is a facies change into more dolomitic rocks in this area. There are insufficient data at present to make reliable distinctions between the Lismore and Islay Limestones, and although it is likely that the above limestones belong to the Blair Atholl Subgroup, definite correlations at formation level are not yet appropriate.

15.4 Sedimentology

Only in recent years has a sedimentological description of the Dalradian been attempted. Harris and his co-workers (Harris *et al.*, 1978) and Anderton (1982, 1985) have developed a generalized sequence of events for the whole Dalradian Supergroup, suggesting that a major gulf opened into the Laurentian landmass, with the gradual deepening of this basin and with changes in the provenance of the sediment also taking place through the sequence. However, the interpretation of the Appin Group is mostly derived from Hickman's work (1972, 1975). There have been very few previous or subsequent papers on rocks of the Appin Group which give any details of the sedimentology, but a study of the Islay area has recently been completed (Stedman, 1986). The two theses (Hickman, 1972; Stedman, 1986) and the author's personal observations are the basis of this account.

15.4.1 *Lochaber Subgroup*

The quartzites of this unit are well bedded, and all three show cross-bedding in almost every outcrop. Individual beds are from 5 to 20 cm thick, averaging about 30 cm, and are generally coarse-grained sands. Although recrystallization has destroyed much of the original grain shapes, where the feldspars are set in quartz, and where there are original grit-sized particles and pebbles, these are subrounded to well-rounded. The cross-sets vary from 10 to 30 cm in thickness and are usually several metres in length, with planar bounding surfaces and curved foresets—the omikron type of Allen's (1963) classification. Xi-cross-bedding also occurs, though much less frequently, and occasionally the lower bounding surfaces are erosional—the gamma-cross-stratification type. Slumped, or at least internally deformed, cross-bedding is quite common in the Eilde and Binnein Quartzites (Hickman, 1975, plate 2), and although this area is one of locally intense deformation, there is no doubt that this is a primary sedimentary feature. Ripple marks have also been observed in each of the formations and are usually straight, asymmet-

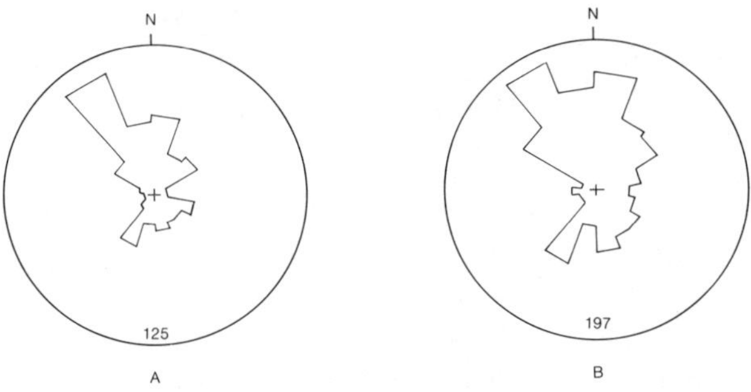

Figure 15.6 Cross-bedding foreset orientations, Lochaber (from Hickman, 1975). (*A*) Binnein Quartzite, Kinlochleven; (*B*) Eilde Quartzite, Binnein Quartzite and Glencoe Quartzite, Lochaber. Number of readings is given at the base of each circle.

rical, 3–15 cm in wavelength and 1–3 cm in amplitude (Hickman, 1975, plate 3).

The current directions, determined by Hickman (1975) from the foreset dip azimuths, are generally north–south with a spread from the northwest to ENE (Fig. 15.6). One of the most significant features noted by Hickman (1975) was the variation in composition of the Glen Coe Quartzite. In the Glen Creran area it contains a fair proportion of pebbly horizons. As it is traced through the Loch Leven area to the north-east, it changes from a feldspathic and rather impure quartzite to a very fine-grained white quartzite in the Glen Nevis, Glen Spean area, before wedging out completely.

Both the Glen Coe and Eilde Quartzites are feldspathic quartzites in the major part of their outcrop. The Eilde Quartzite also disappears north-eastwards, but at the point where it disappears there is apparently a tectonic boundary. However, to the north-east, as recent mapping shows (Hickman, 1975, 1978), this unit does not reappear, and there seems little doubt that it is the least persistent of these three formations. The intervening schist units, the Eilde Schist and Binnein Schist, and the lower part of the Leven Schist above the Glen Coe Quartzite, are very much more heterogeneous formations than the quartzites. Rock types vary from pelitic through semi-pelitic to psammitic, with quite frequent thin (< 10 m) bands of pure quartzite, particularly in the transition zones at the base and top of each formation. The only regional facies changes of any significance are the general decrease in quartzite beds and the increase in the volume of pelitic rocks as the formations are traced north-eastwards. Many of these small psammitic units show graded bedding and some of the quartzites are cross-bedded.

Even within the Binnein Schist there are indications of the onset off a change in sedimentary environment, with black slates, gritty ironstones and calcareous pelites developed near the top. This change also occurs more markedly in the Leven Schist, and probably marks a major change in sedimentary environment.

The extent, tabular nature, and bimodal (rather than dispersed) palaeocurrent patterns of the quartzites, indicate that they are tidal sand bodies with the cross-bedding produced by wave ripples. Such currents are typically coast-parallel, and the apparently south-eastwards slumping in the Eilde Quartzite would suggest that the major slope was in that direction, with longshore currents running north or north-east. The uppermost quartzite may well have been deposited a considerable distance to the north or north-west of the lowest, and it is possible that such sands were widely dispersed as separate bodies along a shallow platform margin. South-westwards they probably merged into truly deltaic deposits, while they passed north-eastwards into pure quartzites and then dominantly pelitic rocks. South-eastwards there is also a tendency for the units to become more pelitic, and this seems also to have been the distal direction.

The only other point of major importance is the occurrence of a boulder bed probably towards the top of this subgroup at Kinlochlaggan (Treagus, 1969), although numerous tectonic slides render precise stratigraphic positioning almost impossible. There are two unbedded psammitic units, each about 7 m thick, within a 43 m sequence of bedded psammites and semi-pelites. The larger clasts are concentrated in the lower unbedded psammite, and range from about 3 cm to 10 cm, with the largest seen to be 33 cm in diameter. Most consist of alkali granite, but psammites, semi-pelites and pelites also occur. No carbonate clasts were found. As such extrabasinal clasts are also found on Islay, in a conglomeratic bed within the Maol an Fhithich Quartzite (Rast and Litherland, 1970; Stedman, 1986) and a single granite clast within the Leven Schists of Glen Creran (Litherland, 1980), there is the possibility that a glacial episode occurred towards the end of Lochaber times.

The stratigraphy of this subgroup elsewhere reveals that one or two quartzite units, sometimes varying from impure feldspathic to pure quartzite within a relatively limited area, are generally developed between Grampian Group facies rocks and Ballachulish Subgroup facies rocks. The only other area with great thicknesses of psammites is the Mayo area but, as was noted above, the whole sequence does seem to thicken into Ireland. Anderton (1985) suggests that the basin of deposition, which opened up in Lochaber times, was north-easterly facing and closed towards the south-west. The evidence from the Lochaber area indicates that sediment supply lay dominantly to the south-west, but this could have been detritus supplied by a river system from a western or north-western landmass. In the absence of a detailed sedimentological investigation it is impossible to be certain, but the many thick psammites and very much lower pelite content of the Mayo area may indicate that there was a second river system flowing into the seas of that area.

15.4.2 *Ballachulish Subgroup*

The transition from Leven Schist to Ballachulish Limestone facies is nowhere well exposed in an area of low tectonism. Towards the top of the Leven Schist there are some calcareous horizons and some darker pelites, and in Ireland and in the Spean area there may well be some distinct limestone horizons. The Ballachulish Limestone Formation, however, is a distinctive blue-grey banded carbonate, with a generally high proportion of dolostone. The lowest member is a white limestone (Bailey, 1960). Although associated with black shales, both within the formation and above and below, there is a higher proportion of siliceous detritus in the impure dolostones than in other formations. The level of Sr is, however, typical of deep-water saline carbonates and some scarce sedimentological indications also suggest that it is in fact a deep-water limestone. The bedding units are uniform and mostly parallel-bedded, with a little ripple-drift bedding but no indication of shallow-water features. There are very occasional chert nodules in this formation. An unusual rock type has been recently exposed at Ballachulish (in the new road-cutting at the River Laroch). This is a chlorite schist with globular (now ellipsoidal) bodies of quartz and pyrite. Its origin is uncertain, but the very high chlorite content might indicate a volcanic origin.

The overlying Ballachulish Slate Formation is a typical black pyritiferous slate with, in places, abundant evidence of bedding. Although black shales are generally regarded as euxinic deposits in areas of low sedimentation, there is no certainty that they are necessarily deep-water deposits or distant from the shore. The sedimentary variation shown by the bedding is between siliceous silty horizons, sometimes showing cross-lamination, and clays (now mica). Sometimes a grading between the two is visible. Quartz clasts, as large grains, are found within the clay horizons, and these are sufficiently common (they are found in thin-sections in every locality so far investigated) for it to seem likely that they are of sedimentary origin. Perhaps they were small 'dropstones' of glacial origin, although if so, the ice was probably quite distant, as the only other indication of glacial conditions below the Ballachulish Limestone may be the Kinlochlaggan Boulder Bed.

The black clays are rather high in alumina, and this could indicate a predominance of chemical weathering in the hinterland, which suggests a hot, humid climate rather than glacial conditions. The major-element chemistry is suggestive of a more montmorillonitic clay (rather than illite) suggesting remoteness from terrigenous supply. The probable environment for these slates is thus a prodelta clay, with the incoming of the quartzites of the 'striped transition' of the Appin Quartzite above indicating the encroachment of the fine quartz sands of the delta into the deeper water.

The Appin Quartzite itself is a coarse, cross-bedded, feldspathic quartzite. Grain size and feldspar content

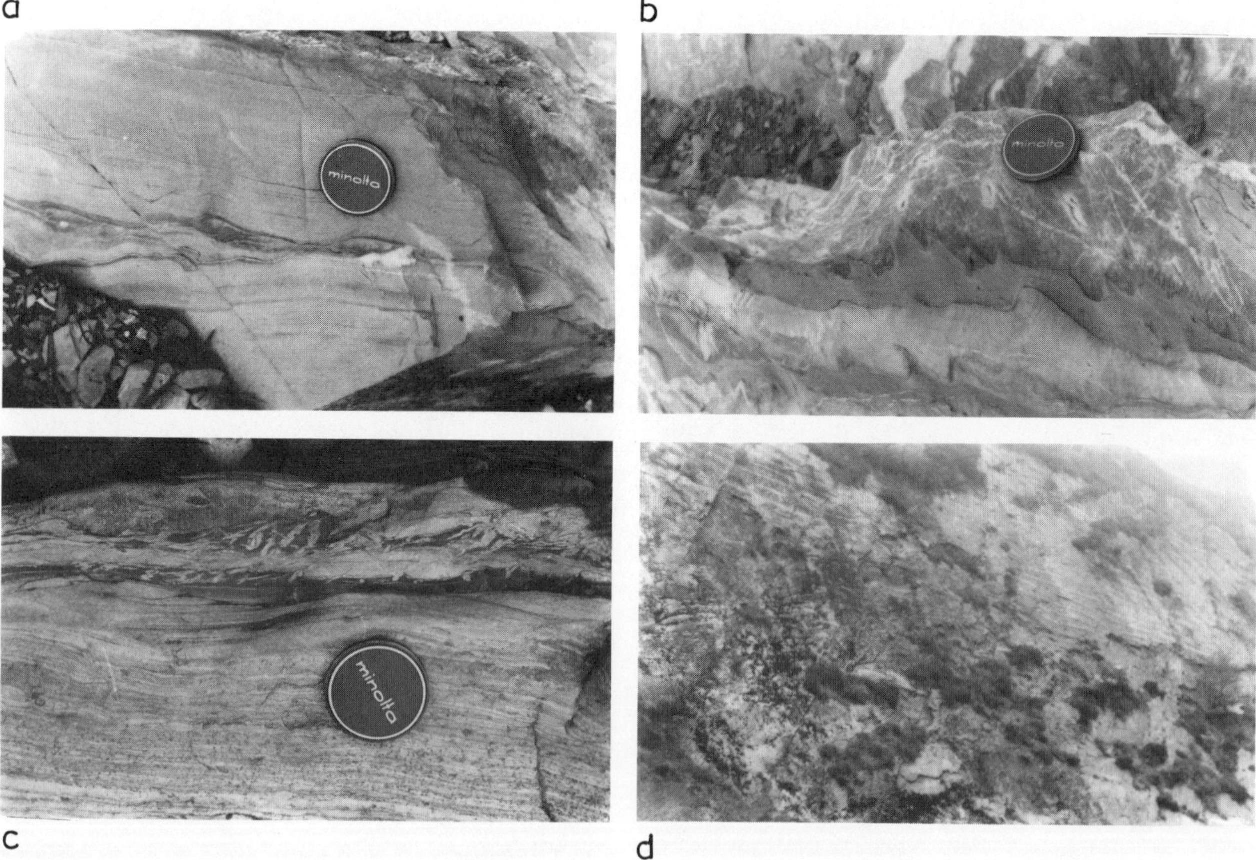

Figure 15.7 Sedimentary structures in Ballachulish Limestone and Appin Quartzite. (*a*) Finely bedded limestone with fine ripple drift bedding (in dark band just below lens cap). West Laroch road cutting, Lochaber. (*b*) Load casts, pure limestone above dolomitic silt (accentuated by deformation). West Laroch road cutting, Lochaber. (*c*) Synaeresis cracks in interbedded quartzites and black slates, of the transition beds of the Appin Quartzite, Laroch Burn, Lochaber. (*d*) Large rippled surfaces of Appin Quartzite beds, Onich, Lochaber. The ripples on this surface have a wavelength of 1.5 m.

increase upwards and away from the type area of Onich towards the south-west where, in the Lynn of Lorn, the homogeneous quartzite gives way to a very variable pebbly quartzite with interbedded thin silty pelites and fine sandstones. It seems to have thickened south-westwards and petered out north-eastwards and south-eastwards. The chemistry indicates that this is a far more immature quartzite than those of the Lochaber Subgroup, becoming a feldspathic sandstone in the Lynn of Lorn.

Sedimentary structures within the quartzite units include cross-bedding, washout structures, ripple marks, graded bedding and slump folding, while in the pelitic interbeds of the striped transition beds there are synaeresis cracks. Xi-, omikron- and trough-cross-bedding are all encountered and are the commonest form of sedimentary structure, being present in a high proportion of the beds. Grading is also fairly common, with the base of some beds being distinctly pebbly. Washout structures and other evidence of channelling are found in the striped transition beds (Bowes and Wright, 1962), and suggest that the beginning of quartzite deposition involved the encroachment of sand masses into deeper water. Ripple marks on a small scale

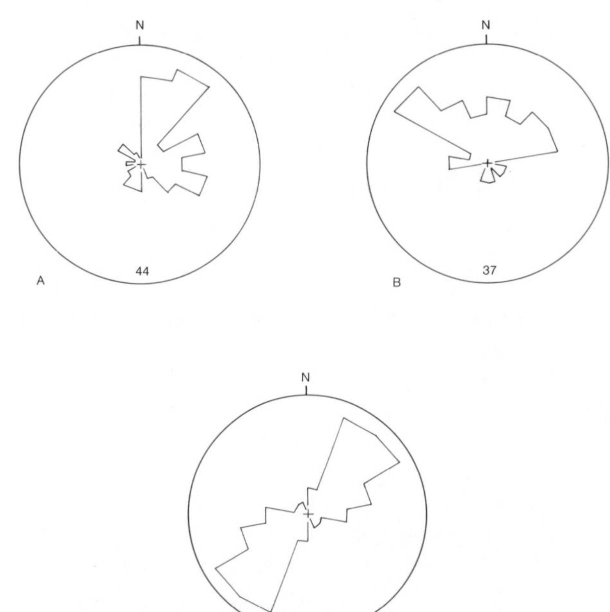

Figure 15.8 (*A* and *B*): cross-bedding foreset orientations for Appin Quartzite (from Hickman, 1975). (*A*) Onich; (*B*) Lynn of Lorn; (*C*) trend of ripple marks, Appin Quartzite, Onich. Number of readings is given at the base of each circle.

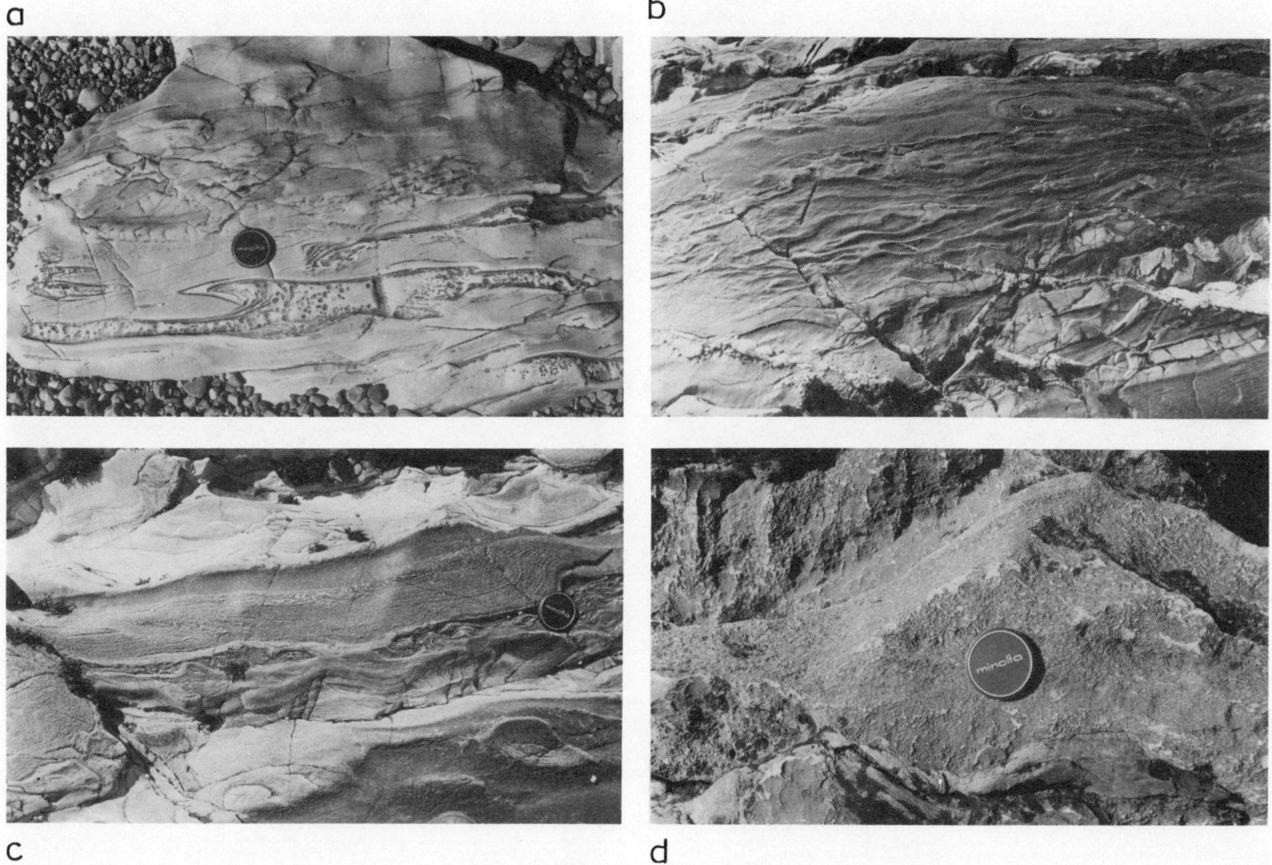

Figure 15.9 Sedimentary structures in the Appin Limestone and Phyllite (*a*) Yellow crystalline limestone of the Appin Limestone highly recrystallized by still preserving slump folds and detached bedded masses in a homogeneous limestone. Onich shore, Lochaber. (*b*) Large scale slump folds within interbedded dolomitic and phyllitic beds. Onich shore, Lochaber. (*c*) Finely bedded dolomitic siltstones in largely arenaceous sequence with open and isoclinal slump folds and ripple drift lamination cut off by the overlying dolomitic horizon. Onich shore, Lochaber. (*d*) Flake breccia of dolomitic fragments, derived by disruption of laminated beds similar to (*c*). Onich shore, Lochaber.

(5–30 cm) are fairly frequent, but mega-ripples seen at Onich (shown in Fig. 15.7*d*) are possibly the type that produced the xi- and omikron-cross-stratification. The ripples, where asymmetrical, possess a steep north-western side, and their axial trend is parallel to the mean foreset inclination of the cross-bedded units (Fig. 15.8) in a north-easterly direction. The only difference noted areally is that the palaeocurrent directions in the Lynn of Lorn area are radial rather than bipolar, suggesting a position closer to the head of a delta (Fig. 15.8*B*).

The synaeresis cracks are very frequent in the pelitic horizons of the striped transition of the Laroch area. Such structures are generally considered to be due to salinity changes, which lends support to the interpretation of the Appin Quartzite as a deltaic unit extending into a marine basin. Slump folding of the cross-bedding is not as frequent in the Appin Quartzite as in the Lochaber quartzites, but its presence also indicates a degree of sedimentary instability.

The transition from quartzite to interbedded carbonate and phyllite is quite sharp, although the lower part of the Appin Limestone Formation contains a fair number of quartzite horizons. The lower marker is a white crystalline limestone (the Onich Limestone Member), and this is followed by interbedded sandy quartzites and green phyllites. The quartzites show channelling at their bases, and often show wave-forms on their upper surfaces. Sandstone dykes, extending several feet down below the sandstones, are frequent. On the Onich shore a wide variety of sedimentary structures can be seen in the phyllites (which are largely silts and sands with very little pure argillite present). Slump folds and large slump breccias occur, and many horizons of laminated dolostone, often slumped and brecciated into fine flake breccias, are also found. These features indicate a high degree of instability in the sedimentary regime of this formation. The limestone which is characteristic of this formation is a banded rock of alternating pure yellow carbonate and brown sandy carbonate, termed 'Tiger Rock' by Bailey (1960). This occurs in at least two horizons—within a great thickness of phyllite with rather more green pelitic material. No exposures of the limestone give any indication of the sedimentary environment. At Glen Creran an upper dolomite is more homogeneous (Litherland, 1980). On

a

b

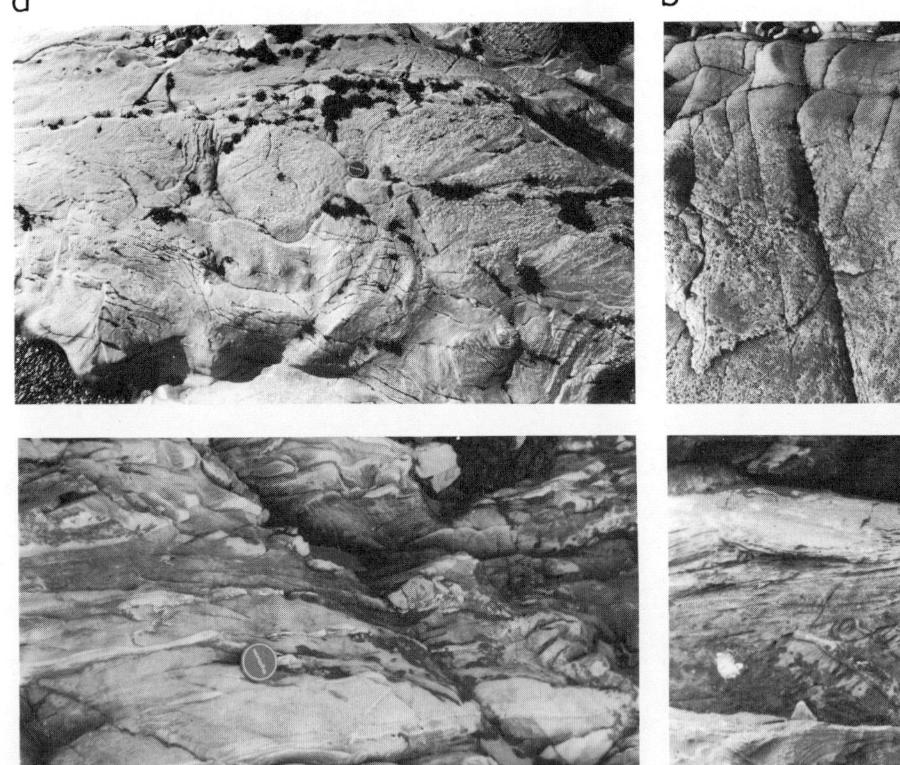

c

d

Figure 15.10 Sedimentary structures in Appin Limestone and Phyllite and Lismore Limestone formations (*a*) Chaotic slumped breccia horizon composed largely of quartzitic and silty horizons in a silt matrix. Onich shore, Lochaber (very close stratigraphically to Fig. 15.9*b*). (*b*) Channel, filled with homogeneous quartzite, cutting bedded quartzite in lower part of the Appin Limestone and Phyllite formation. Onich shore, Lochaber. (*c*) Sandstone dykes in interbedded quartzite and phyllite near base of Appin Limestone and Phyllite. Two small dykes (just above and to the left of the lens cap) are derived from a thin quartzite bed, the major dyke on the right cuts down sequence for many tens of centimetres below a massive quartzite (out of view). Onich shore, Lochaber. (*d*) Banded limestones with intraformational angular unconformity. This is within the horizon showing slump folds (Hickman 1975, plate 8) in the Bagh Clach an Dobhrain transitional member at the very top of the Lismore Limestone Formation, at Bagh Clach an Dobhrain on the island of Lismore.

Islay only a single cream limestone with sandy horizons followed by dolomitic sandstones and interbedded dolomitic siltstones and dolostones occur between the Appin Quartzite and black slates equivalent to the Cuil Bay Slate, with no green phyllites at all. At Glen Creran the Appin Limestone changes facies to the south-east and cuts out the underlying Appin Quartzite (Fig. 15.11).

The Marble Hill Limestone in Donegal is one of at least three carbonate horizons at about this level. Structures within this formation have been interpreted as desiccation cracks (Bliss *et al.*, 1978), while odd rod-like structures and nodules lower in the Sessiagh–Clonmass Formation are regarded as of diagenetic origin. The limestones in Scotland are everywhere too recrystallized to preserve either sedimentary or biogenic structures.

The lack of detailed sedimentological studies in Ireland means that the palaeogeography of this subgroup is not known with much certainty, but the picture of a southward closing bay of an elongate sea, as suggested by Anderton (1985), is the most likely scenario indicated by the present information.

15.4.3 *Blair Atholl Subgroup*

This unit is composed, in the type area, almost entirely of black slates and grey limestones which alternate on scales varying from a few centimetres to formations of several hundred metres. The Cuil Bay Slate at the base is a well-bedded black slate, with the bedding mostly being emphasized by slight variations in colour. Unlike the Ballachulish Slate there is not much silty material deposited.

The geochemistry suggests a more montmorillonitic clay than those higher in the sequence which were probably richer in kaolinite and illite. The trace-element chemistry also suggests a more neritic environment for the higher members of the sequence. The Cuil Bay Slate passes up into the very strongly interbedded Lismore Limestone Formation, with thin shale units and thicker interbedded shales and limestones between major thick limestones. The limestones are all thoroughly recrystallized and no sedimentary structures are preserved. The interbedded members have slumped units exposed both in Lismore and Islay. There are occasional sandstone units both on Lismore and Islay, representing small influxes of detrital sediments.

It is not certain how much of the succession above the Ballygrant Limestone on Islay represents beds above the Lismore Limestone Formation of Appin. The succession does, however, contain a much more varied sequence of rocks although it is still predominantly limestones with interbedded black shales. The Mullach Dubh Phyllite is composed of black slates with sandstones and siltstones and, at the Mull of Oa, a mud-

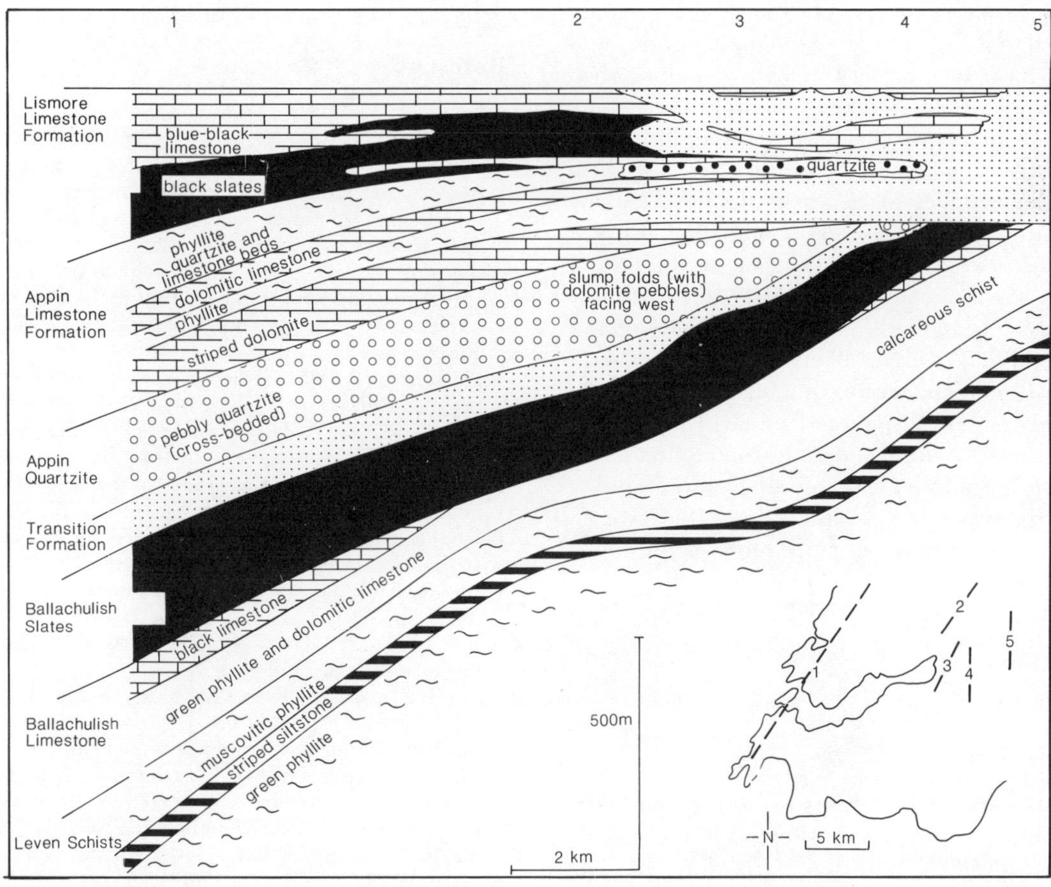

Figure 15.11 Facies changes in the Appin Group across the strike at Loch Creran. Thicknesses are pre-tectonic estimates (from Litherland, 1980).

pellet conglomerate. It is regarded as a deep-water sediment with turbiditic influxes. The Kilslevan Limestone (Fig. 15.1) has some pelitic horizons, some again conglomeratic, while the limestone is often oolitic and has some erosive bedding contacts; it is thought to be of shallow-water origin. The Creag nan Gabhar Slates, black slates with some graded sandstones and limestones, represent a return to deeper water. The Keills Limestone, with oolitic, peloidal and stromatolitic patches, is of shallow-water origin. The Balulive Limestone Conglomerate is a subtidal sequence of lensoid channel-fill conglomerates with clasts of limestone, followed by festoon-cross-bedded then planar-bedded sandstones, and dolomitic siltstones. The Beannean Buidhe Formation comprises dolomitic siltstones and dolomites up to 75 m thick in the core of the anticline. These are interbedded on the SW limb of the Islay Anticline with the Balulive Limestone Conglomerate, and all are overridden by the Port Askaig Tillite. Within the Beannean Buidhe Formation, synaeresis cracks, load casts, sandstone dykes and flake breccias indicate a shallow-water, saline sequence with a great deal of instability. The overall environment of the Blair Atholl Subgroup is thus generally one of shallower-water sequences, probably of lagoonal limestones and shales. There is little evidence of high clastic input for most of this period, but there are substantial indications of instability, including quite small-scale variations in sequences.

On a wider scale these types of rocks (blue limestones and black slates) occur from Banff to Donegal, and there must have been a very extensive shallow shelf parallel to the shoreline, with considerable along-shore continuity of facies. In Northern Ireland the stratigraphy seems to indicate, however, that a radical change in facies takes place across the strike of the orogen, with semi-pelitic rocks comprising a thick sequence at this time in Central Donegal. As this type of change can also be seen along a fairly short transect from Lismore Island to Glen Creran in Scotland (Litherland, 1980) it seems likely that the shelf sedimentation, although extensive in a NE–SW direction, was relatively narrow at right angles to it (Fig. 15.11). The lowest recorded appearance of stromatolites in the Dalradian occurs at the top of this subgroup. This may be because the environment was unfavourable lower in the sequence, or perhaps because most of the lower limestones are too recrystallized to preserve such structures. None have been identified with sufficient accuracy to assist in determining the stratigraphic age of this group.

15.5 Conclusions

The general picture of a southward-closing embayment in a continental margin, proposed by Anderton (1985), is largely based on the data summarized herein. The general northwards trend of the palaeocurrents low in the sequence perhaps suggests that the influx of detritus for the Lochaber quartzites came from the south or south-west, but the presence of a similar thickness of psammitic rocks in western Ireland, though not as well documented, may suggest that several major river systems fed this gulf along its length. After the initial more estuarine conditions, an open-marine environment became established and deep-water limestones and shales were eventually deposited, followed by shallower-water mixed carbonate–clastic deposits, and finally lagoonal limestones and shales.

Instability of the tectonic environment is suggested by slump horizons at various levels ranging as low as the phyllites of the Appin Limestone. This may suggest that the much more definite evidence of syndepositional faulting recorded higher in the Dalradian (Anderton, 1979), may have been initiated much earlier than has previously been suggested. The extensive lateral correlation of certain horizons suggests that towards the end of the Appin Group times there was a stable continental margin along much of the distance from Donegal to Banffshire. This is also borne out by the similar outcrop pattern for the succeeding lowest unit of the Argyll Group–the Port Askaig Boulder Bed, a tillite of enormous extent, which seems to indicate a dramatic climatic change, but which also indicates derivation of detritus from a granitic source unlike the Lewisian or Moinian rocks to the north-west, and thus has generally been regarded as being derived from a southerly or southeasterly source region.

References

Allen, J. R. L. (1963) The classification of cross-stratified units, with notes on their origin. *Sedimentology* **2**, 93–114.

Anderton, R. (1979) Slopes, submarine fans, and syn-depositional faults: sedimentology of parts of the Middle and Upper Dalradian in the SW Highlands of Scotland. *Spec. Publ. geol. Soc. London* **8**, 483–488.

Anderton, R. (1982) Dalradian deposition and the Late Precambrian–Cambrian history of the N. Atlantic region: a review of the early evolution of the Iapetus Ocean. *J. geol. Soc. London* **139**, 421–431.

Anderton, R. (1985) Sedimentation and tectonics in the Scottish Dalradian. *Scott. J. Geol.* **21**, 407–436.

Bailey, E. B. (1917) The Islay Anticline (Inner Hebrides). *Q. J. geol. Soc. London* **72**, 132–159.

Bailey, E. B. (1922) The structure of the South-West Highlands of Scotland. *Q. J. geol. Soc. London* **78**, 82–127.

Bailey, E. B. (1930) New light on sedimentation and tectonics. *Geol. Mag.* **65**, 77–92.

Bailey, E. B. (1934) West Highland tectonics: Loch Leven to Glen Roy. *Q. J. geol. Soc. London* **90**, 462–523.

Bailey, E. B. (1960) Geology of Ben Nevis and Glen Coe. *Mem. geol. Surv. Gt. Br.* 307 pp.

Bliss, G. M., Grant, P. R., Max, M. D. and Diver, W. L. (1978) Sedimentary and diagenetic features in the Sessiagh–Clonmass Formation of the Dalradian Ballachulish Sub-group in Donegal. *Geol. Surv. Ire. Bull.* **2**, 189–204.

Bowes, D. R. and Wright, A. E. (1962) Washout structures in the Dalradian near Kentallen, Argyll. *Geol. Mag.* **99**, 53–56.

Church, W. R. (1969) Metamorphic rocks of the Burlington Peninsula and adjoining areas of Newfoundland and their bearing on continental drift in the North Atlantic. *Mem. Am. Ass. Pet. Geol.* **12**, 212–233.

Clarke, F. W. (1924) The data of geochemistry, 5th edn., *Bull. US geol. Surv.* **770**.

Crow, M. J., Max, M. D. and Sutton, J. S. (1971) Structure and stratigraphy of the metamorphic rocks in part of northwest County Mayo, Ireland. *J. geol. Soc. London* **127**, 579–584.

Flinn, D., May, F., Roberts, J. L. and Treagus, J. E. (1972) A revision of the stratigraphic succession of the East Mainland of Shetland. *Scott. J. Geol.* **8**, 335–343.

Harris, A. L., Baldwin, C. T., Bradbury, H. J., Johnson, H. D. and Smith, R. A. (1978) Ensialic basin sedimentation: the Dalradian Supergroup. *Geol. J. Spec. Issue* **10**, 115–138.

Harris, A. L. and Pitcher, W. S. (1975) The Dalradian Supergroup. *Spec. Rep. geol. Soc. London* **6**, 42–75.

Hickman, A. H. (1972) The Stratigraphy, Geochemistry and Structure of the Balappel Foundation, South-West Highlands of Scotland. Unpublished Ph.D. Thesis, University of Birmingham.

Hickman, A. H. (1975) The stratigraphy of late Precambrian metasediments between Glen Roy and Lismore. *Scott. J. Geol.* **11**, 117–142.

Hickman, A. H. (1978) Recumbent folds between Glen Roy and Lismore. *Scott. J. Geol.* **14**, 191–212.

Hickman, A. H. and Wright, A. E. (1983) Geochemistry and chemostratigraphic correlation of slates, marbles and quartzites of the Appin Group, Argyll, Scotland. *Trans. R. Soc. Edinburgh: Earth Sci.* **73**, 251–278.

Kilburn, C., Pitcher, W. S. and Shackleton, R. M. (1965) The stratigraphy and origin of the Port Askaig Boulder Bed Series (Dalradian). *Geol. J.* **4**, 343–360.

Kruhl, J. and Voll, G. (1975) Large-scale premetamorphic and precleavage inversion at Loch Leven, Scottish Highlands *Neues Jb. Miner. Mh.* **H.2**, 71–78.

Lambert, R. StJ., Winchester, J. A. and Holland, J. G. (1981) Comparative geochemistry of pelites from the Moinian and Appin Group (Dalradian) of Scotland. *Geol. Mag.* **118**, 477–490.

Lambert, R. StJ., Holland, J. G. and Winchester, J. A. (1982) A geochemical comparison of the Dalradian Leven Schists and the Grampian Division Monadhliath Schists of Scotland. *J. geol. Soc. London* **139**, 71–84.

Leake, B. E., Tanner, P. W. G. and Senior, A. (1975) The composition and origin of the Connemara dolomitic marbles and ophicalcites. *J. Pet.* **16**, 237–277.

Litherland, M. (1980) The stratigraphy of the Dalradian rocks around Loch Crenan, Argyll. *Scott. J. Geol.* **16**, 105–123.

McCall, G. J. H. (1954) The Dalradian geology of the Creeslough area, Co. Donegal. *Q. J. geol. Soc. London* **110**, 153–175.

Phillips, W. E. A. (1973) The pre-Silurian rocks of Clare Island, Co. Mayo, Ireland, and the age of the metamorphism of the Dalradian in Ireland. *J. geol. Soc. London* **129**, 585–606.

Phillips, W. E. A. (1981) The orthotectonic Caledonides. In Holland, C. H. (ed.), *A Geology of Ireland.* Scottish Academic Press, Edinburgh, 17–39.

Pitcher, W. S. and Berger, A. R. (1972) *The Geology of Donegal.* Wiley, New York, 435 pp.

Pitcher, W. S., Elwell, R. W. D., Tozer, C. F. and Cambray, F. W. (1964) The Leannan Fault. *Q. J. geol. Soc. London* **120**, 241–273.

Rast, N. and Litherland, M. (1970) The correlation of the Ballachulish and Perthshire (Iltay) Dalradian successions. *Geol. Mag.* **107**, 259–272.

Rickard, M. J. (1962) The stratigraphy and structure of the Errigal area, County Donegal, Ireland. *Q. J. geol. Soc. London* **118**, 207–238.

Roberts, J. L. (1976) The structure of the Dalradian rocks in the North Ballachulish district of Scotland. *J. geol. Soc. London* **132**, 139–154.

Roberts, J. L. and Treagus, J. E. (1975) The structure of the Moine and Dalradian rocks in the Dalmally District of Argyllshire, Scotland. *Geol. J.* **10**, 59–74.

Roberts, J. L. and Treagus, J. E. (1977*a*) The Dalradian rocks of the Southwest Highlands—introduction. *Scott. J. Geol.* **13**, 87–99.

Roberts, J. L. and Treagus, J. E. (1977*b*) The Dalradian rocks of the Loch Leven area. *Scott. J. Geol.* **13**, 165–184.

Roberts, J. L. and Treagus, J. E. (1979) Stratigraphical and structural correlation between the Dalradian rocks of the SW and Central Highlands of Scotland. *Spec. Publ. geol. Soc. London* **8**, 199–204.

Rock, N. M. S. (1985*a*) Value of chemostratigraphical correlation in metamorphic terranes: an illustration from the Colonsay Limestone, Inner Hebrides, Scotland. *Trans. Roy. Soc. Edinburgh: Earth Sci.* **76**, 515–517.

Rock, N. M. S. (1985*b*) A compilation of analytical data for metamorphic limestones from the Scottish Highlands and Islands. *Pet. Miner. Rep. Br. Geol. Surv.* **85/5** (open file report).

Rock, N. M. S., Macdonald, R., Szucs, T. and Bower, J. (1986) The comparative geochemistry of some Highland pelites. (Anomalous local limestone–pelite successions within the Moine outcrop; II.) *Scott. J. Geol.* **22**, 107–126.

Rock, N. M. S. (1986) Chemistry of the Dalradian (Vendian-Cambrian) metalimestones, British Isles. *Chem. Geol.* **56**, 289–311.

Smith, R. A. and Harris, A. L. (1976) The Ballachulish rocks of the Blair Atholl District. *Scott. J. Geol.* **12**, 153–157.

Spencer, A. M. (1971) Late Pre-Cambrian glaciation in Scotland. *Mem. geol. Soc. London* **6**, 100 pp.

Spencer, A. M. (1975) Late Precambrian glaciation in the North Atlantic region. *Geol. J. Spec. Issue* **6**, 217–240.

Stedman, D. (1986) An Examination of the Geology of the Appin Group of the Dalradian on Islay. Unpublished M.Sc. Thesis, University of Birmingham.

Tanner, P. W. G. and Shackleton, R. M. (1979) Structure and stratigraphy of the Dalradian rocks of the Bennabeola area, Connemara, Eire. *Spec. Publ. geol. Soc. London* **8**, 243–256.

Tanton, T. L. (1930) Determination of age-relations in folded rocks. *Geol. Mag.* **67**, 73–76.

Thomas, P. R. and Treagus, J. E. (1968) The stratigraphy and structure of the Glen Orchy area, Argyllshire, Scotland. *Scott. J. Geol.* **4**, 121–134.

Treagus, J. E. (1969) The Kinlochlaggan Boulder Bed. *Proc. geol. Soc. London* **1654**, 55–60.

Treagus, J. E. (1974) A structural cross-section of the Moine and Dalradian rocks of the Kinlochleven area, Scotland. *J. geol. Soc. London* **130**, 525–544.

Treagus, J. E. and King, G. (1978) A complete Lower Dalradian succession in the Schiehallion district, Central Perthshire. *Scott. J. Geol.* **14**, 157–166.

Turekian, K. K. and Wedepohl, K. H. (1961) Distribution of the elements in some major units of the Earth's crust. *Bull. geol. Soc. Am.* **72**, 175–192.

Veizer, J. and Demovič, R. (1974) Strontium as a tool in facies analysis. *J. sedim. Pet.* **44**, 93–115.

Vogt, T. (1930) On the chronological order of deposition of the Highland Schists. *Geol. Mag.* **67**, 68–73.

16
Stratigraphy of the Fleur de Lys Belt, northwest Newfoundland

JAMES HIBBARD

16.1 Introduction

The Fleur de Lys Belt (Hibbard, 1983*a*) comprises predominantly polyphase deformed metamorphosed clastic rocks. The belt is named after one of its major constituents, the Fleur de Lys Supergroup and defines a buffer zone between two major tectonostratigraphic zones in the Newfoundland Appalachians. Rocks of the belt form the eastern margin of the Humber Zone (Williams, 1978) (Fig. 16.1), which towards the west contains autochthonous sedimentary sequences and allochthonous ophiolite suites, that record the evolution of the late Hadrynian to Palaeozoic continental margin of eastern North America. In contrast, to the east of the Fleur de Lys Belt lie ophiolite suites and volcanic complexes of the Dunnage Zone (Williams, 1978) (Fig. 16.1) that collectively represent the vestiges of an early Palaeozoic Iapetus ocean. The evolution of geological relationships between these two tectonostratigraphic realms is recorded in the stratigraphy of the intervening Fleur de Lys Belt.

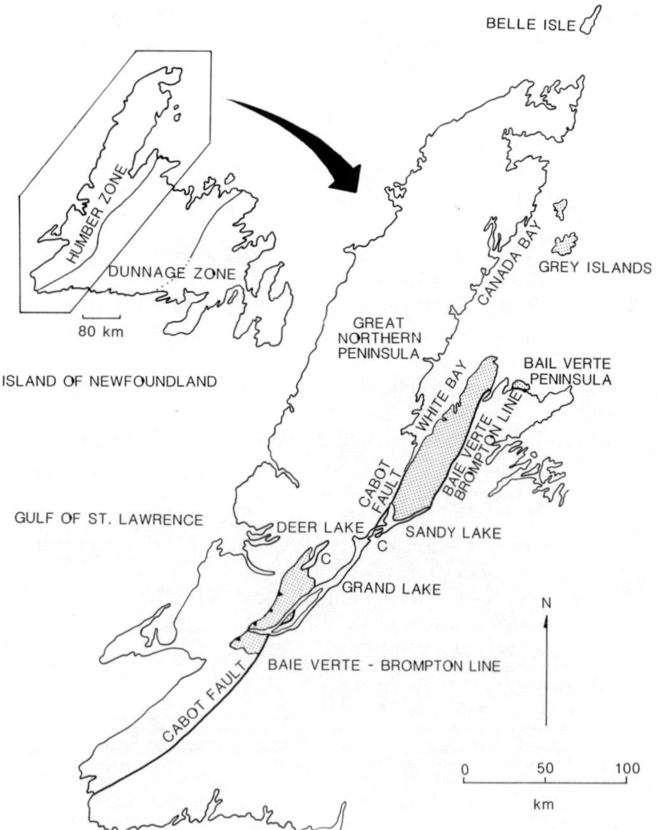

Figure 16.1 Distribution and setting of the Fleur de Lys Belt in Newfoundland; C—Carboniferous cover.

The belt is discontinuously exposed for more than 250 km along strike, extending from the Grey Islands in the north, southward across the Baie Verte Peninsula and into the Grand Lake area in west-central Newfoundland (Fig. 16.1). The continuity of the belt is disrupted by a cover of Carboniferous rocks that unconformably overlie the belt between Grand Lake and Sandy Lake (Fig. 16.1). A steep, narrow structural zone termed the Baie Verte–Brompton Line (Williams and St Julien, 1982) separates the Fleur de Lys Belt from Dunnage Zone rocks to the east. The belt is separated from other Humber Zone rocks by the Cabot Fault in the area south of White Bay and by thrust faults and younger steep faults in the Deer Lake–Grand Lake area (Fig. 16.1).

Two major lithostratigraphic subdivisions are recognized in the belt, including (i) an infrastructure of gneiss and subordinate schists, and (ii) a metaclastic cover sequence with intercalations of metavolcanic rocks and marble and tectonic slivers of ophiolitic rocks, all of which are included in the redefined Fleur de Lys Supergroup (Hibbard, 1983*a*). The infrastructure is a structural unit that exhibits intense, heterogeneous deformational features and hence the stratigraphic position of these rocks is somewhat ambiguous. In contrast, the Fleur de Lys Supergroup represents a relatively less-deformed unit than the infrastructure, and thus the stratigraphy of these rocks can be interpreted with more confidence.

Up until the past decade, the stratigraphic framework of the Fleur de Lys Belt has been largely conjectural, despite the host of workers that have studied the belt at different times since the 1860s. Even now, only a rough outline of the stratigraphy has been established, mainly by extrapolation of local detailed maps (e.g. de Wit, 1972; Bursnall, 1975; D. P. Kennedy, 1981) by regional mapping (Hibbard, 1983*a*, *b*). Unravelling the Fleur de Lys stratigraphy has led to controversy, and thus, this review begins with a brief summary of the major advances in stratigraphic thought on the Fleur de Lys Belt, before delving into a description and contemporary interpretation of the stratigraphy of the belt.

16.2 Evolution of stratigraphic thought

Most of our modern ideas concerning the Fleur de Lys Belt have arisen from investigations undertaken on the Baie Verte Peninsula, which were later extended along strike, both north and south, to other portions of the belt. Hence, the following summary pertains mainly to the peninsula. It encapsulates a short history of precisely which rocks have been termed 'Fleur de Lys', considerations of their ages, the recognition of the basement to these rocks, as well as the recognition of stratigraphically significant units within the Fleur de Lys rocks.

The initial reconnaissance observations of Sir Alexander Murray (Murray and Howley, 1881) of rocks now recognized as Fleur de Lys Belt, have proved accurate and consistent with modern concepts of the belt. Essentially, Murray envisaged these rocks as forming a broad anticlinorium, with schists on the limbs of the structure representing tectonized Lower Palaeozoic strata, and rocks in the core representing Precambrian gneisses. Subsequent detailed studies did not begin until the 1940s, and with these studies began a thirty-year period of confusion concerning the stratigraphy of Fleur de Lys rocks. Ironically, in the past decade the consensus of thought has returned to Murray's original ideas, albeit in more detail than Murray had undertaken.

Fuller (1941) first proposed the name 'Fleur de Lys' for the series of schists and gneisses that outcrop near Fleur de Lys Harbour (Fig. 16.2). Subsequently, Baird (1951) included the rocks in the Fleur de Lys Group, and extended the unit south to the latitude of Baie Verte community (Fig. 16.2). In view of the intense deformation and metamorphism of the group, both workers, as well as a fellow worker, Watson (1947), considered these rocks to be Precambrian in age.

The full extent of the group on the Baie Verte Peninsula was first outlined by Neale and Nash in 1963 during 1:250 000 scale mapping for the Geological Survey of Canada. They recognized an extensive metaconglomerate unit within the group, and reverted back to Murray's (Murray and Howley, 1881) original ideas on the structure and stratigraphy of these rocks.

Church (1969) first compiled the full extent of the rocks now included in the Fleur de Lys Belt in western Newfoundland. He raised the Fleur de Lys Group to supergroup status and added a new eastern division to the supergroup. This eastern division comprised mainly tectonized volcanic rocks, that were structurally similar to, yet lithologically distinct from, the Fleur de Lys Group, and hitherto had not been considered as related to the group. Church (1966, 1969) also interpreted the Fleur de Lys metaconglomerate as a tilloid, possibly correlative with other Precambrian tilloids of the Caledonides; however, he also thought it was possible that this tilloid could be correlative with the Lower Proterozoic Gowganda tillite of the Canadian Shield. Cogently, Neale (1967) noted the intense deformation of the metaconglomerate and stated '... it is highly unlikely that any evidence is preserved to support the recent proposal that these rocks are tilloids'. Also during this time period, Church (1965, 1966, 1969) and M. J. Kennedy (1971, 1973, 1975*a*, *b*) correlated the Fleur de Lys rocks with the Moine and Dalradian assemblages of Scotland and Ireland. Conditioned by the lowermost Ordovician deformation of the latter rocks and by local stratigraphic relationships, they considered the supergroup to have been deformed during or before the early Ordovician, and hence concluded that Fleur de Lys deposition ceased by the Tremadocian.

Many of these earlier ideas were challenged by later

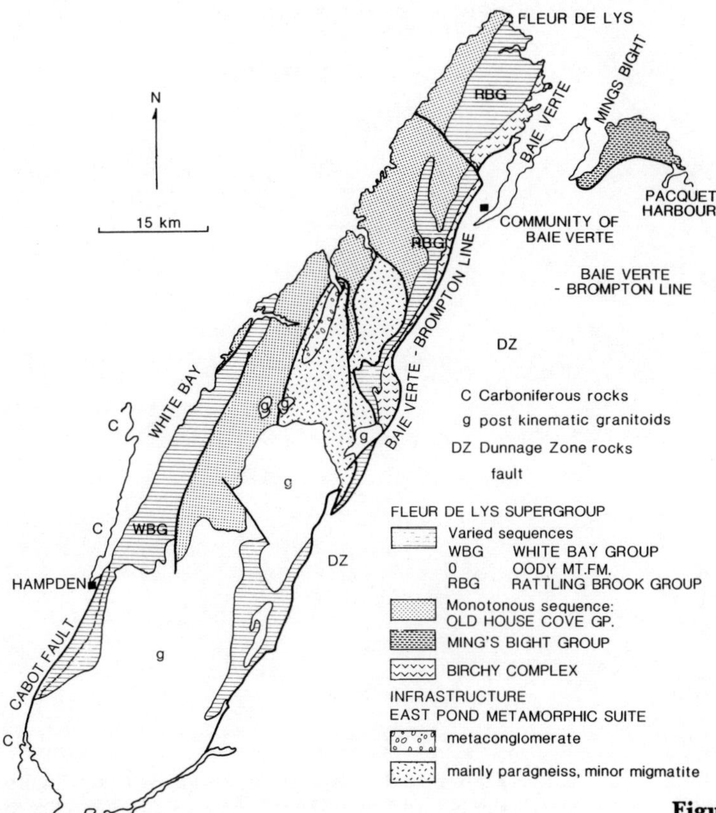

Figure 16.2 Geology of the Fleur de Lys Belt on the Baie Verte Peninsula; see Fig. 16.1 for location.

workers. The validity of Church's (1969) eastern division of the Fleur de Lys Supergroup has largely been refuted (Neale *et al.*, 1975; Degrace *et al.*, 1976; Williams *et al.*, 1977; Hibbard, 1982, 1983*a*; Dallmeyer and Hibbard, 1984). These workers have shown that the eastern division metavolcanics belong to the Dunnage Zone, although a metaclastic portion of the eastern division—the Ming's Bight Group—is tectonically severed from these metavolcanics, and rooted to the main Fleur de Lys outcrop belt in the subsurface (Hibbard, 1982, 1983*a*; Dallmeyer and Hibbard, 1984).

Two Cambridge students, M. J. de Wit and J. Bursnall, also challenged some earlier-held concepts, and established some new stratigraphic ideas concerning the supergroup. Based upon both regional correlation of the supergroup with less-deformed rocks of the western Humber Zone and regional structural considerations, Bursnall and de Wit (1975) concluded that the supergroup could be as young as Llanvirnian, and thus considered the unit to be late Hadrynian to early Ordovician in age. De Wit (1972, 1974, 1980) separated a central core of structurally complicated gneisses from the remainder of the supergroup, and informally designated them as 'the basement'; he considered the core to be the Grenvillian sialic basement to the supercrustal Fleur de Lys Supergroup. He also noted that the metaconglomerate was structurally atop the basement and contained pre-deformed clasts that were strikingly similar to the basement. He thus contended that the metaconglomerate was not a tilloid, but rather represented a basal, probably fluviatile, conglomerate lving above a major unconformity between the basement and the Fleur de Lys Supergroup. De Wit and Strong (1975) also recognized a suite of rift-related amphibolitic dykes in the supergroup, on the basis of petrochemical data and regional correlation (de Wit 1972, 1974).

In contrast to de Wit, Bursnall (1975) recognized a dismembered ophiolitic basement and ophiolitic *mélanges* just to the east of de Wit's area. He concluded that at least part of the supergroup was deposited on oceanic crust, and subsequently the *mélanges* have been related to the regional deformation of the Fleur de Lys Belt and western Newfoundland (Williams, 1977; Williams *et al.*, 1977; Hibbard, 1982).

In recent years, these fundamental stratigraphic concepts have been recognized and applied elsewhere along the length of the belt. Although there are local variations and controversies—e.g. the extent of granitic basement—the belt as a whole portrays a surprisingly uniform stratigraphic package, as described below.

16.3 Infrastructure

Structurally complicated gneisses and schists are exposed both in major anticlinoria on the Baie Verte Peninsula and along major faults affecting the western margin of the tectonite belt in the Deer Lake–Grand Lake area (Fig. 16.2, Fig. 16.3). Intense tectonism has obliterated most primary features in these rocks; consequently, interpretations concerning their origins have been controversial. Locally, as in the Grand Lake area, these rocks are known to be Grenville basement, remobilized during Palaeozoic orogenesis (Martineau, 1980); in other places these gneisses are thought to represent intensely deformed, structurally lower parts of the cover sequence (Hibbard, 1983*a*). Still larger areas of these gneisses are of uncertain origin. There-

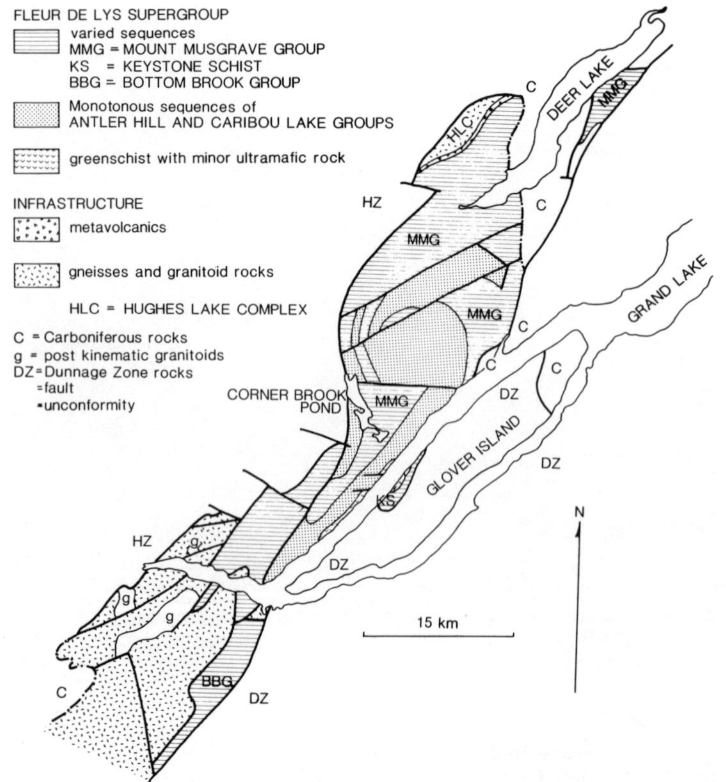

Figure 16.3 Geology of the Fleur de Lys Belt in the Grand Lake–Deer Lake area; see Fig. 16.1 for location.

fore, the tectonic term 'infrastructure' is used here for this collection of intensely deformed, formerly deep-seated rocks, instead of the more confining stratigraphic term 'basement'. The variability in stratigraphic content of the infrastructure mainly reflects the heterogeneity of deformation throughout the belt.

The infrastructure forms less than 20% of the entire Fleur de Lys Belt. It is most extensive and varied on the Baie Verte Peninsula, although smaller slivers of the infrastructure are recognized elsewhere. In all areas, the contact with the cover sequence is generally tectonic, although locally it has been interpreted as a nonconformity. Rocks of the Fleur de Lys Supergroup, immediately adjacent to exposures of the infrastructure, have direct bearing on the interpretation of the gneisses, and thus are included in the following description and discussion.

At the south end of the Baie Verte Peninsula (Fig. 16.2), immediately east of the Cabot Fault, the infrastructure is represented by large screens (km scale) of granitic gneiss that are surrounded by amphibolite. The gneissosity in the granitic rocks is obviously truncated by both thin dykes and more substantial portions of the amphibolite. To the east, the amphibolite is interlayered with metaconglomerate and gritty metaclastic schists, and consistently youngs eastwards; this stratigraphic package has been assigned to the Oody Mountain Formation (Hibbard, 1983*a*). The metaconglomerate contains granitic gneiss cobbles. In view of these relationships, and since the amphibolite and metaclastics show the same deformational features, the granitic gneiss of the infrastructure has been interpreted as a pre-deformed basement to a younger deformed cover sequence: the Oody Mountain Formation. The nature of the contact between the infrastructure and cover, here, is uncertain because Palaeozoic deformation has obscured the primary nature of the Oody Mountain amphibolite; the contact could be either an intrusive or a non-conformable contact.

Similar rock types and relationships are exposed in the area north of Deer Lake, on Glover Island, and at the south-west corner of Grand Lake (Fig. 16.3). However, in each of these areas, there is no evidence to show that deformation of the gneisses predated that of other nearby rocks. In the Deer Lake area, the infrastructure is composed of gneissic and foliated granitic rocks and metavolcanic rocks of the north-east-trending Hughes Lake Complex (Williams *et al.*, 1985). These are exposed beneath a westward-dipping, presumably overturned, thrust fault (Williams *et al.*, 1982). In the complex, granitoid gneiss surrounds outcrops of foliated to massive granite, but the relationship between rock types is unexposed. South-eastwards, amphibolite and chlorite schists form conspicuous melanocratic units in the granite; these units grade south-eastward into north-west-dipping, mixed mafic schist–amphibolite and magnetite-quartz-feldspar schists that structurally overlie a sequence of south-east-younging coarse metaclastic schists.

This sequence was originally interpreted as representing basement gneisses and granite, with zones of intensely mobilized and retrograded gneiss (magnetite-quartz-feldspar schist), all cut by pre-kinematic mafic dykes (melanocratic units) and stratigraphically overlain by a metaclastic cover sequence (Williams *et al.*, 1982). However, recent U/Pb zircon data and geochemical analyses indicate that the foliated granite is 602 ± 10 Ma (latest Hadrynian) in age, and probably related to the mafic rocks (Williams *et al.*, 1985). In addition, field studies indicate that the magnetite-quartz-feldspar schists are intensely deformed metavolcanic rocks, which are associated with the granite and the mafic units (Williams *et al.*, 1985). Thus, part of the infrastructure, the foliated granite, is genetically related to the cover sequence, and the nature of the gneiss is uncertain.

On Glover Island and south-west of Grand Lake, the nature of infrastructural gneisses is equally uncertain. On Glover Island, granitic gneiss, exposed in the core of a major anticline, is concordant with structurally overlying amphibolite, not found elsewhere in the cover sequence. At the SW corner of Grand Lake, similar relationships are found, although the thick mafic unit found directly above the infrastructure elsewhere is absent; instead, thin amphibolite bands are interlayered with the gneisses and these are structurally overlain by a sequence of epidosites and quartzites (D. P. Kennedy, 1981). The epidosites may represent highly deformed felsic volcanics in this area (Hibbard, 1983*b*). South of Grand Lake, gneisses similar to those at the lake merge with the regional Grenvillian Long Range Complex (Martineau, 1980), suggesting that gneisses at the south-west corner of the lake also represent basement.

The largest outcrop area of the infrastructure forms the cores of two major anticlinoria in the centre of the belt, on the Baie Verte Peninsula. In contrast to other occurrences, the infrastructure here is more varied; it consists of migmatite, banded gneiss, quartzo-feldspathic paragneiss, quartzite, epidosite, amphibolite, and metaconglomerate, all of which have recently been assigned to the East Pond Metamorphic Suite (Hibbard, 1983*a*). Pods and concordant layers of amphibolite, most likely remnants of mafic dykes and sills, occur throughout the suite. Rocks of the suite are separated from the surrounding cover sequences by a steep tectonic zone characterized by coarse mica schists (de Wit, 1972, 1980). Within the suite, the migmatite and banded gneisses form small outcrop areas ($< 1\ \text{km}^2$) in the core of each anticlinorium. In the western outcrop area, the migmatite is surrounded by the metaconglomerate, which contains gneissic clasts (Fig. 16.4); the metaconglomerate is, in turn, surrounded by the paragneisses. In the eastern area the banded gneiss is surrounded by paragneiss.

Based on structural evidence, the migmatite has been interpreted as representing an anatectic front within a basement sequence of paragneisses (de Wit, 1972, 1980); since the metaconglomerate contains gneissic clasts, it has been considered as representing the

Figure 16.4 Mildly deformed metaconglomerate of the East Pond Metamorphic Suite; note gneissic clast in foreground.

basal cover deposit on the basement (de Wit, 1972, 1974). Recently, the interpretation of the extent of basement has been challenged (Hibbard, 1979, 1983*a*). The migmatites and banded gneisses display structures that apparently predate any tectonism in the remainder of the suite and the cover rocks surrounding the suite (de Wit, 1972, 1980; Hibbard, 1983*a*). The migmatite layering is steeply dipping and trends north-westward, askew to the typical north-easterly trend of other rocks in the orthotectonic belt. This complex layering is overprinted by all of the deformations evident in cover rocks surrounding the suite. Likewise, the banded gneisses display complex intrafolial fold-interference patterns that are overprinted by the same deformations. The surrounding metaconglomerate and paragneiss units are devoid of any such complex early structures, and display only the regional events recorded in nearby cover rocks, albeit that the deformation is generally much more intense in the suite. These observations suggest that the migmatites and banded gneisses may represent small windows of pre-deformed basement within an intensely deformed, paragneissic equivalent of the cover sequence (Hibbard, 1983*a*).

Regional lithological correlation also supports this newer interpretation. The paragneisses, in the vicinity of the migmatite gneisses, are interlayered with the metaconglomerate, and consist of amphibolite, epidosite, quartzite and psammitic gneiss. This stratigraphic package is seeminly an amalgamation of the lithological associations of the core sequences adjacent to the infrastructure in other parts of Newfoundland. The metaconglomerate of the suite is identical to that found further to the south-west on the peninsula, in the Oody Mountain Formation, which directly overlies infrastructure interpreted to be basement. The amphibolite is a feature of all the other infrastructure–cover boundaries, except that at Grand Lake. At the lake, epidosites and quartzites replace the amphibolite; these too, are part of the lithological association in the suite. It appears then, from both structural and stratigraphic evidence, that the infrastructure, represented by the East Pond Metamorphic Suite, comprises mainly intensely deformed equivalents of the cover sequences, and embraces a basement–cover relationship. The nature of this relationship has been tectonically destroyed, but the character and position of the metaconglomerate suggests that it was a major nonconformity.

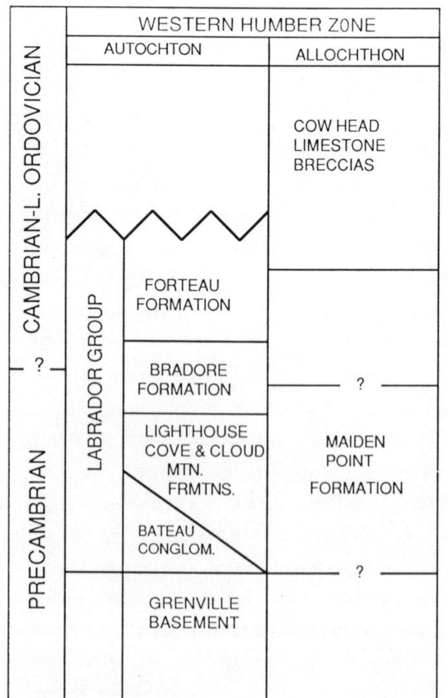

Figure 16.5 Tentative lithostratigraphic correlation chart for the Fleur de Lys Belt and adjacent Humber Zone rocks to the west.

16.3.1 *Age and correlations*

The absolute age of rocks forming the infrastructure is uncertain; however, regional stratigraphic correlation provides reasonable indirect evidence for their age. The granitoid gneisses of the infrastructure, southeast of White Bay, are lithologically identical to nearby Grenville granite gneisses that form the basement to miogeoclinal rocks on the west side of White Bay; likewise, gneisses at the south-west corner of Grand Lake merge with Grenville gneisses of the Long Range Complex on the Great Northern Peninsula. These lithological correlations suggest that the infrastructure in these areas represents Grenville basement.

The gneisses of uncertain origin at Deer Lake, Glover Island, and in the core of the East Pond Metamorphic Suite, provide little compelling evidence for regional correlation; however, the associated cover sequences and paragneisses of the suite consistently display a lithological association of metavolcanics–coarse metaclastic rocks ± metaconglomerate, that is similar to less-tectonized sequences of the miogeocline to the west (Fig. 16.5). Specifically, this assemblage is remarkably reminiscent of the basal part of the Labrador Group (James *et al.*, pers. comm.) on Belle Isle and on the Great Northern Peninsula. Here, a variable sequence of mafic rocks and coarse clastic rocks overlies Grenville basement gneisses. The mafic volcanics, locally termed the Lighthouse Cove Formation and Cloud Mountain Basalt, are generally tholeiitic massive flows that either directly overlie basement, or overlie a basal conglomerate. Mafic dykes that cross-cut the basement are feeders to the basalts. The basal conglomerate, part of the Bateau Formation, is present only locally, and contains gneissic clasts. The associated clastic rocks of the Bradore Formation are coarse quartz-feldspar grits and sandstones.

Most strikingly, metaconglomerate at White Bay and in the East Pond Metamorphic Suite is lithologically identical to the Bateau Formation (Williams and Stevens, 1969). Its close association with amphibolite is similar to the Bateau–Lighthouse Cove association of the Labrador Group. Not surprisingly, amphibolite from the East Pond Metamorphic Suite, the Oody Mountain Formation, and the Deer Lake area are petrochemical correlatives of the tholeiitic Lighthouse Cove Basalts (de Wit and Strong, 1975; Hibbard, 1983*a*; Williams *et al.*, 1985). Epidosites and metafelsites associated with the infrastructure have no direct correlatives within the Labrador Group, but they appear to represent an easterly component equivalent to the Lighthouse Cove volcanism in the west. Coarse metaclastic rocks and minor quartzites of the East Pond Metamorphic Suite, and similar rocks associated with the infrastructure in other areas, bear a resemblance, both stratigraphically and lithologically, to the Bradore Formation (Williams *et al.*, 1982; Hibbard, 1983*a*).

The basal Labrador Group immediately underlies the Lower Cambrian Devil's Cove Limestone in Canada Bay (Betz, 1939) and thus, has been considered to be uppermost Hadrynian to lowermost Cambrian in age. This suggests that paragneisses of the East Pond Metamorphic Suite, and cover sequences associated with the infrastructure, elsewhere, are probably also latest Hadrynian to earliest Cambrian in age. Where infrastructural gneisses show complex structures that predate the Palaeozoic tectonism, the correlations noted above suggest that these rocks are Grenvillian gneisses; this also corroborates correlation of the White Bay infrastructure with Grenville gneisses. Where infrastructural gneisses do not display these complex early structures, they may either be intensely deformed rocks related to the Hadrynian–Cambrian cover sequence, or represent intensely deformed Grenville gneisses in which any evidence of earlier tectonism has been totally obliterated.

16.4 Cover sequences

The cover sequence constitutes the bulk of the Fleur de Lys Belt. It is composed of mainly metaclastic schists with interlayered marble, marble breccia, amphibolite and greenschist. In contrast to the infrastructure of the belt, layering in most of the cover sequence is thought to represent bedding.

Along the eastern margin of the belt, near its contact with the Dunnage Zone, remnants of ophiolitic suites are tectonically interlaced, deformed and metamorphosed with the cover sequence. In at least one location, the cover sequence is in conformable contact with greenschists of these ophiolitic remnants. Thus, the ophiolitic rocks are described here, within the framework of the cover sequence.

All of the rocks of the cover sequence have been assigned to the recently redefined Fleur de Lys Supergroup (Hibbard, 1983*a*) (Figs 16.2, 16.3). This includes the following major lithostratigraphic units: the White Bay, Rattling Brook, Old House Cove and Mings Bight Groups and the Birchy Complex on the Baie Verte Peninsula; the insular Horse Islands and Grey Islands Groups; the Mount Musgrave Group in the Deer Lake–Grand Lake area; and informal units (Bottom Brook Group) south of Grand Lake.

The total thickness of the supergroup is uncertain due to the internal structural complexity; the maximum observed outcrop width of the unit is approximately 13 km, just south of Baie Verte, although the maximum width is probably much greater in the offshore area north of the Baie Verte Peninsula. The contact relationships between constituent units of the cover are complex, and are discussed following a description of rock types.

The units of the cover sequence can be assigned to three divisions according to their lithological content. The White Bay, Rattling Brook, and Mt Musgrave Groups, as well as some informal sequences in the Grand Lake area, are composed of a variety of rock types, and hence will be termed the 'varied sequences' here. The Old House Cove Group and some informal sequences at Grand Lake are less diverse in composition, and are here termed the 'monotonous sequences'. The third division consists of ophiolitic rocks of the Birchy Complex. The insular Horse Islands and Grey Islands Groups, and the Mings Bight Group, display characteristics of both the varied and monotonous groups.

In outcrop belts on the Baie Verte Peninsula and in the Grand Lake area, the cover sequence is disposed in major anticlinoria. In each area, the monotonous sequences tend to form the core areas of these structures, whereas the varied sequences define the flanks.

The components and stratigraphy of the cover sequence have been documented best on the Baie Verte Peninsula; the following description will focus mainly on rocks of the peninsula and draw only lightly on information from other areas.

16.4.1 *Varied sequence*

These sequences include the White Bay and Rattling Brook Groups on the Baie Verte Peninsula, the South Brook Group as well as the informal Mt Musgrave Group in the Grand Lake area, and the informal Keystone schists on Glover Island (Knapp, 1980). They are composed of metaclastic schists, marble, greenschist and amphibolite. The White Bay Group (Betz, 1948) is the best preserved and studied of these sequences, and serves as a model for the varied sequences.

On the west side of the Baie Verte Peninsula, the White Bay Group is disposed in a horizontal, north-north-east trending synclinorium that is reclined steeply to the west. The metaclastic schists are mainly semi-pelitic and pelitic schists with abundant intercalations of sulphurous graphitic schists. Where psammitic rocks are present, they are most commonly thin bedded and fine grained. Marble and carbonate schist form quantitatively small portions of the group; however, they are the most spectacular and conspicuous rock types in the unit. Most striking are layers and large blocks of marble breccia within the pelitic schists (Figs 16.6 and 16.7). These members are generally characterized by numerous angular rip-up remnants of thin-bedded carbonates, as well as subordinate rounded clasts of carbonate, quartzite and vein quartz.

Amphibolite and greenschist locally form distinct formations up to 1 km thick within the metaclastic rocks. The basal unit of the group, the Oody Mountain Formation (Hibbard, 1983*a*) is mainly massive amphibolite that may represent a thick sill or a group of massive flows. A second major mafic schist that occurs higher in the group is of extrusive origin (Fig. 16.8) (de

Figure 16.6 Pod of marble breccia surrounded by metaclastic rocks; White Bay Group on White Bay.

Figure 16.7 Close-up of marble breccia in Fig. 16.6; note slabby clasts.

Figure 16.8 Fragmental mafic metavolcanic in White Bay Group; this feature, as well as interlayering with nearby metaclastic rocks, suggests an extrusive origin for these metavolcanic rocks.

Wit 1972; Hibbard 1983*a*). Petrochemically, these units most closely resemble rift-type tholeiites (Hibbard, 1983*a*).

Rocks of the Rattling Brook Group (Watson, 1947) closely resemble those of the White Bay Group, although the internal stratigraphy of the Rattling Brook Group is less consistent than that of the White Bay Group. This is probably a result of intense structural transposition, due to the proximity of the unit to the Baie Verte Line. One significant difference between the two units is that the Rattling Brook Group contains tectonic slivers of ultramafic rock.

In the Deer Lake area, the cover sequence has a basal metavolcanic unit like the White Bay Group, but here, it is overlain by a thick sequence (2 km) of cross-bedded, metamorphosed arkose, greywacke, and pebble conglomerate, that forms the base of the Mt Musgrave Group. This in turn is overlain by metaclastics and marble similar to the White Bay Group. Further south, the Keystone Schists and Bottom Brook Group are lithologically similar to the White Bay Group. The varied sequence west of Grand Lake locally contains tectonic slivers of ultramafic rock, similar to those in the Rattling Brook Group.

The White Bay Group contains the best-preserved primary features in all of the Fleur de Lys Supergroup; consequently, it divulges most information concerning the depositional environment of the Fleur de Lys protoliths. In particular, the spectacular calcareous metaconglomerates, large marble blocks, and calcareous schists interlayered with dominantly thinly layered metaclastics and sulphurous graphitic schist closely correspond with descriptions of carbonate slope facies, marginal to carbonate banks (Wilson, 1969; McIlreath and James, 1979). The marble breccia layers most likely represent debris sheets emplaced in a slope environment; the thin, flat nature of many of the carbonate clasts suggests that the deposits were slope derived rather than platformally derived (McIlreath and James, 1979). The large, isolated pods of marble breccia may be tectonized equivalents of either debris-flow channels or submarine slump blocks. In the light of these associations, the carbonate schists most closely resemble hemipelagic deposits of peri-platformal ooze (McIlreath and James, 1979).

Based on these observations, the varied sequences are interpreted as having been deposited in a slope environment, bordering a substantial carbonate bank.

16.4.2 *Monotonous sequences*

The Old House Cove Group (Hibbard, 1983*a*) and the informal Antler Hill and Caribou Lake formations (D. P. Kennedy, 1981) in the Grand Lake area crop out extensively in the cores of major anticlinoria. These units are extensive, extremely monotonous sequences of alternating coarse feldspathic psammite and semi-pelite, with local, minor interlayers of graphitic schist, pelite, pebbly metaconglomerate, garnetiferous semi-pelite, and quartzite. The Old House Cove Group contains numerous pods and sills of black-green amphibolite (Fig. 16.9); and a single pre-kinematic felsic stock, the Crow Head intrusion (de Wit, 1972); locally, all these cross-cut bedding in the group. Petrochemical analysis shows that the mafic rocks are tholeiitic (Hibbard, 1983*a*). These mafic rocks are also present in the Grand Lake area, but are not as abundant as in the Old House Cove Group. Most primary features in the metaclastic rocks have been obliterated by regional tectonism, but the alternating monotony of layered psammite and semi-pelite is reminiscent of a turbidite sequence. This, considered with the medium to thick layering so prevalent in the unit, suggests that the group may represent a medial submarine fan facies. The clean, quartzo-feldspathic composition of these rocks suggests that they had a continental derivation.

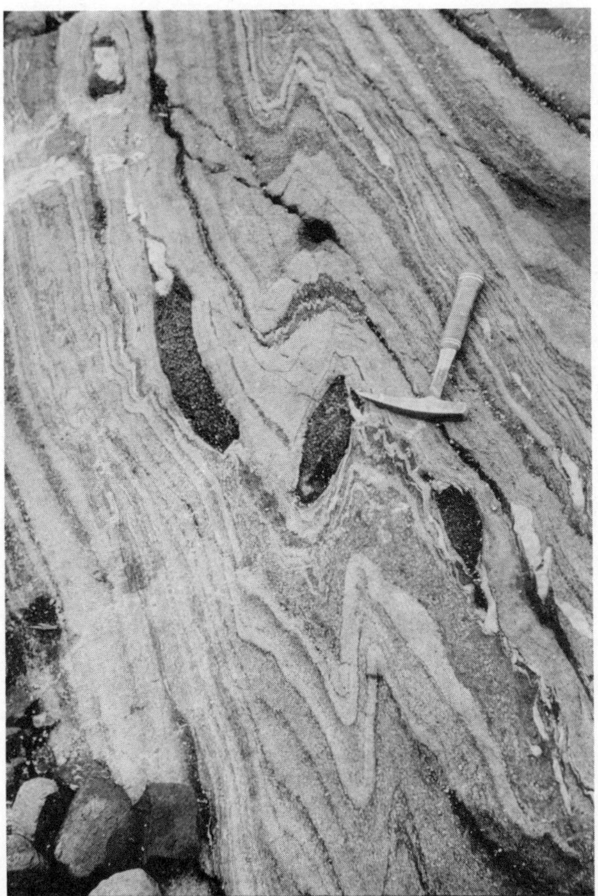

Figure 16.9 Psammite, semi-pelite and amphibolite (dark pods) of the Old House Cove Group.

16.4.3 *Ophiolitic rocks*

The Birchy Complex (Fuller, 1941; Williams *et al.*, 1977) forms the easternmost flank of the supergroup. It is a structural assemblage composed predominantly of greenschist, metagabbro, graphitic schist, *mélange*, and minor ultramafic rocks. The ultramafic rocks, metagabbro, and greenschists have collectively been interpreted as structurally dismembered fragments of an ophiolite (Bursnall, 1975); petrochemical data support this interpretation (Hibbard, 1983*a*). Graphitic schist and *mélange* are interlayered with greenschist; Bursnall (1975) first recognized the *mélanges* and significantly noted that all of the occurrences are characterized by the inclusion of either serpentinized ultramafic rocks or bright-green coarse actinolitic schist of ultramafic derivation.

In direct contrast to other parts of the supergroup that appear to have a continental origin, the Birchy Complex is of oceanic affinity. *Mélanges* in the unit display all of the structures evident in surrounding rocks; hence it has been concluded that they mark the nascent stages of tectonism in the supergroup (Bursnall, 1975; Williams, 1977; Williams *et al.*, 1977).

16.4.4 *Other sequences*

The Horse Island, Grey Islands, and Mings Bight Groups (Watson, 1947; Baird, 1951; Kennedy, 1975*b*; Hibbard and Bursnall, 1979; Hibbard, 1983*a*) are all isolated units that display characteristics of other sequences. They are all composed mainly of monotonous metaclastic rocks; however, they also contain minor variations not found in the monotonous sequences.

The Horse Islands Group contains greenschist and metafelsite infolded with the metaclastics; the metafelsites are locally fragmental. In the Grey Islands Group, one psammitic unit contains amphibolite pods similar to the monotonous sequences, but it also contains greenschist and *mélange* zones similar to those of the Birchy Complex. In addition, graphitic schist and marble occur locally in the group. Likewise, the Ming's Bight Group contains amphibolite with *mélange* zones like those in the Birchy Complex (Hibbard, 1982, 1983*a*).

The depositional sites of protoliths to these sequences are difficult to establish. They appear to represent turbidite sequences that have a volcanic influence; the *mélanges*, similar to those in the Birchy Complex, may indicate that the volcanism is ophiolite-related.

16.4.5 *Contact relationships*

The relationships between sequences appear to be complex (Fig. 16.10); contacts are exposed in many places, although stratigraphic order is difficult to decipher, due to the lack of primary structure in many areas.

The varied sequence overlies a variety of different substrates. The White Bay Group, disposed in a major synclinorium, overlies different units on opposing flanks of the fold. To the southwest, the group overlies and includes screens of infrastructural gneisses, that have been interpreted as representing the regional Grenville basement (Hibbard, 1983*a*). To the east, the group conformably overlies the monotonous sequence: the Old House Cove Group (de Wit, 1972; Hibbard, 1983*a*). Similar relationships are found at Deer Lake and Grand Lake. The Mt Musgrave Group has been interpreted as overlying the infrastructure at Deer Lake (Williams *et al.*, 1982); the same relationship has been documented between the Keystone Schists and the infrastructure on Glover Island (Knapp, 1980).

In contrast to these relationships, the Rattling Brook Group is conformably interlayered to the east with greenschists of the ophiolitic Birchy Complex; in addition, the group contains tectonic slivers of ultramafic rock. These relationships strongly infer that the eastern part, if not all of the Rattling Brook Group, overlies an ophiolitic basement. To the west, the Rattling Brook Group conformably interdigitates with the Old House Cove Group, although the stratigraphic order is uncertain (Bursnall, 1979). Similarly, west of Grand Lake, the eastern portions of the varied sequence contain tectonic slivers of ultramafic rock, suggesting that the sequence here may have originally had an ophiolitic substrate; to the west D. P. Kennedy (1981) has interpreted the varied sequences to overlie the monotonous sequence. This relationship, as well as the relationship of the varied White Bay Group to the monotonous Old House Cove Group, suggests that the Rattling Brook Group may, in part, overlie the Old House Cove Group.

The substrate to the monotonous sequences is uncertain; both on the Baie Verte Peninsula and near Grand Lake, these rocks are faulted against the infrastructure (de Wit, 1972, 1980; D. P. Kennedy, 1981). Relationships outlined above strongly suggest that the monotonous sequence may be partially equivalent to the lowest portions of the varied sequences and underlie the reminder of these units. Since the varied sequences overlie continental basement to the west, and oceanic basement to the east, it would appear that the monotonous sequences may have originally overlain the transition of one substrate to the other (Fig. 16.10).

The relationships of the Mings Bight, Grey Islands, and Horse Islands Groups are uncertain. However,

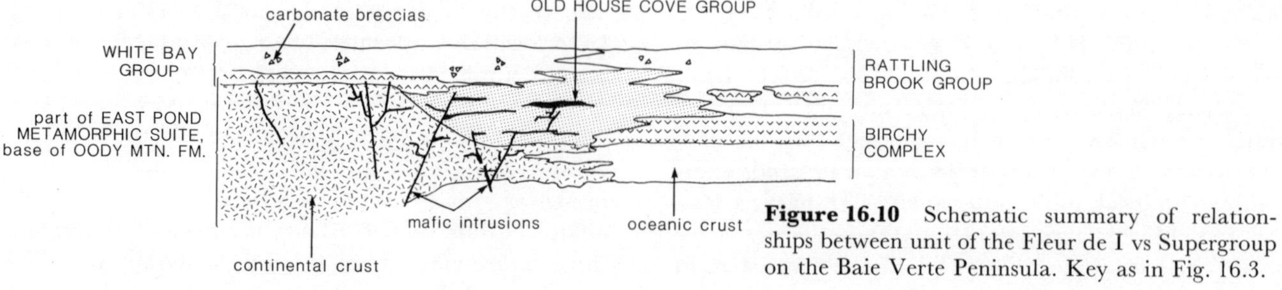

Figure 16.10 Schematic summary of relationships between unit of the Fleur de I vs Supergroup on the Baie Verte Peninsula. Key as in Fig. 16.3.

both the Mings Bight and Grey Islands contain mafic schist members that are correlative with the Birchy Complex; this suggests that these more easterly units may also overlie an ophiolitic substrate, analogous to the Rattling Brook Group.

In summary, the Fleur de Lys Supergroup overlaps two distinct substrates; one is of continental affinity and represented by portions of the infrastructure, whereas the other is the easterly ophiolitic Birchy Complex. The varied sequences of the supergroup overlap these units in the eastern and western parts of the belt, whereas in the middle, they may interdigitate with and overlap the monotonous sequences.

16.4.6 *Age and correlation*

Direct evidence for the age of the Fleur de Lys Supergroup is lacking; a single brachiopod valve recovered from marble in the White Bay Group is the lone fossil occurrence known in the supergroup; it has an affinity with the Order Acrotretida and indicates that at least part of the group is of Palaeozoic age (S. Stouge, pers. comm., 1979, reported in Hibbard, 1983*a*).

Regional correlation provides indirect age constraints on parts of the supergroup. As described above with the infrastructure, lower portions of the supergroup (associated with outcrops of the infrastructure) are lithological correlatives of the lower portion of the Labrador Group to the west (Fig. 16.5). This indicates that stratigraphically lower portions of the supergroup are probably of similar age.

Stratigraphically higher units within the varied sequences are mainly non-diagnostic for regional correlation; however, the carbonate breccias, prevalent in parts of the White Bay Group and in the Grand Lake area, are metamorphosed equivalents of carbonates of the Cow Head Group in western Newfoundland (de Wit, 1972; Bursnall and de Wit, 1975). The group ranges in age from middle Cambrian to middle Ordovician. This does not provide tight age constraints on parts of the supergroup, but lithological correlation provides an important link to the Lower Palaeozoic miogeocline to the west.

The monotonous sequences laterally interdigitate with, and are overlain by, stratigraphically higher parts of varied sequences; this suggests that the monotonous sequences may overlap in age with the lower Labrador Group. Significantly, amphibolite sills and dykes in the Old House Cove Group are petrochemically identical to mafic rocks of the White Bay Group and basalts of the Lighthouse Cove Formation (Hibbard, 1983*a*). Thus the bulk of the monotonous sequences are probably slightly older than the Labrador Group. The monotonous sequences are also lithologically similar to allochthonous clastic rocks and mafic volcanics of the Maiden Point Formation on the Great Northern Peninsula; the formation is inferred to be Precambrian to Cambrian in age (Smyth, 1973), similar to the inferred age for monotonous sequences.

Rocks of the Birchy Complex, at the eastern margin of the supergroup, are correlative with rocks both to the east and west. Ophiolitic *mélanges* of the complex have been correlated with unmetamorphosed *mélanges* related to allochthonous ophiolites on the miogeocline to the west (Williams, 1977; Williams *et al.*, 1977). Since these allochthons were emplaced by the middle Ordovician, the *mélanges* both under the allochthons, and their lithological correlatives in the Birchy Complex are probably older than middle Ordovician. In addition, Bursnall (1975) demonstrated that the bulk of the complex is correlative with ophiolitic rocks of the Dunnage Zone, to the east. The best age constraint on the nearby Dunnage ophiolites is early Ordovician (Dunning and Krough, 1985), suggesting a similar age for the Birchy Complex.

Based on regional correlation, then, the Fleur de Lys Supergroup appears to range in age from late Hadrynian to possibly as young as early Ordovician. Regional tectonic constraints put a probable upper age limit of early Ordovician on the supergroup (Bursnall and de Wit, 1975; Williams *et al.*, 1977; Hibbard, 1982, 1983*a*).

16.5 Significance and interpretation of the Fleur de Lys Belt

Pre-kinematic rocks of the Fleur de Lys Belt appear to form a coherent depositional and magmatic unit at the eastern margin of the Humber Zone. Even though the effects of the intense tectonism are apparent in these rocks, i.e. transposition of original stratigraphy and obliteration of many primary features, it is still possible to reconstruct the broad scheme of deposition and intrusion in the belt prior to the regional tectonism. The recognition that the belt intervenes between platformal deposits of the westerly Humber Zone and the oceanic rocks of the easterly Dunnage Zone is a decisive aid in reconstructing the pre-kinematic Fleur de Lys stratigraphy.

The infrastructure is a structural unit that recorded the tectonism experienced by some of the most deep-seated rocks in the Humber Zone. Logically, rocks lowermost in the original stratigraphic succession have more potential for deeper burial during orogenesis than those higher in the succession. Thus, it is not surprising that the infrastructure incorporates elements of remobilized Grenville basement, as well as intensely deformed equivalents of lower parts of the cover succession.

Cover rocks of the Fleur de Lys Supergroup appear to have been deposited on both continental and oceanic substrates, from late Hadrynian to early Ordovician times. The varied sequences locally overlie granitic gneisses of probable Grenvillian age, and the metasediments of the East Pond Metamorphic Suite—which are probably correlative with those of the supergroup, appear to overlie reworked migmatite and gneiss, also probably of Grenvillian age. In contrast, the Birchy Complex is interpreted to be ophiolitic, and thus originally oceanic crust. However, the extent of this crust is uncertain; pre-kinematic ultramafic bodies within the varied sequences may indicate an original oceanic basement to these units. The nature of the basement to the Ming's Bight, Grey Islands, and Horse Islands Groups is uncertain.

Metasedimentary rocks of the cover sequences appear to represent both slope and submarine basinal deposits. The varied sequences appear to represent the slope deposits, which seem to conformably interfinger with basinal deposits of the monotonous sequences. Marble breccia blocks within the varied sequences indicate that the slope–basin geometry formed along the flank of a substantial carbonate bank. The Ming's Bight, Horse Islands and Grey Islands Groups may represent distal basinal facies.

Mafic meta-igneous rocks occur throughout the cover sequence, and are both extrusive and intrusive. Based on geochemical data and geochemical correlation, most of the amphibolites (except those of the Birchy Complex) appear to the rift-facies igneous rocks. The Birchy Complex rocks appear to be ophiolitic and geochemically distinct from these rift-related rocks.

Thus, it appears that deposition and accompanying intrusion in the Fleur de Lys Belt during the late Precambrian–early Palaeozoic took place along the interface between continental and oceanic crust. Most reasonably, the cover sequence of the Fleur de Lys Supergroup reflects the evolution of a continental rift and the subsequent formation of oceanic crust. The metasedimentary rocks record the establishment of a slope and basin environment on the west side of the rift zone, and establishment of a stable carbonate bank further to the west. The meta-igneous rocks were generated during the rifting process and during the formation of new oceanic crust. These characteristics, in conjunction with the regional setting of the belt, between platformal rocks of the Humber Zone and oceanic rocks of the Dunnage Zone, indicate that the belt represents the ancient early Palaeozoic continental margin of eastern North America.

Acknowledgements

This study was completed while the author worked for the Newfoundland Dept. of Mines and Energy, Mineral Development Division; I thank the director, B. Greene, for permission to publish these data. I also thank John Bursnall and Harold Williams for introductions to, and different perspectives on, the Fleur de Lys rocks. I especially thank Tammy Babcock for the preparation of the manuscript.

References

Baird, D. M. (1951) The geology of the Burlington Peninsula, Newfoundland. *Geol. Surv. Can. Pap.* **51–21**, 70 pp.

Betz, F. Jr. (1939) Geology and mineral deposits of the Canada Bay area, northern Newfoundland. *Newfoundland Geol. Surv. Bull.* **16**, 49 pp.

Betz, F. Jr. (1948) Geology and mineral deposits of Southern White Bay. *Newfoundland Geol. Surv. Bull.* **24**, 24 pp.

Bursnall, J. T. (1975) Stratigraphy. Structure and Metamorphism West of Baie Verte, Burlington Peninsula, Newfoundland. Ph.D. Thesis, Cambridge University.

Bursnall, J. T. (1979) Geology of part of the Fleur de Lys map area (12I/1E) Newfoundland. In Gibbons, R. V. (ed.), *Report of Activities for 1978, Newfoundland Dept. Mines & Energy, Miner. Dev. Div., Rep.* **79–1**, 68–74.

Bursnall, J. T. and de Wit, M. J. (1975) Timing and development of the orthotectonic zone in the Appalachian orogen of northwest Newfoundland. *Can. J. Earth Sci.* **12**, 1712–1722.

Church, W. R. (1965) Structural evolution of northeast Newfoundland—comparison with that of the British Caledonides. *Maritime Seds.* **1**, 10–14.

Church, W. R. (1966) Geology of the Burlington Peninsula, northeast Newfoundland. *Geol. Ass. Can. & Min. Ass. Can. Tech. Prog.* 11–12.

Church, W. R. (1969) Metamorphic rocks of Burlington Peninsula and adjoining areas of Newfoundland and their bearing on continental drift of the North Atlantic. In Kay, M. (ed.), North Atlantic Geology and Continental Drift. *Am. Ass. Pet. Geol. Mem.* **12**, 212–233.

Dallmeyer, R. D. and Hibbard, J. (1984) Geochronology of the Baie Verte Peninsula, Newfoundland: implications for the tectonic evolution of the Humber and Dunnage Zones of the Appalachian Orogen. *J. Geol.* **92**, 489–512.

Degrace, J. R., Kean, B. F., Hsu, E. and Green, T. (1976) Geology of the Nippers Harbour map area (2E/13), Newfoundland. *Newfoundland Dept. Mines Energy, Miner. Dev. Div., Rep.* **76–3**, 73 pp.

de Wit, M. J. (1972) The Geology around Bear Cove, Eastern White Bay, Newfoundland. Ph.D. Thesis, University of Cambridge.

de Wit, M. J. (1974) On the origin and deformation of the Fleur de Lys metaconglomerate, Appalachian fold belt, northwest Newfoundland. *Can. J. Earth Sci.* **11**, 1168–1180.

de Wit, M. J. (1980) Structural and metamorphic relationships of the pre-Fleur de Lys and Fleur de Lys rocks of the Baie Verte Peninsula, Newfoundland, *Can. J. Earth Sci.* **17**, 1559–1575.

de Wit, M. J. and Strong, D. F. (1975) Eclogite-bearing amphibolites from the Appalachian mobile belt, northwest Newfoundland: dry versus wet metamorphism. *J. Geol.* **83**, 609–627.

Dunning, G. and Krough, T. (1985) Geochronology of ophiolites of the Newfoundland Appalachians. *Can. J. Earth Sci.* **22**, 1659–1670.

Fuller, J. O. (1941) Geology and mineral deposits of the Fleur de Lys area, Newfoundland. *Geol. Surv. Newfoundland., Bull.* **15**, 41 pp.

Hibbard, J. (1979) Geology of the Baie Verte map area, west (12H/16W), Newfoundland. In R. V. Gibbons. (ed.), *Report of Activities for 1978, Newfoundland Dept. Mines & Energy, Miner. Dev. Div., Rep.* **79–1**, 58–63.

Hibbard, J. (1982) Significance of the Baie Verte flexure, Newfoundland. *Bull. geol. Soc. Amer.* **93**, 790–797.

Hibbard, J. (1983*a*) Geology of the Baie Verte Peninsula, Newfoundland. *Newfoundland Dept. Mines & Energy, St John's, Newfoundland, Mem.* **2**, 279 pp.

Hibbard, J. (1983*b*) Notes on the metamorphic rocks in the Corner Brook area (12A/13) and regional correlation of the Fleur de Lys belt, Newfoundland. In Murray, J., Saunders, P., Boyce, D. and Gibbons, R. V. (eds.), Current Research. *Newfoundland Dept. Mines Energy Rep.* **83–1**, 41–50.

Hibbard, J. and Bursnall, J. T. (1979) Geology of the Horse Islands, Newfoundland. In R. Gibbons (ed.), *Report of Activities for 1978, Newfoundland Dept. Mines & Energy, Min. Dev. Div., Rep.* **79–1**, 64–68.

Kennedy, D. P. (1981) Geology of the Corner Brook Lake Area, Newfoundland. M.Sc. Thesis, Memorial University of Newfoundland, St John's.

Kennedy, M. J. (1971) Structure and stratigraphy of the Fleur de Lys Supergroup in the Fleur de Lys area, Burlington Peninsula, Newfoundland. *Geol. Ass. Can. Proc.* **24**, 59–71.

Kennedy, M. J. (1973) Pre-Ordovician polyphase structures

in the Burlington Peninsula of the Newfoundland Appalachians. *Nature* **241**, 114–116.

Kennedy, M. J. (1975a) Repetitive orogeny of the northeastern Appalachians—New plate models based upon Appalachian examples. *Tectonophysics* **28**, 39–87.

Kennedy, M. J. (1975*b*) The Fleur de Lys Supergroup: stratigraphic comparison of Moine and Dalradian equivalents in Newfoundland with the British Caledonides. *Bull. geol. Soc. London* **131**, 305–310.

Knapp, D. (1980) The stratigraphy, structure, and metamorphism of central Glover Island, western Newfoundland. *Geol. Surv. Can. Pap.* **81–1B**, 89–96.

Martineau, Y. (1980) The Relationship among Rock Groups between the Grand Lake Thrust and Cabot Fault, West Newfoundland. M.Sc. Thesis, Memorial University of Newfoundland, St Johns, 132 pp.

McIlreath, I. A. and James, N. P. (1979) Carbonate slopes. In Walker, R. G. (ed.), Facies Models. *Geoscience Canada Reprint Ser.* **1**, 133–143.

Murray, A. and Howley, J. P. (1881) Report of the Geological Survey of Newfoundland from 1864 to 1880. *Geol. Surv. Newfoundland*, 536 pp.

Neale, E. R. W. (1967) Burlington (Baie Verte) Peninsula, Newfoundland. *Geol. Surv. Can., Pap.* **67–1A**, 183–186.

Neale, E. R. W., Kean, B. F. and Upadhyay, H. D. (1975) Post-ophiolite unconformity, Tilt Cove–Betts Cove area, Newfoundland. *Can. J. Earth Sci.* **12**, 800–886.

Smyth, W. R. (1973) The Stratigraphy and Structure of the Southern Part of the Hare Bay Allochthon, Northwest Newfoundland, Ph.D. Thesis, Memorial University of Newfoundland, St John's, 172 pp.

Watson, K. de P. (1947) Geology and mineral deposits of the Baie Verte–Mings Bight area, Newfoundland. *Geol. Surv. Newfoundland, Bull.* **21**, 48 pp.

Williams, H. (1977) Ophiolite mélange and its significance in the Fleur de Lys Supergroup, northern Appalachians. *Can. J. Earth Sci.* **14**, 987–1003.

Williams, H. (1978) Tectonic lithofacies map of the Appalachian orogen. *Memorial University of Newfoundland, map 1.*

Williams, H. and Stevens, R. (1969) Geology of Belle Isle, northern extremity of the deformed Appalachian miogeoclinal belt. *Can. J. Earth Sci.* **6**, 1145–1157.

Williams, H. and St Julien, P. (1982) The Baie Verte–Brompton Line: early Paleozoic continent–ocean interface in the Canadian Appalachians. In St P. Julien and Béland (eds.), Major Structural Zones and Faults of the Northern Appalachians. *Geol. Ass. Can. Spec. Pap.* **24**, 177–207.

Williams, H., Hibbard, J. and Bursnall, J. T. (1977) Geologic setting of asbestos-bearing ultramafic rocks along the Baie Verte Lineament, Newfoundland. *Geol. Surv. Can. Pap.* **77–1A**, 351–360.

Williams, H., Gillespie, R. T. and Knapp, D. (1982) Geology of the Pasadena map area, Newfoundland. In Current Research, Part A. *Geol. Surv. Can., Pap.* **82–1A**, 281–288.

Williams, H., Gillespie, R. and van Breemen, O. (1985) A Late Precambrian rift related igneous suite in Western Newfoundland. *Can. J. Earth Sci.* **22**, 1727–1735.

Wilson, J. L. (1969) Microfacies and sedimentary structures in 'deeper water' lime mudstones. In Friedman, G. M. (ed.), Depositional Environments in Carbonate Rocks. *Soc. Econ. Miner. Paleont. Spec. Pub.* **9**, 56–68.

17
The Eleonore Bay Group (central East Greenland)

R. CABY and JANINE BERTRAND-SARFATI

17.1 Introduction

The Eleonore Bay Group crops out in central East Greenland between 72 and 75°N (Fig. 17.1). The most complete sections are found in the outer fjord zone, and incomplete sections occur in the nunatak zone and in Canning Land.

The complete succession is about 15 000 m thick, and its base is not seen. At the top, it is separated by a paraconformity from the overlying Tillite Group (*c.* 500–800 m) (Katz, 1981), which is itself conformably overlain by the fossiliferous Cambro-Ordovician which is about 3000 m thick.

No angular unconformity was recorded in this thick sedimentary succession, which was folded during the Caledonian orogeny and was also intruded by syn- and post-orogenic granites. These granites are particularly concentrated in the Stauning Alps domal structure, where they are accompanied by thermal metamorphism. A detachment zone always separates the Eleonore Bay Supergroup from underlying gneisses of the Central Metamorphic Complex, which are considered as its high-grade metamorphic equivalent, as proposed by Wegmann (1935), Haller (1956, 1958, 1971) and Peucat *et al.* (1985).

Different nomenclatures have been proposed for this series. Koch (1929) Backlund (1930) and Teichert (1933) established the traditional lithostratigraphic nomenclature of the unit, subsequently made more precise and elaborated by Fränckl (1953*a* and *b*); Haller (1953, 1955, 1958, 1971) and Katz (1952*a*, *b*, 1954).

In agreement with modern rules, and following Katz (1961), we have divided the Eleonore Bay Group into (from base to top): the Alpefjord Formation, the Agardhsbjerg Formation, the Brogetdal Formation and the Nøkkefossen Formation (Table 17.1).

17.2 Alpefjord Formation

The Alpefjord Formation (Fig. 17.2) is more than 11 000 m thick. It essentially comprises grey, impure feldspathic quartzites, black and green siltstone silty shales, orthoquartzites and a few carbonates (in decreasing abundance). The Formation has been subdivided into three members, according to the different lithologies (Fig. 17.3). The lowermost part was affected by progressive regional Caledonian metamorphism and cut by various granitoids. A detachment zone everywhere obscures relationships with the central metamorphic complex underneath, which includes high-grade and migmatized gneisses derived from rocks lithologically similar to the Alpefjord Formation.

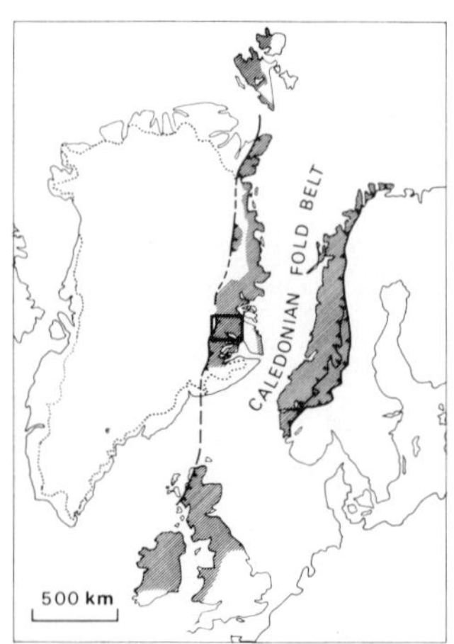

Figure 17.1 Eleonore Bay Group outcrops in central East Greenland (71°40–76°00′ N). Inset: the Caledonian fold belt in a pre-drift reconstruction.

Table 17.1 Provisional chronology of Proterozoic III to Vendian lithostratigraphic units distinguished in different areas of central East and north-east Greenland.

	Nunatak zone	Fjord zone	Canning Land	NE Greenland
			Cambrian	
Vendian	?			
		Tillite Group Nøkkefossen Formation	Tillite Group Nøkkefossen Formation	Fin Sø Formation
~ 700 Ma				
		Brogetdal Formation	Brogetdal Formation	Hagen Fjord Group
	? Synclinal S.			
		Agardhsbjerg Formation	Agardhsbjerg Formation	
	Summit S.			
			Upper member	
Proterozoic III	Petermann Bjerg Group	Alpefjord Formation: Middle member Stauning Alps migmatites Krummedal	? Pelitic gneisses?	
		Lower member Forsblads Fjord gneisses		
~ 950 Ma?				

17.2.1 *The lower member*

A measured section which is superbly exposed along the western side of Alpefjord with a regular dip to the E or ENE (Fig. 17.4) is about 10 000 m thick. However, the actual thickness can perhaps be reduced by 5%, in agreement with the overall extensional cleavage and also because of the presence of some normal faults oblique to the section (Caby, 1976*a*).

The lowermost unit occupies the cores and normal limbs of east-vergent recumbent folds which can be traced at Schaffhauserdalen and farther south. This unit is cut by various synkinematic Caledonian granites which are often leucocratic, and by foliated diorites—both occurring as axial-planar sheets within and below the hinge zones of large folds.

The unit comprises thinly layered metapelites; compact mica schists; impure metaquartzites; orthoquartzites with ripple marks and high-angle cross-bedding and overturned bedding; marble and calc-silicate lenses. Mobilization has selectively affected some lithologies, which develop a gneissose fabric close to a dense granitic network. Although ductile deformation was incipient in many layers, metamorphic mineral assemblages are of lower amphibolite facies: biotite ± muscovite + oligoclase + orthoclase + sillimanite + almandine + hercynite + graphite in metapelites; and diopside + Ca-garnet + calcite + Ca-plagioclase + quartz in impure marbles. In metapelites, the planar disposition of micas and sillimanite defines a metamorphic cleavage oblique to compositional sedimentary layering, and results from a synkinematic growth. Synkinematic fibrolitic sillimanite grew after biotite, together with prograde muscovite and complex reactions involving blastesis of hercynite.

It is thus well established that high-temperature Caledonian metamorphism reached temperatures > 600 °C in the lowermost part of the sedimentary pile, and that the grade increased with depth. General considerations imply the development of a similar thermal regime below and point to the necessarily Caledonian age of the metamorphism in the underlying gneisses.

17.2.1.1 *Transition zone with the Central Metamorphic Complex* In Schaffhauserdalen, a deeper unit is exposed on the eastern flank of the valley (Fig. 17.4). It is cut by a voluminous network (> 50%) of aluminous granites. This dismembered unit was affected by east-vergent recumbent folding and boudinage. Lithologies are identical to those above, but mineral associations include prismatic sillimanite and K-feldspar neoblasts in metapelites. Cross-bedding and overturned bedding is, however, well preserved in metaquartzites.

In Forsblads Fjord, isograds are related to the emplacement on the northern side, of a two-mica 'ballooning' granite. Foliated two-mica hornfelses are developed, with giant porphyroblasts of garnet and staurolite several centimetres across, and less frequently of cordierite and/or andalusite. Below the root of the granite, we are dealing with a lower stratigraphic unit of mica schists and metaquartzites (*c.* 500 m) with garnet porphyroblasts and occasional fibrolitic sillimanite, cut by a spectacular synkinematic network of pegmatites and leucocratic granite affected by syn-to post-emplacement boudinage in the sillimanite field. At Randenøes, preserved sedimentary structures in these metamorphic rocks include ripple marks, mud-cracks, channels, cross-bedding and load structures. To the

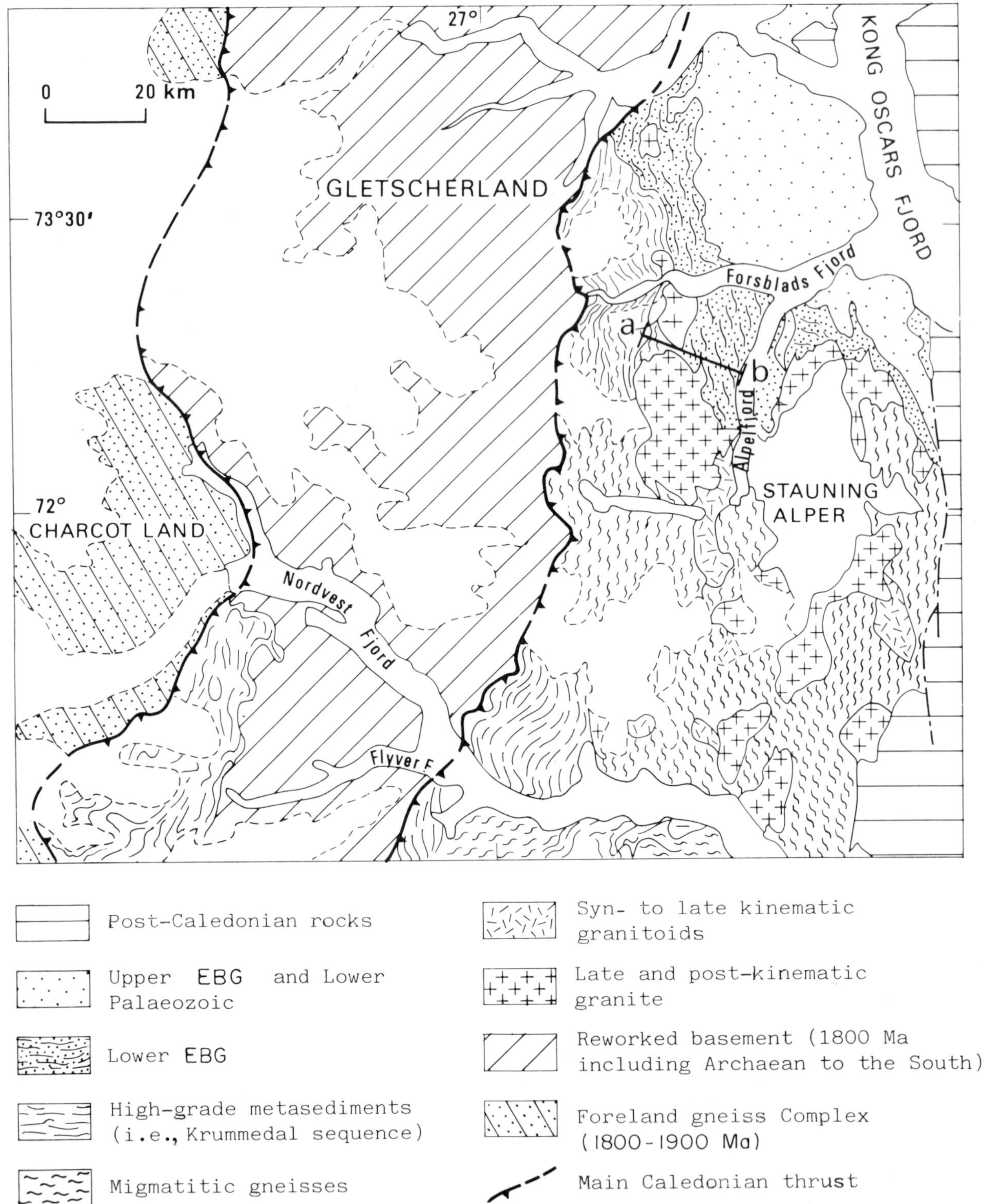

Figure 17.2 Simplified geological map of the western part of the Caledonian fold belt between 71°30′ and 73° N.

north-west, below these mica schists, grain size progessively increases and mineral associations comprise large almandine garnet, prismatic sillimanite, and K-feldspar. Complex nodules, with sillimanite, muscovite pseudomorphs after cordierite(?) and garnet, are very common in impure metaquartzites with well-preserved cross-bedding. (Fig. 17.5) Post-cleavage folding at high temperature also affected the numerous leucogranite sheets and dykes, which become more abundant downwards, and gradually built a banded sheet 2 km thick, including pendants of orthoquartzite and calc-silicates, which forms the boundary with the underlying Central Metamorphic Complex.

The Central Metamorphic Complex in inner Forsblads Fjord conformably underlies the composite granite sheet. It mainly comprises pelitic gneisses,

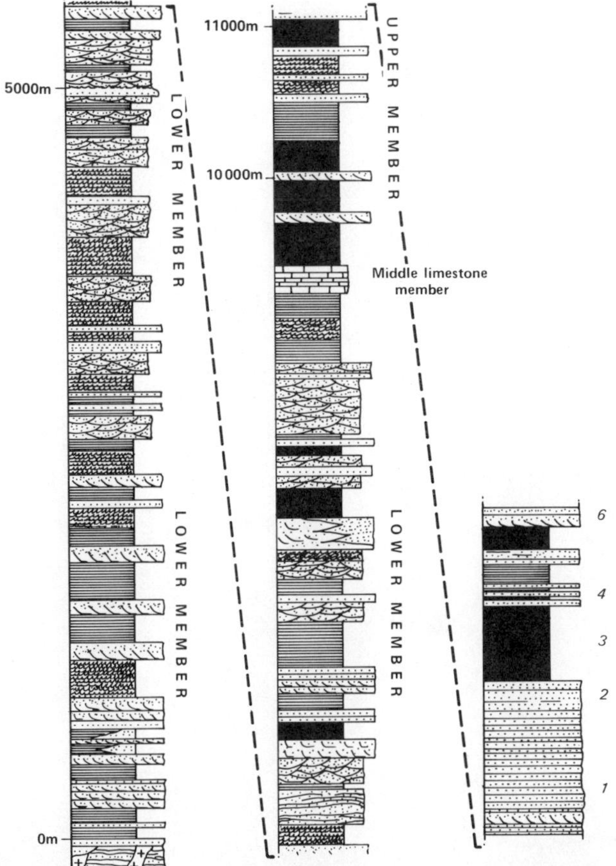

Figure 17.3 Schematic log of the Alpefjord Formation in the type locality and the Agardhsbjerg Formation in Canning Land.

displaying compositional layering on a decimetre to metre scale, with conformable veins and sheets of leucocratic mobilizates, calc-silicate rocks and various metaquartzites. Mineral associations to the east are similar to those of the highest-grade rocks of the Alpefjord Formation, but post-foliation folding is ubiquitous. The only preserved sedimentary structures are cross-bedding and overturned bedding in orthoquartzites. There is a marked compositional layering parallel to the planar disposition of biotite and sillimanite, which is affected by a strong boudinage.

Heavy-mineral bands, containing ilmenite and rutile, but mainly zircon, have been recorded in both pelitic gneisses and quartzites. The U–Pb zircon ages are slightly more discordant than those from detrital zircons from undoubted Alpefjord Formation. They also give evidence of a pre-1030 Ma source terrain (Peucat *et al.*, 1985, p. 16).

In the middle part of Forsblads Fjord, migmatization of gneisses becomes more pronounced, and the metasediments display transitional contacts with sheets of foliated aluminous granitoids (sillimanite + garnet bearing). Dark lenses and dispersed boudins of any size (decimetre, metre and 100 metre) include both calc-silicate gneiss (brown amphibole + diopside, Ca-garnet and quartz) and mafic to ultramafic rocks (garnet-plagioclase amphibolites, pyribolites, phlogopite-pyroxenites, etc.) which can be considered as synmetamorphic intrusives. In inner Forsblads Fjord, higher-pressure conditions are indicated by the abundance of kyanite both in metasediments and mobilizates, and by the occurrence of rutile instead of sphene. Calc-silicate layers and boudins occur throughout the succession, together with rare layers of impure marbles. Some pelitic gneisses have a high percentage of both graphite and sulphides (pyrite and pyrrhotite).

In the Alpefjord region at least, geochronological results (Peucat *et al.*, 1985) and unpublished petrological data furnish strong evidence for a gradual transition between the metamorphic lower part of the lower member and the underlying metasedimentary gneisses.

17.2.1.2 *Description of the lithofacies of the lower member.* Orthoquartzites make up several mappable units which are entirely quartzitic and some several tens of metres thick, and some less-frequent thinner intercalations in other lithotypes. Thick units are generally traceable for several kilometres, with beds ranging mostly between 0.5 and 2 metres in thickness and generally displaying high-angle cross-laminations and frequent overturned bedding, both of which are emphasized by discrete concentrations of heavy minerals (Figs 17.6 and 17.7). The bases of units are generally sharply defined, whereas tongue-like protrusions of beds and lenses of orthoquartzites within grey arenites or silts are frequent at the tops (Fig. 17.8). In contrast to the Agardsbjerg Formation, pelitic intercalations in orthoquartzite members are lacking, and mud pebbles are also rare.

Petrographically, the orthoquartzites derive from recrystallized rather pure quartz sand, without noticeable primary matrix. When not metamorphosed, some carbonate concretions allow the preservation of subspherical grains with a mean diameter of 0.8 to 1.5 mm, which suggests a well-sorted and purified feldspar-free and mica-free sand of continental origin. Heavy minerals include predominant zircon, with ilmenite, tourmaline and secondary pyrite. All the sedimentary features observed are consistent with fluviatile deposition.

Grey psammitic feldspathic quartzites are the predominant lithofacies. They make up several composite units, either regularly layered with silt-shale alternating layers or more compact units with internal lenticular bedding.

In the regularly layered units ripple marks are widely developed, and mud-cracks in alternating black silts and shales show that these shallow water deposits were repeatedly subjected to subaerial conditions. These desiccation features are mostly recognizable by the sandy infill of cracks, some centimetres deep and 2 to 5 mm thick (Fig. 17.9) and detrital muscovite films concentrated at the top of the strata subjected to desiccation. While these cracks are frequently polygonal, others appear to represent 'shrinkage' cracks formed subaqueously by synaeresis (Fig. 17.10).

In the lenticular bedded facies, there is often com-

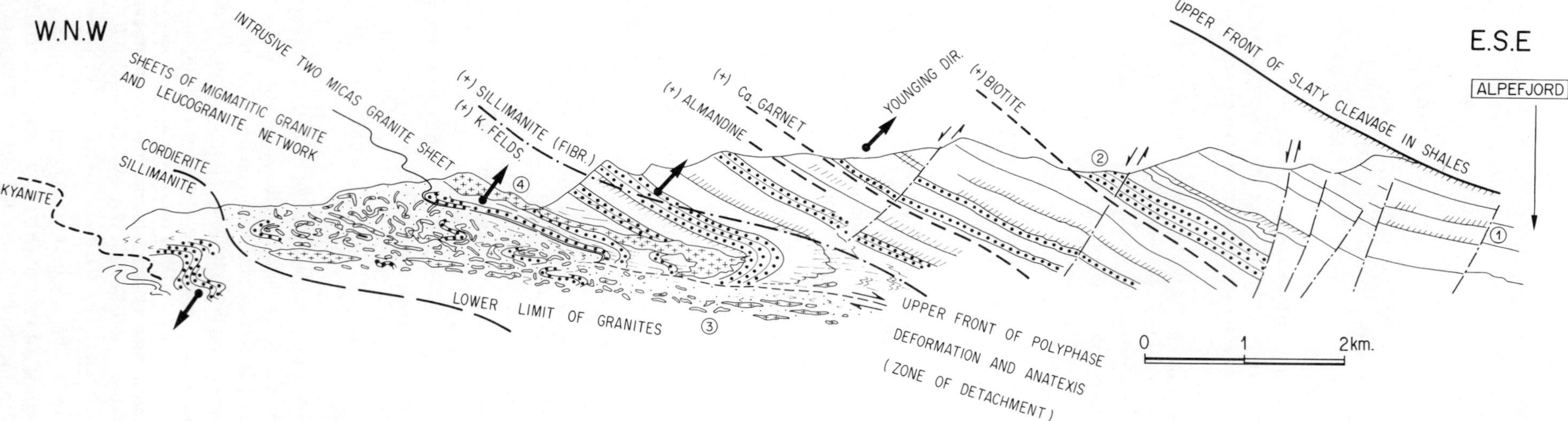

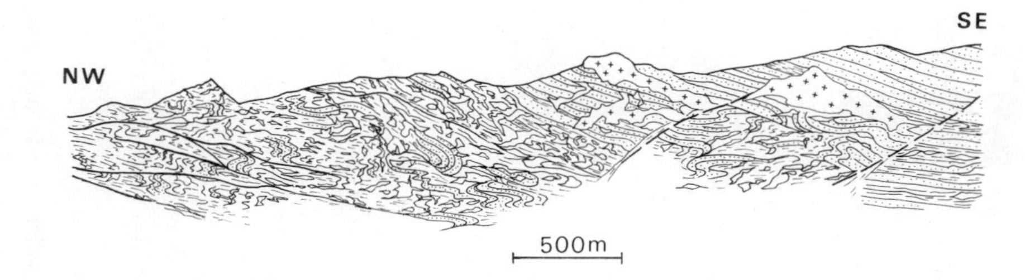

Figure 17.4 (*a*) Geological cross-section along Schaffhauserdalen (a–b of Fig. 17.2) showing prograde metamorphic zones in the lowermost unit of the Alpefjord Formation. (1) silty shales; (2) orthoquartzites; (3) migmatized metasediments migmatitic granite, and granite network with abundant roof pendants and 'schollen' of metasediments and sillimanite-garnet-bearing pelitic gneisses; (4) intrusive two-mica granite. The different mapped isograds are indicated; hatched pattern is the slaty cleavage, grading downwards into a foliation defined by muscovite and brown biotite. (*b*) Detailed section along the northern side of Schaffhauserdalen, showing the transition of sillimanite-bearing metasediments into pelitic gneisses cut by sheets, veins and pods of peraluminous granitoids (crosses for the larger bodies). Note the recumbent folds outlined by metaquartzite layers (dotted.) (Drawing after oblique photographs).

Figure 17.5 Impure, cross-bedded metaquartzite with sillimanite nodules, cut by pegmatite veins, from the lowermost Alpefjord unit, Schaffhauserdalen.

Figure 17.6 Base of an orthoquartzite unit, lower member. Note the north-east-trending channel. In the background is the Murchison Bjerg (2072 m). Note the marked large-scale parallel layers of the upper member of the Alpefjord Formation, paraconformably overlain by the Agardhsbjerg Formation, sharply delimited at its base by the white orthoquartzite layer of member 1 of the Agardhsbjerg Formation.

Figure 17.7 Orthoquartzite with cross-bedding and well-developed overturned laminations (middle part of the lower member).

Figure 17.8 High-angle cross-bedding in orthoquartzite layers alternating with parallel-laminated silty arenite.

Figure 17.9 Cross-section of small-scale cracks with sandy infill, in a silt layer interlayered with ripple-mark facies with lenticular bedding. Note the compaction effect and the incipient rotation of cracks.

plete separation of pure sand (building ripples) and impure sandy-silty material, on a centimetre scale. (Fig. 17.11). Other layers display a regular alternation of 5–10 cm thick, grey quartzite and silty shale, always with very progressive transitions, though some pseudo-graded bedding may also occur. Both upward coarsening and downward coarsening sequences have been observed, which result from progressive variations of silty matrix content, without an increase in the grain size of sand. Ripple marks generally are moderately flattened, but their frequent asymmetry, if sometimes primary, may also be due to deformation.

In the massive grey quartzitic layers, devoid of silty shale layers, channels of tens of metres to hundreds of metres in size, and low-angle cross-bedding, are the

Figure 17.10 Small-scale cracks in black silt of unknown origin (synercesis cracks).

Figure 17.11 Impure quartzite with a parallel-laminated structure overlain by current ripples. The darker, spotted zone with internal ghost layering (below hammer) is a carbonate concretion recrystallized into grossular-amphibole calc-silicate (lower member, biotite zone).

dominant features. Channels 2.5 m deep, with steep walls (30° to 90°) and 10 m wide have been observed. Sandy infills of channels display cross-bedding and overturned laminae indicative of high-energy currents. Such features suggest channel systems with both vertical and lateral accretion (Collinson, 1986). Accumulations several tens or hundreds of metres thick suggest a fluviatile environment, also in agreement with abundant Ca-Fe-Mg carbonate concentrations.

In the massive grey quartzites, load structures and

Figure 17.12 Pseudonodules of grey quartzite in dark silt, formed by load and injection features of sandy material in water-saturated mudstone (upper member).

slump-balls are frequent, while spectacular pseudonodules (Fig. 17.12), generated by downward injection of sand, characterize some regular layers.

Petrographically, impure quartzites contain a bimodal population of grains. Coarser quartz grains (1, or more rarely 2 to 3 mm in diameter) are mostly well rounded. Smaller grains (*c.* 0.5 mm) are more frequently angular. Both K-feldspar and plagioclase are abundant (quartz = K-feldspar = plagioclase) and muscovite flakes, mostly very thin but up to 2 mm across, form up to 20% of some thin layers. Most of these rocks are thus arenites or wackes, according to their matrix content. Carbon content is lower than in the black silty units.

The dark chloritic quartzites differ by the presence of an abundant (15–30%) primary matrix of pure Fe-chlorite. This phase is possibly derived from altered glauconite, or chamosite. With increasing metamorphic grade, chlorite has entirely recrystallized into Fe-rich dark-green biotite, and Fe-oxide granules.

Black silts and silty shales make up several regular units of variable thickness (decimetre to 300 m). They were very sensitive to metamorphism, and now are often spotted with porphyroblasts of brown biotite and muscovitized andalusite. They generally display a thinly layered structure on a very small scale (150 μm to few millimetres), and this layering is lenticular or micro-cross-bedded. In thin-section, the original silty nature of rocks is well preserved, with a minor clay component—if present—and a high proportion of detrital micas (muscovite, chloritized biotite). Microlaminations are marked by a higher content of carbon; Fe-sulphides (pyrite + marcasite + arsenopyrite) are common.

Small-scale channels (Fig. 17.13) with steep walls are ubiquitous in the succession, especially in the thinly banded alternations of silty quartzite–silty shale with mud-cracks. They sometimes appear as branching and meandering rill-marks on bedding surfaces. Sandy infill of channels cuts across the silt shales, and vice versa.

Carbonates represent a minor component. They consist of:

(i) Impure marble layers: 1 to 3 m thick, which repeatedly occur in the metamorphic lowermost unit. They are highly recrystallized and contain variable amounts of grossular, diopside, amphibole and biotite,

Figure 17.13 Selection of channels with sandy infill (right) and black silty infill (left) cutting parallel-laminated silty arenite. Note the important compaction effects.

and gradually pass into calc-silicates and skarns, with local development of Pb-Zn mineralization

(ii) Iron-magnesium carbonate lenses and ovoid concretions generally 20 to 50 cm in length, which occur throughout the succession. They are of early diagenetic origin and invade and replace selected layers of impure quartzite, silt and more rarely orthoquartzite. Ghost layering of enclosing rocks is sometimes still visible in the concretions, with a concentric zonation. When unmetamorphosed, they consist of coarse-grained ankerite ± calcite ± siderite replacing detrital grains. They frequently contain disseminated sulphides (pyrite and chalcopyrite), and in the lower part are thoroughly recrystallized into calcium silicates (grossular + green amphibole + biotite), with the development of diopside in the deepest units.

Heavy-mineral layers have been recorded at all levels of both the upper and lower members. Ilmenite is the chief mineral, magnetite being much less abundant. Zircon is more abundant than apatite, sphene and tourmaline. In orthoquartzites, heavy-mineral layers occur as thin bands which emphasize the internal structure of the layers (oblique or overturned laminae), but also as regular layers up to some cm thick in parallel-laminated metaquartzite units from the deepest part of the lower member. In silty units rich in organic matter, the heavy-mineral layers also form regular layers up to 20 cm thick, with a variable content of tiny opaque grains (ilmenite and/or magnetite) and smaller amounts of zircon and apatite than in the quartzites.

In the high-grade lowermost unit, as well as in the mobilized garnet-sillimanite paragneisses of outer Forsblads Fjord, some layers up to 20 cm thick (Fig. 17.14) include ilmenite, rutile and zircon as the chief minerals, set in a biotite matrix. The Ti-bearing minerals are euhedral, indicative of a complete synmetamorphic recrystallization and/or nucleation (rutile). The zircon grains ($\sim 100\,\mu m$) have maintained their subspherical shape of sedimentary origin, although discrete, newly formed faces are noticeable (Peucat *et al.*, 1985). Xenoliths of heavy-mineral layers have also been frequently

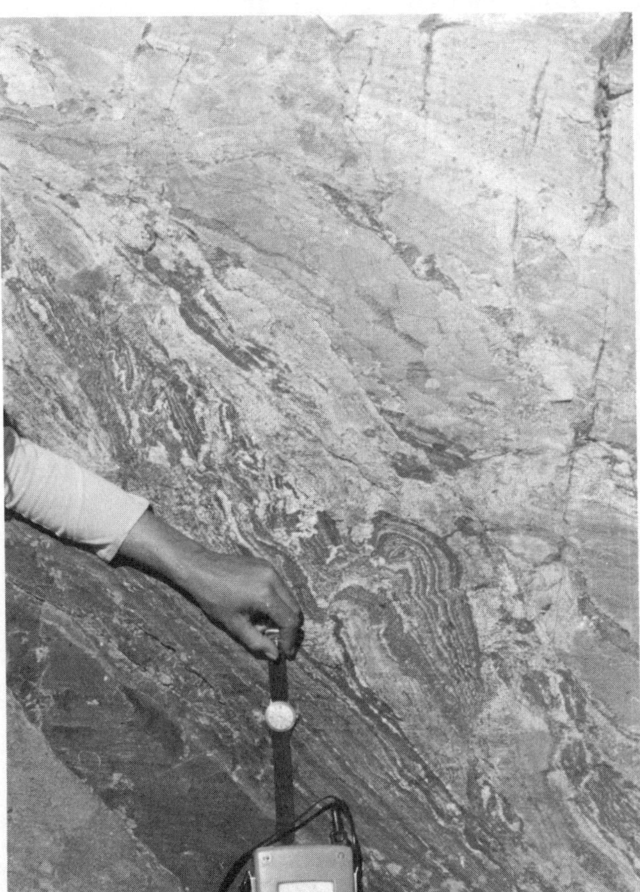

Figure 17.14 Heavy-mineral layer (in dark, close to the Geiger-Müller counter) in migmatized sillimanite-garnet-orthoclase paragneiss, upper Schaffhauserdalen.

found in the cordierite migmatites and granitoids of the western Stauning Alps.

17.2.2 *The middle carbonate member*

This member is exposed in Alpefjord as a regularly layered unit *c.* 120 m thick. It includes impure carbonates, limestones and silty layers subsequently affected by spectacular desiccation features and complex redeposition processes.

Alternating dark silty shales and silty limestones with a cm-scale layering and rarely preserved ripples represent the oldest carbonate deposits. Parallel-laminated carbonates with a micritic microstructure and secondary ankerite and pyrite may derive from carbonate muds deposited in quiet-water conditions. Other higher-energy deposits include sandy limestones with cross-bedding and recurrent levels of polygenetic conglomerates with flat pebbles of limestone and siltstone predominant.

The slightly transported elements may derive from adjacent fragmented and desiccated beds, possibly reworked during stormy periods. Throughout the unit, complex cracks with carbonate infill, and pseudobreccias, may represent the result of repeated periods of desiccation and/or evaporitic conditions. Undisputable sun-cracks with a quadrangular to polygonal shape

Figure 17.15 Large desiccation cracks in a silty layer of the middle carbonate member.

(Fig. 17.15) have been observed at several levels. The cracks are up to several cm in width, with both sandy and carbonate infill and up to one metre in depth. Undeformed in the siltstones, they are ptygmatically folded in limestone layers, with well-developed slaty cleavage at a low angle to bedding and associated disharmonic recumbent folds. Dolomites have been reported in this unit north of 73°N.

17.2.3 *The upper member* ($\geqslant$ *1300 m thick*)

This member (Fig. 17.16) differs from the lower member by its greater development of green silty units, by the appearance of red hematitic layers, and by a higher carbonate content in the matrix of various lithologies. In Forsblads Fjord it lies above the upper front of slaty cleavage, whereas pelitic schists intruded by foliated granites are considered to be their counterparts in the northern Stauning Alps.

17.2.3.1 *Description of lithofacies* (i) Orthoquartzites make up units several tens of metres thick, with 20–40-cm thick beds. They display less-frequent high-angle laminae than in the lower member, and no overturned features. Silty interbands are present, which are frequently reworked as mud pebbles. Illite and calcite cement is common, and ankerite spots and elliptical concretions occur. The Grey-green quartzite units are

Figure 17.16 Upper part of the lower member of Alpefjord Formation. Two orthoquartzite units (clear bands) overlie monotonous impure quartzite-arenites which make up the lower half of the mountain (1947 m high) on the right. In the background is the upper member, capped by a outlier of basal quartzite of the Agardhsbjerg Formation.

Figure 17.17 Alternating ripple marks and parting lineation in grey impure quartzite, upper member.

massive, with common metre-scale channels filled with coarse-grained lithologies (grain size up to 3 mm, dispersed quartz gravels with clasts up to 1 cm across and mud pebbles) displaying cross-bedding and contorted to overturned bedding. Local conglomeratic lenses with rounded pebbles of siltstones and quartzites from underlying beds occur.

Asymmetrical current ripples, (locally 10 cm high and with up to 50 cm wavelengths) have been observed (Fig. 17.17). Internal erosion features are conspicuous (Figs 17.18 and 17.19). Petrographically, both feldspar-free and arkosic and micaceous quartzites occur. Coarser grains of quartz frequently have a grey-blue colour and contain abundant rutile needles, suggestive of a granulitic source terrain.

Other feldspar-free and muscovite-free quartzites have a cement of illite, chlorite and carbonates, and some dark-green varieties may contain up to 20% of chlorite matrix, possibly derived from montmorillonite and glauconite; the spherical shape of the latter mineral is recognizable in some rock types.

(ii) Black silt–silty shale units display a thinly layered structure, with alternating sand-rich and sand-free, carbonate-rich microcycles. Sun-cracked surfaces, ripple-marks, and small-scale channels are extremely common in this lithofacies (Fig. 17.20). Both carbonate-free units with large development of mud-

Figure 17.18 Steep erosion surface sharply cutting white quartzite with diffuse carbonates. Subsequent burial of the topography took place by deposition of thinly laminated silty arenite. Synform–antiform structures are related to differential compaction.

Figure 17.19 Deep erosion on top of an impure sandstone, with parallel laminations in the form of numerous channels and small canyons. These features were passively buried by low-energy deposition of sandy and silty material with lenticular bedding and ripple marks (middle part of the upper member).

Figure 17.20 Small channels of carbonate-rich sandstone in black silt. Note the carbonate layer and the diagenetic concretion (containing ankerite and siderite) below.

cracks on a cm and dm scale and with sandy infill, and ripple-marked units with carbonate in the matrix and less-frequent mud-cracks, are present.

(iii) Composite units of black silt, shale and pure quartzite, all with various amounts of carbonate, include rippled facies, and channels passively filled by overlying sediments. Other features with local steep erosion surfaces close to mud-cracked surfaces may also indicate repeated sub-aerial exposure.

In thin-section, 10 to 30 microcycles/cm can be recognized in thinly banded dark silty layers. Each microcycle of 0.15 to 0.5 mm thickness is defined by thin (200–500 μm) films highly enriched in organic matter, and very thin detrital muscovite flakes and chlorite pseudomorphs after biotite. Alternating pale sandy layers (mm to several cm thick) which form the ripples with an internal cross-bedding, are composed of angular grains ($\sim$ 200 μm in size) of quartz, plagioclase and K-feldspar in equal proportions, with lesser amounts of muscovite, tourmaline and apatite. Above the biotite isograd, cryptocrystalline illitic mica may derive from minor amounts of clay-rich cement. It is noteworthy that heavy minerals are highly concentrated ($\sim$ 2000 grains per cm^2) in selected microcycles some 100 μm thick within or close to the thinly layered silty bands. These features were possibly formed by residual enrichment from an already deposited sand removed by a regular slow bottom current.

The features observed suggest a tidal flat environment: the carbonate-rich microcycles of silt represent immersion periods of the low intertidal zone; the mud-crack-rich layers represent temporary emerged areas above high tide; and the ripple marks represent sandy layers higher-energy deposit in subtidal areas.

Intense compaction of rocks is indicated by microkinks and stylolites, developed along the silty layers which deformed the muscovite flakes, by pressure-solution processes around feldspar grains, and by ptygmatically folded mud-cracks.

(iv) Green rhythmic silty shale units up to 300 m thick display a thin parallel and lenticular microstructure with rare tiny ripple marks. They comprise two types of sediment free of organic matter:

(a) Dark-green layers with quartz and feldspar grains ($\sim$ 50 μm), with fresh detrital muscovite and biotite, set in a secondary recrystallized abundant matrix (20–30%) of illite $\pm$ chlorite $\pm$ carbonate

(b) Pale-green regular silty layers with an obviously parallel-laminated structure. These display recurrent graded bedding at the scale of 100 to 300 μm, grading from silt into a cryptocrystalline illitic groundmass clearly derived from recrystallized clay. Quartzo-feldspathic layers abruptly appear above the parallel-laminated layers and contain load structures and even synsedimentary microfaults. Some laminae are also defined by heavy-mineral enrichment. Other compact non-laminated beds may also derive from more homogeneous clay deposits.

With increasing grain size, silty rocks gradually pass into green psammitic quartzites with parallel-bedding and diffuse carbonate cement. Rare current marks resembling flute casts are locally preserved on flaggy surfaces, which are mostly flat or display minute ripples. They include carbonate-rich varieties, occasional surfaces with mud-cracks and some red layers with hematite pigment affected by large desiccation cracks. Small channels with coarser-grained sandy infill with frequent ankeritic cement are common. Diffuse pyrite $\pm$ chalcopyrite primary impregnations occur repeatedly.

These rhythmites may represent low-energy subtidal or lacustrine deposits in shallow water, with episodic emergent periods. They are lithologically similar to the green units of the Agardhsbjerg Formation. Note that no oblique cleavage is developed in these units, because compaction and flattening acted parallel to bedding. Some units display a cherty appearance due to regional thermal metamorphism.

17.2.3.2 *Interpretation of palaeoenvironments* (i) Lower member: the presence at many levels of the succession of orthoquartzite units, with sedimentary features characteristic of fluviatile deposits, may provide a guide for the depositional environment of this succession. The abrupt recurrent deposition of well-rounded feldspar-free sand within both silty units and grey impure and unsorted clastic rocks suggests that the former may derive from redeposited aeolian sand, which may have invaded a different environment characterized by deposition of unaltered, mica-rich sand and silt rich in organic matter.

Grey arenites with low-angle cross-bedding and channels may represent channel deposits formed both by lateral and vertical accretion in a fluvial system. Silty units with ripple marks and desiccation cracks may represent shallow-water flood-plain deposits, whereas thinly layered units with rhythmic microcycles may be lake deposits. The abundant presence of Ca-Fe-Mg carbonate, mainly in the form of diagenetic concretions, the abundance of fresh detrital micas and the high carbon content are arguments for a non-marine environment (Courel *et al.*, 1984; Allen and Collinson, 1986; Collinson, 1986).

The permanent unaltered, clay-poor sandy and silty material may indicate a mostly rainy climate in the source area and also a negligible meteoric and/or pedogenetic alteration (and/or a rather rapid erosion following renewed uplift).

Thus, following criteria developed by Collinson (1986), we interpret the unoxidized, micaceous, carbonate-rich units as ancient mudstones and siltstones deposited in shallow interchannel areas, flood-plains and lakes. Shallow-marine conditions and tidal flats are also likely, but no evidence for shorelines has been found. Heavy-mineral layers are frequent along modern shorelines and their presence within silty units must be pointed out.

Thus we propose that the lower member was deposited in a flat, mainly non-marine environment, with sediment derived from a huge fluviatile system originating in an area of high rainfall where muscovite-rich metamorphic rocks were being eroded.

Current directions occasionally measured (channels, parting lineations, axes of overturned laminae) always indicate a mean eastward direction of transport.

(ii) Middle member: the appearance of carbonate layers, together with green and black rhythmites, is strong evidence for open lake deposition. Carbonate muds may represent cyclic precipitates and are associated with carbon-rich silts and muds, which would also be offshore deposits. Repeated desiccation features indicate many emergent periods. Conglomeratic beds and brecciated carbonates may represent marginal facies of temporary closed lakes, with possible evaporitic conditions.

(iii) Upper member: the great abundance of silts and carbonate-cemented rocks, together with the occurrence of a prominent clay component and possible glauconite, support the interpretation of both lacustrine and shallow-marine conditions with possible tidal flats including carbon-rich microbial mats. Orthoquartzites do not have the typical high-angle and overturned laminae, indicating a reduced current and possible marine conditions. Current directions measured in channels, however, again indicate east-north-east to eastward transport. It is noteworthy that the Canning Land black units underlying the Agardhsbjerg Formation 130 km to the south-east (Fig. 17.3) share similar facies. Orthoquartzites tend, however, to build thinner units. On the other hand, the clay content is possibly higher in silty shales, and a better separation of clay, silt and sand can be deciphered in many rock units with slaty cleavage, in which the typical small-scale channels and cracks of unknown origin have also been observed (Caby, 1972).

17.3 Agardhsbjerg Formation

This formation ($\sim$ 2000 m thick) (Fig. 17.3) is sharply delimited at its base by a regular pure orthoquartzite layer of great lateral extent, that rests with a paraconformity or with a locally slight angular unconformity (the only one recorded in the whole group) on black and green silty units. This lower white orthoquartzite layer (25 m thick), studied in northern Alpefjord, displays parallel-bedding and includes interlayers of green silty shale at its base. This abrupt change in sedimentation recorded in Alpefjord and other areas suggests a marine transgression through marine dunes with rather pure and well-sorted sand. This layer grades upward into medium-bedded sandstones to quartzites (70 m), with the thickness of the beds ranging from 3 to 20 cm, with internal cross-bedding and flat, eroded surfaces with thin silty layers, ripple marks and local giant mud-cracks. Red and green muscovite-bearing siltstone interbands are also reworked as mud-pebbles. Several layers of carbonate-cemented sandstones with parallel laminations occur at the top of the unit.

Overlying beds include pink to violet hematitic sand

stones to quartzites, with cross-laminations and ripple marks on top of beds. Red mud-pebbles in sandstone are frequent in the channel infills. Mud-cracks occur in several red and green silty to shaly layers (< 1 cm), which separate individual sandy beds. Carbonate cement is present in a few layers.

According to several authors (Katz, 1952*b*, Eha, 1953; Fränkl, 1953*a*, *b*; Sommer, 1957*a*, *b*), the last units of the central Fjord zone include pure quartzites, and semi-pelitic to pelitic and psammitic layers. Contacts between individual beds are often sharp. Thus non-quartzitic layers apparently share lithological and petrographic similarities with many units of the upper member of Alpefjord Formation.

17.3.1 *Agardhsbjerg Formation in Canning Land*

Here, the thickness of this formation is only about 1500 m, with fewer quartzites than in the Alpefjord region (Caby, 1972, 1976*b*). Unit 1 (270 m thick) also includes a basal massive, white quartzite layer (30 m thick), with quartz gravels, grading into green chloritic siltstones and glauconitic psammites (20 m thick), and overlain by black rhythmic silty shales with mud-cracks and small sandy channels (220 m). Unit 2 (170 m) begins with ripple-marked quartzite with oblique laminae (50 m), capped by 100 m of green silty shales with a varved structure and small channels of ankeritic sandstone. Unit 3 (420 m) is composed of layered, often carbonate-rich quartzites with oblique bedding and ripple-marks, alternating with black silty shales. Unit 4 (480 m) begins with a massive quartzite (100 m), overlain by 150 m of alternating black and green silty shales and a 2–5-m thick quartzite layer, grading up into purple and green rhythmic silty shales (200 m) (Fig. 37) and ankeritic sandstone layers. Unit 5 (150 m) includes green rhythmic silty rocks (Fig. 17.21) and overlying black silty shales with ankeritic sandstones. Diffuse Cu mineralization repeatedly occurs in green silty units (Caby, 1972). Unit 6 is a white cross-bedded sandstone.

Figure 17.21 Red silty shale with lenticular bedding underlain by white sandstone (member 4 of the Agardhsbjerg Formation, Canning Land).

17.3.1.1 *Interpretation of palaeoenvironments* The occurrence of typical desiccation cracks, repeatedly developed in thin red silty shale interbands, indicates that very shallow-water deposits of continental affinity exist in Alpefjord. Compared with the upper member of the Alpefjord Formation, the clay component in the silty shale is higher, and the small amount of both detrital feldspar and mica in the sandstones is consistent with a previously altered sand, following chemical and/or pedogenetic alteration. East to north-north-east current directions have been observed. Thus, most of the quartzites are interpreted as low-energy fluvial sediments.

On the other hand, if the correlations between the units proposed above are correct, the environment at Canning Land was more similar to that of the upper member of the Alpefjord Formation, with alternating of lacustrine and shallow-marine conditions. However, the presence of copper in green rhythmic units is reminiscent of those generally associated with red-type series, which actually exist in some of the oxidized red deposits in the upper members.

17.4 Brogetdal Formation

Using the lithological subdivisions proposed by Katz (1952), seven members, or less formally, beds 6 to 13 (Bertrand-Sarfati and Caby, 1974, 1976) we will try here to interpret the sedimentological features of the Brogetdal Formation. Four 'major cycles', capped by deposits showing subaerial exposure features can be recognized (Fig. 17.22*a*). Among them, some specific characters allow parts of these cycles to be assigned to specific environments, and it is therefore possible to propose a sedimentological model for the Brogetdal Formation. All the examples are taken in Canning Land, where the measured section is located. Equivalent facies have been recognized in Alpefjord and are described in this chapter.

17.4.1 *Description of the major cycles*

Besides the lithological subdivisions, we can recognize four major cycles, which are characterized by features of subaerial exposure in the uppermost beds, but which are not equivalent. The four major cycles are preceded by a sedimentological break, for example: cycle B finishes with desiccated beds deposited in the supratidal zone and cycle C begins with limestones, deposited in a quiet, subtidal environment. This can be considered as a sequence boundary. These four major cycles are, from base to top:

17.4.1.1 *Major Cycle A* mainly comprises multicoloured shales and siltstones, with intercalations of

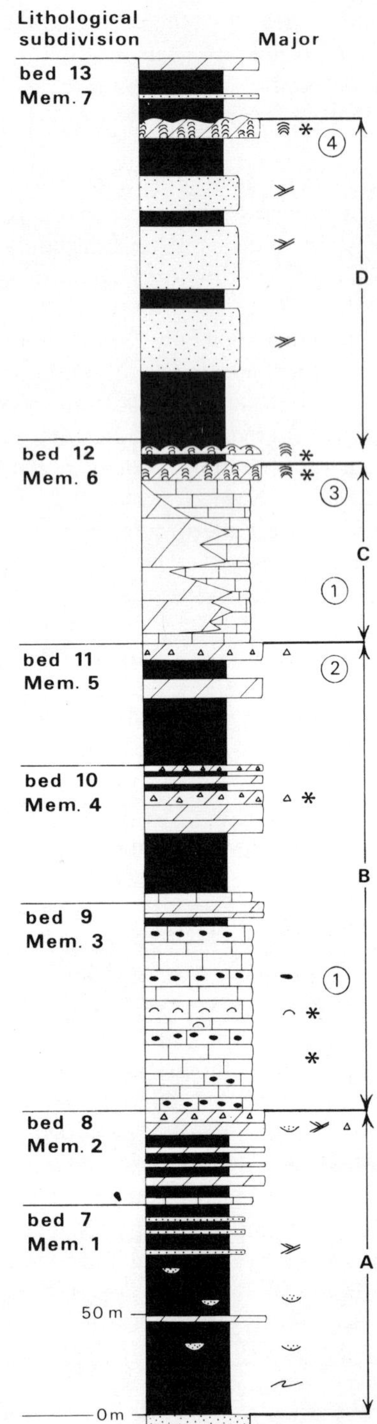

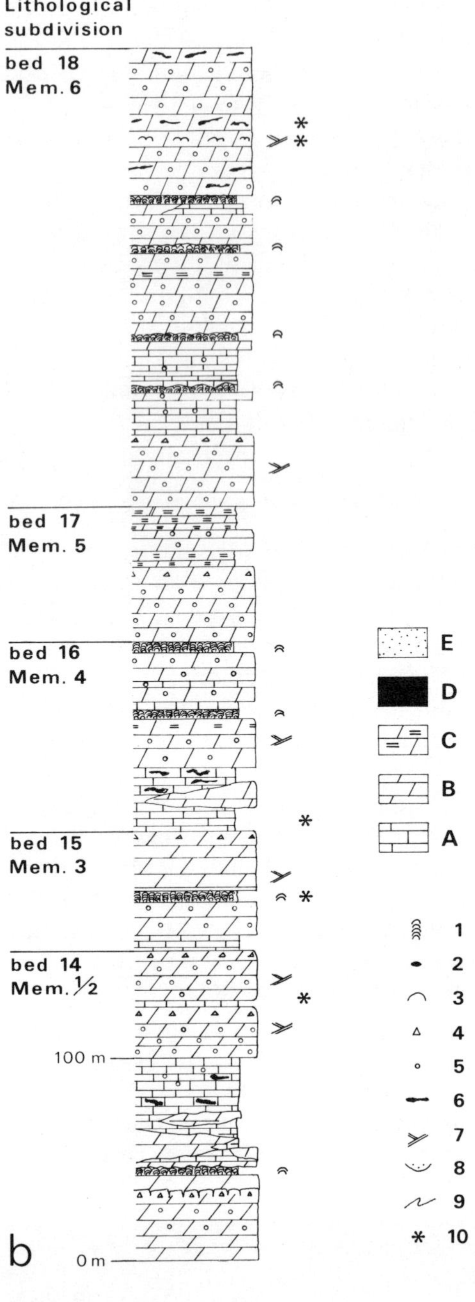

Figure 17.22 Schematic sections of (*a*) Brogetdal Formation; (*b*) Nøkkefossen Formation. *A*—limestones; *B*—dolomites; *C*—shaly dolomite; *D*—shales; *E*—quartzites; 1—stromatolites; 2—siliceous dolomite concretions; 3—mud mounds; 4—breccias; 5—oncoids; 6—cherts; 7—cross-bedding; 8—channels; 9—ripple marks; 10—localities mentioned in the text.

thin sandstone beds with small-scale cross-bedding and channel-like structures (Member 1, bed 7). Dolomitic intervals are very rare. The upper part of the cycle (Member 2, bed 8) lacks sandy intercalations, and, on the contrary, shows an increase in dolomitic layers, first as dolomitic shales or siltstone, and then as intercalations of pure-yellow clastic dolomite with occasional mud pebble breccias.

17.4.1.2 *Major Cycle B* has a thick limestone unit at the base, (Member 3, bed 9) that implies a sedimentological break with the preceding cycle. The limestones are rhythmites (facies 1) with thin layers of laminated or unlaminated mudstone and wackestone limestones. They are overlain by red shales and clastic dolomite layers, which are predominant in the upper part of the cycle. The middle part of the major cycle, after a very thin limestone layer, is mainly composed of multicoloured shales and siltstones, intercalated with clastic dolomites (Member 4, bed 10) and laminated dolomites. The upper part of the major cycle (Member 5, bed 11) is essentially made of black shales or siltstones with few quartzitic beds, and the uppermost part comprises laminated and clastic dolomite (facies 2).

17.4.1.3 *Major cycle C.* This also begins with a thick limestone unit. In Canning Land this unit is entirely or partially dolomitized secondarily (Member 6, bed 12). The cycle changes, above the clastic dolomitic beds, with the growth of stromatolites in sequences of decreasing energy associated with cross-bedded clastic dolomites (facies 3). They are interrupted by the deposition of black shales, forming the base of the next major cycle.

17.4.1.4 *Major cycle D.* This begins with stromatolites in patch-reefs or discrete bioherms embedded in the black shales. The stromatolites may be absent, with black shales alone deposited. This first unit is overlain either by alternating layers of black shales and quartzites or laterally by massive pure quartzites. These siliciclastic beds are lithologically identical to those from the underlying formations. The quartzites are made up of pure quartz sand with large-scale cross-bedding, and they contain—in the basal and uppermost parts—channels up to 2 m deep and conglomerates of quartz pebbles (2 cm across), which can represent high-energy environments, passing up, in the uppermost bed, to the stromatolitic sequences. The latter are a succession of shallowing-upwards sequences with stromatolites (facies 4), in which features indicative of subaerial exposure can be found. The siliciclastic interval may be fluviatile in a regressive sequence or marine deltaic in a transgressive sequence. At present we are unable to decide which.

17.4.2 *Description of the characteristic facies*

17.4.2.1 *Black, thinly layered limestones.* These rocks correspond with facies usually called rhythmites (Reineck and Singh, 1973; Jackson, 1985; Grotzinger, 1986) and are composed essentially of thinly layered lime mudstone. Two main facies are present: small mud mounds (between 10 and 15 cm) and laminated layers and thinly laminated lime mudstones with 'molar tooth' structures.

The facies of the unlaminated mudstones contain scattered, minute organic spheres, pyrite framboids, and tiny peloids in a dark micritic matrix. They sometimes resemble mounds in cushions (Fig. 17.23), outlined by lines of siliceous dolomitic spherical concretions (less than 1 cm in diameter) like threads of beads. Laminated layers are draped over these mounds. We interpret these mounds as being like Devonian mud mounds, caused by the early cementation of mud by microbial activity (Monty *et al.*, 1982; Monty, 1984).

The second facies of thinly laminated limy mudstones associated with 'molar tooth' structures is made of dark and clear microlayers (1 cm, maximum thickness) of mudstone. Each layer represents a composite cycle,

Figure 17.23 Thinly layered limestone (facies 1): (*a*) mud mound; (*b*) layers of thinly laminated limestone draped over the mounds; (bed 9, Canning Land). Scale bar: 4 cm.

composed of thinner clear and dark doublets usually 1 to 4–5 mm thick (Fig. 17.24). The dark layers are made of micrite-size crystals of calcite, and a faint lamination is outlined by clay/organic matter flakes. Change in colour results from changes in the amount of dark flakes and the intercalation of thin and rare clear layers. Generally, they grade upwards into a thick, clear layer presenting a nodular aspect. Composed mainly of clear micrite, the nodular texture is due to interlayering with stylolite seams outlined by dark films or flakes. The clear layer contains spherical concretions of silica and dolomite identical to those outlining the previous facies, only smaller. The uppermost boundary seems to be sharp and the microcycles can be interpreted as follows: dark mudstone with decreasing numbers of flakes and clear nodular micrite. The second-order cycles are due to an increase in thickness of one of the two components of the microcycles. These rhythmites are deformed and cross-cut by contorted cracks, probably as an effect of compaction (Fig. 17.24*b*). The cracks are filled by very fine-grained micrite, differing from both micritic layers found in the rhythmites and probably representing a micritic cement in which darker lines describe convolutions. The cracks are generally tapered at both ends and they seem to deform the rhythmites in two opposite directions. Our hypothesis is that they have resisted compaction, while the rhythmites, as shown by stylolites and nodular micrite, have been deformed. They are interpreted as synaeresis cracks because they are oriented regardless of bedding, they thin at both ends and are not accompanied by reworked mud clasts. Similar features have been abundantly recorded in Precambrian beds and interpreted as products of dewatering processes. They can be found both in siliciclastic and carbonate sediments (Smith, 1968; Donovan and Foster, 1972; Horodyski, 1976, 1983;

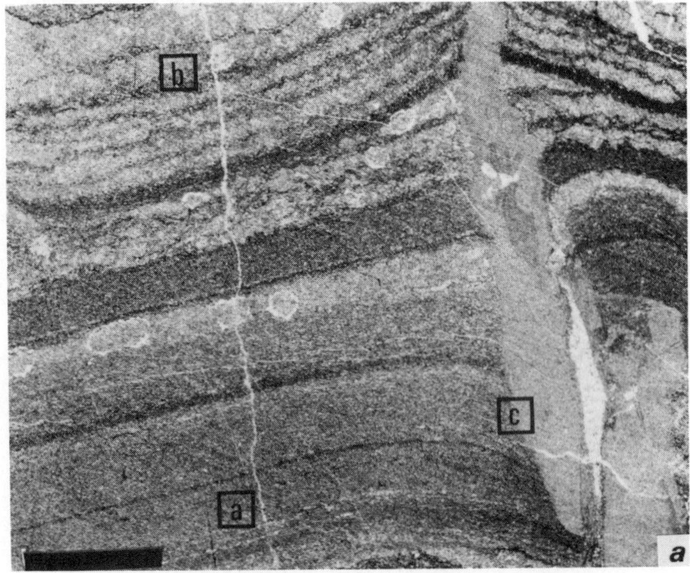

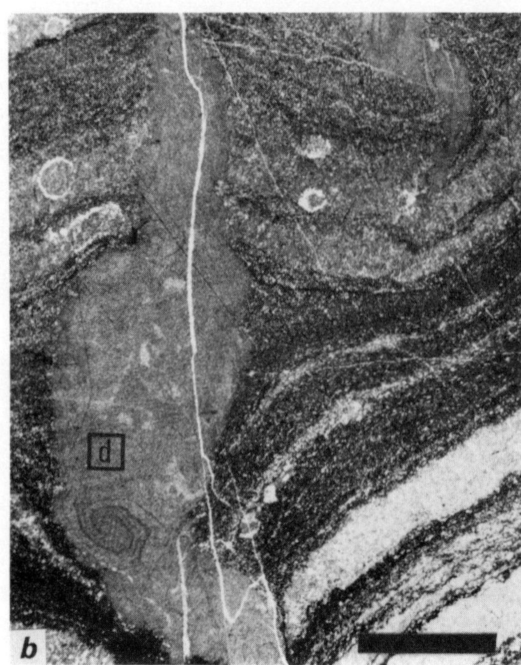

Figure 17.24 Microcycles in rhythmites and 'synaeresis cracks': (*a*) dark layers; (*b*) clear nodular layers with spherical concretions; (*c*) cracks filled with micritic equigranular cement. In (*a*) note the secondary crack filled by sparite. In (*b*) note the coloration rings (*d*) and the stylolite which does not affect the crack cement (bed 9, Canning Land, SGU 145 352). Scale bar: 3 mm.

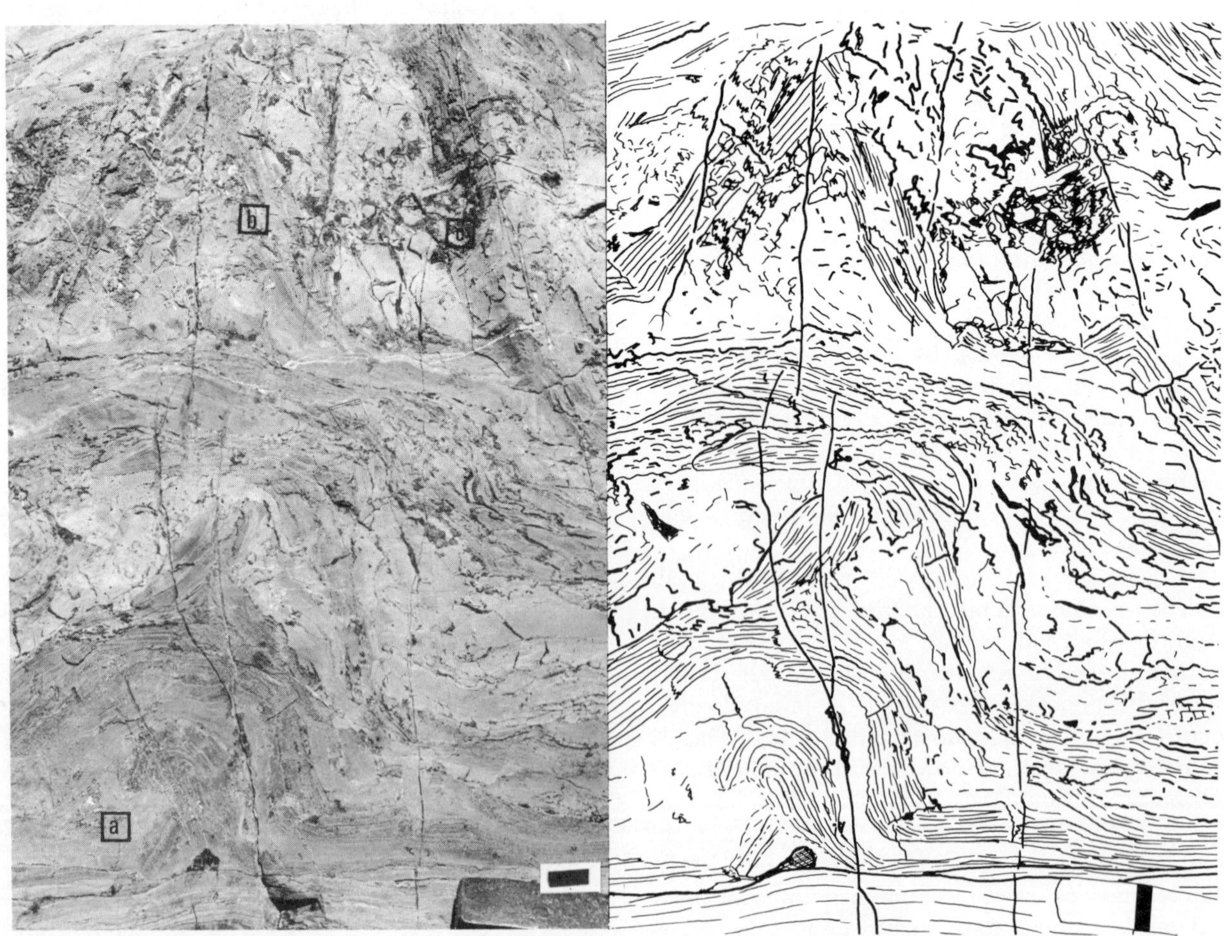

Figure 17.25 Dolomitic brecciated facies: (left), over a basal, laminated bed. (*a*) desiccated surface, (*b*) folded and tepeelike beds, (*c*) collapse breccia in less-competent beds (bed 10, Canning Land); (right) id. drawing. Scale bar: 4 cm.

Siedlecka, 1978; Fairchild, 1980; Fairchild and Hambrey, 1984), and have sometimes been attributed to organic activity of unknown origin (Bertrand-Sarfati, 1972; O' Connor, 1972). In some limestones in Alpefjord (bed 12), desiccation cracks have also been recorded, indicating occasional subaerial exposure.

Rhythmites are found in a wide range of modern and past environmental settings; deep- or shallow-marine (Grotzinger, 1986*a*, *b*), marine, intertidal estuarine (Reineck and Singh, 1973) and from continental seas, ponds and lakes (Jackson, 1985). One possible distinction between tidal flats and stagnation basins (Reineck and Singh 1973, p. 106) lies in the size of the detrital material, which is very much smaller in the latter examples. If we compare 'varves' described by Jackson (1985), they also appear to be deformed and cut through by secondary features, for example, spherical nodules attributed to evaporite growth. The Eleonore Bay facies of thinly banded and laminated limestone is interpreted as having been deposited in subtidal, quiet-water, protected environments, rich in organic matter. However, it is not yet clear whether they are lacustrine or shallow marine.

17.4.2.2 *Dolomitic laminated and brecciated facies*. One of the most typical lithofacies is a laminated dolomite which displays—above an undisturbed laminated layer—a sudden irregularity of the bedding resembling a desiccated torn-off pebble; the structure develops rapidly in a folded, and then tepee-like structure (Fig. 17.25). The overlying beds of massive unlaminated dolomite continue to be distorted by the tepee-like structure, and are deeply brecciated, with early cementation of sparite. The interpretation we suggest is that desiccation and evaporite growth might be responsible for the tepee structure (Asseretto and Kendall, 1977; Enos, 1983) and that dissolution of the evaporite might generate collapse breccias. Smaller desiccation sheet-cracks are visible in other dolomite beds, with an internal growth of botryoidal calcite, perhaps guided by microbial rods and growing as endostromatolites (Monty, 1982). These facies represent subaerially exposed sediments in alternating arid and humid seasons (no evaporite clasts are recorded).

17.4.2.3 *First stromatolite sequences*. The first stromatolites have been described (Bertrand-Sarfati and Caby, 1976) as *Poludia boreica* and *Poludia tyrrellina*. They appear in a succession of sequences (Fig. 17.26). The sequence begins with clastic dolomite deposited on the eroded surface of the underlying sequence. The deposit consists of 3–5-cm thick beds, containing aggregated grains (botryoids, grapestone, microbial lumps) which are generally round grains with a micritic envelope and an empty centre (now sparite) and coated by a micritic common envelope. These envelopes are usually considered to be of microbial origin and generated in relatively low-energy settings. Such structures have been defined as microphytolites in Russian literature, and their possible origin is discussed in Swett and Knoll (1985).

Figure 17.26 First stromatolitic sequences: (*a*) erosional base of microconglomerate: mainly grapestone; (*b*) flat pebble conglomerate, in *grainstone*, then in *packstone*; (*c*) stromatolite head in mudstone (bed 12, Canning Land). Scale bar: 4 cm.

The sediment is gradually enriched in flat pebbles. Flat-lying at first, they become imbricated. The top of the detrital layer displays a change in texture from grainstone to wackestone, but still containing flat pebbles. Over this layer, the stromatolite laminae grow rapidly in large discrete heads showing a high synoptic relief over the carbonate mud that fills the interspaces. Uppermost parts of the stromatolites show branching columns of a smaller size covered by carbonate mud, representing the uppermost layer of the sequence. We interpret this sequence as reflecting a decrease in the energy of deposition in a subtidal environment; the sequence is repeated many times. Higher up in the bed, the energy increased and cross-bedded clastic dolomites were deposited without stromatolites (Fig. 17.27). When the energy decreased, shales were deposited and stromatolites were also prevented from growing. Higher up, stromatolites which belong to *Eleonora ramosa* suggest a shallower environment of growth with the

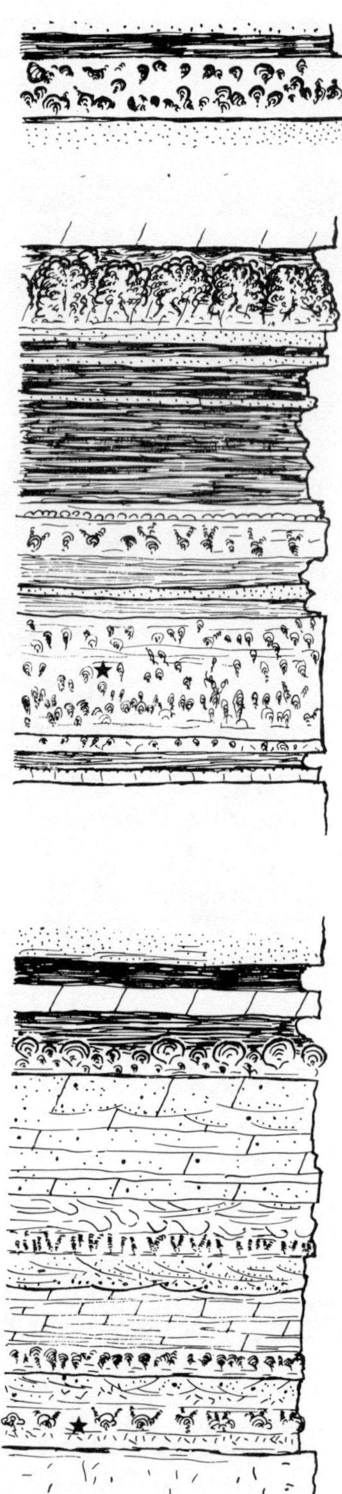

Figure 17.27 Schematic section of the first stromatolite sequences (from Bertrand-Sarfati and Caby, 1976).

existence of internal erosion surfaces, brecciation and desiccation of the internal sediment, together with cyclical growth of the stromatolite bioherms. Later, the indurated sediments were cyclically submitted to dissolution (Fig. 17.28). The last stromatolite bioherms are embedded in shales, and higher up, the terrigenous sediment changes from shales to quartzites. Stromatolites still grew occasionally in subtidal settings, but

Figure 17.28 Synsedimentary erosion, dissolution and sedimentation–cementation in a stromatolite, before and after a second period of stromatolite growth (bed 12, Canning Land). Scale bar: 4 cm.

showed frequent column erosion—due to wave action rather than subaerial exposure:

17.4.2.4 *Stromatolites in shallowing-upward sequences.* After a period of purely sand deposition with local intercalations of shales, a change in the environment allowed stromatolites to grow again in repetitive sequences approximately one metre thick. The most complete sequence comprises laminated dolomites overgrown by stromatolitic laminae (Fig. 17.29*a*), quickly growing in ramified columns assigned to the form-species *Inzeria groenlandica* (Bertrand-Sarfati and Caby, 1976). These columns reflect periodic changes in the environment: laminae erosion, probably desiccation, then colonization of the whole column by new stromatolite laminae unconformably capping the previous ones, and followed by the deposition of sediment in the interspaces. Above that bed, the stromatolites form discrete short heads with highly variable shape, buried in layered dolomite where desiccation features can be seen: such as reworked mud polygons (Fig. 17.29*b*). Such sequences have been described in the Proterozoic of Australia by Grey and Thorne (1985). The sequence of deposition is interpreted as beginning in the shallow subtidal zone, moving to low intertidal (with stromatolite columns) then to the intertidal zone with evidence of subaerial exposure.

17.4.3 *Interpretation of the palaeoenvironments*

The four major cycles represent complex superimposed sequences which we have not detailed. They are

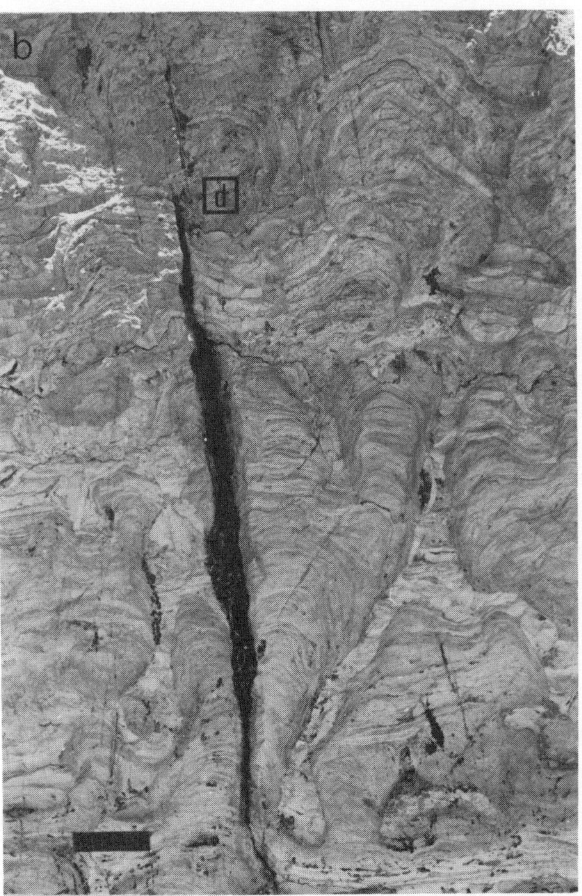

Figure 17.29 Stromatolites in shallowing-upwards sequences: (*a*) laminated dolomitic mudstones, columnar stromatolites (*Inzeria groenlandical*) in cyclic growth (bed 13, Canning Land). Scale bar: 4 cm; (*b*) stromatolite columns overlain by intertidal to supratidal stromatolites (desiccation polygons: (*d*) bed 13, Canning Land). Scale bar: 4 cm.

characterized by uppermost beds indicating a subaerial exposure in the supratidal zone (facies 2) or in the high subtidal to intertidal (facies 3), and low intertidal to supratidal zone (facies 4). They often (except cycle D) begin with limestones implying subtidal, quiet-water, protected environments (facies 1). The cause of such a change from supratidal environments to subtidal environments involving transgression of a large body of water, may be due to irregular subsidence, occasional sea-level rises or a sediment supply/subsidence ratio varying from base to top. The presence of tepee structures and the absence of evaporite casts testify to alternating arid and humid warm seasons. The Brogetdal Formation thus corresponds with the shoal zone of a carbonate shelf.

17.5 Nøkkefossen Formation

This formation mainly comprises carbonates with a very small proportion of siltstone and shale. Stromatolites are frequent among the carbonates as columnar biostromes or as flat-lying stromatolite laminae. Another characteristic feature of the entire formation is the abundance of cherts in all facies, both dolomites and limestones (except in stromatolites), either as extensive silicified layers or as discrete cherts. Some of the most typical facies are described, and can be grouped in an ideal sequence which is found repeatedly in the section but which is never complete.

17.5.1 *Description of some characteristic facies*

There are five distinct facies recognizable in the Nøkkefossen Formation.

17.5.1.1 *Black, thinly layered limestones.* They resemble those previously described, except that they are in sequential succession and do not show synaeresis cracks. Like the other limestone facies they contain large amounts of organic matter. Small-scale sequences (Fig. 17.30) begin with thin layers of black bituminous shales, which may be sometimes interpreted as boghead coals. (Schidlowski *et al.*, 1975). Carbonates can form unlaminated cores (10 cm in size), on which thinly laminated carbonates are draped in more or less discontinuous layers. They can be compared with the mud mounds in the Brogetdal Formation. They may also be in layers and overlain by laminated limestones

Figure 17.30 Thinly layered limestone in small-scale sequences: (*a*) black shales; (*b*) unlaminated lime mudstone in mound; (*c*) laminated mudstones draped over the mounds; (*d*) storm deposit of dolomitic mud breccia; (*e*) laminated limestone with planar fenestrae (bed 16, Canning Land). Scale bar: 4 cm.

with planar sheet-cracks. Occasionally, one sequence may display a scoured surface filled by an allochthonous breccia of dolomite mud clasts. The whole sequence seems to be subtidal, with the sheet-cracks testifying to the influence of a period of relative aridity or microbial activity (gas trapping). The mud clast breccia may reflect wave action in the surf zone (James, 1983), depositing dolomite debris from a surrounding tidal flat. Dominance of lime mudstones indicates a protected environment.

17.5.1.2 *Dolomitic, oncolitic and detrital facies*
Representing a large part of the sedimentary pile, this facies consists of 2–5-cm thick layers of dolomite sand, sometimes cross-stratified, successive layers of unlaminated or laminated organic-rich mudstones, wackestones and grainstones, where the grains are mainly oncoids, redeposited oncoids of various sizes (up to 1 cm) and reworked oncolitic clasts. In places monogenetic or polygenetic mud clasts are also found. The oncoids are asymmetrical, with microcolumnar laminae which are obviously microbial. Such oncoids have been analysed by Grotzinger (1986*b*) as being pedogenetic, but we have not seen any evidence of pedogenesis. Layers of spherical ooids of various sizes are also present but never mixed with oncoids. Both ooids and oncoids are frequently silicified. Among these ooids, the first endoliths of late Proterozoic age have been found (Campbell, 1982; Campbell *et al.*, 1982). In places, very small, bushy columnar stromatolites (2–3 cm high) display a curious microstructure

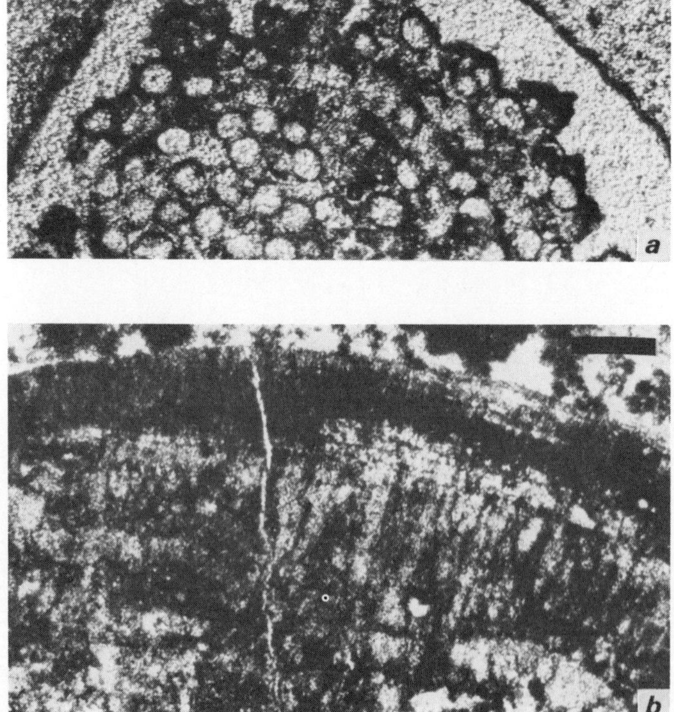

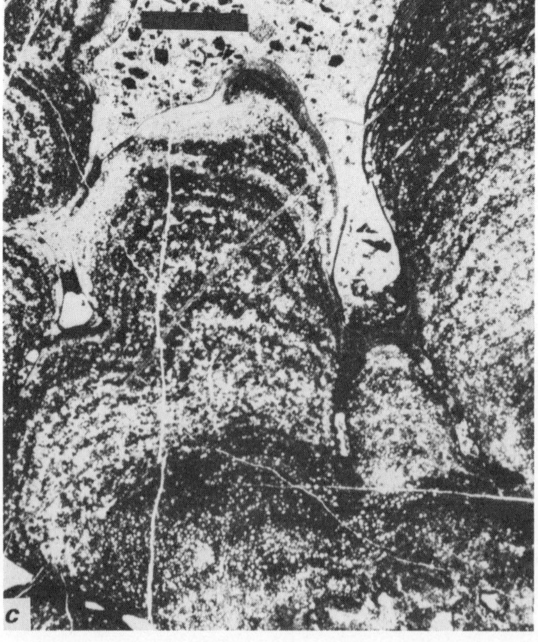

Figure 17.31 Microstromatolites built by closely packed spheres (bed 18, Canning Land, GGU 145 442); (*a*) details of a sphere layer in transverse section. Scale bar: 1 mm; (*b*) detail of a filamentous layer prior to silicification (see the same outer layer in (*a*), silicified). Scale bar: 1 mm; (*c*) general view of the small columns. Scale bar: 10 mm.

(Fig. 17.31*c*), which is of great interest in understanding the various spherical bodies widespread in both stromatolites and detrital layers. They are composed of thick laminae of micrite with radial filament rods (Fig. 17.31*b*). These peculiar laminae are subject to an intense and specific silicification. Alternating with them are thick laminae made of adjacent spheres, closely packed, with a thick micrite wall and an empty (now sparite) centre which seems to be more resistant to silicification. These spheres appear to have been generated *in situ* (Fig. 17.31*a*).

17.5.1.3 *Columnar stromatolites with mudstone or wackestone sediment.* The columnar stromatolites, forming biostromes 1 or 2 metres high, belong to different groups but have not been described at form level (Bertrand-Sarfati and Caby, 1976). The columns are vertical or slightly oblique, and the interspace filling is usually a carbonate mud with very thin layers of silt and a few stromatolite clasts. Rounded hollows similar to borings in the stromatolite columns, and strange bodies of equigranular micrite sheltering a spar-filled cavity, might be interpreted as indicating formerly living organisms (Fig. 17.32) of unclear origin.

17.5.1.4 *Columnar or pseudocolumnar stromatolites with cyclical growth.* The stromatolite growth was periodic; after 10 cm of undisturbed growth, the columns showed surface erosion, perhaps desiccation, and dissolution cavities filled with internal sediment, which in turn displayed desiccation sheet-cracks. These stromatolites are interpreted as growing in the intertidal zone and submitting to periods of subaerial exposure after lithification (Fig. 17.33).

17.5.1.5 *Small stromatolites and associated breccia.* These small-scale sequences begin with a layer of imbricated intraclast–flat pebble breccia, probably resulting from desiccation and deposited together with smaller detrital grains by wave or storm action. Stromatolites in small columns (1–3 cm high) have grown directly on top of the breccia and are also covered by flat pebbles. They in turn are overlain by laminated stromatolitic dolostones, showing desiccation features, sheet-cracks and angular flat intraclasts (Fig. 17.33*b*). This reflects deposition in low intertidal to high intertidal or supratidal zones. In another sequence, mud polygons are also reworked among storm deposits (Fig. 17.34), similar to those deposited on the present-day tidal flats of Florida or to the Cretaceous carbonates of Texas (Shinn, 1983; Figs 11 and 13).

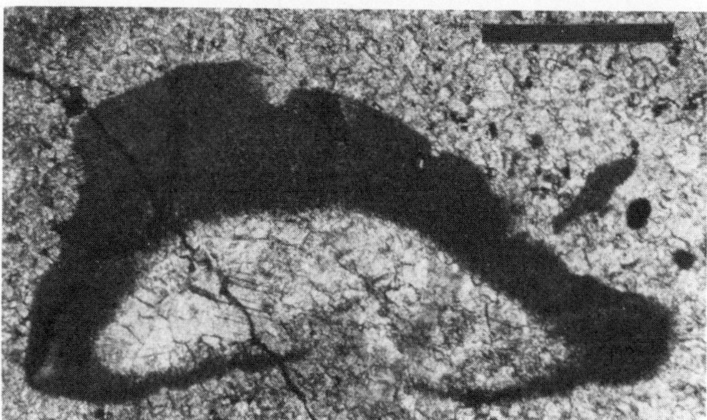

Figure 17.32 Micrite body attributed to an unknown organism, in the mud filling the interspaces of stromatolite columns (bed 15, Canning Land; GGU 145 424). Scale bar: 1 mm.

17.5.2 *Interpretation of the palaeoenvironments*

The Nøkkefossen Formation is a cyclical unit. From specific facies analysis and from field data, we consider that the ideal sequence may be (regardless of thickness of the beds):

(i) limestones and rare black shales, probably subtidal in small-scale sequences, with rare planar fenestrae
(ii) Detrital dolomites with cross-bedding, oncoids or ooids, which could have been deposited in shallow subtidal, relatively low-energy environments; columnar stromatolites are usually associated with these oncolitic dolomites, either above or below them. They pass upward to:

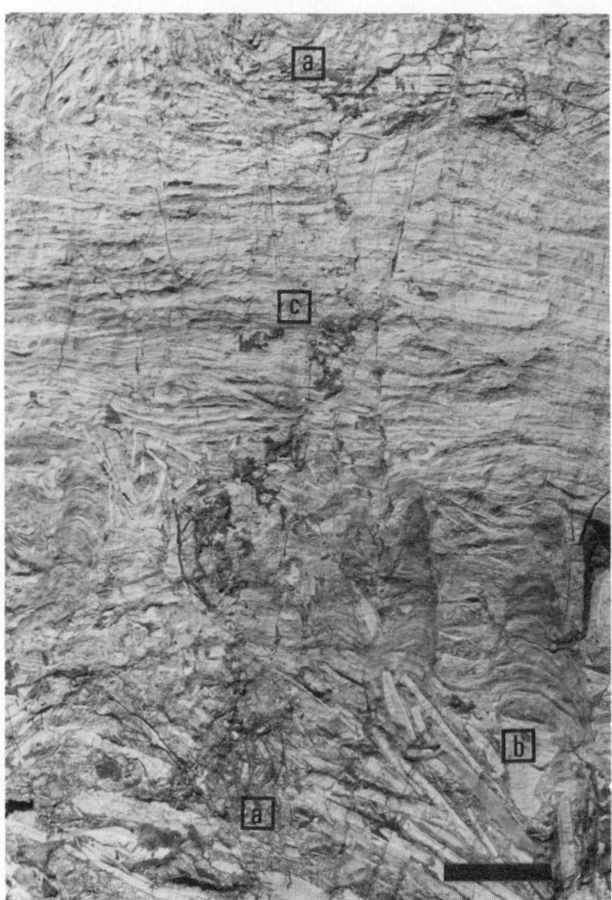

Figure 17.33 Shallowing-upwards sequences: (*a*) conglomerate of flat pebbles (surf zone?); (*b*) small stromatolites also overlain by flat pebbles; (*c*) laminated dolomites with desiccation features (bed 14, Canning Land). Scale bar: 4 cm.

Figure 17.34 Reworked mud-cracks (m) in the intertidal to supratidal zone (laminated dolomites: (*d*); (bed 18, Canning Land). Scale bar: 4 cm.

(iii) Laminated dolomites with short stromatolites which are in the low intertidal to high intertidal facies, and are locally associated with shaly dolomites

(iv) Desiccation is obvious in the laminated dolomites and is responsible for the dolomite brecciation with large flat pebbles or mud polygons. The cyclical sedimentation of the Nøkkefossen Formation reflects shallowing-upward sequences which differ from the Brogetdal sequences in their frequent repetition and small scale. We consider that the entire formation has been deposited in an inner-shelf environment, probably under a tidal regime. The alternations of limestones and dolomites of probable early diagenetic origin also provide evidence of deposition in the inner-shelf zone. It is clear that sediment supply in the Nøkkefossen Formation was more important than the rate of subsidence.

The uppermost beds of the formation, entirely dolomitic, are invaded by cherts and pass upward to a more or less continuous layer of laminated black chert crust, which can be interpreted as a silcrete crust (Friedmann and Sanders, 1978; Ross and Chiarenzelli, 1985) indicative of a period of prolonged weathering and pedogenetic alteration, prior to the deposition of a new group: the Tillite Group.

17.6 The age of the Eleonore Bay Group

17.6.1 *Geochronological data*

The U–Pb dating of detrital zircon from placer deposits of the Alpefjord Formation gives ages of 1162 ± 36 Ma (Fig. 17.35), thought to be pre-depositional, whereas K–Ar ages of detrital muscovites are in the range of 1030 ± 22 Ma when sampled above the Caledonian biotite isograd (Peucat *et al.*, 1985). A U–Pb age of 1060 ± 37 Ma was also obtained on zircon from a placer deposit interlayered with high-grade pelitic gneisses of the Central Metamorphic Complex in Schaffhauserdalen (Fig. 17.14). This is also interpreted as a pre-depositional age. If so, the gneisses may be the metamorphic equivalents of the Alpefjord Formation. Poorly defined Rb–Sr isochrons and errochrons obtained on high-grade gneisses of the Krummedal sequence and of inner Forsblad Fjord (Higgins, 1974, 1976; Henriksen and Higgins, 1976; Rex *et al.*, 1977; Hansen *et al.*, 1978; Higgins *et al.*, 1978; Steiger *et al.*, 1979) are here interpreted as inherited ages from the source, which according to our data, was also represented by unaltered and unweathered muscovite and feldspar-rich arenites.

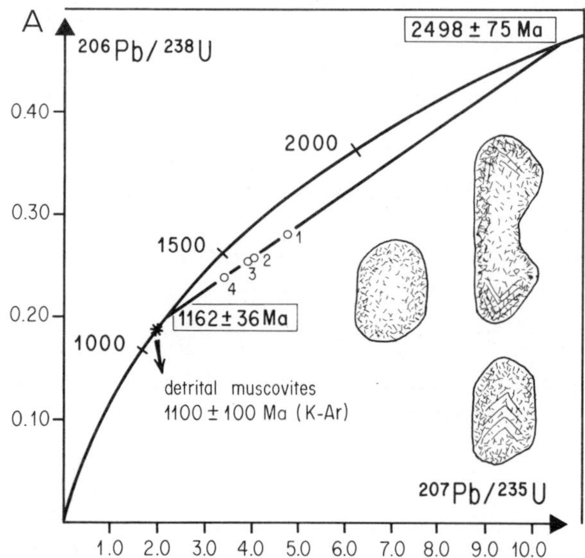

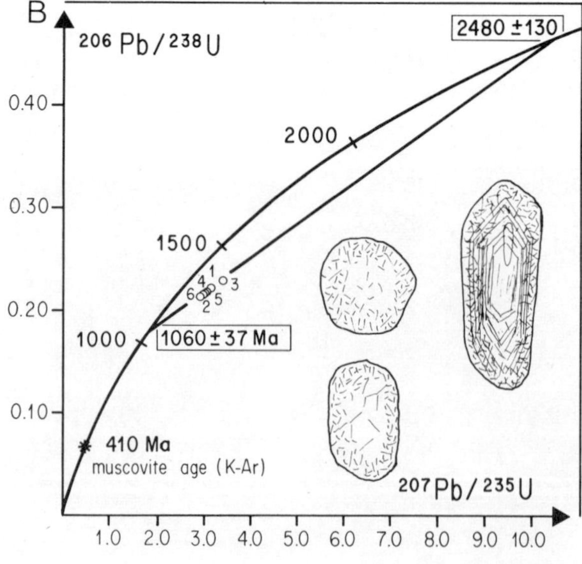

Figure 17.35 Concordia diagrams of detrital zircons from placer deposits. *A*—Alpefjord Formation (lower member, chlorite zone); *B*—garnet-sillimanite pelitic gneisses of Fig. 17.14 (after Peucat *et al.*, 1985).

Thus, pre-depositional ages are in the same range, within the error limits, as those reported from the inner Grenville Province of eastern Labrador, *c.* 1030 Ma old (Schärer *et al.*, 1986). The Alpefjord detrital zircons, however, differ in their initial Archaean age, whereas the Labrador zircons derive from protoliths of early Proterozoic age. The Alpefjord sediments thus originated from a recycled Archaean terrain that is not presently exposed, since reworked Archaean units of the frontal part of the belt suffered regional metamorphism at *c.* 970 Ma (Dallmeyer and Rivers, 1983; Schärer *et al.*, 1986).

It is thus likely that a persistent supply of unaltered clastics during deposition of the Alpefjord Formation was dependent on renewed uplift of the Grenville Province, which lasted until *c.* 850 Ma. We therefore propose to consider the Alpefjord Formation as a possible grey molasse (*s.l.*) to the Grenvillian Belt, with possible depositional ages in the range of 950–850 Ma.

Dating of clay-derived sediments of the Brogetdal Formation above the slaty-cleavage front gave Caledonian ages by both Rb–Sr and K–Ar methods (Bonhomme and Caby, in press).

17.6.1.1 *Stromatolites.* The stromatolites described and analysed by Bertrand-Sarfati and Caby (1976) are relatively abundant and diversified, but differ from those recorded in the late Riphean (Proterozoic III) of other areas (Bertrand-Sarfati, 1972): they are less abundant (with regard to the carbonate thickness) and morphologically more variable. They have been interpreted as indicating a Vendian age.

17.6.1.2 *Geochemistry.* Samples from the upper carbonate formations have been analysed by Schidlowski and the results published (Schidlowski *et al.*, 1975). These carbonates are 'heavy' with respect to the mean isotopic content in C and O of other Precambrian carbonates of the same age (Table 17.2). The implications for the depositional environment are not very significant, the deviation from normal ratios being very small.

17.6.1.3 *Acritarchs.* These have been studied by Vidal (1975, 1979) and organic matter is apparently not preserved in the lower Eleonore Bay, Alpefjord Formation as microfossils, except in the middle and upper members where *Trachysphaeridium levis* and *Chuaria circularis* are found. In the Aghardhsbjerg Formation, *Kildinosphaera* sp. and *Chuaria circularis* have been found in one bed only (bed 3). The Brogetdal Formation is still relatively poor in microfossils; it contains only *P. densicoronata*, *Trachisphaeridium levis*, and fragments of *Chuaria circularis*. The harvest is better in the Nøkkefossen Formation where all beds contain microfossils, 'the assemblage is characterized by the abundance of *Chuaria circularis*, *Leiosphaeridia asperata*, *Kildinosphaera chagrinata*. *K. verljanica* cf. *Stictosphaeridium* sp., *Synsphaeridium* sp., *Trachisphaeridium laminaritum*, *T. levis*, *Pterospermopsimorpha? densicoronata*.' This assemblage is interpreted as late Riphean in age by Vidal, in apparent

Table 17.2 $\delta^{13}C$ and $\delta^{18}O$ values of late Precambrian (assumed age 1.10 Ma) sedimentary carbonates from the Eleonore Bay Formation, East Greenland; the samples are from the upper part of the formation (Bed Group 8–18 of 'Multicoloured' and 'Limestone-Dolomite' Series; from Schidlowski *et al.*, 1975, p. 36).

No.	Description of sample	Total carbonate (%)	Dolomite (%) in total carbonate	$\delta^{13}C$ (‰ PDB)	$\delta^{18}O$ (‰ SMOW)
1.	Stromatolitic dolomite of Bed Group 18	98.5	> 99.0	+ 5.2	+ 16.4
2.	Dark bituminous dolomite (containing carbonaceous material) of Bed Group 16	98.0	> 99.0	+ 5.6	+ 20.4
3.	Dark bituminous magnesian limestone (siliceous) of Bed Group 16	83.0	10.0	+ 7.2	+ 20.9
4.	Dark carbonaceous dolomite associated with chert of Bed Group 15	97.0	> 99.0	+ 4.9	+ 16.7
5	Black carbonaceous limestone (siliceous) of Bed Group 14	79.0	< 1.0	+ 5.2	+ 17.6
6.	Grey dolomitic chert with stromatolites of Bed Group 13	40.0	97.0	+ 5.1	+ 21.7
7.	Dark stromatolitic dolomite of Bed Group 13	92.0	> 99.0	+ 3.7	+ 26.1
8.	Light-grey stromatolitic dolomite of Bed Group 12	73.0	> 99.0	+ 4.6	+ 22.6
9.	Grey dolomitic limestone of Bed Group 9	96.0	22.5	+ 5.4	+ 15.1
10.	Dark magnesian limestone (bituminous and oncolite-bearing) of Bed Group 9	94.5	7.5	+ 5.1	+ 19.3

disagreement with the indications given by the stromatolites. We consider that all microfossils of the assemblage cover a long time span from the late Riphean to the Vendian (Vidal and Knoll, 1983). The difference between these assemblages and those found in the Tillite Group (Vidal, 1979), emphasized by the presence of a weathering surface between the two groups, does not imply a major age difference in terms of palaeontology, and it is insufficient evidence to demonstrate that the upper part of the Eleonore Bay Group might be considered as latest Riphean or early Vendian.

17.6.2 *Comparisons with the Phanerozoic*

The palaeogeography of southern Western Europe during the period 310–200 Ma (from late Carboniferous to late Triassic) can be roughly compared with that during the late Proterozoic. During the late Carboniferous, 4 to 8 km of grey clastic deposits were laid down in continental subsiding areas during a rather wet period. Conditions were mostly fluviatile to lacustrine. Palaeosols with coal accumulations are abundant. We therefore suggest that the carbon-rich unoxidized mudcrack facies, so common in the lower member of the Alpefjord Formation, may represent the Proterozoic counterparts of Carboniferous palaeosols. Permian red beds and Lower Triassic quartzite deposits can be lithologically compared with the Agardhsbjerg and the Brogetdal Formations, whereas Middle to Upper Triassic carbonates can be compared with the Nøkkefossen Formation. This comparison may suggest a similar climatic change during deposition of the Eleonore Bay Group to that well-established in palaeo-Europe from the Carboniferous to the Permian, and which was related to complete peneplanation of the Variscan Belt.

17.7 Conclusions

Although major westward thrusting of the East Greenland Caledonides on to the Greenland Shield occurred, the steadily subsiding area where the Eleonore Bay Group was deposited can be considered as part of the same Greenland Shield.

Palaeoenvironmental interpretation suggests shallow, subaqueous non-marine conditions for the deposition of the Alpefjord Formation, under the influence of a huge fluviatile system draining from the west-southwest.

According to geochronological data, the source terrain proposed is the Grenvillian Belt. Thus, the Alpefjord Formation may be considered as a distal grey molasse (*s.l.*) of the Grenvillian Belt, with depositional ages similar to that of the uplift of the Grenville Province (950–850 Ma).

Deposition of the Agardhsbjerg Formation also took place in shallow-water conditions. The predominantly sandy material possibly originated from continental dunes accumulated on the continent.

The two carbonate formations (Brogetdal and Nøkkefossen) show an increase in the number and thickness of carbonate beds and the amount of desiccation features, and a relative decrease in the number of shale beds. The unique quartzitic episode is indicative of a sudden change in sediment supply, implying a new erosional phase on the continent. The two Formations are included in a prograding sedimentary system on a carbonate shelf: the Brogetdal major cycles being deposited in the shoal zone, and the Nøkkefossen cyclical sediments in the inner shelf.

The uppermost surface of the Nøkkefossen Formation is interpreted as a silcrete (a pedogenetic surface), and seems to indicate a period of weathering and pedogenetic evolution prior to the deposition of the Tillite Group. It thus represents an important diastem in the Eleonore Bay Group shelf evolution.

This thick sedimentary pile accumulated in a steadily subsiding area of the Greenland Shield. Persistence of shallow-water conditions until the Cambro-Ordovician indicates that subsidence exactly matched the sedimentary supply. Since no magmatic activity took place before the Caledonian events, it is assumed that deposition of this huge pile was possibly connected with a passive marginal facies of the Iapetus ocean.

References

Allen, P. A. and Collinson, J. D. (1986) Lakes. In Reading (ed.), H. G. *Sedimentary Environments and Facies*. Blackwell Scientific, Oxford, 63–94.

Asseretto, R. L. A. M. and Kendall, C. G. StC. (1977) Nature, origin and classification of peritidal tepee structures and related breccias. *Sedimentology* **24**, 153–210.

Backlund, H. G. (1930) Das alter des 'Metamorphen komplexes' von Franz Josef fjord in Ost-Grönland. *Meddlr Grønland*, 87, p. 119.

Bertrand-Sarfati, J. (1972) *Les Stromatolites du Précambrien Supérieur du Sahara Nord-occidental. Inventaire, Morphologie et Microstructure des Laminations, Corrélations Stratigraphiques.* Thèse d'Etat, Montpellier. *Edit. CNRS-CRZA* **14**, 245 pp.

Bertrand-Sarfati, J. and Caby, R. (1974) Précisions sur l'âge Précambrien terminal (Vendien) de la série carbonatée à stromatolites du Groupe d'Eleonore Bay (Groenland oriental). *C. R. Acad. Sci. Paris* **278**, 2267–2270.

Bertrand-Sarfati, J. and Caby, R. (1976) Carbonates et stromatolites du sommet du Groupe d'Eleonore Bay (Précambrien terminal) au Canning Land (Groenland oriental). *Grønlands Geolog. Unders. Bull.* **119**, 51.

Bonhomme, M. and Caby, R. (1987) Rb/Sr Caledonian ages of Upper Eleonore Bay Group and Cambrian metasediments in the East Greenland fold belt. *Grønlands Geol. Unders. Rapport* (in press).

Caby, R. (1972) Preliminary results of mapping in the Caledonian rocks of Canning Land and Wegener Halvø, East Greenland. *Grønlands Geol. Unders. Rapport* **48**, 21–38.

Caby, R. (1976*a*) Investigations on the lower Eleonore Bay Group in the Alpefjord region, central East Greenland. *Grønlands Geol. Unders. Rapport* **80**, 102–106.

Caby, R. (1976*b*) Tension structures related to gliding tectonics in the Caledonian structure of Canning Land and Wegener Halvø, central East Greenland. *Grønlands Geol. Unders. Rapport* **72**, 1–24.

Campbell, S. E. (1982) Precambrian endoliths discovered. *Nature* **299**, 429–431.

Cambpell, S. E., Bertrand-Sarfati, J. and Simone, L. (1982) Earliest endolithic borings in Late Precambrian silicified ooids and lowermost Cambrian ooids. *Geol. Soc. Am. Bull.* Abstract.

Collinson, J. D. (1986) Alluvial sediments. In Reading, H. G. (ed.), *Sedimentary Environments and Facies*. Blackwell Scientific, Oxford, 20–62.

Courel, L.; Laurent, P. and Vetter, P. (1984) Coal position in the coal-bearing sequences of the limnic basins: importance of the subsidence rate. *Proc. 27th Intern. Geol. Congr.* **4**, 87–98.

Dallmeyer, R. D. and Rivers, T. N. (1983) Recognition of excess ^{40}Ar through incremental-release ^{40}Ar–^{39}Ar analysis on biotite and hornblende across the Grenvillian metamorphic gradient in southwestern Labrador. *Geochim. Cosmochim. Acta* **47**, 413–428.

Donovan, R. N. and Foster, R. J. (1972) Subaqueous shrinkage cracks from the Caithness flagstone series (middle Devonian) of northeast Scotland. *J. sedim. Pet.* **42**, 309–317.

Eha, S. (1953) The pre-Devonian sediments on Ymersφ, Suess Land, and Ellaφ (East Greenland) and their tectonics. *Meddlr Grφnland* **111**, 2, 115 pp.

Enos, P. (1983) Shelf environment. In Scholle, P. A., Bebout, D. G. and Moore, C. H. (eds.), Carbonate Depositional Environments. *AAPG Mem.* **63**, 267–296.

Fairchild, I. J. (1980) Sedimentation and origin of a late Precambrian 'Dolomite' from Scotland. *J. sedim. Pet.* **50**, 423–446.

Fairchild, I. J. and Hambrey, M. J. (1984) The Vendian succession of Northeastern Spitzbergen: petrogenesis of a dolomite–tillite association. *Precambr. Res.* **26**, 111–167.

Fränkl, E. (1951) Die untere Eleonore Bay Formation im Alpefjord. *Meddlr Grφnland* **151**, 15.

Fränkl, E. (1953*a*) Geologische Untersuchungen in Ost-Andrées Land (NE-Grönland). *Meddlr Grφnland* **113**, 160.

Fränkl, E. (1953*b*) Die geologische Karte von Nord-Scoresby Land (NE-Grönland). *Meddlr Grφnland* **113**, 56 pp.

Friedmann, G. M. and Sanders, J. E. (1978) *Principles of Sedimentology*. Wiley, New York, 160 pp.

Grey, R. and Thorne, A. M. (1985) Biostratigraphic significance of stromatolites in upward shallowing sequences of the early Proterozoic Duck Creek dolomite, Western Australia. *Precambr. Res.* **29**, 183–206.

Grotzinger, J. P. (1986*a*) Cyclicity and paleoenvironmental dynamics of an Early Proterozoic passive-margin carbonate platform, Rocknest Formation (*c.* 1.9 Ga), Wopmay Orogen, N. W. T., Canada, *Geol. Soc. Am. Bull.* **97**, 1208–1231.

Grotzinger, J. P. (1986*b*) Evolution of early Proterozoic passive-margin carbonate platform, Rocknest Formation, Wopmay Orogen, N. W. T., Canada, *J. sedim. Pet.* **56**, 831–847.

Haller, J. (1953) Geologie und Petrographie von West-Andrées Land und Ost-Frænkels Land (NE-Grönland). *Meddlr Grφnland* **113**, 196 pp.

Haller, J. (1955) Der 'Zentrale Metamorphe Komplexe' von NE-Grönland. Teil I. Die geologische Karte von Suess Land, Glestscherland und Goodenoughs Land. *Meddlr Grφnland* **73**, 174 pp.

Haller, J. (1956) Geologie der Nunatakker Region von Zentral-Ostgrönland. *Meddlr Grφnland* **154**, 1721.

Haller, J. (1958) Der 'Zentrale Metamorphe Komplexe' von NE-Grönland. Teil II. Die geologische Karte der Staunings Alper und des Forsblads Fjordes. *Meddlr Grφnland* **154**, 27.

Haller, J. (1971) *Geology of East Greenland Caledonides*. Interscience, New York, 413 pp.

Hansen, B. J., Higgins, A. K. and Bar, M. T. (1978) Rb–Sr and U–Pb age patterns in polymetamorphic sediments from the southern part of the East Greenland Caledonides. *Bull. Geol. Soc. Denmark* **27**, 55–62.

Henriksen, N. and Higgins, A. K. (1976) East Greenland Caledonian fold belt. In Escher, A. and Watt, W. S. (eds.), *Geology of Greenland*. Grφnland Geol. Unders. 182–246.

Higgins, A. K. (1974) The Krummedal supracrustal sequence around inner Nordvestfjord, Scoresby Sund, East Greenland. *Grφnlands Geol. Unders. Rapport* **67**, 34.

Higgins, A. K. (1976) Pre-Caledonian metamorphic complexes within the southern part of the east Greenland Caledonides. *J. geol. Soc. London* **132**, 289–305.

Higgins, A. K. Friderichsen, J. D., Rex, D. C. and Gledhill, A. R. (1978) Early Proterozoic isotopic ages in the East Greenland Caledonides. *Contrib. Miner. Pet.* **67**, 87–94.

Horodyski, R. J. (1976) Stromatolites of the Upper Siyeh limestone (Middle Proterozoic), Belt supergroup, Glacier National Park, Montana. *Precambr. Res.* **3**, 51–536.

Horodyski, R. J. (1983) Sedimentary geology and stromatolites of the Middle Proterozoic Belt supergroup, Glacier National Park Montana. *Precambr. Res.* **20**, 391–425.

Jackson, M. P. (1985) Middle-Proterozoic dolomitic varves and microcycles from the McArthur basin, Northern Australia. *Sedim. Geol.* **44**, 301–326.

James, N. P. (1984) Shallowing upward sequences. In Walken, R. (ed.), Carbonate Facies Models. *Geoscience Can.* **2**, 213–228.

Katz, H. R. (1952*a*) Zur Geologie von Strindbergs Land (NE-Grönland). *Meddlr Grφnland* **111**, 150 pp.

Katz, H. R. (1952*b*) Ein Querschnitt durch die Nunatakzone Ostgrönlands. *Meddlr Grφnland* **144**, 65.

Katz, H. R. (1954) Einige Bemerkungen zur Lithologie und Stratigraphie der Tellitprofile im Gebiet des Kejser Franz Josephs Fjord Ostgrönland. *Meddlr Grφnland* **72**, 64.

Katz, H. R. (1961) Late Precambrian to Cambrian stratigraphy in East Greenland. In Raasch, G. O. (ed.), *Geology of Arctic*. Vol. 1, Toronto, 299–328.

Koch, L. (1929) The geology of East Greenland. *Meddlr Grφnland* **73**, 209 pp.

Monty, Cl. (1982) Cavity fissure dwelling stromatolites (endostromatolites) from Belgian Devonian mud-mounds (extended abstract). *Ann. Soc. géol. Belgique* **105**, 343–344.

Monty, Cl. (1984) Mud-mounds: geology and palaeoecology. In Geister, J. and Herb, R. (eds.), *Géologie et Paléoécologie des Récifs*. 23.1–23.8.

Monty, Cl. Bernet-Rollande, M. C. and Maurin, A. F. (1982) Reinterpretation of the Frasnian classical 'reefs' of the southern Ardennes, Belgium. *Ann. Soc. geol. Belgique* **105**, 339–341.

O'Connor, M. P. (1972) Classification and environmental interpretation of the cryptalgal organosedimentary 'molar tooth' structure from the late Precambrian Belt–Purcell supergroup. *J. Geol.* **80**, 592–610.

Peucat, J. J., Tisserant, D., Caby, R. and Clauer, N. (1985) Resistance of zircons to resetting in a prograde metamorphic sequence of Caledonian age, East Greenland. *Can. J. Earth Sci.* **22**, 330–338.

Reineck, H. E. and Singh, I. B. (1973). *Depositional Sedimentary Environments*. Springer–Verlag, New York.

Rex, D. C., Gledhill, A., Higgins, A. K. (1977) Precambrian Rb–Sr isochron ages from the crystalline complexes of inner Forsblads Fjord, East Greenland fold belt. *Grφnlands Geol. Unders. Rapport* **85**, 122–126.

Ross, G. M. and Chiarenzelli, J. R. (1985) Paleoclimatic significance of widespread Proterozoic silcretes in the Bear and Churchill provinces of the Northwestern Canadian Shield. *J. sedim. Pet.* **55**, 196–204.

Schidlowski, M., Eichmann, R. and Junge, C. (1975) Precambrian sedimentary carbonates: carbon and oxygen isotope geochemistry and implications for the terrestrial oxygen budget. *Precambr. Res.* **2**, 1–69.

Siedlecka, A. (1978) Late Precambrian tidal flat deposits and algal stromatolites in the Batsfjord formation, East Finnmark, North Norway. *Sedim. Geol.* **21**, 277–310.

Schärer, U., Krogh, T. E. and Gower, C. F. (1986) Age and evolution of the Grenville Province in eastern Labrador from U–Pb systematics in accessory minerals. *Contrib. Miner. Pet.* **94**, 438–451.

Shinn, E. A. (1983) Tidal flat. In Scholle, P. A., Bebout, D. G. and Moore, C. H. (eds.), Carbonate Depositional Environments, *AAPG Mem.* **63**, 17–210.

Smith, G. (1968) The origin and deformation of some 'molar-tooth' structures in the Precambrian Belt–Purcell Supergroup. *J. Geol.* **76**, 426–443.

Sommer, M. (1957*a*) Geologie von Lyells Land (NE-Grönland). *Meddlr Grønland* **155**, 157.

Sommer, M. (1957*b*) Geologische Untersuchungen in den präekambrischen Sedimenten zwischen Grandjeans Fjord und Bessels Fjord in NE-Grönland. *Meddlr Grønland* **160**, 56.

Steiger, R. H., Hansen, B. T., Schuler, Ch., Bar, M. T. and Henriksen, N. (1979) Polyorogenic nature of the southern Caledonian fold belt in East Greenland: an isotopic age study. *J. Geol.* **86**, 475–495.

Swett, K. and Knoll, A. H. (1985) Stromatolite bioherms and microphytolites from the Late Proterozoic Draken Conglomerate formation, Spitsbergen. *Precambr. Res.* **28**, 327–347.

Teichert, C. (1933) Untersuchungen zum Bau des kaledonischen Gebirges in Ostgrönland. *Meddlr Grønland* **95**, 121.

Vidal, G. (1975) Late Precambrian acritarchs from the Eleonore Bay Group and Tillite Group in East Greenland. *Grønland Geol. Unders. Rapport* **78**, 19.

Vidal, G. (1979) Acritarchs from the Upper Proterozoic and Lower Cambrian of East Greenland. *Grønlands Geol. Unders. Rapport* **134**, 40.

Vidal, G. and Knoll, A. H. (1983) Proterozoic Plankton. *Geol. Soc. Am. Mem.* **161**, 265–277.

Wegmann, C. E. (1935) Preliminary report on the Caledonian orogeny in Christian X's Land (NE-Greenland). *Meddlr Grønland* **103**, 59.

18
The 'Sparagmites' of Norway

J. P. NYSTUEN and ANNA SIEDLECKA

18.1 Introduction

The Upper Proterozoic sequences of late Riphaean to Vendian and early Cambrian age in southern and northern Norway and adjacent parts of Sweden (Fig. 18.1) have been studied since the early decades of last century. The successions of this age in Scandinavia became known as the 'sparagmites', after the term 'sparagmite' was introduced by Esmark (1829) for the dominant, feldspathic sandstone type of these sequences. At present the 'sparagmites' are divided into formal lithostratigraphic units, and the old term is only retained as a trivial designation, and in the name 'sparagmite region' for the outcrop area of these rocks in South Norway and neighbouring districts of Sweden (Fig. 18.1).

The Upper Riphaean to Lower Cambrian (*c.* 900–600 Ma) sequences occur

(i) Autochthonously, resting unconformably on the crystalline rocks of the Fennoscandian Shield

(ii) In nappes of the lower and middle allochthon of the Scandinavian Caledonides (Fig. 18.1). Strongly deformed and highly metamorphosed Upper Proterozoic to Lower Cambrian rocks of the higher nappes have an uncertain stratigraphy and will not be dealt with here. The rocks of the Barents Sea Region in East Finnmark (Fig. 18.1) form an allochthonous terrane which may be a north-western continuation of the Timanian fold belt of the northern Soviet Union.

Recent reviews of the stratigraphy and structural geology of these sequences in South Norway are given by Nystuen (1981, 1982, 1983), Bockelie and Nystuen (1985), Kumpulainen and Nystuen (1985) and Nickelsen *et al.* (1985), and in North Norway by Siedlecka (1975, 1985), Vidal (1981*a*), Vidal and Siedlecka (1983), Føyn (1985), Ramsay *et al.* (1985) and Roberts (1985). A very brief review of late Proterozoic sedimentation in Scandinavia is also given by Kumpulainen (1986). A summary correlation diagram for the Upper Proterozoic to Lower Cambrian Sequences in the North Atlantic region is given in Fig. 18.2.

18.2 South Norway

Upper Proterozoic to Lower Cambrian sequences with fairly well-known stratigraphy occur in three major nappe units: the Osen–Røa Nappe Complex, including the Synnfjell Nappe (lower allochthon), and the Valdres and Kvitvola Nappe Complexes (middle allochthon). The nappes were emplaced in late Silurian time and displaced a distance in the range of 130 to *c.* 400 km (Oftedahl, 1943; Gee, 1978; Hossack, 1978; Nystuen, 1981; Hossack *et al.*, 1985; Kumpulainen and Nystuen, 1985; Morley, 1986).

The Osen–Røa Nappe Complex and the Synnfjell Nappe contain the *Hedmark Group* (Table 18.1), the Valdres Nappe Complex includes the *Valdres Group* and the *Mellsenn Group* (Table 18.2) and the Kvitvola Nappe Complex includes the *Engerdalen Group* (Table 18.3). The Hedmark Group also occurs as a thin and partly discontinuous autochthonous sequence at the eroded nappe front and on basement rocks in tectonic windows (Fig. 18.1). The Hedmark Group (Bjørlykke *et al.*, 1967) includes the classical 'sparagmites' of Esmark (1829). The group comprises formations accumulated in (i) western and (ii) eastern depositional provinces separated by the Imsdalen Fault which is suggested to be of synsedimentary origin (Sæther and Nystuen 1981) (Table 18.1).

The *c.* 2500-m thick *Brøttum Formation* is thought to be the oldest unit and consists of grey turbiditic sandstones and intercalated black shales (Fig. 18.3). The *Elstad Formation* that locally underlies the Brøttum Formation in Gudbrandsdalen is of uncertain tectonic and stratigraphic position (Englund, 1973). Acritarchs of early Vendian age occurring in the upper part of the Brøttum Formation (Vidal, 1981*a*) show that a marine environment prevailed in the depositional basin. In the eastern province, the >2000-m thick *Storskarven Formation* is a pink and white fluvial sandstone laid down on a distal alluvial plain (Nystuen and Ilebekk, 1981). This formation may be a lateral equivalent to, or older than, the Brøttum Formation (Nystuen, in press). It is overlain by the *Rendalen Formation* which consists of coarse-grained, red arkosic sandstones and conglomerates. This major formation is preserved with depositional contacts on thrust granitic basement rocks. The *Litlesjøberget Conglomerate* occurs here as alluvial-fan bodies along basement escarpments. The Rendalen Formation, being 2000–3000 m thick, passes westwards into coarse-grained turbidites in the upper part of the Brøttum Formation, through a transitional zone of fan-delta facies (Sæther and Nystuen, 1981; Nystuen, 1982).

The *Atna Formation* in the eastern province is a transgressive shallow-marine quartzite (Nystuen, 1982). Locally this formation contains tholeiitic basalt flows or is overlain by thin basalt sheets, the *Svarttjørnkampen Basalt* (Sæther and Nystuen, 1981; Nystuen, 1982, Furnes *et al.*, 1983). Further transgression in the east and west is reflected by grey limestones, dolomitic limestones, siltstones and black shales of the *Biri Formation* (Fig. 18.4) which also contains Vendian microfossils (Vidal, 1981*a*). The first recorded microfossils from the late Proterozoic in South Norway were

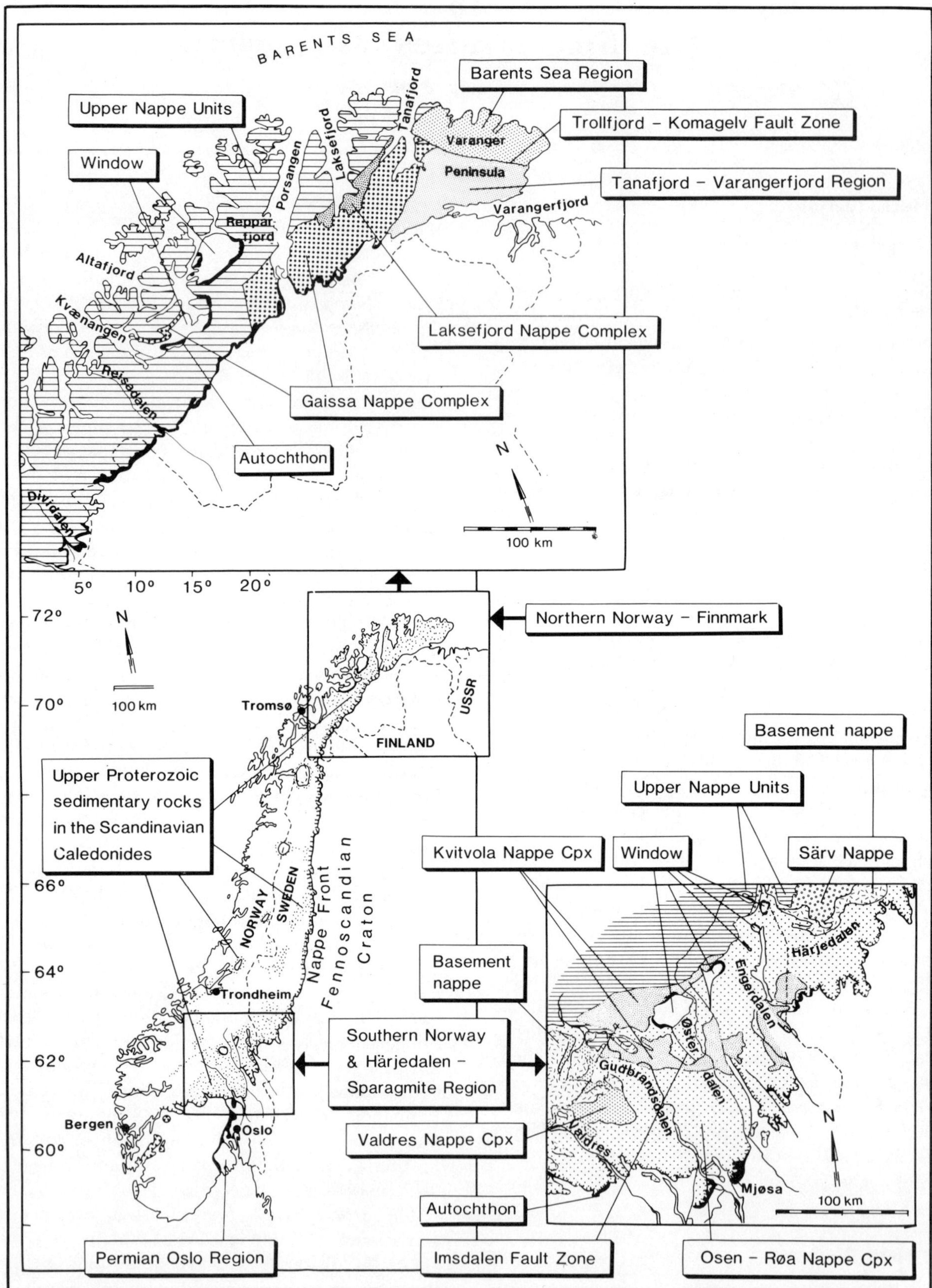

Figure 18.1 Key maps showing occurrence of Upper Proterozoic to Lower Cambrian sedimentary rocks in the Scandinavian Caledonides (central map), in the sparagmite region, South Norway and Härjedalen, Sweden (lower right map), and in Finnmark, North Norway (upper left map). The key maps are modified from Gee *et al.* (1985).

Table 18.1 Lithostratigraphy of the Hedmark Group, Osen-Røa Nappe Complex and autochthon, South Norway. After Englund (1972, 1973); Bjørlykke *et al.* (1976); Nystuen and Ilebekk (1981); Sæther and Nystuen (1981); Vidal (1981*a*); Nystuen (1982). A = acritarch, S = stromatolite, T = trace fossil.

Age	Group	Western depositional province: Formation thickness	Western: Lithology	Western: Depositional environment	Eastern depositional province: Formation thickness	Eastern: Lithology	Eastern: Depositional environment
Early Cambrian		Vangsås, (A, T) 50–200 m	Conglomerate, sandstone, quartzite	Fluvial, deltaic, shallow-marine	Vangsås 4–1000 m	Conglomerate, sandstone, quartzite	Subaqueous fan, fluvial deltaic, shallow-marine
Vendian	Hedmark Group 4000–5000 m	Ekre 5–50 m	Siltstone, shale	Marine	Ekre 5–200 m	Siltstone, shale	Marine (lacustrine)
		Moelv 1–10 m	Diamictite, mudstone	Glacial and glaciomarine	Moelv 5 (min.)–160 m	Diamictite, mudstone	Glacial and glacio-marine
		Ring (A) ≤ 300 m	Conglomerate	Submarine part of fan delta	Osdalen ≤ 300 m	Conglomerate	Alluvial fan
		Biri (A, S) 50–200 m	Shale, carbonate	Marine	Biri and BjøRånes (S)	Shale, carbonate, sandstone	Marine
		Biskopåsen (A) and Imsdalen ≤ 700 m	Conglomerate	Submarine part of fan delta	Svarttjørn–Kampen ≤ 15 m	Basalt	Volcanic
					Atna ≤ 100 m	Quartzite	Shallow-marine
Late Riphean		Brøttum (A) minimum of 2500 m	Sandstone, shale	Submarine fan	Rendalen and Litlesjøberget	Sandstone, conglomerate	Alluvial plain Alluvial fan
					Storskarven minimum of 2000 m	Sandstone	Alluvial plain

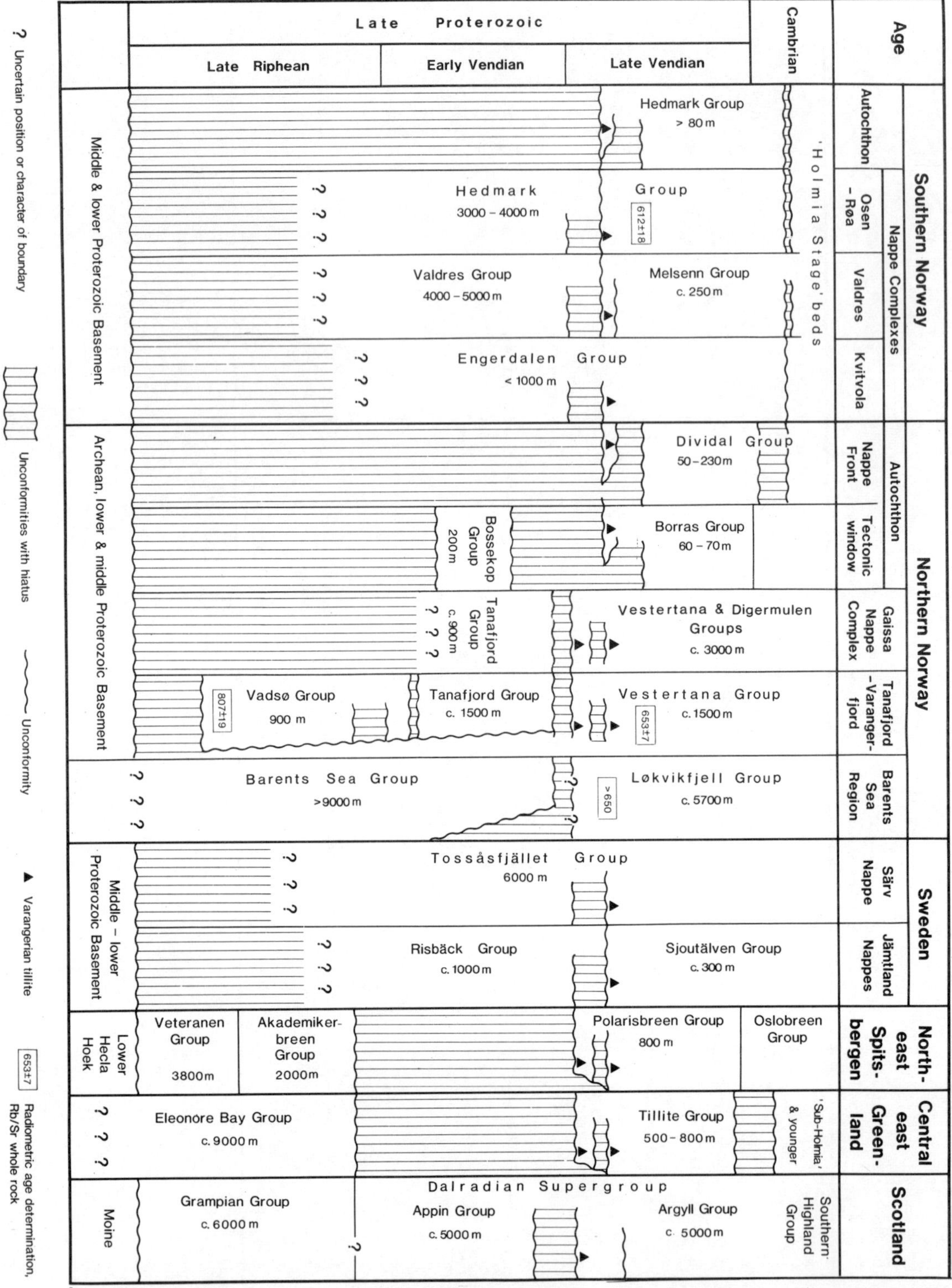

Figure 18.2 Correlation diagram for Upper Proterozoic to Lower Cambrian sequences in the North Atlantic region. Sources: southern and northern Norway: Tables 18.1–18.5 of present chapter and Vidal (1981*a*); Sweden: Gee *et al.* (1974), Kumpulainen (1980, 1982); NE Spitsbergen: Harland (1959, 1985), Knoll (1982*a*, *b*); central East Greenland: Henriksen and Higgins (1976), Henriksen (1985); Scotland: Harris and Pitcher (1975), Johnson (1983).

Table 18.2 Lithostratigraphy of the Valdres Group and late Vendian to Cambrian part of the Mellsenn Group, Valdres Nappe Complex, South Norway. After Nickelsen (1974); Nickelsen *et al.* (1985); Siedlecka *et al.* (1986).

Age	Group	Formation thickness	Lithology	Sedimentary environment
Cambrian	Mellsen Group *c.* 240 m	Informal unit 50–60 m	Slate	Shallow-marine
		Vangsås 90–200 m	Sandstone, slate, quartzite	Fluvial, deltaic, shallow-marine
Vendian		Ekre 5 m	Slate	Deltaic–marine (lacustrine?)
		Moelv 5 m	Tillite	Glacial
	Valdres Group 3000–5000 m	Olefjell	Arkose, conglomerate	Fluvial
		Bygdin	Quartzite-conglomerate	Fluvial
Late Riphean		Krusgrav	Arkose, conglomerate	Fluvial
		Ormtjernskampen *c.* 700 m	Conglomerate	Alluvial fan

Table 18.3 Lithostratigraphy of the Engerdalen Group, Kvitvola Nappe Complex, South Norway. After Nystuen (1980). S = stromatolite.

Age	Group	Formation thickness	Lithology	Sedimentary environment
Cambrian	Engerdalen Group 1000–1500 m			
Vendian		Høyberget 800–1000 m	Sandstone, calcareous sandstone, quartzite	Fluvial, shallow-marine
		Koppang 10 m	Diamictite	Glacial
		Hylleråsen (S) 15–50 m	Dolomite, magnesite phyllite	Shallow-marine, tidal
?		Fron	Sandstone, calcareous sandstone	Fluvial, shallow-marine

found in the Biri Formation, in part as clasts in the Biskopåsen Conglomerate (Manum, 1967; Spjeldnæs, 1967). Small columnar stromatolites and oncolites are also recorded from the Biri Formation.

Figure 18.3 Turbidites in the Brøttum Formation, Hedmark Group, younging to the right. The section exhibits a coarsening-upward unit of a submarine fan lobe. Lillehammer, South Norway.

The *Biskopåsen* and *Imsdalen Conglomerates* (Fig. 18.5) occur in several subaqueously formed fan bodies, deposited from pebble- to boulder-bearing gravity flows, wedging into Biri shales and Brøttum sandstones (Løberg, 1970; Englund, 1972; 1973; Bjørlykke *et al.*, 1976; Nystuen, 1982). The *Osdalen Conglomerate* originated as alluvial fans (Nystuen and Sæther, 1979; Nystuen, 1982). Erosion of the transgressive Biri carbonates took place during the deposition of these conglomerates.

The *Moelv Tillite* is a mojor marker bed within the Hedmark Group and was formed during the middle Vendian Varangerian glaciation. This is the oldest unit of the Hedmark Group which is preserved in the authochthonous sequence on the Fennoscandian basement (Nystuen, 1976 Bjørlykke and Nystuen, 1981; Siedlecka and Ilebekk, 1982). It rests with an erosional unconformity on several of the older formations. A complete sequence of the glacial unit consists of subglacially deposited tillite (diamictite), overlain by laminated mudstone with dropstones which grades upwards into the post-glacial *Ekre Shale*. This formation is

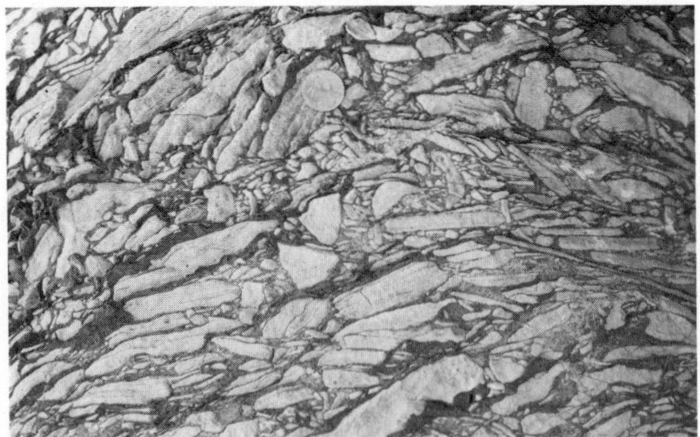

Figure 18.4 Intraformational limestone breccia from the Biri Formation, Hedmark Group. This breccia probably formed in a tidal channel. Jordet, South Norway.

Figure 18.5 Conglomerate and sandstone beds in the Biskopasen Conglomerate, Hedmark Group. The sandstone bed in the centre contains imbricated shale clasts. This section youngs to the right and is thought to be a subaqueous gravity flow facies. Biri, South Norway.

dominated by finely laminated or massive siltstone and represents deltaic to offshore marine environments.

The *Vangsås Formation* consists of the Vardal and Ringsaker Members. Deltaic, braided-fluvial and marginal-marine facies dominate the Vardal Member and shallow-marine orthoquartzites the Ringsaker Member. The latter unit contains, in its upper part, *Skolithos* and *Diplocraterion* trace fossils (Skjeseth, 1963). The Vangsås Formation onlaps the cratonic basement and is overlain by 'Holmia-stage' Lower Cambrian shales and sandstones that contain macroscopic body fossils. The Precambrian–Cambrian boundary is present within the Vangsås Formation (Vidal, 1981*b*, *c*; Bergström and Gee, 1985).

Schiøtz (1902) and several later authors (Nystuen, 1981; Bjørlykke, 1983; Bockelie and Nystuen, 1985) suggested that the Hedmark Group was deposited in a local 'sparagmite basin' within the outcrop area in the sparagmite region. The term *Hedmark Basin* was applied by Kumpulainen and Nystuen (1985) for the thrust palaeobasin.

18.2.1 *Valdres and Mellsenn Groups*

The Valdres Nappe Complex consists of Precambrian basement rocks overlain by the Upper Riphaean(?) –Vendian *Valdres Group* (Strand, 1964; Nickelsen, 1974), at least 4 km thick, which is succeeded by the uppermost Vendian (or early Cambrian–Middle Ordovician *Mellsenn Group* (Strand, 1938; Loeschke, 1967) (Table 18.2).

The basement rocks consist of high-grade metamorphic gneisses, gabbro and anorthosites of types similar to the rocks of the overlying Jotun Nappe. The sedimentary rocks were deposited on the basement, and clasts and mineral debris from the local basement rock types characterize the sandstones and conglomerates of the Valdres Group.

Loeschke and Nickelsen (1968) subdivided the sandstones and conglomerates of the Valdres Group into several formations of mixed-lithostratigraphic–petrographic character. The formations presented in Table 18.2 are all shown on the recent map of Sieldecka *et al.* (1986), and should be considered as formal units.

The *Ormtjernskampen Conglomerate*, up to 700 m thick, is a boulder-bearing alluvial-fan conglomerate that wedges into the braided-stream facies feldspathic sandstones of the Valdres Group (Nickelsen, 1974). Similar thick basin-margin conglomerates also occur elsewhere within the Valdres Group (Nickelsen *et al.*, 1985).

The *Krusgrav Formation* is a major thick sandstone unit with conglomerate beds which were deposited in proximal alluvial fans by braided streams. The *Bygdin Conglomerate* is a fluvial quartz and quartzite conglomerate, and the *Olefjell Formation* is a feldspathic sandstone, also of fluvial origin. A *dolomite unit* is also present locally above the Bygdin Conglomerate in the upper part of the Valdres Group (Hossack, 1976). This dolomite unit has no formal lithostratigraphic name, but may correspond with the Biri Formation in the Hedmark Group.

The Varangerian Moelv Tillite overlies, with an erosional unconformity, either the older part of the Valdres Group or the crystalline basement rocks included in the Valdres Nappe Complex (Nickelsen, 1974; Nickelsen *et al.*, 1985). The Ekre Shale forms the top of the Valdres Group.

The Mellsenn Group unconformably overlies the Valdres Group rocks. The *Vangsås Formation* is developed here as a lower arkosic or subarkosic member (Hornskjøkampen Member) and upper quartzite member with intercalated shales (Obleikhaugen Member). The overlying green slates and quartz arenites are probably of 'Holmia-stage' early Cambrian age (Nickelsen, 1974; Nickelsen *et al.*, 1985).

According to Heim (in Kumpulainen and Nystuen, 1985) dykes and sills of tholeiitic dolerite in the Valdres Nappe Complex were possibly intruded during the formation of the Valdres rift basin along marginal faults. Detritus in conglomerates and sandstones of mafic composition were suggested to have been derived from such eroded intrusive and extrusive rocks.

18.2.2 *Engerdalen Group*

The Kvitvola Nappe Complex may consist of the most far-travelled basinal sequences of the groups described here from South Norway. The thrust complex includes the Engerdalen Group (Nystuen, 1980) which mostly has a well-defined stratigraphy, with the exception of some lithostratigraphic units of uncertain stratigraphic position (Table 18.3). Precambrian crystalline basement rocks are also included in the nappe complex.

The lowest preserved unit in the Engerdalen Group is the *Fron Formation*, which consists of grey, micaceous feldspathic sandstones, calcareous sandstones and quartzites. The depositional environment probably included coastal plain to shallow-marine settings (Englund, 1973). The *Hylleråsen Formation* (Nystuen, 1980) has a wide lateral extent and consists of dolomite, magnesitic dolomite (Nystuen, 1969), black and grey phyllite and dolomitic arenites. Pseudomorphs after evaporite minerals and non-columnar stromatolites are also present. A shallow-marine to coastal sabkha environment has been suggested for this unit (Nystuen, 1980). An erosional unconformity separates the *Koppang Formation* from underlying carbonate rocks. The Koppang Formation is a diamictite of glacial origin that is correlated with the Varangerian tillites in Finnmark and the Moelv Tillite (Nystuen, 1980).

The upper part of the Engerdalen Group consists of the *Høyberget Formation* which includes fluvial feldspathic sandstones, marginal-marine or marine quartzites and shallow-marine calcareous sandstones (Nystuen, 1980). This is the laterally dominant formation in the Engerdalen Group. It is overlain by grey and greyish green phyllite of Cambrian to Ordovician age.

The *Rosten Formation* is a diamictitic to clast-supported conglomerate that occurs in the northwestern outcrop area of the Kvitvola Nappe Complex. Its stratigraphic position is uncertain, but it probably occurs above a carbonate equivalent to the Hylleråsen Formation and beneath a diamictite bed corresponding with the Koppang Formation. The conglomeratic unit, up to *c.* 300–400 m thick, is probably an alluvial-fan body.

The *Rondeslottet Formation* (Siedlecka *et al.*, 1986) is a 3000–4000-m thick unit of grey and greenish grey metasandstones of subarkosic to arkosic composition with intercalated thin conglomerate beds. The unit occurs in a separate thrust sheet in the northernmost part of the Kvitvola Nappe Complex, and its stratigraphic relationship to the Engerdalen Group is uncertain. The formation is suggested to be mainly of fluvial origin.

18.3 North Norway

Sedimentary rocks of late Proterozoic to early Cambrian age occur in an authochthonous position

(i) In the Tanafjord–Varangerfjord region
(ii) In a narrow belt at the Caledonian nappe front
(iii) In the tectonic windows in the Altafjord–Kvænangen area (Fig. 18.1). The bulk of the Gaissa Nappe Complex (lower allochthon) is also composed of unmetamorphosed to very low-grade metamorphic, Upper Proterozoic sedimentary sequences. The stratigraphy of the rocks in the Laksefjord Nappe Complex (middle allochthon) is uncertain; however, as a whole, the sequence is considered to be late Proterozoic in age. The Barents Sea region is entirely composed of Upper Proterozoic sedimentary to very low-grade metasedimentary formations.

The Tanafjord–Varangerfjord region contains the *Vadsø, Tanafjord* and *Vestertana Groups* with a total maximum thickness of *c.* 4000 m (Table 18.4). Complete sequences of the Tanafjord and Vestertana Groups also occur in the Gaissa Nappe Complex, whereas only partial autochthonous stratigraphic equivalents of these groups are present as the *Bossekop, Dividal* and *Borras Groups*. The Barents Sea region includes the *Barents Sea Group* and the *Løkvikfjell Group* (Table 18.5). The latter possibly has lithostratigraphic equivalents in the *Laksefjord Group* of the *Laksefjord Nappe Complex* (Fig. 18.1). Correlations between the groups mentioned here are shown in Fig. 18.2.

18.3.1 *Vadsø Group*

The Vadsø Group (Banks *et al.*, 1974) rests unconformably on Archaean gneisses. The group is subdivided into seven formations (Table 18.4), all exposed along the northern side of the Varangerfjorden. The group is dominated by sandstones.

The lower part of the *Veidnesbotn Formation* and the *Fugleberget, Paddeby* and *Golneselv Formations* consist of subarkosic–arkosic sandstones with abundant unidirectional cross-bedding. The sandstones are interpreted as fluvial (Banks *et al.*, 1974; Banks and Røe 1974). The upper part of the Veidnesbotn Formation consists of shallow-marine quartz-arenites, passing upwards into shales of the *Klubbnes Formation*, which is interpreted as a regressive, probably deltaic, sequence. The shales and mudstones of the *Andersby Formation* are either interpreted similarly to the Klubbnes Formation or as lacustrine, interfluvial deposits (Banks *et al.*, 1974). The *Ekkerøy Formation* is thought to represent a

Table 18.4 Lithostratigraphy of the Vadsø, Tanafjord and Vestertana Groups, Tanafjord–Varanger region and Gaissa Nappe Complex, Finnmark, North Norway. Modified from Johnson *et al.* (1978); Edwards (1984). A = acritarch, S = stromatolite, P = *Platysolenites* sp., T = trace fossil.

Age	Group	Formation thickness	Lithology	Sedimentary environment
Early Cambrian	Vestertana Group *c.* 1500 m	Breivik (A, P, T) 600 m	Silty sandstone, shale	Shallow-marine
- - - -		Stappogiedde (A, T) 505–545 m	Siltstone, sandstone	Shallow-marine and fluvial
		Mortensnes (A) 10–60 m	Diamictite, mudstone	Glacial and glacio-marine
		Nyborg (A, S) 200–400 m	Shale, siltstone, sandstone, dolomite	Marine
		Smalfjord 2–50 m	Diamictite, conglomerate, sandstone, siltstone	Glacial and glacio-fluvial
Vendian	Tanafjord Group *c.* 1500 m	Grasdal (A, S) 280 m	Dolomite, shale	Shallow-marine
		Hanglecærro 200 m	Sandstone	Shallow-marine
		Vagge 80 m	Shale	Shallow-marine
		Gamasfjell 280–300 m	Sandstone	Shallow-marine
		Dakkovarre (A) 275–350 m	Sandstone, silt-stone, shale	Shallow-marine
		Stangenes (A) 205–255 m	Shale, mudstone	Shallow-marine
		Grønnes 130–200 m	Conglomerate sandstone, mudstone	Shallow-marine
Late Riphaean	Vadsø Group ≤ 900 m	Ekkerøy (A) 15–190 m	Shale, sandstone	Shallow-marine to coastal plain
		Golneselv (A) 50–135 m	Sandstone	Fluvial
		Paddeby 25–120 m	Sandstone	Fluvial
		Andersby (A) 25–40 m	Shale, mudstone	Deltaic or lacustrine
		Fugleberget 125 m	Sandstone	Fluvial
		Klubbnes (A) 50 m	Shale	Deltaic
		Veidnesbotn 300 m	Sandstone	Shallow-marine fluvial

transgressive–regressive sequence. Marine shales in the lower part of the formation were deposited after a non-depositional or erosional hiatus, and are succeeded by regressive coastal-plain sandstones of the upper Ekkerøy Formation (Banks *et al.*, 1974; Røe, 1975; Johnson, 1975*a*, 1978).

18.3.2 *Tanafjord Group*

The Tanafjord Group (Siedlecka and Siedlecki, 1971) starts with the *Gønnes Formation*, a conglomerate–sandstone–mudstone unit, which marks a new transgression after the emergence event that occurred above the Ekkerøy Formation (Table 18.4). The sandstone facies predominates in the Grønnes Formation, which is interpreted in part as a shallow-marine and in part as a beach deposit. The *Stangenes Formation* is dominated by grey and variegated shale and mudstone which accumulated offshore as a blanket mud. Shallowing conditions are marked by a tidal sand complex, represented by quartzitic sandstones in the lower part of the *Dakkovarre Formation*. The remainder of this formation consists of interbedded quartzitic, in places ferruginous sandstone and arenaceous to muddy shale. The formation is thought to be of shallow-marine origin, characterized by regressive nearshore and transgressive offshore facies sequences (Siedlecka and Siedlecki, 1971; Johnson, 1975*a*, 1975*b*).

The overlying *Gamasfjell* and *Hanglecærro Formations* are homogeneous quartzitic sandstones, the former pink, the latter white, both interpreted as shallow-marine sheet sands. They are separated from each other by the sandy–silty shales of the *Vagge Formation* of offshore shallow-marine origin. The *Grasdal Formation*

Table 18.5 Lithostratigraphy of the Barents Sea Group and the Løkvikfjell Group, Barents Sea region, Finnmark, North Norway. After Siedlecka and Siedlecki (1967); Siedlecka (1978 and unpublished data); Siedlecki and Levell (1978); Siedlecka and Edwards (1980); Vidal (1981*a*). A = acritarch, S = stromatolite.

Age	Group	Formation thickness	Lithology	Sedimentary environment
Vendian	Løkvikfjell Group 5700–5800 m	Skidnefjell > 800 m	Sandstone	Shallow-marine
		Stordalselva 1200 m	Shale, sandstone	Shallow-marine
		Skjærgårdsnes 210 m	Sandstone	Shallow-marine
		Styret (A) 1500–1600 m	Sandstone, shale	Fluvial to shallow-marine
		Sandfjord 2000 m	Sandstone	Shallow-marine
	Barents Sea Group > 9000 m	Tyvjofjell 1500 m	Sandstone	Shallow-marine and fluvial
		Båtsfjord (A, S) 1400–1600 m	Sandstone, mudstone, dolomite, limestone	Shallow-marine and tidal
Late Riphaean		Båsnæring (A) 2500–3500 m	Shale, siltstone, sandstone	Prodelta and delta plain
		Kongsfjord (A)	Sandstone, shale	Submarine fan

which caps the Tanafjord Group is shale-dominated in the lower part, whereas the upper part consists of dolomite containing columnar stromatolites (Siedlecka and Siedlecki, 1969, 1971; Bertrand-Sarfati and Siedlecka, 1980; Siedlecki, 1980).

The Tanafjord Group can be mapped in the Gaissa Nappe Complex, with no major lithostratigraphic changes as far westward as the area south of Laksefjorden (Edwards *et al.*, 1973). In fact, according to a new tectonic interpretation, the Gaissa Nappe Complex continues to the eastern side of Tanafjorden, and the type section of the Tanafjord Group is thus located within this allochthonous nappe complex (e.g. Chapman *et al.*, 1985). The Tanafjord Group thins westwards, and in the Porsanger area there occur two major changes in the unit:

(i) The lower part of the group is thinner, and correlation with the formations defined in the Tanafjord–Varangerfjord region is not straightforward

(ii) The uppermost, dolomite-dominated part of the group: the *Porsanger Formation*, is thicker and much more extensive than its eastern equivalent: the upper part of the Grasdal Formation.

The Porsanger Formation contains various columnar stromatolites (Fig. 18.6) and evaporite mineral pseudomorphs (Bertrand-Sarfati and Siedlecka, 1980; Siedlecka, 1976; Tucker, 1976; 1977). The shaly *Stabbursdal Formation* in the Porsangen area is equivalent to the lower part of the Grasdal Formation. The three key formations in the Tanafjord Group, the Gamasfjell, Vagge and Hanglecærro Formations (Table 18.4) are recognized in the Porsanger area with no difficulty, so the former name *Porsangerfjord Group* has been abandoned in this area.

18.3.3 *Bossekop Group*

The Bossekop Group (Føyn 1964) rests unconformably on the Svecokarelian Raipas supracrustals in the Alta-Kvænangen and Repparfjord tectonic windows (see Figs. 18.1, 18.2). The group has been subdivided into three informal stratigraphic units south-east of the head of Altafjorden: (i) Quartzite (lowest, *c.* 140 m), (ii) Shale and thin-bedded sandstone (*c.* 35 m) and (iii) Quartzite (*c.* 10 m). These units have been correlated

Figure 18.6 Columnar stromatolites in the Porsanger Formation, North Norway. Coastal section, NW of Porsanger.

with the Gamasfjell, Vagge and Hanglecærro Formations of the Tanafjord Group (Zwaan and Gautier, 1980).

18.3.4 *Vestertana Group*

The Vestertana Group (Reading, 1965) rests with a slight angular unconformity on either the Vadsϕ or Tanafjord Group (Table 18.4). The group comprises five formations and is characterized by the presence of two tillite formations and of one major stratigraphic gap (Fϕyn, 1937; Banks *et al.*, 1971; Vidal, 1981*a*). The Precambrian-Cambrian boundary is located within the upper part of the Vestertana Group (Fϕyn, 1967; Banks, 1970; Vidal, 1981*a*).

The *Smalfjord Formation* at the base of the group includes (i) glacial diamictites, including the famous Bigganjarga tillite (Fig. 18.7), and (ii) glacio-fluvial sandstones and conglomerates (Bjϕrlykke, 1967; Banks *et al.*, 1974; Edwards, 1975; 1984; Fϕyn and Siedlecki, 1980). The *Nyborg Formation* rests with a slight unconformity on the Smalfjord Formation. It is an interglacial unit of grey, purple and variegated marine sandstone, shale and light-grey dolomite. The Nyborg Formation is unconformably overlain by the *Mortensnes Formation*. This latter unit consists primarily of glacial diamictite overlain by laminated mudstone carrying ice-dropped stones. The outcrops of these three formations in the Tanafjord–Varangerfjord region (Fig. 18.1) comprise the type area and type sections for the middle Vendian *Varanger Ice Age* (Kulling, 1942).

The Mortensnes Formation is succeeded by the post-glacial *Stappogiedde Formation*. Mudstone and feldspathic, conglomeratic sandstone in the lower part of the formation are of fluvial and deltaic origin (Edwards, 1984). Blue-green and purple laminated mudstone with trace fossils constitutes the bulk of the formation and has been interpreted as a shelf deposit (Banks, 1973*a*). The formation is capped by red or grey quartzitic sandstones and turbidites (Reading, 1965). The *Breivik Formation*, dominated by sandstones in the lower part and mudstones higher up, is of early Cambrian age (Fϕyn, 1967; Banks 1970; Vidal, 1981*a*).

Figure 18.7 Tillite of the Smalfjord Formation resting upon the Veidnesbton Formation (lower Vadsϕ Group). Coastal exposure at Bigganjarrga, Varangerfjorden, Finnmark.

18.3.5 *Borras Group*

The Borras Group (Fϕyn, 1964) rests with an unconformity either on the Bossekop Group or on the Svecokarelian basement. The lowermost *c*. 10 thick *Alta Tillite Formation* is unconformably overlain by a *c*. 50-m thick sequence starting with a conglomerate that is succeeded by quartzite units and blue-green and purple shale (Fϕyn, 1964; Zwaan and Gautier, 1980). The Borras Group can be correlated with the Dividal Group at the nappe front.

18.3.6 *Dividal Group*

The Dividal Group (Fϕyn, 1967) extends as a narrow belt along the eroded margin of the Caledonian nappes, and thins from 230 m in the north-east to *c*. 75–80 m in the Dividalen area in the south-east (Fig. 18.1). The group starts with a conglomerate that rests unconformably on the Svecokarelian basement rocks. This basal conglomerate, although in places thought to be alluvial, generally marks a widespread transgression of Vendian age (Fϕyn, 1967). Locally, discontinuous erosional remnants of tillite occur beneath the transgressive conglomerate and represent basal till deposits laid down on the metamorphic basement from a Varangerian ice sheet.

The middle and upper parts of the group consist of alternating sandstone and shale units which are laterally persistent. Blue-green and purple coloured shales are a characteristic feature. Trace and body fossils in the Dividal Group contain both elements of the Ediacara fauna and early Cambrian species, particularly of the genus *Platysolenites* (Hamar, 1967; Fϕyn and Glaessner, 1979; Banks, 1973*b*). Thus, the Precambrian–Cambrian boundary is thought to be present within the upper part of the group. It is uncertain whether the Dividal sequence is continuous in the north-east (south-east of Porsanger); further south-west a hiatus has been suggested at the Precambrian–Cambrian boundary, primarily on the basis of biostratigraphic evidence (Fϕyn and Glaessner, 1979).

The Dividal Group continues along the Caledonian nappe front into Finland and northern Sweden, and links up with equivalent autochthonous sequences further south in Central Sweden and South Norway (e.g. Willdén, 1980; Nystuen, 1982; Thelander, 1982; Wallin, 1982; Lehtovaara, 1986) (Fig. 18.1).

18.3.7 *Barents Sea Group*

The Barents Sea Group (Siedlecka and Siedlecki, 1967, 1969) (Table 18.5) starts with the *c*. 2500-m thick *Kongsfjord Formation*, which consists of turbidites with subordinate channel-fill, slump and pelagic deposits (Fig. 18.8). The formation has been interpreted as a flysch-like sequence accumulated as a submarine fan (Siedlecka, 1972; Pickering, 1981, 1985). The Kongsfjord Formation is followed by the *Båsnæring Formation*, a succession of greenish grey or red sandstones, mudstones

Figure 18.8 Turbidites of the Kongsfjord Formation, Barents Sea Group. The striated sole marks in the upper half of the picture are groove casts at the base of a thick turbiditic sandstone. Veineset in Kongsfjorden, NE coast of the Varanger Peninsula, Finnmark.

Figure 18.9 Stromatolites in the lower Båtsfjord Formation of the Barents Sea Group. Syltefjorden, NE coast of the Varanger Peninsula, Finnmark.

and shales more than 2500 m thick. The lithological transition between these two formations was interpreted as a gradual shallowing sequence from the submarine fan to a prodelta environment; it has been suggested that the Båsnæring Formation as a whole was a major prograding deltaic deposit (Siedlecka and Edwards, 1980; Pickering, 1982).

The next unit, the *c.* 1500 m thick *Båtsfjord Formation*, consists in its lower part of grey sandstone, mudstone and claystone interbedded with yellowish grey dolomite and grey limestone. The limestone contains non-columnar stromatolites in places (Fig. 18.9). This part of the formation has been interpreted as a tidal-flat deposit (Siedlecka, 1978). Acritarch assemblages found in these beds indicate that the Riphean–Vendian boundary occurs within the lower part of the Båtsfjord Formation (Vidal and Siedlecka, 1983). The upper part of the Båtsfjord Formation is multicoloured, and consists of grey and pink sandstone, purple mudstone and shale and subordinate yellowish grey dolomite, and is believed to have been accumulated in a shallow-marine, mostly coastal area.

The shallow-marine sedimentation with fluvial incursions continued throughout the *Tyvjofjell Formation*. This unit is about 1500 m thick and is dominated by pinkish grey sandstone interbedded with purple mudstone. The latter facies is mostly present in the lower part of the formation.

18.3.8 *Løkvikfjell Group*

The Løkvikfjell (Siedlecki and Levell, 1978) rests with a slight angular unconformity on the Barents Sea Group (Table 18.5). A transgressive conglomerate occurs at the bottom of the *Sandfjord Formation* which consists of coarse, pink feldspathic sandstone and granule conglomerate, interpreted as high-energy shallow-marine sands (Levell, 1980). The Sandfjord Formation grades upwards into the *Styret Formation* which consists of grey feldspathic sandstone interbedded with siltstone. This formation is interpreted as a progradational fluvial sequence. The two formations which follow—the *Skjærgården Formation* and the *Stordalselva Formation*—both consist of interbedded grey sandstone, siltstone and mudstone, interpreted as shallow-marine sediments. The sandstones and siltstones in the Stordalselva Formation are arranged in tens to hundreds of metres thick mappable units. The Løkvikfjell Group is capped by the *Skidnefjell Formation*, a unit of coarse-grained, white to buff sandstone with subordinate conglomerate, remarkably similar to the Sandfjord Formation (Siedlecki and Levell, 1978).

18.3.9 *Laksefjord Group*

The stratigraphic position of the Laksefjord Group, which occurs in the Laksefjord Nappe Complex, is uncertain. The group consists, in stratigraphic order, of the *Ifjord, Landersfjord and Friarfjord Formations* (Føyn, 1960). The Ifjord Formation consists of a diamictite which has been interpreted as a glacial deposit and correlated with the Vendian Tillites in the Tanafjord–Varangerfjord region (Laird, 1972*a*). However, the glacial origin of the Ifjord diamictite has been questioned. A correlation between the Laksefjord and Løkvikfjell Groups has been proposed by several authors (e.g. Føyn, 1969; Laird, 1927*b*), but as yet there is no conclusive evidence of the age of the Laksefjord Group.

18.4 Palaeogeographical environment and sedimentary evolution

The Upper Proterozoic to lowermost Cambrian sequences now present in southern and northern Norway originally formed a sedimentary wedge on the western and northern margins of the Fennoscandian craton. The sequences formed in four major environments which are, from the sea towards the craton:

(i) Shelf–slope and deep-marine base-of-slope
(ii) Shelf to coastal plain
(iii) Intracratonic rift basin
(iv) Coastal plain to epicontinental shallow-sea.

The lowermost part of the Barents Sea Group, with the turbiditic Kongsfjord Formation, represents a shelf–slope to base-of-slope environment of a very thick sequence that was built out into a north-west–south-east-running seaway, bordering the Fennoscandian craton towards the north-east. Progradation of the shelf sequence gave rise to shallowing and deposition in deltaic, shore and coastal plain areas for the middle and upper parts of the Barents Sea Group. Tilting and erosion resulted in removal of parts of the Barents Sea Group prior to accumulation of the predominantly shallow-marine Løkvikfjell Group. Siedlecka (1975, 1985) proposed that the basin in which the Barents Sea Group and the Løkvikfjell Group accumulated formed a part of the Timanian aulacogen. Kjøde *et al.* (1978) have inferred that this basin was located adjacent to the East Greenland part of the Laurentian continent.

The Vadsø, Tanafjord and Vestertana Groups also represent shelf to shallow-marine, shore and coastal plain settings. The structural position of the basin was on a slightly subsiding north-eastern margin of the Fennoscandian Shield (Fig. 18.2). Repeated stratigraphic breaks in these sequences reflect relative sea-level changes and slight tectonic tilting of the cratonic basement (Siedlecka, 1975, 1985).

The depositional basin of the Engerdalen Group in South Norway can be compared with that of the Vadsø and Tanafjord Groups. The Engerdalen Basin was formed in a shelf setting, where shallow-marine and coastal plain environments alternated as a result of variations in the sea-level. The basin reveals a trend from dominantly marine environments in the lower part of the sequence towards dominantly coastal plain or alluvial plain settings in the late Vendian upper part of the succession (Kumpulainen and Nystuen, 1985).

Intracratonic rift basins are represented by the Hedmark and Valdres Groups in South Norway. Both basins probably consisted of one major elongated graben and several subsidiary fault basins formed by structural basin expansion (Kumpulainen and Nystuen, 1985; Nystuen, in press). The Hedmark Basin was open to the sea in the north-west like an aulacogen (Nystuen, 1982). A marine incursion in the Valdres Basin is thought to be marked by a thin dolomite unit of an age probably corresponding with the Vendian Biri Formation in the Hedmark Group.

In middle Vendian to early Cambrian times the marginal cratonic areas of the Fennoscandian Shield had been effectively denuded and Varangerian glacial till and glacio-fluvial sediments were deposited and preserved in shallow depressions and palaeovalleys (e.g. Føyn and Siedlecki, 1980; Nystuen, 1985). Post-Varangerian fluvial, deltaic and shallow-marine sedimentation took place during a slow subsidence of the cratonic marginal areas, and gave rise to the onlapping sequences in the Vestertana, Borras and Dividal Groups in North Norway and in the upper part of the Hedmark Group in South Norway. An epicontinental sea transgressed over the peneplaned Fennoscandian Shield during Cambrian times, probably in response to a tectono-eustatic sea-level rise due to a high spreading rate in the new-born Iapetus ocean.

18.5 Correlation

The upper Proterozoic sequences of Norway are, in Fig. 18.2, correlated with successions of corresponding age in the Swedish Caledonides, in Ny Friesland of north-eastern Spitsbergen, in central East Greenland and in Scotland.

The *Tossåsfjället Group* (Kumpulainen, 1980) in the Särv Nappe of the middle allochthon in the Caledonides of Sweden has been tectonically displaced from a structural position beyond and west of that of the Engerdalen Group in the Kvitvola Nappe Complex (Gee, 1978; Kumpulainen and Nystuen, 1985). Fluvial and shallow-marine sandstones predominated, and the palaeobasin represented a coastal to shelf area. The Tossåfjället Group is penetrated by the Ottfjället dolerite dyke swarm, that may have intruded during the stage of crustal rupture which led to the opening of the Iapetus ocean (Gee, 1975).

The *Risbäck Group* in the lower allochthon of the central Swedish Caledonides is probably derived from a fault-bounded basin dominated by coarse-clastic alluvial facies, similar to the eastern sedimentary province of the Hedmark Group. The *Sjoutälven Group* reveals a lithostratigraphy very similar to the Varangerian and post-Varangerian part of the Hedmark Group (Gee *et al.*, 1974; Kumpulainen, 1982). This succession represents a segment of the middle Vendian–early Cambrian onlap sequence of the western region of the Fennoscandian craton.

The middle Vendian tillites from the Varanger Ice Age can be correlated with sequences in Scandinavia, Spitsbergen, Greenland and Scotland (Spencer, 1975; Hambrey, 1983). The shelf to shelf–slope sequences in Finnmark exhibit considerable similarities with the successions of Spitsbergen and central East Greenland, as regards thickness and facies associations (Vidal and Siedlecka, 1983; Siedlecka, 1985). On the other hand, the Upper Riphean to Vendian sequences in southern Norway do not resemble the successions of corresponding age either in Scotland or in East Greenland (Kumpulainen and Nystuen, 1985).

However, in all these areas, tillite formations are

underlain by dolomitic carbonates of shallow-marine to tidal flat origin. In the Scandinavian sequences, these dolomites appear to be of early Vendian age, according to acritarch biostratigraphy (Vidal, 1981*a*), whereas pre-Varangerian dolomites in north-east Spitsbergen and Nordaustlandet in Svalbard are suggested to be of late Riphaean age (Knoll, 1982*a*, *b*), and likewise the dolomites beneath the Tillite Group in the upper part of the Eleonore Bay Group in East Greenland (Vidal, 1979). The lack of biostratigraphic correspondence between the tops of these dolomite formations may result from variable amounts of glacial erosion, or regional facies changes and possibly also lack of deposition.

determinations suggest that the deposition of the Upper Proterozoic sequences of Norway commenced in the late Riphaean, more than 800 Ma ago (Vidal, 1985). The oldest units are represented by the lower part of the Barents Sea Group, which may be about 900 Ma old, or older. The Barents Sea Group comprises, by its age, biostratigraphy, thickness and lithofacies, a major link between the Russian Timanian palaeo-seaway in the east and the pre-Iapetus seaways that must have existed in the North Atlantic region farther west (Siedlecka, 1975).

Acknowledgements

We thank Liv Ravdal and Gudbrand Framgården for drawing the figures, and Jill Renee Mφrk and Fride W. Bråten for typing the manuscript.

References

Banks, N. L. (1970) Trace fossils from the late Precambrian and Lower Cambrian of Finnmark, Norway. In Crimes, T. P. and Harper, J. C. (eds.), Trace Fossils. Geol. J. Spec. Issue **3**, 19–34.

Banks, N. L. (1973*a*) Innerelv Member: Late Precambrian marine shelf deposit, East Finnmark. *Norges Geol. Unders.* **288**, 7–25.

Banks, N. L. (1973*b*) Trace fossils in the Halkkavarre section of the Dividal Group (? late Precambrian–Lower Cambrian), Finnmark. *Norges Geol. Unders.* **288**, 1–16.

Banks, N. L., Edwards, M. B., Geddes, W. P., Hobday, D. K. and Reading, H. G. (1971) Late Precambrian and Cambro-Ordovican sedimentation in East Finnmark. *Norges Geol. Unders.* **269**, 197–236.

Banks, N. L. and Rφe, S.-L. (1974) Sedimentology of the Late Precambrian Golneselv Formation, Varangerfjorden, Finnmark. *Norges Geol. Unders.* **303**, 17–38.

Banks, N. L., Hobday, D. K. Reading, H. G. and Taylor, P. N. (1974) Stratigraphy of the Late Precambrian 'Older Sandstone Series' of the Varangerfjord area, Finnmark. *Norges Geol. Unders.* **303**, 1–15.

Bergström, J. and Gee, D. G. (1985) The Cambrian in Scandinavia. In Gee, D. G. and Sturt, B. A. (eds.), *The Caledonide Orogen—Scandinavia and Related Areas. Part 1.* Wiley, Chichester, 247–271.

Bertrand-Sarfati, J. and Siedlecka, A. (1980) Columnar stromatolites of the terminal Precambrian Porsanger Dolomite and Grasdal Formation of Finnmark, North Norway. *Norsk Geol. Tidsskr.* **60**, 1–27.

Bjφrlykke, K. (1967) The Eocambrian 'Reusch Moraine' at Bigganjargga and the geology around Varangerfjorden, Northern Norway. *Norges Geol. Unders.* **251**, 18–44.

Bjφrlykke, K. (1983) Subsidence and tectonics in Late Precambrian and Palaeozoic sedimentation basins of southern Norway. *Norges Geol. Unders.* **380**, 159–172.

Bjφrlykke, K. and Nystuen, J. P. (1981) Late Precambrian tillites of South Norway. In Hambrey, M. J. and Harland, W. B. (eds.), *Earth's Pre-Pleistocene Glacial Record.* Cambridge University Press, 624–628.

Bjφrlykke, K., Englund, J.-O. and Kirkhusmo, L. (1967) Latest Precambrian and Eocambrian stratigraphy of Norway. *Norges Geol. Unders.* **251**, 5–17.

Bjφrlykke, K., Elvsborg, A. and Hφy, T. (1976) Late Precambrian sedimentation in the central sparagmite basin of South Norway. *Norsk Geol. Tidsskr.* **56**, 233–290.

Bockelie, J. F. and Nystuen, J. P. (1985) The southeastern part of the Scandinavian Caledonides. In Gee, D. G. and Sturt, B. A. (eds.), *The Caledonide Orogen—Scandinavia and Related Areas. Part 1.* Wiley, Chichester, 69–88.

Chapman, T. J., Gayer, R. A. and Williams, G. D. (1985) Structural cross-sections through the Finnmark Caledonides and timing of the Finnmarkian event. In Gee, D. G. and Sturt, B. A. (eds.), *The Caledonide Orogen—Scandinavia and Related Areas. Part 1.* Wiley, Chichester, 593–609.

Edwards, M. B. (1975) Glacial retreat sedimentation in the Smalfjord Formation, Late Precambrian, North Norway. *Sedimentology* **22**, 75–94.

Edwards, M. B. (1984) Sedimentology of the Upper Proterozoic glacial record, Vestertana Group, Finnmark, North Norway. *Norges Geol. Unders. Bull.* **394**, 1–76.

Edwards, M. B., Bayliss, P., Gibling, M., Goffe, W., Potter, M. and Suthern, R. J. (1973) Stratigraphy of the 'Older Sandstone Series' (Tanafjord Group) and Vestertana Group north of Stallogaissa, Laksefjord district, Finnmark. *Norges Geol. Unders.* **294**, 25–41.

Englund, J.-O. (1972) Sedimentological and structural investigations of the Hedmark Group in the Tretten-Øyer–Fåberg district, Gudbrandsdalen. *Norges Geol. Unders.* **276**, 1–59.

Englund, J.-O. (1973) Stratigraphy and structure of the Ringebu–Vinstra district, Gudbrandsdalen; with a short analysis of the western part of the sparagmite region in southern Norway. *Norges Geol. Unders.* **293**, 1–58.

Esmark, J. (1829) *Reise fra Christiania til Throndhjem.* Christiania, Oslo, 1–81, 1 map.

Fφyn, S. (1937) The Eocambrian series of the Tana district, northern Norway. *Norsk Geol. Tidsskr.* **17**, 65–164.

Fφyn, S. (1960) Tanafjord to Laksefjord. In Holtedahl, O., Fφyn, S. and Reitan, P. H. (eds.), Guide to Excursion No. A3, Intern. Geol. Congr. XXI, Norden, 1960. *Norges Geol. Unders.* **212**, 45–54.

Fφyn, S. (1964) Den tillittfφrende formasjonsgruppe i Alta—enjevnfφring med Øst-Finnmark og med indre Finnmark. *Norges Geol. Unders.* **228**, 139–150.

Fφyn, S. (1967) Dividal-gruppen ('*Hyolithus*-sonen') i Finnmark og dens forhold til de eokambrisk–kambriske formasjoner. [Summary: the Dividal Group ('the *Hyolithus* zone') in Finnmark and its relations to the Eocambrian–Cambrian formations.] *Norges Geol. Unders.* **249**, 1–84.

Fφyn, S. (1969) Laksefjord-gruppen ved Tanafjorden. *Norges Geol. Unders.* **258**, 5–16.

Fφyn, S. (1985) The Late Precambrian in northern Scandinavia. In Gee, D. G. and Sturt, B. A. (eds.), *The Caledonide Orogen—Scandinavia and Related Areas. Part 1.* Wiley, Chichester, 233–245.

Fφyn, S. and Glaessner, M. F. (1979) *Platysolenites*, other animal fossils, and the Precambrian–Cambrian transition in Norway. *Norsk Geol. Tidsskr.* **59**, 25–46.

Fφyn, S. and Siedlecki, S. (1980) Glacial stadials and interstadials of the Late Precambrian Smalfjord Tillite on Laksefjordvidda, Finnmark, North Norway. *Norges Geol. Unders.* **358**, 31–45.

Furnes, H., Nystuen, J. P., Brunfelt, A. O. and Solheim, S. (1983) Geochemistry of Upper Riphean–Vendian basalts associated with the 'sparagmites' of southern Norway. *Geol. Mag.* **120**, 349–361.

Gee, D. G. (1975) A tectonic model for the central part of the Scandinavian Caledonides. *Am. J. Sci.* **275-A**, 468–515.

Gee, D. G. (1978) Nappe displacements in the Scandinavian Caledonides. *Tectonophysics* **47**, 393–419.

Gee, D. G., Karis, L., Kumpulainen, R. and Thelander, T. (1974) A summary of Caledonian front stratigraphy, northern Jämtland/southern Västerbotten, central Swedish Caledonides. *Geol. Fören. Stockh. Förh.* **96**, 389–397.

Gee, D. G., Kumpulainen, R., Roberts, D., Stephens, M. B., Thon, A. and Zachrisson, E. (1985) Scandinavian Caledonides—tectonostratigraphic map, scale 1:2 mill. In Gee, D. G. and Sturt, B. A. (eds.), *The Caledonide Orogen—Scandinavia and Related Areas.* Wiley, Chichester (IGCP Project no. 27).

Hamar, G. (1967) *Platysolenites antiquissimus* Eichw. (Vermes) from the Lower Cambrian of northern Norway. *Norges Geol. Unders.* **249II**, 89–95.

Hambrey, M. J. (1983) Correlation of Late Proterozoic tillites in the North Atlantic region and Europe. *Geol. Mag.* **120**, 209–320.

Harland, W. B. (1959) The Caledonian sequence in Nŷ Friesland, Spitsbergen. *Q. J. geol. Soc. London* **114**, 307–342.

Harland, W. B. (1985) Caledonide Svalbard. In Gee, D. G. and Sturt, B. A. (eds.), *The Caledonide Orogen—Scandinavia and Related Areas. Part 2.* Wiley, Chichester, 999–1016.

Harris, A. L. and Pitcher, W. S. (1975) The Dalradian Supergroup. In Harris, A. L. *et al.* (eds.), A correlation of Precambrian rocks in the British Isles. *Geol. Soc. London Spec. Rep.* **6**, 52–75.

Henriksen, N. (1985) The Caledonides of central East Greenland 70°–76°N. In Gee, D. G. and Sturt, B. A. (eds.), *The Caledonide Orogen—Scandinavia and Related Areas. Part 2.* Wiley, Chichester, 1095–1113.

Henriksen, N. and Higgins, A. K. (1976) East Greenland Caledonian fold belt. In Escher, A. and Watt, W. S. (eds.), *Geology of Greenland*, Geological Survey of Greenland, Copenhagen, 182–246.

Hossack, J. (1976) Geology and structure of the Beito window, Valdres. *Norges Geol. Unders.* **327**, 1–33.

Hossack, J. R. (1978) The correlation of stratigraphic sections for tectonic finite strain in the Bygdin area, Norway. *J. geol. Soc. London* **135**, 229–241.

Hossack, J. R., Garton, M. R. and Nickelsen, R. P. (1985) The geological section from the foreland up to the Jotun thrust sheet in the Valdres area, south Norway. In Gee, D. G. and Sturt, B. A. (eds.), *The Caledonide Orogen—Scandinavia and Related Areas. Part 1.* Wiley, Chichester, 443–456.

Johnson, H. D. (1975*a*) Sedimentological Studies in Late Precambrian Upper Vadsφ Group and Lower Tanafjord Group of East Finnmark, North Norway. Unpublished D. Phil. Thesis, University of Oxford.

Johnson, H. D. (1975*b*) Tide- and wave-dominated inshore and shoreline sequences from the late Precambrian, Finnmark, North Norway. *Sedimentology* **22**, 45–74.

Johnson, H. D. (1978) Facies distributions and lithostratigraphic correlation in the late Precambrian Ekkerφy Formation, East Finnmark, Norway. *Norsk Geol. Tidsskr.* **58**, 175–190.

Johnson, H. D., Levell, B. K. and Siedlecki, S. (1978) Late Precambrian sedimentary rocks in East Finnmark, North Norway, and their relationship to the Trollfjord–Komagelv Fault. *J. geol. Soc. London* **135**, 517–533.

Johnson, M. R. W. (1983) Dalradian. In Craig, G. Y. (ed.), *Geology of Scotland.* Scottish Academic Press, Edinburgh, 77–104.

Kjφde, J., Storetvedt, K. M., Roberts, D. and Gidskehaug, A. (1978) Palaeomagnetic evidence for large-scale dextral movement along the Trollfjord–Komagelv Fault, Finnmark, North Norway. *Phys. Earth Planet Inter.* **161**, 132–144.

Knoll, A. H. (1982*a*) Microfossil-based biostratigraphy of the Precambrian Hecla Hoek Sequence, Nordaustlandet, Svalbard. *Geol. Mag.* **119**, 269–279.

Knoll, A. H. (1982*b*) Microfossils from the Late Precambrian Draken Conglomerate, Ny Friesland, Svalbard. *J. Palaeont.* **56**, 755–790.

Kulling, O., (1942) Grunddragen av fjällkedjerandens bergbyggnad inom Västerbottens län. *Sver. Geol. Unders.* **C445**, 1–319.

Kumpulainen, R. (1980) Upper Proterozoic stratigraphy and depositional environments of the Tossåsfjället Group, Särv Nappe, southern Swedish Caledonides. *Geol. Fören. Stockholm Förh.* **102**, 531–550.

Kumpulainen, R. (1982) The Upper Proterozoic Risbäck Group, northern Västerbotten, central Swedish Caledonides. *Univ. Uppsala, Dept. Miner. Pet. Res. Rep.* **28**, 1–60.

Kumpulainen, R. (1986) Outlines of the Late Proterozoic sedimentation in western and northern Baltoscandia. *Geol. Fören. Stockholm Förh.* **108**, 289–291.

Kumpulainen, R. and Nystuen, J. P. (1985) Late Proterozoic basin evolution and sedimentation in the westernmost part of Baltoscandia. In Gee, D. G. and Sturt, B. A. (eds.), *The Caledonide Orogen—Scandinavia and Related Areas. Part 1.* Wiley, Chichester, 213–232.

Laird, M. G. (1972*a*) The stratigraphy and sedimentology of the Laksefjord Group, Finnmark. *Norges Geol. Unders.* **278**, 13–40.

Laird, M. G. (1972*b*) Sedimentation of the ?Late Precambrian Raggo Group, Varanger Peninsula. *Norges Geol. Unders.* **278**, 1–11.

Lehtovaara, J. J. (1986). Tectonostratigraphical outline of the Finnish Caledonides. *Geol. Fören. Stockholm Förh.* **108**, 291–294.

Levell, B. K. (1980) A late Precambrian tidal shelf deposit, the lower Sandfjord Formation, Finnmark, North Norway. *Sedimentology* **27**, 539–557.

Lφberg, B. (1970) Investigations at the south-western border of the sparagmite basin (Gausdal Vestfjell and Fåberg Vestfjell), southern Norway. *Norges Geol. Unders.* **266**, 160–205.

Loeschke, J. (1967) Zur Stratigraphic and Petrographie des Valdres-Sparagmites und der Mellsenn-Gruppe bei Mellane/Valdres (Süd-Norwegen). *Norges Geol. Unders.* **243**, 5–66.

Loeschke, J. and Nickelsen, R. P. (1968) On the age and tectonic position of the Valdres Sparagmite in Slidre (Southern Norway). *Neues Jb. Geol. Paläont. Abh.* **131**, 337–367.

Manum, S. (1967) Microfossils from Late Precambrian sediments around Lake Mjφsa, southern Norway. *Norges Geol. Unders.* **251**, 45–52.

Morley, C. K. (1986) The Caledonian thrust front and palinspastic restorations in the southern Norwegian Caledonides. *J. struct. Geol.* **8**, 753–765.

Nickelsen, R. P. (1974) Geology of the Rφssjφkollan–Dokkvatn area, Oppland. *Norges Geol. Unders.* **314**, 53–100.

Nickelsen, R. P., Hossack, J. R., Garton, M. and with a note on conodont identification by Repetsky, J. (1985) Late Precambrian to Ordovician stratigraphy and correlation in the Valdres and Synnfjell thrust sheets of the Valdres area, southern Norwegian Caledonides; with some comments on sedimentation. In Gee, D. G. and Sturt, B. A. (eds.), *The Caledonide Orogen—Scandinavia and Related Areas. Part 1.* Wiley Chichester, 369–378.

Nystuen, J. P. (1969) On the paragenesis of chert and carbonate minerals in chert-bearing magnesitic dolomite from the Kvitvola Nappe, southern Norway. *Norges Geol. Unders.* **258**, 66–78.

Nystuen, J. P. (1976) Facies and sedimentation of the Late Precambrian Moelv Tillite in the eastern part of the Sparagmite Region, southern Norway. *Norges Geol. Unders.* **329**, 1–70.

Nystuen, J. P. (1980) Stratigraphy of the Upper Proterozoic Engerdalen Group, Kvitvola Nappe, southeastern Scandinavian Caledonides. *Geol. Fören Stockholm Förh.* **102**, 551–560.

Nystuen, J. P. (1981) The Late Precambrian 'sparagmites' of southern Norway: a major Caledonian allochthon—the Osen–Røa Nappe Complex. *Am. J. Sci.* **281**, 69–94.

Nystuen, J. P. (1982) Late Proterozoic basin evolution on the Baltoscandian craton: the Hedmark Group, southern Norway. *Norges Geol. Unders.* **375**, 1–74.

Nystuen, J. P. (1983) Nappe and thrust structures in the Sparagmite Region, southern Norway. *Norges Geol. Unders.* **380**, 67–83.

Nystuen, J. P. (1985) Facies and preservation of glaciogenic sequences from the Varanger Ice Age in Scandinavia and other parts of the North Atlantic Region. *Palaegeogr., Palaeoclim., Palaeoecol.* **51**, 209–229.

Nystuen, J. P. (1987) Synthesis of the tectonic and sedimentological evolution of the late Proterozoic–early Cambrian Hedmark Basin, the Caledonian thrust belt, southern Norway. *Norsk Geol. Tidsskr.* (in press).

Nystuen, J. P. and Ilebekk, S. (1981) Stratigraphy and Caledonian structures in the area between the Atnsjøen and Spekedalen windows, Sparagmite Region, southern Norway. *Norsk Geol. Tidsskr.* **61**, 17–24.

Nystuen, J. P. and Sæther, T. (1979) Clast studies in the Late Precambrian Moelv Tillite and Osdal Conglomerate, Sparagmite Region, South Norway. *Norsk Geol. Tidsskr.* **59**, 239–254.

Oftedahl, C. (1943) Overskyvninger i den norske fjellkjede. *Naturen (Oslo)* **5**, 143–150.

Pickering, K. T. (1981) The Kongsfjord Formation—a Late Precambrian submarine fan in north-east Finnmark, North Norway. *Norges Geol. Unders.* **367**, 77–104.

Pickering, K. T. (1982) A Precambrian upper basin-slope and prodelta in northeast Finnmark, North Norway—a fossil ancient upper continental slope. *J. sed. Pet.* **52**, 171–186.

Pickering, K. T. (1985) Kongsfjord turbidite system. In Bouma, A. H., Normark, W. R. and Barnes, N. E. (eds.), *Submarine Fans and Related Turbidite Systems.* Springer-Verlag, New York, 237–244.

Ramsay, D. M., Sturt, B. A., Zwaan, K. B. and Roberts, D. (1985) Caledonides of northern Norway. In Gee, D. G. and Sturt, B. A. (Eds.), *The Caledonide Orogen—Scandinavia and Related Areas. Part 1.* Wiley, Chichester, 163–184.

Reading, H. G. (1965) Eocambrian and lower Palaeozoic geology of the Digermul Peninsula, Tanafjord, Finnmark. *Norges Geol. Unders.* **234**, 167–191.

Roberts, D. (1985) The Caledonian fold belt in Finnmark: a synopsis. *Norges Geol. Unders.* **403**, 161–177.

Røe, S. L. (1975) Correlation between the Late Precambrian Older Sandstone Series of the Varangerfjord and Tanafjord areas. *Norges Geol. Unders.* **266**, 230–245.

Sæther, T. and Nystuen, J. P. (1981) Tectonic framework, stratigraphy, sedimentation and volcanism of the Late Precambrian Hedmark Group, Østerdalen, South Norway. *Norsk Geol. Tidsskr.* **61**, 193–211.

Schiøtz, O. E. (1902) Den sydøstlige del af Sparagmit-Kvarts-Fjeldet i Norge. *Norges Geol. Unders.* **35**, 1–131.

Siedlecka, A. (1972) Kongsfjord Formation—a Late Precambrian flysch sequence from the Varanger Peninsula, Finnmark. *Norges Geol. Unders.* **278**, 41–80.

Siedlecka, A. (1975) Late Precambrian stratigraphy and structure of the north-eastern margin of the Fennoscandian shield (East Finnmark—Timan Region). *Norges Geol. Unders.* **316**, 313–348.

Siedlecka, A. (1976) Silicified Precambrian evaporite nodules from Northern Norway: a preliminary report. *Sedim. Geol.* **16**, 161–175.

Siedlecka, A. (1978) Late Precambrian tidal-flat deposits and algal stromatolites in the Båtsfjord Formation, East Finnmark, North Norway. *Sedim. Geol.* **21**, 177–310.

Siedlecka, A. (1985) Development of the Upper Proterozoic basins of the Varanger Peninsula, East Finnmark, North Norway. *Geol. Surv. Finland Bull.* **331**, 175–185.

Siedlecka, A. and Edwards, M. B. (1980) Lithostratigraphy and sedimentation of the Riphean Båsnæring Formation, Varanger Peninsula, North Norway. *Norges Geol. Unders.* **355**, 27–47.

Siedlecka, A. and Ilebekk, S. (1982) Forekomster av tillitt på nordsiden av Atnsjøen-vinduet, Syd-Norge. *Norges Geol. Unders.* **373**, 33–37.

Siedlecka, A. and Siedlecki, S. (1967) Some new aspects of the geology of Varanger Peninsula (Northern Norway) *Norges Geol. Unders.* **247**, 288–306

Siedlecka, A. and Siedlecki, S. (1969) Some new geological observations from the inner part of Varanger Peninsula, northern Norway. *Program og Resumeer af Foredag, IX Nordiske Geologiske Vintermøde, Lyngby* (Denmark), 54–55.

Siedlecka, A. and Siedlecki, S. (1971) Late Precambrian sedimentary rocks of the Tanafjord–Varangerfjord region of Varanger Peninsula, northern Norway. *Norges Geol. Unders.* **269**, 246–294.

Siedlecka, A., Nystuen, J. P., Englund, J.-O. and Hossack, J. R. (1986) Lillehammer—berggrunnskart M. 1:250 000 (geological map). *Norges Geol. Unders.*

Siedlecki, S. (1980) Geologisk kart over Norge, berggrunnskart Vadsø—M. 1:250 000, *Norges Geol. Unders.*

Siedlecki, S. and Levell, B. K. (1978) Lithostratigraphy of the Late Precambrian Løkvikfjell Group on Varanger Peninsula, East Finnmark, North Norway. *Norges Geol. Unders.* **343**, 73–85.

Skjeseth, S. (1963) Contributions to the geology of the Mjøsa districts and the classical sparagmite area in southern Norway. *Norges Geol. Unders.* **220**, 1–126.

Spencer, A. M. (1975) Late Precambrian glaciation in the North Atlantic region. In Wright, A. E. and Mosely, F. (eds.), Ice Ages: Ancient and Modern. *Geol. J. Spec. Issue* **6**, 217–240.

Spjeldnæs, N. (1967) Fossils from pebbles in the Biskopåsen Formation in southern Norway. *Norges Geol. Unders.* **251**, 53–82.

Strand, T. (1938) Nordre Etnedal. Beskrivelse til det geologiske gradteigskartet. *Norges Geol. Unders.* **152**, 1–71.

Strand, T. (1964) Otta-dekket og Valdres-gruppen i strøkene langs Bøverdalen og Leirdalen. *Norges Geol. Unders.* **228**, 280–288.

Thelander, T. (1982) The Torneträsk Formation of the Dividal Group, northern Swedish Caledonides. *Sver. Geol. Unders.* **C789**, 1–49.

Tucker, M. E. (1976) Replaced evaporites from the Late Precambrian of Finnmark, Arctic Norway, *Sedim. Geol.* **16**, 193–204.

Tucker, M. E. (1977) Stromatolite biostromes and associated facies in the Late Precambrian Porsanger Dolomite Formation of Finnmark, Arctic Norway. *Palaeogeogr., Palaeoclimatol., Palaeoecol.* **21**, 55–83.

Vidal, G. (1979) Acritarchs from the Upper Proterozoic and Lower Cambrian of East Greenland. *Grønlands Geol. Unders. Bull.* **134**, 1–40.

Vidal, G. (1981*a*) Micropalaeontology and biostratigraphy of the Upper Proterozoic and Lower Cambrian sequence in East Finnmark, northern Norway. *Norges Geol. Unders.* **362**, 1–53.

Vidal, G. (1981*b*) Micropalaeontology and biostratigraphy of the Lower Cambrian sequence in Scandinavia. In Taylor, M. E. (eds.), Short Papers for the Second International

J

Symposium on the Cambrian System, 1981. US *Geol. Surv. Open-File Rep.* **81–743**, 232–235.

Vidal, G. (1981*c*) Lower Cambrian acritarch stratigraphy in Scandinavia. *Geol. Fören. Stockholm Förh.* **103**, 183–192.

Vidal, G. (1985) Biostratigraphic correlation of the Upper Proterozoic and Lower Cambrian of the Fennoscandian Shield and the Caledonides of East Greenland and Svalbard. In Gee, D. G. and Sturt, B. A. (eds.), *The Caledonide Orogen—Scandinavia and Related Areas. Part* 1. Wiley, Chichester, 331–338.

Vidal, G. and Siedlecka, A. (1983) Planktonic, acid-resistant microfossils from the Upper Proterozoic strata of the Barents Sea Region of Varanger Peninsula, East Finnmark, Northern Norway. *Norges Geol. Unders.* **382**, 45–79.

Wallin, B. (1982) Sedimentology of the Lower Cambrian sequence at Vassbo, Sweden. *Stockholm Contrib. Geol.* **39**, 1–111

Willden, M. Y. (1980) Paleoenvironment of the autochthonous sedimentary rock sequence at Laisvall, Swedish Caledonides. *Stockholm Contrib Geol.* **33**, 1–100.

Zwaan, K. B. and Gautter, A. M. (1980) Alta and Gargia. Description of the geological maps (AMS-M 711) 1834 I and 1934 IV—M:50 000. *Norges Geol. Unders.* **357**, 1–47.

19
Later Proterozoic environments and tectonic evolution in the northern Atlantic lands

J. A. WINCHESTER

19.1 Introduction

Enough information is now available for an attempt to reconstruct the original depositional environments and their tectonic controls in the Northern Atlantic lands during the later Proterozoic. In recent years several maps showing the postulated relative positions of Laurentia, Greenland and Scandinavia have been produced, much of the supporting evidence being based upon palaeomagnetic data (Patchett and Bylund, 1977; Patchett *et al.*, 1978; Piper, 1980, 1982). Using both this information, and much new evidence first described in this volume, maps have been drawn detailing the later Proterozoic evolution of this area. The first map shows the positions of the 'basement' blocks around 1150 Ma BP, at the time of the deposition of the Moine Assemblage and before the main Grenville–Sveconorwegian tectonometamorphic event. On this map the areas of basement are separated into five categories:

1. Archaean cratons, essentially unmodified during the Proterozoic.
2. Slightly modified Archaean rocks, with some Proterzoic cover, as represented by the Amer Lake Zone (Van Schmus *et al.*, 1987).
3. Archaean rocks intensely reworked in Early Proterzoic mobile belts, possibly with early Proterozoic supracrustal rocks infolded.
4. Early Proterozoic mobile belts with no Archaean rocks, probably representing Proterozoic continental accretion.
5. Major Mid-Proterozoic granitoid belts.

Because little relative motion between the main continental blocks appears to have occurred during the mid-Proterozoic, when they all seem to have formed part of the 'Proterozoic Supercontinent' (Piper, 1982), the main features formed in the basement before 1600 Ma still seem to match in a reconstruction of their relative positions around 1150 Ma BP.

However, any attempt to match the continental blocks during the Proterozoic must take account of the later phanerozoic movements, in particular recent plate tectonic movements and the major crustal shortening and strike slip movements which accompanied the Lower Palaeozoic Caledonide movements. Accordingly, the outlines of the main continental blocks have been modified, and these maps are thus an attempt to piece together palinspastic reconstructions.

19.2 Correlating Canada and West Greenland

A correlation between NW Greenland and the Arctic Islands of Canada (van Schmus *et al.*, 1987) shows both the Rinkian and Nagssugtoqidian Belts of West Greenland continuing west into the Churchill Province of Canada as part of the Trans-Hudson Orogen. The Trans-Hudson Orogen is devided by them into different zones, and the Rinkian Belt (Escher and Pulvertaft, 1976) is equated with the Foxe Zone of Central Baffin Island and the Cree Lake Zone SW of Hudson Bay, all areas characterized by early Proterozoic reworking of earlier rocks. The full breadth of the Rinkian Belt in West Greenland is not well known, hence lines indicating its northern limits remain speculative. To the south, the Nagssugtoqidian Belt is equated with the Baffin and Labrador Zones (van Schmus *et al.*, 1987), and possibly with the Reindeer Zone collage of Saskatchewan, although in the latter area evidence of 'new' Proterozoic crustal material is present. By contrast, in the Nagssugtoqidian Belt extensive reworking of Archaean gneisses has occurred. Van Schmus *et al.* (1987) also acknowledge that extensive reworked Archaean basement may also be present in the Baffin and Labrador Zones. In Labrador, links with west Greenland are now clearly established (Collerson, 1982), and the Archaean Nain Province is accepted as a detached part of the North Atlantic Craton, which is mostly exposed in South Greenland. In the extreme S of Greenland the Ketilidian Belt, which apparently represents Proterozoic continental accretion, links with the Makkovik Province of Labrador which in turn may connect S of the Superior Craton with the contemporary Penokean Zone rocks S of the Great Lakes. Later overprinting during the Grenville event has made the latter link less clear. Finally, in the far N of Canada, the Amer Lake Zone of the Churchill Province, consisting of mildly reworked Archaean rocks, which is exposed in NW Baffin Island, may extend E to form most of northern Greenland, although this area is still very poorly understood. This correlation may prove to be less speculative than those linking lands across the Atlantic because the relative displacement of Greenland and North America is well established and no Phanerozoic

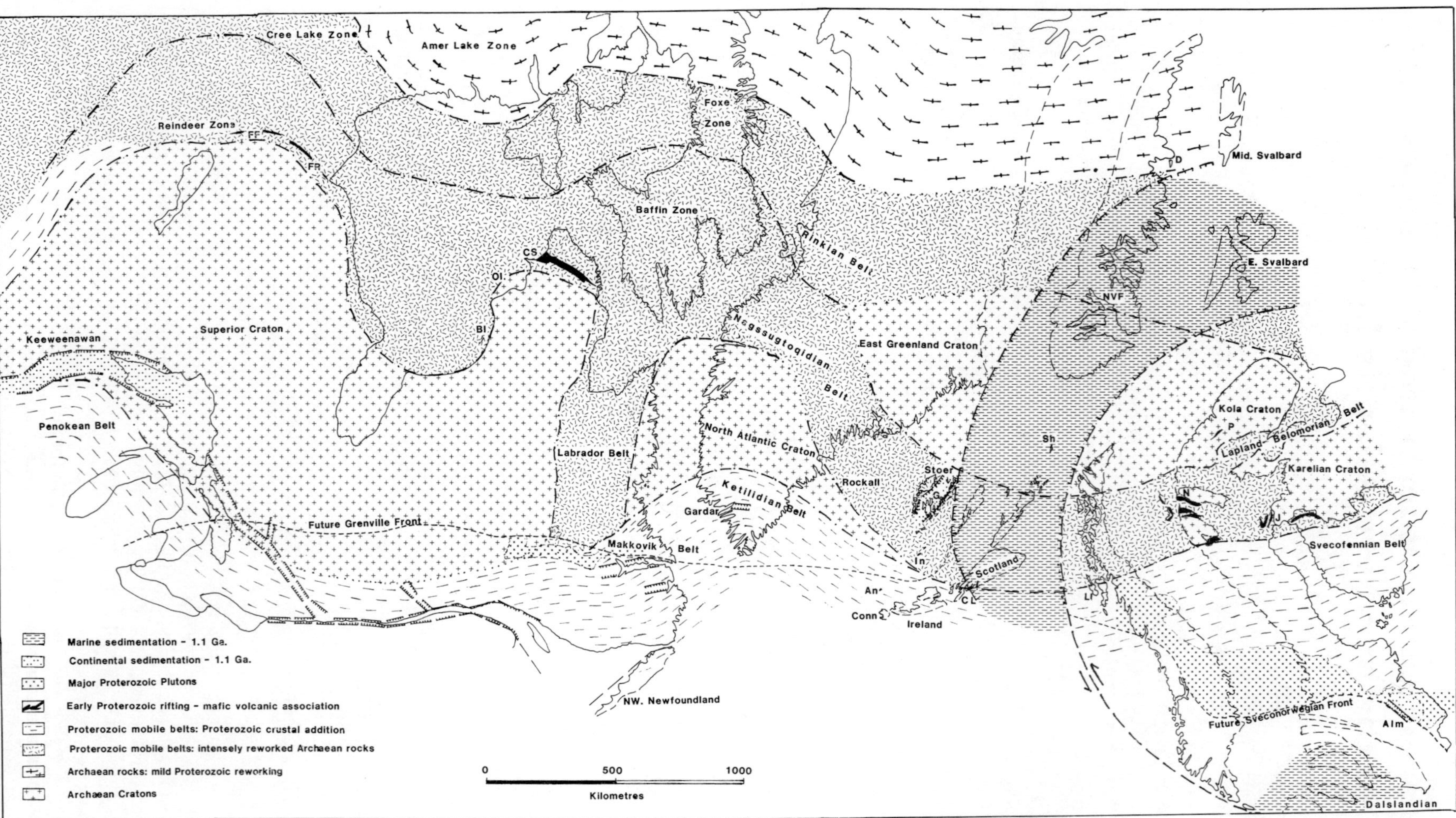

Figure 19.1 A simplified palinspastic reconstruction of the Laurentian–Scandinavian portion of the Proterozoic Super-continent at approximately 1150 Ma BP. Early Proterozoic mobile belts and Archaean cratons are distinguished, together with the location of intracratonic rifts and basins developing at the time. Abbreviations: *Alm*, Almesåkra; *An*, Annagh Division; *BI*, Belcher Islands; *CL*, Cruachan Lineament; *Conn*, Connemara; *CS*, Cape Smith; *FF*, Flin Flon; *FR*, Fox River; *G*, Gairloch; *In*, Inishtrahull; *J*, Jormua metavolcanics; *LI*, Lofoten Islands; *N*, Nussir metavolcanics; *NVF*, Nordvestfjord; *P*, Pechenga metavolcanics; *Sh*, Shetland, *OI*, Ottawa Islands.

movements have produced extensive crustal shortening or strike-slip movements to complicate the pattern.

19.3 Correlating West and East Greenland

The extent of the Greenland icecap is so large that correlation between the W and E coasts of the island is not always clear. In the S of the island the broad links are plain: the Archaean North Atlantic Craton is well-exposed on both coasts and the Nagssugtoqidian Belt is likewise recognized in both W and E Greenland. Further N, however, correlation is less clear, partly because the nature and age of the Precambrian basement is somewhat obscured on the E coast by the Caledonian fold belt and by later deposits, and partly because of the breadth of the ice-cap. However, enough is seen to show that the match between W and E coasts becomes less simple further N. On the W coast the Rinkian Belt, which is distinguished from the contemporaneous Nagssugtoqidian Belt by its very different tectonic style (Escher and Pulvertaft, 1976), is recognized flanking the Nagssugtoqidian Belt to the N. On the E coast north of Angmagssalik, however, the Nagssugtoqidian Belt is bounded to the NE by essentially unmodified Archaean rocks, and similar Archaean ages have been obtained from basement windows within the Caledonides of East Greenland from various localities south of Nordvestfjord (Higgins and Phillips, 1979; Rex and Gledhill, 1974; Rex *et al.*, 1977). From the scanty evidence available, therefore, it seems that an Archaean craton is present on the E coast of Greenland which is not seen in the W. North of Nordvestfjord, however, early Proterozoic ages have been obtained from basement windows and Archaean ages are absent in a zone at least 200 km wide (Rex and Gledhill, 1981). Although no clear correlation has been made, perhaps this area of basement, reworked in early Proterozoic time, represents the Rinkian Belt on the E Coast. Further to the N, Archaean ages have been obtained from banded gneisses at Danmarkshavn (Higgins and Phillips, 1979) and these could represent the northern limit of possible Rinkian reworking. Perhaps these Archaean ages, taken together with the early Proterozoic ages obtained from the basement of NE Greenland (Jepsen and Kalsbeek, 1985) form part of a belt which could be interpreted as an eastward extension of the Amer Lake Zone of mildly reworked Archaean (Fig. 19.1). Clearly such correlations must remain highly speculative until far more isotopic dating has been done.

19.4 Correlation between SE Greenland and the NW British Isles

Previous reconstructions of the relative positions of the main cratonic blocks during the middle Proterozoic (Patchett *et al.*, 1978; Piper, 1982) ignore the presence of a substantial area of ancient continental material beneath Scotland, NW Ireland and the Rockall micro-continent. Perhaps it was omitted because it was considered too small to include. However, during the Caledonide event considerable crustal shortening occurred in this area, and if a palinspastic reconstruction is made, showing its pre-Caledonide form, it emerges as a substantial crustal relic occupying an area of perhaps 500 000 km^2. To accommodate this block, the position of the Baltic Shield relative to Laurentia must be relocated 'eastward' to the limit of the arrow indicated in Patchett *et al.* (1978). On the reconstruction attempted here, adjustments to the outlines of the present-day landmasses in this block have been made in an attempt to accomodate some of the movement which occurred during the Caledonide event (Figs 19.1, 19.2). The following displacements are represented within the terrane lying NW of the Highland Boundary Fault, and its continuation in Ireland, the Fair Head—Clew Bay Line (Riddihough and Max, 1975). On the Moine Thrust Zone a total movement of 100 km is indicated (Coward, 1980; 1983; Coward *et al.*, 1980; Elliott and Johnson, 1980), without complicating the pattern by attempting to include all the varying movements on different thrust planes within the zone. A further displacement of 140 km is inferred for the ductile Sgurr Beag Thrust in Scotland (Tanner, 1970; Tanner *et al.*, 1970) which separates the Moine Nappe (consisting of the Morar Division described in Chapter 2) from the Sgurr Beag Nappe (consisting of the Glenfinnan and Loch Eil Divisions described in Chapter 3) (Winchester, 1985) (Fig. 19.2), while the smaller Knoydart and Naver Nappes are also shown schematically. Even without additional calculations of further Caledonide shortening measured along other minor ductile thrusts and by folding, the amount of Caledonian nappe displacement in Scotland is comparable with that in Scandinavia. Finally, a sinistral displacement of 160 km is inferred along the Great Glen Fault (after Winchester, 1973), based upon correlations of Caledonide metamorphic pattern and more recent information from the Central Highlands, including in particular the recognition of an older (probably Moine) basement in the Central Highlands, termed by Piasecki (1980) the Central Highland Division (Piasecki and Temperley, this volume) (Fig. 19.2). The resemblance between the Central Highland Division and the Glenfinnan Division of the Northern Highlands on lithological, geochemical and geochronological evidence is so pronounced that larger strike-skip movements postulated for the Great Glen Fault (for example Van der Voo and Scotese, 1982) are more likely to have occurred along the major terrane boundaries such as the Highland Boundary Fault, which shows clear evidence of major strike-slip displacement (Bluck, 1984) on a scale which precludes correlation of early Devonian and pre-Devonian rocks across it. The Great Glen Fault is thus considered to be a relatively minor splay of the major Caledonide strike-slip faults or terrane boundaries which cross Scotland, such as the Highland Boundary Fault, the Southern Uplands Fault and the Iapetus Suture.

In a similar manner, relative displacement of Wes-

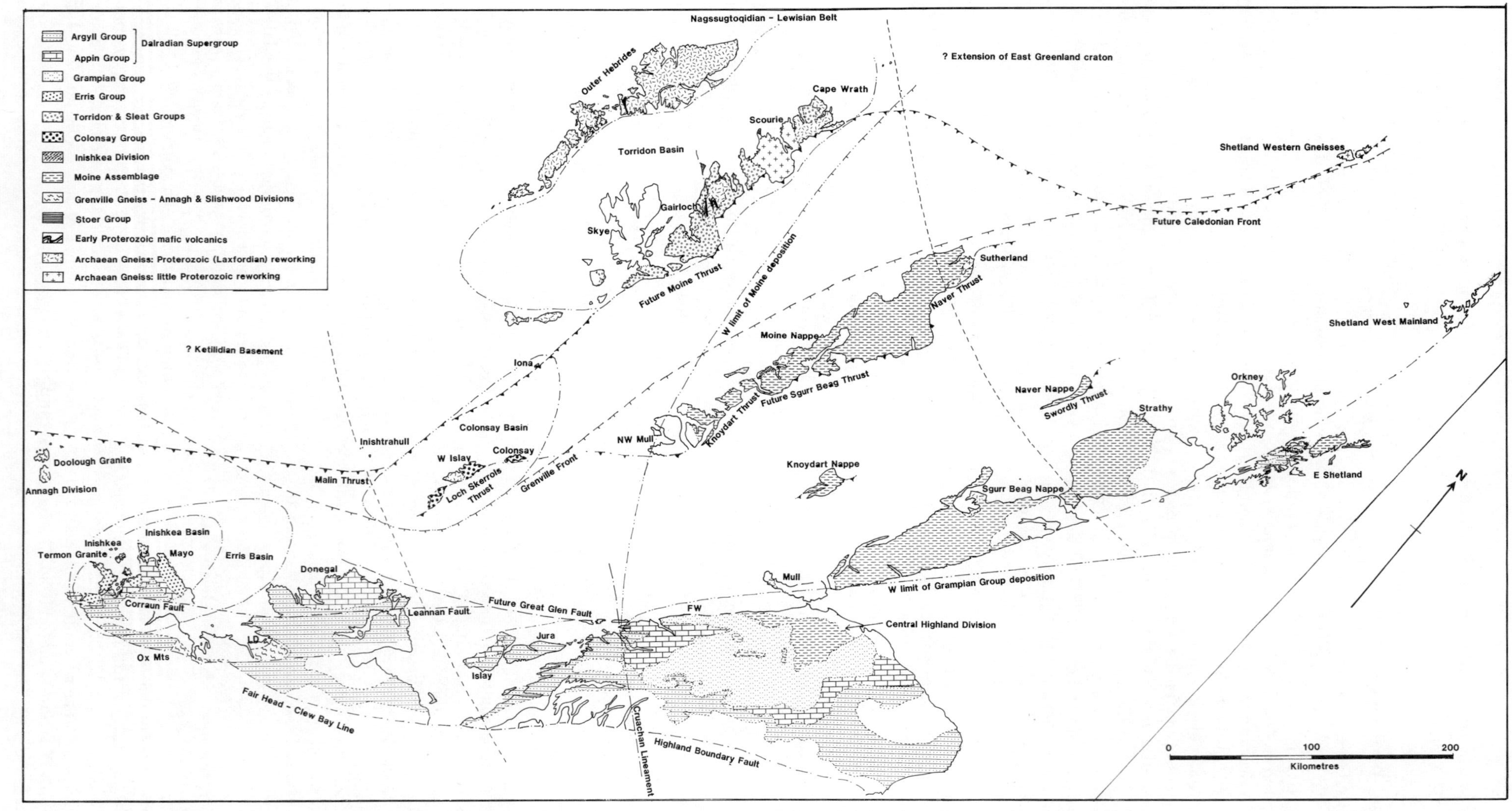

Figure 19.2 A pre-Caledonian reconstruction of the NW British Isles in Late Proterozoic time. Direct evidence of the full extent of Late Proterozoic sedimentary basins is scarce, and hence their boundaries shown on this map are speculative. Abbreviations: *FW*, Fort William; *LD*, Lough Derg inlier.

tern Islay and Colonsay is shown, compared to their present position (Fig. 19.2). This reflects westerly transport on the Loch Skerrols Thrust, which has been somewhat speculatively linked with the Sgurr Beag Thrust (a link with the Moine Thrust previously proposed (Kennedy, 1946) is not supported by Bentley, this volume). Likewise the 40 km sinistral offset along the Leannan Fault (Pitcher *et al.*, 1964) and its probable continuation into Co. Mayo, the Corraun Fault, accounts for the distortion of the NW portion of Ireland shown. The position of Connemara, which was probably displaced sinistrally along the Clew Bay Fault and overthrust to the south (Leake *et al.*, 1983; Leake and Singh, 1986) is speculative as its precise former location is unknown. In Shetland the Walls Boundary Fault is interpreted as a northward continuation of the Great Glen Fault: hence eastern Shetland is shown displaced so that it is situated E of Orkney on this reconstruction. Likewise, the eastward curvature of the Moine Thrust N of Scotland and its continuation to NW Shetland is based on seismic reflection profiling (Andrews, 1985; Brewer and Smythe, 1984; Smythe, 1987). Accordingly the Archaean Western Gneisses of NW Shetland are shown palinspastically restored using the 100 km displacement on the Moine Thrust Zone on the Scottish mainland, although the movement on the Thrust in Shetland has not been measured. West of Scotland, space is also left to accommodate the Rockall microcontinent, its area and outline adapted from Lefort (1984). Relative positions of the NW Foreland of Scotland, including the Outer Hebrides, and SE Greenland, are based upon correlation of the Archaean/Lower Proterozoic rocks in both areas (Le Pichon *et al.*, 1977; Myers, 1987) (Fig. 19.1).

The Nagssugtoqidian Belt in SE Greenland is equated with the Lewisian gneisses of the NW Scottish Foreland in their entirety (Myers, 1987). Previous suggestions of a link between the North Atlantic Craton and the Scourie Complex in the Lewisian (Fig. 19.2) are not supported by the nature of the Laxfordian gneisses to the S, which contain reworked Archaean gneisses (Giletti *et al.*, 1961; Park and Tarney, 1987), and are thus quite different from the entirely Proterozoic rocks of the Ketilidian Belt of southernmost Greenland. Furthermore, early Proterozoic Scourie mafic dykes cut the Laxfordian gneisses in a relationship analogous to that seen in the Nagssugtoqidian Belt (Evans, 1965; Chapman, 1979). Similar gneisses, mostly yielding Early Proterozoic ages, crop out as far S as the island of Inishtrahull, off the north coast of Ireland (Roddick and Max, 1983), which is shown here palinspastically restored to its former position before movement occurred on the Malin Thrust, which is assumed to be on the same scale as that of other major thrusts in the area. Both in structure and age the entire Lewisian Complex on the NW Foreland and the Nagssugtoqidian Belt are well matched, and the Scourian Complex, once interpreted as part of the North Atlantic Craton (Watterson, 1978; Bridgwater *et al.*, 1973) is better viewed as a preserved nucleus of Archaean rocks resembling many others identified within the Nagssugtoqidian Belt of Greenland (Myers, 1987). This early Proterozoic Nagssugtoqidian—Lewisian 'orogenic belt' is now interpreted as the result of shear-zone displacements caused by the relative motion of more stable blocks (Park and Tarney, 1987), in contrast to the coeval Ketilidian Belt of southernmost Greenland, which may reflect an early Proterozoic plate margin passing 'south' of the present Lewisian basement in Scotland. However, the Nagssugtoqidian Belt which is only 240 km wide in SE Greenland (Myers, 1987) must increase in breadth to over 450 km in Scotland (the distance between Inishtrahull and Cape Wrath on the palinspastic reconstruction). In SE Greenland the Nagssugtoqidian Belt is flanked on both sides by Archaean cratons. The craton to the SW, usually termed the North Atlantic craton (Korstgård *et al.*, 1987; van Schmus *et al.*, 1987) separates the Nagssugtoqidian Belt from the Ketilidian Belt. In the British Isles no equivalent block is seen lying SW of the Laxfordian gneisses of Inishtrahull, and no evidence for an Archaean basement has been obtained from the pre-Caledonian basement rocks of NW Ireland in Co. Mayo. Indeed two lines of evidence suggest instead that NW Mayo is underlain by an early Proterozoic basement, possibly akin to the rocks of the Ketilidian Belt. The oldest rocks exposed, termed the Annagh Division, described in Chapter 12 of this volume, have yielded upper intercept U–Pb zircon ages of $1297 + 164 - 101$ Ma (Aftalion and Max, 1987) and are intruded at Doolough Point by oversaturated peralkaline granite bodies which have yielded an age of $1093 + 8 - 7$ Ma (Winchester and Max, 1987), similar to the late Gardar Tugtutoq dykes in the Ketilidian Belt (Martin, 1985). In addition, zircon dating of the Caledonian Termon Granite on the Mullet Peninsula of Co. Mayo yielded an upper intercept age of 1956 ± 125 Ma (Aftalion and Max, 1987), which could be interpreted as either a Lewisian or Ketilidian age. However, the absence of evidence for Archaean crust beneath NW Ireland may suggest that the early Proterozoic ages obtained relate to the presence of an accretionary belt in the basement which may be an extension of the Ketilidian Belt. If so, the Ketilidian and Nagssugtoqidian Belts must converge west of the NW British Isles, which are thus nowhere underlain by the North Atlantic Craton. Further information from the Rockall microcontinent is needed to clarify this projection.

It is also likely that the Archaean craton lying NE of the Nagssugtoqidian Belt is also present beneath the northernmost part of the NW British Isles block. The tendency to describe all sub-Caledonide basement as 'Lewisian' has confused this picture, but in the northernmost part of mainland Scotland, Moorhouse and Moorhouse (this volume) describe the 'Lewisian' inliers which occur in the structurally higher parts of the Moine Nappe, and in the Naver Nappe as similar to the Archaean Scourie gneiss. Also present in that area is the enigmatic Strathy Complex, which includes high-grade metamorphic rocks chemically and lithologically dissimilar to Lewisian rocks (Moorhouse and Moorhouse, 1983). Attempts to date these rocks have so far proved unsatisfactory, but in Shetland the Western Gneisses ('Lewisian') have yielded Archaean ages (Flinn, 1985, and this volume). This evidence thus suggests that the basement to the northernmost part of Scotland is not an extension of the Nagssugtoqidian Belt, but may instead be an extension of the East Greenland Archaean craton (Fig. 19.1).

19.5 Correlating East Greenland and Scandinavia

Establishing a match between the basements of East Greenland and Scandinavia has proved difficult. This is partly because major displacements during the Caledonide event have to be considered, and partly because a substantial amount of later supracrustal material is present. In Scandinavia a displacement of 300 km is inferred for the eastward transport of Caledonide nappes, based crudely upon existing estimates of a 290 km transport of the Jotun Nappe (Roberts and Sturt, 1980), although much greater movements on the higher nappes have been inferred (Hossack *et al.*, 1985; Kumpulainen and Nystuen, 1985; Morley, 1986). It is thus possible that the original position of the 'Western Gneisses' may have lain much further to the 'west'. No attempt has been made to distinguish the individual travel of each nappe, as has been done in Scotland. In East Greenland, in view of the generally symmetrical disposition of the Caledonide nappes, a similar westerly transport of 300 km, has been postulated, although the few measurements so far undertaken have only indicated nappe travel of up to 100 km (Henriksen and Higgins, 1976).

On their reconstructions, Patchett *et al.* (1978) and Piper (1982) do not modify the outline of the continental blocks they match to allow for Caledonide displacements, nor do they include the crucially important NW British Isles–Rockall block. For these reasons, the relative positions of Scandinavia and Greenland shown on their models need modification, within the constraints imposed by the palaeomagnetic data. Accordingly Scandinavia has to be positioned further 'east', while maintaining the orientation required to fit the palaeomagnetic data (Fig. 19.1). With this modification the Karelian craton is no longer juxtaposed against the North Atlantic craton of southern Greenland (as shown in Piper, 1982, Fig. 1), but is situated opposite the East Greenland craton (as shown in Piper 1982, Fig. 9), and the evidence from the British Isles also suggests that this is a much more probable match. The East Greenland craton is flanked to the SW by the Nagssugtoqidian Belt, while the Karelian craton is flanked to the 'south' (in its pre-rotation position) by the Karelian lithological province (Gorbatschev, 1985) which, like both the Nagssugtoqidian and Lewisian Belts, consists of large areas of Archaean rock reworked during early Proterozoic movements. Further N, the East Greenland early Proterozoic mobile belt already discussed passes clear of the Kola craton on the Kola Peninsula.

19.6 Correlating Scandinavia and the NW British Isles

Within the Scandinavian Caledonides, Archaean inliers occur in the Lofoten islands, about 450 km away from the Karelian craton. This distance appears to indicate the breadth of the Karelian province, which encloses large areas of unmodified Archaean rocks between zones of Proterozoic reworking. This province corresponds best both in breadth and in its nature to the Lewisian in Scotland. Furthermore, just as the evidence from NW Ireland suggests that the Lewisian may be flanked to the south by a Proterozoic mobile belt devoid of Archaean material analogous to the Ketilidian of southernmost Greenland, so in Scandinavia the similar Proterozoic Svecofennian Belt, essentially devoid of Archaean relics, flanks the Karelian Province to the S in the pre-rotation position. Both in orientation and in the breadth of mobile belts and cratons, this match is surprisingly good. There may even be parallels between the Trans Scandinavian Granite-Porphyry Belt (Gorbatschev, 1985) and the Trans-Labrador Batholith (Dallmeyer *et al.*, 1987), which may have formed contemporaneously as part of the same mobile belt.

19.7 Early Proterozoic mafic volcanic belts

A feature of many of the early Proterozoic mobile belts in which Archaean gneisses are reworked is the presence of sporadically-preserved early Proterozoic supracrustal rocks which include thick sequences of mafic volcanic rocks which may include komatiitic lavas. Of these the best documented are the metavolcanics of the Cape Smith foldbelt in northern Quebec (Francis and Hynes, 1979; Francis *et al.*, 1981, 1983; Hynes and Francis, 1982) which form a prominent part of the Circum-Superior belt together with the mafic volcanics on the Ottawa and Belcher Islands in Hudson Bay, and the Fox River and Flin Flon mafic volcanic suites in Manitoba (Baragar and Scoates, 1981; Chauvel *et al.*, 1987; Watters and Armstrong, 1985). In these belts, Chauvel *et al.* (1987) state that the earlier formations were deposited in a rifting environment which, like Baragar and Scoates (1981) they envisaged as a trough of limited width. Similar mafic suites of early Proterozoic age are also present in the Lewisian Complex in Scotland at Gairloch where the chemistry of the basic metavolcanic rocks suggests a similar environment of formation (Johnson *et al.*, 1987). In this area, although the amphibolites possess an 'oceanic' chemical character, the field evidence does not support the presence of 'ophiolitic' material, and suggests that the rift in which the metavolcanics were emplaced was rapidly filled by clastic sediments after rifting ceased. Chemically similar and contemporaneous metavolcanic suites are also represented in Lapland by the Nussir Group (Pharaoh, 1985; Pearce and Pharaoh, 1984) and in Finland by numerous suites including the Jormua metavolcanics (Kontinen, 1987) (Fig. 19.1). In addition, several other Proterozoic basic suites, such as the Pechenga picrites, have been emplaced in rift environments in the Kola and Karelian cratons (Gaal and Gorbatschev, 1987; Barbey and Martin, 1987).

The same pattern of development seems to characterize all of these belts. Extension and rifting were followed by the extrusion of voluminous mafic volcanics and rapid filling of the sedimentary trough. No convincing evidence for the formation of any 'ocean' at the time has so far been advanced. Hence, if the Proterozoic basement model is correct (Fig. 19.1), it suggests that the early Proterozoic Supercontenent underwent major rifting about 2.0 Ga ago along a line extending at least 5000 km from Manitoba to Finland.

19.8 Sedimentation between 1150 and 1000 Ma BP

The match between the principal cratonic blocks established on the basis of linking early Proterozoic mobile belts still seems to apply during the period around 1150 Ma ago, suggesting that there was little relative displacement of the cratonic blocks during the Middle Proterozoic. Over most of the region there is little evidence of deposition in the 1150–1000 Ma period across much of the Canadian and Baltic Shields. However, in the region between Greenland and Scotland, extensive and rapid sedimentation took place, and elsewhere lesser thicknesses of non-marine sediment were laid down in ensialic rifts. These latter deposits will be described first.

19.9 Continental sedimentation

Within both the Canadian and Baltic Shields intracratonic rifting occurred at this time, together with a reactivation of some pre-existing grabens. The most extensive example of this rifting is filled by the Keweenawan Sequence in North America (Fig. 19.1) which consists dominantly of arid continental fluvial siliciclastic red beds over 5000 m thick (Kalliokoski, 1986), associated with mafic dykes and extrusive rocks emplaced around 1100 Ma (van Schmus *et al.*, 1982; Sutcliffe, 1987). Comparable deposits of similar age occur in southern Sweden, where the Almesåkra Group comprises 1200 m of dominantly fluviatile red shales, feldspathic sandstones and conglomerates (Rodhe, 1986), cut by dolerites emplaced at about 1000 Ma (Patchett and Bylund, 1977). The Jotnian Dala Sandstone of central Sweden, once equated with the Almesåkra Group, is an aeolian deposit shown by isotopic data to be older, yielding an age of 1300 Ma (Pulvertaft, 1985).

More comparable is the age of the Dalslandian Series (Magnusson, 1965; Berthelsen, 1980; Jakobsen *et al.*, 1984) of SW Sweden, which might be as old as 1130 Ma (Magnusson, 1965). Clearly deposited before the Sveconorwegian metamorphic event locally dated at 1030 Ma (Skiold, 1976), it consists of conglomerates, quartzites and slates associated with spilitic pillow lavas which may have been deposited in a small marine intracratonic basin, which could have once been linked to the Moine basin to the NW.

In NW Scotland the dominantly fluvial and lacustrine red beds of the Stoer Group (Stewart, 1962, 1966, 1982, and this volume) were formed in an intracratonic half-graben. Although Stoer Group shales have yielded a Rb–Sr date of 968 Ma (Moorbath, 1969), a depositional age as old as 1100 Ma has been suggested by palaeomagnetic studies (Piper, 1982; Smith *et al.*, 1983) in which Stoer Group palaeopoles coincide with those from formations within the Middle and Upper Keeweenawan.

Further rifting is indicated in the Gardar Province of SW Greenland in which alkaline intrusives were emplaced, and this province may extend into both Labrador and NW Ireland (Winchester and Max, 1983; 1987). In summary, therefore, an extensive pattern of rifting appears to have happened at the same time as the opening of both a major marine basin between Greenland and the Baltic Shield and subsidiary fault-bounded ensialic basins.

19.10 Marine basins 1150–1000 Ma

The Moine Assemblage, which is present in Scotland in the Central Highlands (Central Highland Division) (Piasecki, 1980; Piasecki and Temperley, this volume), the Northern Highlands (Morar, Glenfinnan and Loch Eil Divisions) and in Shetland (Flinn, this volume) consists of a thick sequence of shallow marine clastic rocks. Before metamorphism most of these rocks were arkosic and lithic sandstones: greywackes are less abundant, and the sedimentary structures reveal a northward transport of sediment. Lithological and sedimentological evidence suggests that the main basin margin lay a short distance to the west of existing outcrop, as parts of the Morar Division may have been fluviatile and conglomerates become increasingly abundant in the Ardnamurchan area in the SW of the Moine Nappe. However, the eastward and northward extent of the basin is unknown. To the S, although Moine rocks are concealed beneath later deposits, gravity and seismic evidence has suggested that the Moine basin may be limited by the Cruachan Lineament in the SW Scottish Highlands (Graham, 1986) (Fig. 19.2).

Within the Morar Division, which as part of the Moine Nappe was formerly situated W of the other preserved portions of the Moine Assemblage, tidal environments are postulated by Glendinning (this volume) who argues that deposition was within a N–S trending half-graben with a major growth fault situated to the W. This basin geometry apes that of the possibly contemporary Stoer Group basin, and suggests that the western margin of the Moine basin was characterized by the development of several major faults and tilted blocks with considerable variations in thickness of sediment.

East of the Morar Division a larger marine depositional basin was filled by sediments assigned to the Glenfinnan and Loch Eil Divisions, now exposed within the Sgurr Beag Nappe of the Northern Scottish Highlands (Roberts *et al.*, 1987; Barr *et al.*, this volume). Because the Sgurr Beag Thrust forms the contact with the Morar Division at all mainland localities, the stratigraphic relationship of the Morar and Glenfinnan Divisions is uncertain, but it has been recently claimed that in SW Mull (Fig. 19.2) Glenfinnan Division rocks pass stratigraphically down into the Morar Division (Holdsworth *et al.*, 1987).

Throughout the outcrop of the Moine Assemblage in Scotland, preserved sedimentary structures indicate that current flow was towards the N. This has been interpreted as longitudinal flow along the length of the basin, which was so broad that no evidence of an eastern margin is preserved in Scotland. However, although Moine rocks are present in Shetland (Flinn, 1985; this volume) there is no evidence that any deposition occurred in the formerly adjacent part of Scandinavia, which must therefore have lain beyond the eastern limit of the trough. Apart from locally-derived material near the western margin of the basin, the bulk of the clastic material forming the Moine Assemblage must have been derived from the Proterozoic basement lying to the S of Scotland, which almost certainly consisted of a part of the Ketilidian–Svecofennian Province (Plant *et al.*, 1984), together with later intrusive bodies, such as the Mid-Proterozoic acid-intermediate suite now probably represented by the Annagh Division exposed in NW Ireland (Winchester and Max, 1984). The size of this basin, the dominantly marine sedimentation and the thickness of the deposits, amounting to several kilometres, vastly exceeds that of the contemporary basins already mentioned. The precise age of deposition is not known, but all these rocks appear to have been subjected to a metamorphic event at approximately 1030 Ma (Brook *et al.*, 1976, 1977; Brewer *et al.*, 1979; Powell *et al.*, 1981). Only to the N, in the Caledonides of East Greenland, does there appear to be a contemporary marine succession of comparable size.

Apart from the synorogenic granites represented by the West Highland Granitic Gneiss (Barr *et al.*, 1985) the Moine Assemblage includes minor amounts of basic meta-igneous rock. These bodies, interpreted as highly deformed former dykes and sills, include an early suite which shares the entire deformational history of the adjacent metasediments, and which was thus emplaced before the Grenville event. Such amphibolites occur widely in the Sgurr Beag Nappe, where they are all tholeiitic in composition, but amphibolites of comparable chemistry are resticted to the NE part of the Moine Nappe, where the Ben Hope Sill suite is the best-known group (Moorhouse and Moorhouse, 1979; Smith, 1979; Winchester, 1984, 1985; Winchester and Floyd, 1984). Rare alkali basaltic amphibolites occur within the central part of the Moine Nappe, but the western part of the nappe is devoid of early basic rocks (Winchester, 1976; 1985). The absence of abundant metavolcanics in the Moine Assemblage argues, together with the arkosic nature of the sediments, that the basin never became truly 'oceanic', but this conclusion cannot apply to the eastern portion of the basin, which is entirely concealed.

Links between the Moine Assemblage of Scotland and Proterozoic supracrustal sequences in East Greenland were proposed by Higgins and Phillips (1976), and their former proximity and lithological similarity makes this correlation attractive. In the Krummedal Sequence of East Greenland a considerable thickness of clastic marine deposits is present (Higgins, this volume) although the proportion of metasediments interpreted as deep marine turbidites is greater than in the Moine Assemblage. Both rock groups have been intruded by granitic bodies yielding isotopic dates of approximately 1050 Ma (Rex and Gledhill, 1981) and both have apparently been affected by a metamorphic event at about that time, although there is still some disagreement about the interpretation of the ages from East Greenland (see also Caby and Bertrand-Sarfati, this volume). Higgins (this volume) records the existence of thicker amphibolites within the Krummedal Sequence, and it is possible that the greater igneous activity implied hints at greater extension in the northern part of the Moine–Krummedal basin. Still further N, in NE Greenland, contemporary sedimentary sequences appear to be unaffected by a Grenville-age metamorphic event (Jepsen and Kalsbeek, 1985), implying that later compressive movements did not affect that area.

19.11 Tectonic setting and depositional patterns

The distribution of rift systems and metamorphic belts during the Grenville–Sveconorwegian event has been largely explained by relative plate motion. The main Grenville Belt, extending from Canada to Sweden, has usually been interpreted as a 'collision belt' (e.g. Baer, 1981*b*), although the identity of the colliding continent remains obscure. Palaeomagnetic evidence has indicated that the rotation of Scandinavia relative to Laurentia, which may have been a response to continental collision, occurred between 1190 and 1050 Ma ago (Stearn and Piper, 1984), and this period covers both the 'early' Grenville metamorphism (Baer, 1981*a*, *b*) and the deposition of the Moine Assemblage in Scotland, the Krummedal Sequence in East Greenland, the later Keeweenawan sediments in rifts near Lake Superior, and possibly also the Stoer Group. It therefore seems probable that the Moine–Krummedal intracratonic basin was formed as a direct result of the rotation of Scandinavia. In Scotland, the limits of this basin may be indicated by the position of the Cruachan Lineament, and the consistent current flow towards the N hints that the basin widened northwards. With the simultaneous clockwise rotation of Scandinavia, this basin is thus likely to have developed in a sinistral transtensional regime, while the simultaneous development of subsidiary rifting, such as the Keeweenawan graben, may also be linked with dextral movement within the Grenville Province (McWilliams and Dunlop, 1978). In Scotland and in part of East Greenland, deposition seems to have been followed by a transpressional regime accompanied by regional metamorphism, while further to the N in Greenland there is little evidence of contemporary metamorphism (Jepsen and Kalsbeek, 1985). This lack of subsequent compression may be explained if the rotation of Scandinavia was accompanied by a southward movement of Scandinavia relative to Laurentia. The metamorphism of the Moine Assemblage, which does not appear to have been accompanied at this stage by major overthrusting, may be explained as the result of sinistral transpression in a relatively narrow shear zone, which formed a 'third branch' of the Grenville–Sveconorwegian orogen. By contrast, in NW Ireland the Moine Assemblage seems to be absent (Winchester and Max, this volume): the area formed part of the Laurentian plate and did not form part of the Moine–Krummedal shear zone. In the 'south' the rotation of Scandinavia induced a largely compressional regime, characterized by relatively little deposition (the thickness of the Dalslandian being relatively small), while the subsequent Sveconorwegian metamorphism was accompanied by major nappe translations indicating considerable crustal shortening (Berthelsen, 1980). After 1050 Ma ago, the rotational movement ceased, as subsequently did the metamorphism in both the Moine Assemblage and the Krummedal Sequence. Neither of these rock assemblages, situated in the 'third arm' of the Grenville–Sveconorwegian belt, previously indicated by Ziegler (1986), were affected by the later metamorphism around 950 Ma ago (Baer, 1981*a*) which affected the main belt.

19.12 Post-Grenville basin development

In the time interval 1050–840 Ma, palaeomagnetic evidence suggests that the relative positions of Laurentia and Scandinavia changed little (Stearns and Piper, 1984). During this period there is also little evidence of deposition, unless the palaeopole positions of the Torridon Group are interpreted as evidence of deposition around 1040 Ma (Piper, 1982) (Table 19.1). If so, deposition of the Torridon Group, occurring at a time of uplift of the metamorphic Moine Belt to the E, might be explicable as a product of rapid erosion of the uplifted area, except for the persistent derivation of clastic material from the W. It therefore remains quite possible that the original isotopic date (Moorbath, 1969; Moorbath *et al.*, 1967) of around 777 + 24 Ma may more closely reflect the original depositional date (Table 19.1). Equally, during this period there appears to have been little deposition in either the Grenville Province of Canada or Scandinavia.

19.13 Early restricted basin development

After 840 Ma, deposition was apparently renewed in the NW British Isles segment, in East Greenland and possibly also in the adjacent portions of Scandinavia with the development of small ensialic basins which were probably fault-bounded, and may have been interconnected. In each of these areas, local names have been given to each sedimentary sequence, and an attempt to relate them chronologically is portrayed in Table 19.1. It is possible that in NW Ireland the deposition and subsequent metamorphism of the clastic, probably marine Inishkea Division rocks may have occurred before 800 Ma, on the evidence of a poorly constrained date of 800 + 146 Ma (Winchester and Max, this volume; Winchester and Max, 1987*b*), while the age of the Colonsay Group in the Inner Hebrides may also be comparable, although there is currently little available evidence for the deposition date of the group (Bentley, this volume). Perhaps the metamorphism of the Inishkea Division and the precambrian deformation of the Colonsay Group both testify to pre-Grampian Group local basin formation and subsequent compression, as a result of minor adjustment to the relative positions of Scandinavia and the Laurentian Shield, and any associated metamorphism was local rather than regional in extent.

19.14 Late Proterozoic deformation and metamorphism

Only the Inishkea Division and the Colonsay Group provide clear evidence of post-Grenville, pre-Caledonian metamorphism of deformation. Neither the Grampian Group nor the Erris Group have yielded evidence of pre-Caledonian deformation. This suggests that a late Proterozoic localized deformational event may have occurred within the Grenville mobile belt separating Scandinavia from the Laurentian Shield, possibly as a response to some short-lived differential movement between the two blocks. Support for this hypothesis comes from isotopic evidence for a late Proterozoic event, first recognized in Scotland and generally known as the 'Morarian' event (Lambert, 1969). Dates range from 765–700 Ma and have since been obtained from additional sites in the Scottish Highlands (van Breemen *et al.*, 1974; Piasecki, 1980; Piasecki *et al.*, 1981), Greenland (Rex and Gledhill, 1981), and Denmark (Ziegler, 1986). However, these dates tend to be younger than the assumed depositional dates of the Grampian Group and so a considerable problem of interpretation remains.

19.15 Early widespread subsidence and deposition

After 800 Ma, far more widespread basin development seems to have occurred. In Scotland and NW Ireland the Grampian Group and Erris Group clastic sequences were deposited, although the absence of rocks occupying a comparable stratigraphic position in Donegal, where Dalradian rocks are brought into tectonic contact with the Slishwood Division, suggests that the Grampian and Erris Group basins were not necessarily

Table 19.1 Correlation chart showing the relative ages of events between 1200 Ma and the end of the Proterozoic in the north Atlantic lands.

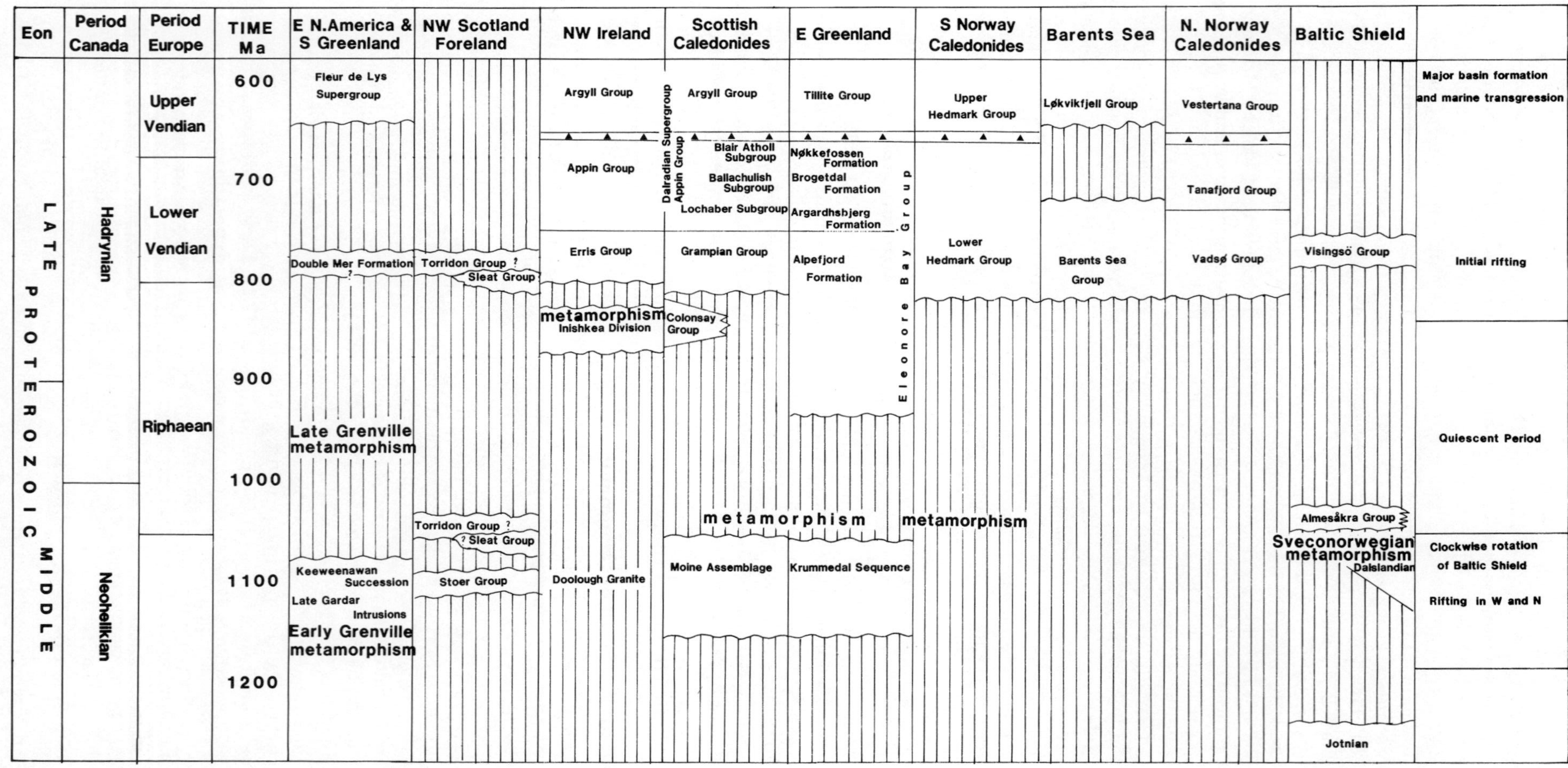

directly connected. In both sequences earlier turbiditic psammites were succeeded by maturer sequences which in the Grampian Group have been interpreted on the evidence of their preserved sedimentary structures as fluviodeltaic in origin (Winchester and Glover, this volume). Rapid changes in the thickness of individual formations in the Grampian Group may record the existence of fault control on the rate of subsidence and deposition. It also suggests that the western margin of the Grampian Group basin may have been close to the westernmost outcrops. Chemical comparisons between the Grampian Group and the Erris Group show that similar chemical changes occur with time in each group, but comparable sedimentological studies have not yet been undertaken in the Erris Group, and thus the environment of deposition is less well understood at present (Winchester *et al.*, this volume). However, the striking chemical similarity of individual formations in each group argues that similar depositional conditions prevailed and the sediment was derived from very similar source areas. However, the absence of Erris Group type sediments in the NE Ox Mountains and Lough Derg inliers argues that the Erris Group and Grampian Group basins were not necessarily connected. In addition, the absence of Grampian Group sediments (which are 'Moine-like' in their geophysical properties) SW of the Cruachan Lineament also suggests that the basins were separate.

In East Greenland parallels have previously been drawn between the lower part of the Eleonore Bay Group (the Alpefjord Formation, Katz, 1961; Caby and Bertrand-Sarfati, this volume) and the Grampian Group in Scotland (Higgins and Phillips, 1979). However, while both consist dominantly of sequences of metamorphosed impure sandstones and silts, recent sedimentological work within the Grampian Group (Winchester and Glover, this volume) has indicated that only the uppermost part (the Glen Spean Subgroup) may be interpreted as fluviodeltaic. By contrast the underlying Corrieyairack Subgroup consists mainly of turbiditic deposits: a clear difference from the Alpefjord Formation which is almost entirely fluviodeltaic or lacustrine, with abundant evidence of emergence and desiccation (Caby, 1976; Caby and Bertrand-Sarfati, this volume). Indeed, the depositional environment of the Alpefjord Formation may have more in common with that of the Sleat Group and Torridon Group in NW Scotland. These two groups incorporate a combined maximum thickness of up to 9000 m, although the Sleat Group is only preserved below the Torridon Group on the Isle of Skye. Like the Alpefjord Formation, both the Sleat and Torridon Groups are interpreted as suites of fluvial, alluvial and lacustrine deposits (Stewart, this volume) while the isotopic age obtained of 777 Ma (Moorbath, 1969) suggests that deposition was contemporaneous with that of the Alpefjord Formation. However, it remains possible that the Torridon and Sleat Groups might be considerably older if the calculated palaeomagnetic pole positions are correct (Piper, 1982; Smith *et al.*, 1983). It is, however, tempting to link the deposition of the Torridon and Sleat Groups with that of the Alpefjord Formation as ensialic continental deposits formed during a major rifting event. Deposition associated with such a model would include alluvial and fluvial sedimentation at the margins and within separate but related minor continental rifts, while deeper water deposits accumulated near the centre of the main rift. However, work in Shetland (Flinn, this volume) suggests that no equivalent of the Grampian Group exists in the archipelago, whereas, if a single Grampian–Torridon–Alpefjord basin has developed, the position of Shetland would argue strongly that deposits akin to the Grampian Group should be present. That they are not illustrates the dangers of long-distance correlation and perhaps suggests that, while widespread rifting may have occurred about 800 Ma ago, the basins that formed were not necessarily interconnected.

In Scandinavia, the lower parts of the Late Precambrian succession were probably deposited at the same time as the Grampian Group in Scotland. In southern Norway deposition occurred simultaneously in several basins, collectively termed the 'western Baltoscandian basins' (Kumpulainen and Nystuen, 1985). Deposition was dominantly non-marine, consisting of fluvial, alluvial or lacustrine deposits in all but the Hedmark basin, in which alluvial deposits prograde westwards into a marine trough in which marine turbiditic sediments were laid down (the Brøttum Formation). Some of the detritus in the Brøttum Formation was derived from the W and S (Kumpulainen and Nystuen, 1985), directions similar to those noted in the Grampian Group, and raising the possibility that the Hedmark Group and Grampian Group may have been laid down on opposite sides of the same basin (Fig. 19.3). Higher formations in the western Hedmark basin show a progressive change to submarine deltaic deposits (Nystuen and Siedlecka, this volume), thus showing some resemblance to the progressive changes seen in the Grampian and Erris Groups in the NW British Isles.

In northern Norway an age of 807 + 19 Ma has been obtained from near the base of the Vadsø Group (Føyn, 1985), suggesting that renewed deposition also started in this area at the same time as in southern Norway and Scotland. In the Vadsø Group fluvial conditions predominate, but in the overlying Tanafjord Group shallow marine clastic deposition is typical, recording that subsequently a widespread marine transgression occurred.

A somewhat different sedimentary succession occurs in the northern part of the Varanger Peninsula. This succession, termed the Barents Sea Group, was deposited at the same time as the Vadsø Group, as attested by an age of 810 + 60 Ma (Føyn, 1985), but consists of a thick sequence of marine turbiditic deposits. It is separated from the Vadsø Group succession by the Trollfjord–Komagelv Fault. Much discussion has centred on whether the Vadsø Group and Barents Sea Group successions can be related (Johnson *et al.*, 1978;

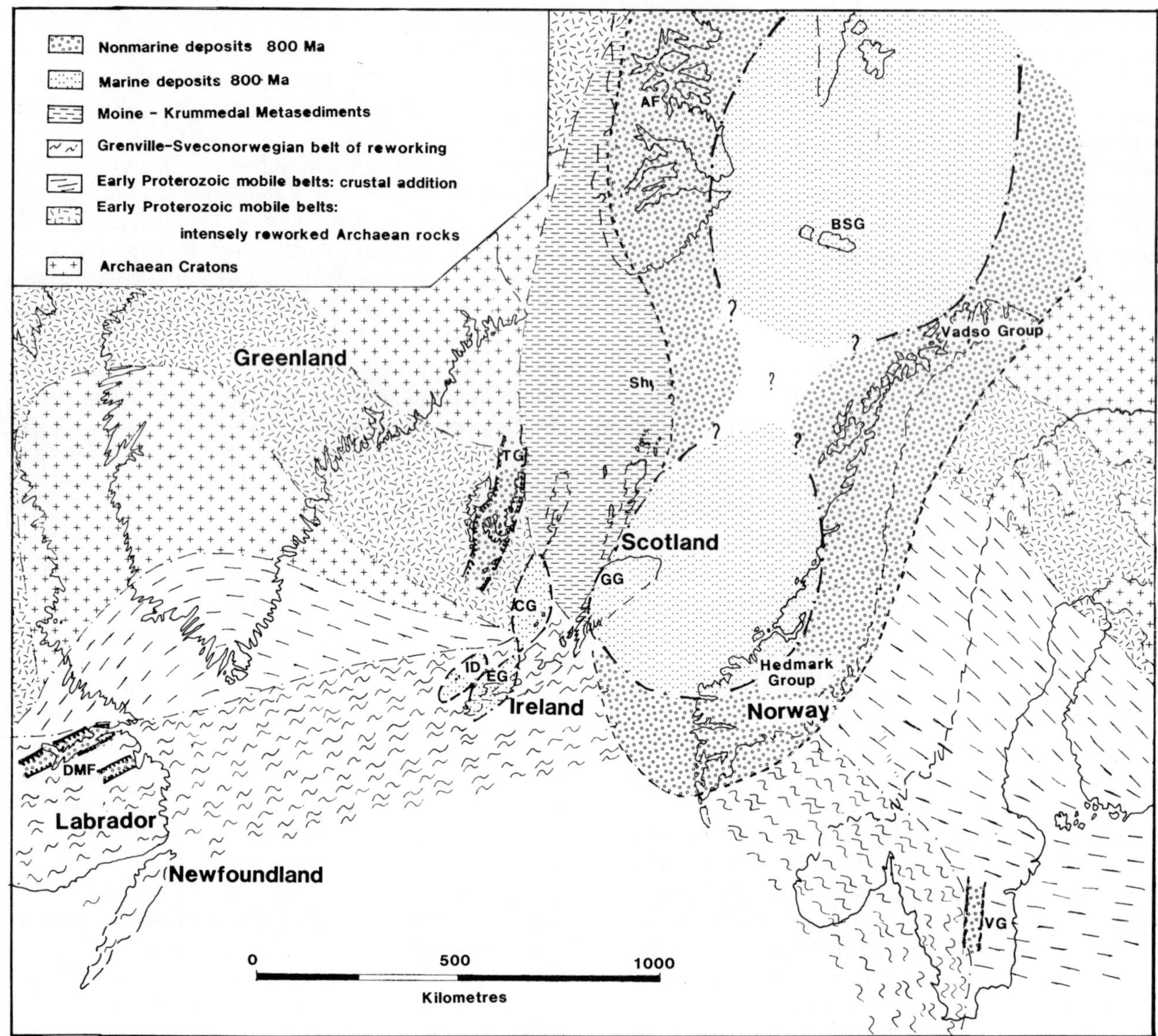

Figure 19.3 A simplified palaeogeographic reconstruction of the Northern Atlantic Lands 800 Ma ago. Possible limits of early post-Grenville basins are indicated. Relative positions of Greenland and Scandinavia are adapted from Piper (1982). Abbreviations: *AF*, Alpefjord Formation; *BSG*, Barents Sea Group; *CG*, Colonsay Group; *DMF*, Double Mer Formation; *EG*, Erris Group; *GG*, Grampian Group; *ID*, Inishkea Division; *Sh*, Shetland; *TG*, Torridon Group; *VG*, Visingsö Group.

Siedlecka, 1975; Harland and Gayer, 1972), but Johnson *et al.* (*op. cit.*) postulate a dextral strike-slip movement exceeding 500 km on the Trollfjord–Komagelv Fault, which is supported by palaeomagnetic evidence (Kjφde *et al.*, 1978). On a palinspastic reconstruction, such a movement places the Barents Sea Group within the axial zone of the Grampian Group–Alpefjord Formation–Vadsφ Group basin (if it was a single basin) (Fig. 19.3), and might account for the presence of thick turbiditic deposits. Thus the series of ensiallic basins developed around 800 Ma between East Greenland and Scandinavia suggest that some crustal extension occurred in this region. This in turn may have been the result of renewed relative motion between Scandinavia and the Laurentian Shield.

On the Baltic craton sediments of comparable age are preserved in the small rift-bounded Vattern basin in southern Sweden. Acritarch assemblages suggested that these sediments, collectively termed the Visingsö Group, were laid down between 750–800 Ma (Vidal, 1974), although isotopic data from shales in the group have suggested a somewhat younger age (Bonhomme and Welin, 1984). They consist of up to 1000 m of dominantly fluvial and deltaic sediments with some marine intercalations. The sedimentary evolution of the Visingsö Group is similar to that of contemporary sedimentary basins in southern Norway (Kumpulainen and Nystuen, 1985) and has also been compared with many platform sequences on the Russian Platform (Vidal, 1985).

19.16 Widespread marine transgression

Succeeding deposits in Greenland and Scotland suggest that environmental changes followed the deposition of the Grampian Group and its likely equivalents. In the NW British Isles, the Grampian Group and Erris Group are usually overlain with apparent conformity by metasediments assigned to the Appin Group, which forms the lower part of the Dalradian Supergroup (Harris and Pitcher, 1975; Kelling *et al.*, 1985). Generally interpreted as marine deposits laid down in an ensialic basin which witnessed increasing instability with time (Harris and Pitcher, 1975; Harris *et al.*, 1978; Anderton, 1979), the Dalradian Supergroup was apparently laid down during a time when Scandinavia separated from its position adjacent to Greenland at the end of the Riphaean (illustrated using palaeomagnetic data in Piper, 1985) to its position in early Cambrian times, when it was separated by a broad ocean from Greenland (Piper, 1985). Appin Group sedimentation in the British Isles was unaccompanied by volcanic activity and records a period of steady crustal subsidence and a marine transgression. In the British Isles the base of the Appin Group is marked by a pure quartzite (the Eilde Quartzite in Scotland; the Srahlaghy Quartzite and its equivalents in Ireland) of considerable lateral extent, while the succeeding Lochaber Subgroup rocks consist of an alternation of more laterally impersistent quartzites and pelitic schists. Close to the present line of the Great Glen Fault, both NE and SW of Fort William, much of the Lochaber Subgroup seems to be missing and, while local tectonic disruption may be responsible in part, 40 km NE of Fort William an apparently complete succession shows the Leven Schist in sharp contact with underlying Grampian Group schists. It seems probable that the lower formations of the Appin Group are missing because of the proximity of the basin margin, which may have been fault-controlled, and hence it is possible that neither the Grampian Group nor overlying Dalradian Supergroup rocks were deposited over much of the NW Highlands of Scotland (Figs 19.3, 19.4).

A similar lithological change is recorded in the Upper

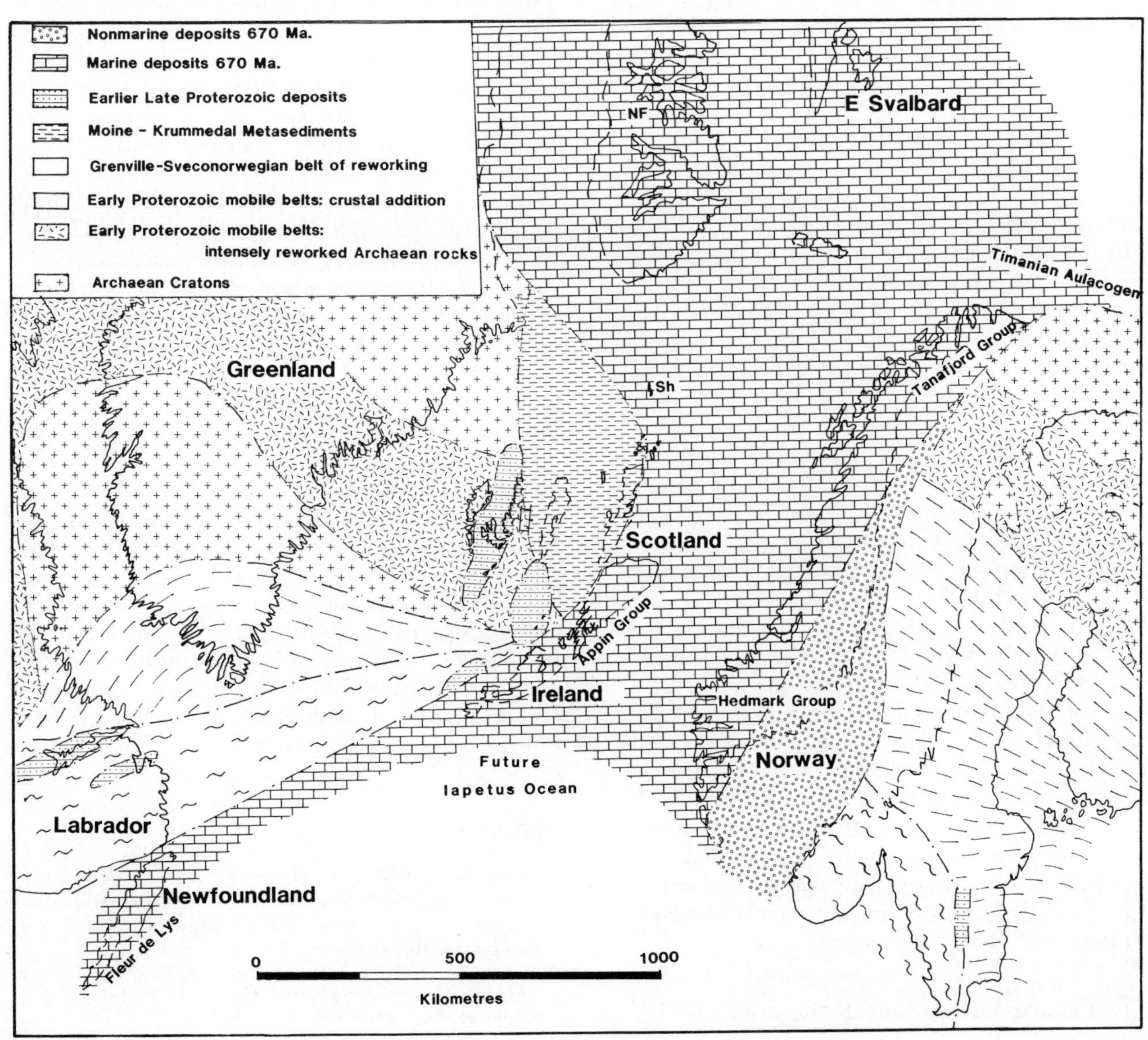

Figure 19.4 A simplified palaeogeographic reconstruction of the Northern Atlantic Lands approximately 680 Ma ago, before the separation of Scandinavia from the Laurentian Shield and the formation of the Iapetus Ocean. Abbreviations: *NF*, Nokkefossen Formation; *Sh*, Shetland.

Eleonore Bay Group of Greenland where the base of the Argardhsbjerg Formation is marked by a laterally extensive orthoquartzite (Caby and Bertrand-Sarfati, this volume), which is succeeded by up to six units, each consisting of a quartzite–sandstone–shale sequence interpreted as lacustrine or very shallow marine (Caby and Bertrand-Sarfati, this volume) (Table 19.1).

In the British Isles the Ballachulish and Blair Atholl Subgroups, which underlie the Port Askaig Tillite, correlated with the Varanger Tillite (Spencer, 1971; Pringle, 1973) (Table 19.1) form a sequence of limestone and black shales which are laterally very continuous. In Greenland the Brogetdal and Nokkefossen Formations of the Upper Eleonore Bay Group are lithologically very similar and consist of a cyclic succession of limestones, dolomites and black shales interpreted as marine supratidal to subtidal deposits (Caby and Bertrand-Sarfati, this volume).

In Northern Norway the Tanafjord Group, which is overlain by the Varanger Tillite, consists largely of a sequence of sandstones and shales, capped by a stromatolite-bearing dolomitic unit (Siedlecka and Siedlecki, 1971; Bertrand-Sarfati and Siedlecka, 1980; Nystuen and Siedlecka, this volume). It has been interpreted as a record of a marine transgression (Nystuen and Siedlecka, this volume), although the derivation of the sediment was more from the E, suggesting that it lay on the opposite side of the basin from Scotland.

In southern Norway, by contrast, shallow marine deposition in the sparagmite basins was of relatively brief duration. It is represented in the Hedmark Basin (Kumpulainen and Nystuen, 1985) by the Atna and Biri Formations (Nystuen and Siedlecka, this volume), which were in turn overlain by the Osdalen Conglomerate, interpreted as alluvial fans (Nystuen and Saether, 1979; Nystuen, 1982), which suggests that the margin of the basin was never far distant. Also, unlike the other areas in which contemporary deposition was in progress, the Hedmark Basin alone contains a record of igneous activity, with the extrusion of basaltic flows, which on analysis have been shown to have the chemical characteristics of continental tholeiites (Saether and Nystuen, 1982; Nystuen, 1982; Furnes *et al.*, 1983; Nystuen and Siedlecka, this volume).

In none of these areas do the deposits predating the Varanger glaciation, dated at approximately 654 Ma (Pringle, 1973), provide evidence of deep-water deposition. Instead, slow crustal extension, accompanied by the marine invasion of a broad gulf separating emergent parts of the Laurentian and Baltic Shields, is indicated. With periods of temporary emergence, and subsequent deposition of glacial deposits, the present-day North Sea seems to be a fitting analogy.

19.17 Postglacial Proterozoic deposits

Correlation between some of the end-Proterozoic deposits in the northern Atlantic lands becomes increasingly difficult. During the 80 million years between the Varanger glaciation episodes and the beginning of the Cambrian Period, crustal extension appears to have accelerated, so that correlation of the deposits on the Laurentian and Scandinavian sides of the basin becomes impossible. In Scotland, detailed sedimentological analysis of the Argyll Group (Anderton, 1975, 1979, 1982, 1985) has shown that the deposition patterns were dominated by extensional tectonics which gave rise to deep fault-bounded basins. Rapid lateral facies changes are typical, and tholeiitic metavolcanics become widespread in both Scotland (Sturt, 1961; Harris and Pitcher, 1975) and Ireland (Max and Long, 1979, 1985; Winchester *et al.*, 1987). By contrast, in East Greenland and in Norway, deposition of shallow marine-shelf sea sediments continued, devoid of accompanying vulcanicity. These areas may have been more remote from the axis of the developing basin, but in NW Newfoundland the nature of the probably contemporaneous (Kennedy, 1975) Fleur de Lys sedimentation is indicative of developing crustal instability. These varied deposits, apparently laid directly upon 'Grenville' basement, consist of clastics, marbles and associated basaltic metavolcanic rocks deposited on an unstable slope environment (Hibbard, this volume). By this stage, therefore, the late Proterozoic basin is taking on 'oceanic' characteristics: hence the end-Proterozoic rocks of Scotland, Ireland and Newfoundland appear to have been formed on the Laurentian continental sheld, while across the widening Iapetus Ocean, the contemporary rocks of Norway were laid down on a continental margin in which the relative scarcity of marine rocks suggests that the continental shelf was relatively narrow. Among the sediments closest to the axial part of the basin, such as the Dalradian Supergroup, there is no clearly identifiable break to mark the start of the Cambrian Period, although the existence of a widespread marine transgression across lands not formerly submerged is evidence of continuing subsidence as the Iapetus Ocean continued to widen. This latter development is beyond the scope of this book, which has thus compiled a story of the stratigraphic changes that accompanied each progressive stage of the disintegration of the Proterozoic Supercontinent in the latter part of the Proterozoic. Only further work will show whether some of the more speculative correlations are valid, and how truly this story records the later Proterozoic history of this region.

References

Aftalion, M. and Max, M. D. (1987) U-Pb zircon geochronology from the Precambrian Annagh Division gneisses and the Termon Granite, NW Co. Mayo, Ireland. *J. geol. Soc. London* **144**, 401–406.

Anderton, R. (1975) Tidal flat and shallow marine sediments from the Craignish Phyllites, Middle Dalradian, Argyll, Scotland. *Geol. Mag.* **112**, 429–458.

Anderton, R. (1979) Slopes, submarine fans, and syn-depositional faults: sedimentology of parts of the Middle and Upper Dalradian in the SW Highlands of Scotland. In Harris, A. L., Holland, C. H. and Leake, B. E. (eds.), The

Caledonides of the British Isles–Reviewed. *Spec. Publ. geol. Soc. London* **8**, 483–488.

Anderton, R. (1982) Dalradian deposition and the late Precambrian–Cambrian history of the North Atlantic region: a review of the early evolution of the Iapetus Ocean. *J. geol. Soc. London* **139**, 421–431.

Anderton, R. (1985) Sedimentation and tectonics in the Scottish Dalradian. *Scott. J. Geol.* **21**, 407–436.

Andrews, I. J. (1985) The deep structure of the Moine Thrust, southwest of Shetland. *Scott. J. Geol.* **21**, 213–217.

Baer, A. J. (1981*a*) Two orogenies in the Grenville Belt. *Nature (London)* **290**, 129–131.

Baer, A. J. (1981*b*) A Grenvillian model of Proterozoic Plate Tectonics. In Kroner, A. (ed.), *Precambrian Plate Tectonics*, Elsevier, Amsterdam, 353–385.

Baragar, W. R. A. and Scoates, R. F. J. (1981) The Circum-Superior Belt: a Proterozoic Plate Margin. In Kroner, A. (ed.), *Precambrian Plate Tectonics*, Elsevier, Amsterdam, 297–330.

Barbey, P. and Martin, H. (1987) The role of komatiites in plate tectonics. Evidence from the Archaean and early Proterozoic crust in the eastern Baltic Shield. *Precambrian Res.* **35**, 1–14.

Barr, D., Roberts, A. M., Highton, A. J., Parson, L. M. and Harris, A. L. (1985) Structural setting and geochronological significance of the West Highland Granitic Gneiss, a deformed early granite within Proterozoic Moine rocks of NW Scotland. *J. geol. Soc. London* **142**, 663–676.

Berthelsen, A. (1980) Towards a palinspastic tectonic analysis of the Baltic Shield. In Géologie de l'Europe du Precambrian aux bassins sedimentaires post-hercyniens. *Colloque Co. Publ. du 26*, CGI, Paris, 5–21.

Bertrand-Sarfati, J. and Siedlecka, A. (1980) Columnar stromatolites of the terminal Precambrian Porsanger Dolomite and Grasdal Formation of Finnmark, North Norway. *Norsk geol. Tidsskr.* **60**, 1–27.

Bluck, B. J. (1984) Pre-Carboniferous history of the Midland Valley of Scotland. *Trans. R. Soc. Edinburgh* **75**, 275–295.

Bonhomme, M. G. and Welin, E. (1984) Rb-Sr and K-Ar isotopic data on shale and siltstone from the Visingsö Group, Lake Vattern basin, Sweden. *Förh. geol. Fören. Stockholm* **105**, 363–366.

Brewer, M. S., Brook, M. and Powell, D. (1979) Dating of the tectonometamorphic history of the southwestern Moine, Scotland. In Harris, A. L., Holland, C. H. and Leake, B. E. (eds.), The Caledonides of the British Isles–Reviewed. *Spec. Publ. geol. Soc. London* **8**, 129–137.

Brewer, J. A. and Smythe, D. K. (1984) MOIST and the continuity of crustal reflector geometry along the Caledonian–Appalachian orogen. *J. geol. Soc. London* **142**, 245–258.

Bridgwater, D., Escher, A. and Watterson, J. (1973) Tectonic displacements and thermal activity in two contrasting Proterozoic mobile belts from Greenland. *Phil. Trans. R. Soc. London* **A273**, 513–533.

Brook, M., Powell, D. and Brewer, M. S. (1976) Grenville age for rocks in the Moine of north-western Scotland. *Nature (London)* **260**, 515–517.

Brook, M., Powell, D. and Brewer, M. S. (1977) Grenville events in Moine rocks of the Northern Highlands, Scotland. *J. geol. Soc. London* **133**, 489–496.

Bylund, G. (1981) Sveconorwegian palaeomagnetism in hyperite dolerites and syenites from Scania, Sweden. *Förh. geol. Fören. Stockholm* **103**, 173–182.

Caby, R. (1976) Investigations in the Lower Eleonore Bay Group in the Alpefjord region, Central East Greenland. *Rapport Grønl. Geol. Unders.* **80**, 102–106.

Chapman, H. J. (1979) 2390 Myr Rb-Sr whole rock age for the Scourie dykes of north-west Scotland. *Nature (London)* **277**, 642–643.

Chauvel, C., Arndt, N. T., Kielinzcuk, S. and Thom, A. (1987) Formation of Canadian 1.9 Ga old continental crust. I: Nd isotopic data. *Can. J. Earth Sci.* **24**, 396–406.

Collerson, K. D. (1982) Geochemistry and Rb-Sr geochronology of associated Proterozoic peralkaline and subalkaline anorogenic granites from Labrador. *Contrib. Mineral. Petrol.* **81**, 126–147.

Coward, M. P. (1980) The Caledonian thrust and shear zones of NW Scotland. *J. Struct. Geol.* **2**, 11–17.

Coward, M. P. (1983) The thrust and shear zones of the Moine Thrust Zone and the NW Scottish Caledonides. *J. geol. Soc. London* **140**, 795–811.

Coward, M. P., Kim, J. H. and Parke, J. (1980) A correlation of Lewisian structures across the lower thrusts of the Moine Thrust zone, NW Scotland. *Proc. geol. Assoc. London* **91**, 327–337.

Dallmeyer, R. D. (1987) $^{40}Ar/^{39}Ar$ mineral age record of variably superimposed Proterozoic tectonothermal events in the Grenville Orogen, Central Labrador. *Can. J. Earth Sci.* **24**, 314–333.

Elliott, D. and Johnson, M. R. W. (1980) Structural evolution in the northern part of the Moine Thrust belt, NW Scotland. *Trans. R. Soc. Edinburgh Earth Sci.* **71**, 69–96.

Escher, A. and Pulvertaft, T. C. R. (1976) Rinkian mobile belt of West Greenland. In Escher, A. and Watt, W. S. (eds.), *Geology of Greenland*, Odense, 105–119.

Evans, C. R. (1965) Geochronology of the Lewisian basement near Lochinver, Sutherland. *Nature (London)* **207**, 54–56.

Flinn, D. (1985) The Caledonides of Shetland. In Gee, D. G. and Sturt, B. A. (eds.), *The Caledonide Orogen—Scandinavia and Related Areas*, John Wiley, New York, 1158–1171.

Føyn, S. (1985) The Late Precambrian in northern Scandinavia. In Gee, D. G. and Sturt, B. A. (eds.), *The Caledonide Orogen—Scandinavia and Related Areas*, John Wiley, New York, 233–245.

Francis, D. M. and Hynes, A. J. (1979) Komatiite-derived tholeiites in the Proterozoic of New Quebec. *Earth Planet. Sci. Lett.* **44**, 473–481.

Francis, D. M., Hynes, A. J., Ludden, J. N. and Bedard, J. (1981) Crystal fractionation and partial melting in the petrogenesis of a Proterozoic high-MgO volcanic suite, Ungava, Quebec. *Contr. Mineral. Petrol.* **78**, 27–36.

Francis, D. M., Ludden, J. N. and Hynes, A. J. (1983) Magma evolution in a Proterozoic rifting environment. *J. Petrol.* **24**, 556–582.

Furnes, H., Nystuen, J. P., Brunfelt, A. O. and Solheim, S. (1983) Geochemistry of Upper Riphaean–Vendian basalts associated with the 'sparagmites' of southern Norway. *Geol. Mag.* **120**, 349–361.

Gaal, G. and Gorbatschev, R. (1987) An outline of the Precambrian evolution of the Baltic Shield. *Precambrian. Res.* **35**, 15–52.

Giletti, B. J., Lambert, R. StJ. and Moorbath, S. (1961) The basement rocks of Scotland and Ireland. *Ann. N.Y. Acad. Sci.* **91**, 464–468.

Gorbatschev, R. (1985) Precambrian basement of the Scandinavian Caledonides. In Gee, D. G. and Sturt, B. A. (eds.), *The Caledonide Orogen—Scandinavia and Related Areas*, John Wiley, New York, 197–212.

Graham, C. M. (1986) The role of the Cruachan Lineament during Dalradian evolution. *Scott. J. Geol.* **22**, 257–270.

Harland, W. B. and Gayer, R. A. (1972) The Arctic Caledonides and earlier oceans. *Geol. Mag.* **109**, 289–384.

Harris, A. L. and Pitcher, W. S. (1975) The Dalradian Supergroup. In Harris, A. L. *et al.* (eds.), A correlation of Precambrian rocks in the British Isles, *Spec. Rep. geol. Soc. London* **6**, 42–75.

Harris, A. L., Baldwin, C. T., Bradbury, H. J., Johnson, H. D. and Smith, R. A. (1978) Ensialic basin sedimentation: the Dalradian Supergroup. In Bowes, D. R. and Leake, B. E. (eds.), Crustal evolution in northwestern

Britain and adjacent areas. *Geol. J. Spec. Issue* **10**, 115–138.

Henriksen, N. and Higgins, A. K. (1976) East Greenland Caledonian fold belt. In Escher, A. and Watt, W. S. (eds.), *Geology of Greenland*, Copenhagen, 182–246.

Higgins, A. K. and Phillips, W. E. A. (1979) East Greenland Caledonides—a continuation of the British Caledonides. In Harris, A. L., Holland, C. H. and Leake, B. E. (eds.), The Caledonides of the British Isles—reviewed, *Spec. Publ. geol. Soc. London* **8**, 19–32.

Holdsworth, R. E., Harris, A. L. and Roberts, A. M. (1987) The stratigraphy structure and regional significance of the Moine rocks of Mull, Argyllshire, W. Scotland. *Geol. J.* **22**, 83–107.

Hossack, J. R., Garton, M. R. and Nickelsen, R. P. (1985) The geological section from the foreland up to the Jotun thrust sheet in the Valdres area, south Norway. In Gee, D. G. and Sturt, B. A. (eds.), *The Caledonide Orogen—Scandinavia and Related Areas*, John Wiley, New York, 443–456.

Hynes, A. D. and Francis, D. M. (1982) A transect of the early Proterozoic Cape Smith foldbelt, New Quebec. *Tectonophysics* **88**, 23–59.

Jakobsen, H., Munksgaard, N. C. and Zeck, H. P. (1984) Pre-Dalslandian deformation and recrystallization in the basement of the Dalslandian supracrustals, Grenvillian (Sveconorwegian) Belt, south-western Sweden. *Förh. Geol. Fören. Stockholm* **105**, 205–212.

Jepsen, H. F. and Kalsbeek, F. (1985) Evidence for non-existence of a Carolinidian fold belt in eastern North Greenland. In Gee, D. G. and Sturt, B. A. (eds.), *The Caledonian Orogen—Scandinavia and Related Areas*, John Wiley, New York, 1071–1076.

Johnson, H. D., Levell, B. K. and Siedlecki, S. (1978) Late Precambrian sedimentary rocks in East Finnmark, North Norway, and their relationship to the Trollfjord-Komagelv Fault. *J. geol. Soc. London* **135**, 517–533.

Johnson, Y. A., Park, R. G. and Winchester, J. A. (1987) Geochemistry petrogenesis and tectonic significance of the early Proterozoic Loch Maree Group amphibolites of the Lewisian Complex, NW Scotland. In Pharaoh, T. C., Beckinsale, R. D. and Rickards, D. (eds.), Geochemistry and mineralization of Proterozoic volcanic suites, *Spec. Publ. geol. Soc. London* **33**, 255–269.

Kalliokoski, J. (1986) Calcium carbonate cement (caliche) in Keeweenawan sedimentary rocks (~ 1.1 Ga), Upper Peninsula of Michigan. *Precambrian Res.* **32**, 243–259.

Katz, H. R. (1961) Late Precambrian to Cambrian stratigraphy in East Greenland. In Raasch, G. O. (ed.), *Geology of the Arctic*, Toronto, 299–328.

Kelling, G., Phillips, W. E. A., Harris, A. L. and Howells, M. F. (1985) The Caledonides of the British Isles: a review and appraisal. In Gee, D. G. and Sturt, B. A. (eds.), *The Caledonide Orogen—Scandinavia and Related Areas*, John Wiley, New York, 1125–1146.

Kennedy, M. J. (1975) The Fleur de Lys Supergroup: stratigraphic comparison of Moine and Dalradian equivalents in Newfoundland with the British Caledonides. *J. geol. Soc. London* **131**, 305–310.

Kennedy, W. Q. (1946) The Great Glen Fault. *Q. J. geol. Soc. London* **102**, 41–76.

Kjøde, J., Storetvedt, K. M., Roberts, D. and Gidskehaug, A. (1978) Palaeomagnetic evidence for large-scale dextral movement along the Trollfjord-Komagelv Fault, Finnmark, North Norway. *Phys. Earth Planet. Inter.* **161**, 132–144.

Kontinen, A. (1987) The Jormua mafic-ultramafic complex, northeastern Finland—an early Proterozoic ophiolite. *Precambrian Res.* **35**, 313–341.

Korstgård, J., Ryan, B. and Wardle, R. (1987) The boundary between Proterozoic and Archaean crustal blocks in central West Greenland and northern Labrador. In Park, R. G. and Tarney, J. (eds.), Evolution of the Lewisian and comparable Precambrian high-grade terrains, *Spec. Publ. geol. Soc. London* **27**, 247–259.

Kumpulainen, R. and Nystuen, J. P. (1985) Late Proterozoic basin evolution and sedimentation in the westernmost part of Baltoscandia. In Gee, D. G. and Sturt, B. A. (eds.), *The Caledonian Orogen–Scandinavia and Related Areas*, John Wiley, New York, 213–232.

Lambert, R. StJ. (1969) Isotopic studies relating to the Precambrian history of the Moinian of Scotland. *Proc. geol. Soc. London* **1652**, 243–245.

Le Pichon, X., Sibuet, J. C. and Francheteau, J. (1977) The fit of the continents around the North Atlantic Ocean. *Tectonophysics* **38**, 169–209.

Leake, B. E. and Singh, D. (1986) The Delaney Dome Formation, Connemara, W. Ireland, and the geochemical distinction of ortho- and para-quartzofeldspathic rocks. *Mineral. Mag.* **50**, 205–215.

Leake, B. E., Tanner, P. W. G. and Singh, D. (1983) Major southward thrusting of the Dalradian rocks of Connemara, western Ireland. *Nature (London)* **305**, 210–213.

Lefort, J. P. (1984) The main basement features recognized in the northern part of the North Atlantic area. In de Gracianski, P.C., Poag, C. W., *et al.* (eds.), *Init. Rep. D.S.D.P.* **80**, US Government Printing Office, Washington DC, 1103–1114.

Magnussen, N. H. (1965) The Precambrian history of Sweden. *Q. J. geol. Soc. London* **121**, 1–30.

Martin, A. R. (1985) The Evolution of the Tugtutoq-Illimaussaq Dyke Swarm, southwest Greenland. Ph.D. Thesis, University of Edinburgh (Unpubl.).

Max, M. D. and Long, C. B. (1979) Basic volcanic rocks in the Dalradian of Ireland. In Harris, A. L., Holland, C. H. and Leake, B. E. (eds.), The Caledonides of the British Isles—reviewed, *Spec. Publ. geol. Soc. London* **8**, 585–589.

Max, M. D. and Long, C. B. (1985) Pre-Caledonian basement in Ireland and its cover relationships. *Geol. J.* **20**, 341–366.

McWilliams, M. O. and Dunlop, D. J. (1978) Grenville palaeomagnetism and tectonics. *Can. J. Earth Sci.* **15**, 687–695.

Moorbath, S. (1969) Evidence for the age of deposition of the Torridonian sediments of north-west Scotland. *Scott. J. Geol.* **5**, 154–170.

Moorbath, S., Stewart, A. D., Lawson, D. E. and Williams, G. E. (1967) Geochronological studies on the Torridonian sediments of north-west Scotland. *Scott. J. Geol.* **3**, 389–412.

Moorhouse, S. J. and Moorhouse, V. E. (1979) The Moine amphibolite suites of central and northern Sutherland, Scotland. *Mineral. Mag.* **43**, 211–225.

Moorhouse, V. E. and Moorhouse, S. J. (1983) The geology and geochemistry of the Strathy Complex of north-east Sutherland, Scotland. *Mineral. Mag.* **47**, 123–137.

Morley, C. K. (1986) The Caledonian Thrust Front and palinspastic restorations in the southern Norwegian Caledonides. *J. struct. Geol.* **8**, 753–765.

Myers, J. S. (1987) The East Greenland Nagssugtoqidian mobile belt compared with the Lewisian Complex. In Park, R. G. and Tarney, J. (eds.), Evolution of the Lewisian and comparable Precambrian high-grade terrains, *Spec. Publ. geol. Soc. London* **27**, 235–246.

Nystuen, J. P. (1982) Late Proterozoic basin evolution on the Baltoscandian craton: the Hedmark Group, southern Norway. *Nor. geol. Unders.* **375**, 1–74.

Nystuen, J. P. and Saether, T. (1979) Clast studies in the Late Precambrian Moelv Tillite and Osdal Conglomerate, Sparagmite Region, south Norway. *Norsk geol. Tidsskr.* **59**, 239–254.

Park, R. G. and Tarney, J. (1987) The Lewisian Complex: a typical Precambrian high-grade terrain. In Park, R. G. and Tarney, J. (eds.), Evolution of the Lewisian and comparable Precambrian high-grade terrains, *Spec. Publ. geol. Soc. London* **27**, 13–25.

Patchett, P. J. and Bylund, G. (1977) Age of Grenville Belt magnetization: Rb-Sr and palaeomagnetic evidence from Swedish dolerites. *Earth Planet. Sci. Lett.* **35**, 92–104.

Patchett, P. J., Bylund, G. and Upton, B. G. J. (1978) Palaeomagnetism and the Grenville Orogeny: new Rb-Sr ages from dolerites in Canada and Greenland. *Earth Planet. Sci. Lett.* **40**, 349–364.

Pharaoh, T. C. (1985) Volcanic and geochemical stratigraphy of the Nussir Group of Arctic Norway—an early Proterozoic greenstone suite. *J. geol. Soc. London* **142**, 259–278.

Pharaoh, T. C. and Pearce, J. A. (1984) Geochemical evidence for the geotectonic setting of early Proterozoic metavolcanic sequences in Lapland. *Precambrian Res.* **10**, 283–309.

Piasecki, M. A. J. (1980) New light on the Moine rocks of the Central Highlands of Scotland. *J. geol. Soc. London* **137**, 41–59.

Piasecki, M. A. J., van Breemen, O. and Wright, A. E. (1981) Late Precambrian Geology of Scotland, England and Wales. *Mem. Can. Soc. Petrol. Geol.* **7**, 57–94.

Piper, J. D. A. (1980) Palaeomagnetic study of the Swedish rapakivi suite: Proterozoic tectonics of the Baltic Shield. *Earth Planet. Sci. Lett.* **46**, 443–461.

Piper, J. D. A. (1982) The Precambrian palaeomagnetic record: the case for the Proterozoic Supercontinent. *Earth Planet. Sci. Lett.* **59**, 61–89.

Piper, J. D. A. (1985) Continental movements and breakup in Late Precambrian–Cambrian times: prelude to Caledonian orogenesis. In Gee, D. G. and Sturt, B. A. (eds.), *The Caledonide Orogen–Scandinavia and Related Areas*, John Wiley, New York, 19–34.

Pitcher, W. S., Elwell, R. W. D., Tozer, C. F. and Cambray, F. W. (1964) The Leannan Fault. *Q. J. geol. Soc. London* **120**, 241–273.

Plant, J. A., Watson, J. V. and Green, P. M. (1984) Moine-Dalradian relationships and their palaeotectonic significance. *Proc. R. Soc. London* **A395**, 185–202.

Powell, D., Baird, A. W., Charnley, N. R. and Jordan, P. J. (1981) The metamorphic environment of the Sgurr Beag Slide: a major crustal displacement zone in Proterozoic, Moine rocks of Scotland. *J. geol. Soc. London* **138**, 661–673.

Pringle, R. (1973) Rb-Sr age determinations on shales associated with the Varanger Ice Age. *Geol. Mag.* **109**, 465–472.

Pulvertaft, T. C. R. (1985) Palaeocurrent directions in the Lower Dala Sandstone, west central Sweden. *Förh. Geol. Fören. Stockholm* **107**, 59–62.

Rex, D. C. and Gledhill, A. R. (1974) Reconnaissance geochronology of the infracrustal rocks of Flyvefjord, Scoresby Sund, East Greenland. *Bull. geol. Soc. Denmark* **23**, 49–54.

Rex, D. C. and Gledhill, A. R. (1981) Isotopic studies in the East Greenland Caledonides (72°–74 °N)—Precambrian and Caledonian ages. *Bull. Grønlands geol. Unders.* **104**, 47–72.

Rex, D. C., Gledhill, A. R. and Higgins, A. K. (1977) Precambrian Rb-Sr isochron ages from the crystalline complexes of inner Forsblads Fjord, East Greenland fold belt. *Bull. Grønlands geol. Unders.* **85**, 122–126.

Riddihough, R. P. and Max, M. D. (1975) Continuation of the Highland Boundary Fault in Ireland. *Geology* **3**, 206–210.

Roberts, D. and Sturt, B. A. (1980) Caledonian deformation in Norway. *J. geol. Soc. London* **137**, 241–250.

Roddick, C. and Max, M. D. (1983) A Laxfordian age from the Inishtrahull Platform, County Donegal, Ireland. *Scott. J. Geol.* **19**, 97–102.

Rodhe, A. (1986) Geochemistry and clay mineralogy of argillites in the Late Proterozoic Almesåkra group, south Sweden. *Förh. Geol. Fören. Stockholm* **107**, 175–182.

Saether, T. and Nystuen, J. P. (1981) Tectonic framework, stratigraphy, sedimentation and volcanism of the Late Precambrian Hedmark Group, Østerdalen, South Norway. *Norsk. geol. Tidsskr.* **61**, 193–211.

Siedlecka, A. (1975) Late Precambrian stratigraphy and structure of the north-eastern margin of the Fennoscandian shield (East Finnmark—Timan Region). *Nor. geol. Unders.* **316**, 313–348.

Siedlecka, A. and Siedlecki, S. (1971) Late Precambrian sedimentary rocks of the Tanafjord—Varangerfjord region of Varanger Peninsula, northern Norway. *Nor. geol. Unders.* **269**, 246–294.

Skiold, T. (1976) The interpretation of the Rb-Sr and K-Ar ages of Late Precambrian rocks in south-western Sweden. *Förh. Geol. Fören. Stockholm*, **98**, 3–29.

Smith, D. I. (1979) Caledonian minor intrusions of the Northern Highlands of Scotland. In Harris, A. L., Holland, C. H. and Leake, B. E. (eds.), The Caledonides of the British Isles—Reviewed. *Spec. Publ. geol. Soc. London* **8**, 683–697.

Smith, R. L., Stearn, J. E. F. and Piper, J. D. A. (1983) Palaeomagnetic studies of the Torridonian sediments, NW Scotland. *Scott. J. Geol.* **19**, 29–45.

Smythe, D. K. (1987) Deep seismic reflection profiling of the Lewisian Foreland. In Park, R. G. and Tarney, J. (eds.), Evolution of the Lewisian and Comparable Precambrian High Grade Terrains. *Spec. Publ. geol. Soc. London* **27**, 193–203.

Spencer, A. M. (1971) Late Precambrian glaciation in Scotland. *Mem. geol. Soc. London* **6**, 98 pp.

Stearn, J. E. F. and Piper, J. D. A. (1984) The palaeomagnetism of the Sveconorwegian mobile belt of the Fennoscandian Shield. *Precambrian Res.* **23**, 201–246.

Steiger, R. H. and Jäger, E. (1977) Subcommission on Geochronology: Convention on the use of decay constants in geo- and cosmochronology. *Earth Planet. Sci. Lett.* **28**, 359–362.

Stewart, A. D. (1962) On the Torridonian sediments of Colonsay and their relationship to the main outcrop in north-west Scotland. *Liverpool Manchester geol. J.* **3**, 121–156.

Stewart, A. D. (1966) An unconformity in the Torridonian. *Geol. Mag.* **103**, 462–465.

Stewart, A. D. (1982) Late Proterozoic rifting in NW Scotland: the genesis of the 'Torridonian'. *J. geol. Soc. London* **139**, 413–420.

Sturt, B. A. (1961) The geological structure of the area south of Loch Tummel. *Q. J. geol. Soc. London* **117**, 131–156.

Sutcliffe, R. H. (1987) Petrology of Middle Proterozoic diabases and picrites from Lake Nipigon, Canada. *Contrib. Mineral. Petrol.* **96**, 201–211.

Tanner, P. W. G. (1970) The Sgurr Beag Slide—a major tectonic break within the Moinian of the western Highlands of Scotland. *Q. J. geol. Soc. London* **126**, 435–463.

Tanner, P. W. G., Johnstone, G. S., Smith, D. I. and Harris, A. L. (1970) Moine stratigraphy and the problem of the Central Ross-shire inliers. *Bull. geol. Soc. Am.* **81**, 299–306.

van Breemen, O., Pidgeon, R. T. and Johnson, M. R. W. (1974) Precambrian and Palaeozoic pegmatites in the Moines of northern Scotland. *J. geol. Soc. London* **130**, 493–507.

van der Voo, R. and Scotese, C. (1981) Palaeomagnetic evidence for a large (2000 km) sinistral offset along the Great Glen Fault during Carboniferous time. *Geology* **9**, 583–589.

van Schmus, W. R., Bickford, M. E., Lewry, J. F. and Macdonald, R. (1987) U-Pb geochronology in the Trans-Hudson Orogen, northern Saskatchewan, Canada. *Can. J. Earth Sci.*, **24**, 407–424.

van Schmus, W. R., Green, J. C. and Halls, H. C. (1982) Geochronology of Keeweenawan rocks of the Lake Superior region: a summary. In Wold, R. J. and Hinze, W. J. (eds.), Geology and Tectonics of the Lake Superior Basin, *Mem. Geol. Soc. Am.* **156**, 165–171.

Vidal, G. (1974) Late Precambrian microfossils from the basal sandstone unit of the Visingso Beds, south Sweden. *Geol. ed Palaeontol.*, **8**, 1–14.

Vidal, G. (1985) Biostratigraphic correlation of the Upper Proterozoic and Lower Cambrian of the Fennoscandian Shield and the Caledonides of East Greenland and Svalbard. In Gee, D. G. and Sturt, B. A. (eds.), *The Caledonide Orogen—Scandinavia and Related Areas* John Wiley, New York, 331–338.

Watters, B. R. and Armstrong, R. L. (1985) Rb-Sr study of metavolcanic rocks from the La Ronge and Flin Flon domains, northern Saskatchewan. *Can. J. Earth Sci.* **22**, 452–463.

Watterson, J. (1978) Proterozoic intraplate deformation in the light of south-east Asian neotectonics. *Nature (London)* **273**, 636–640.

Winchester, J. A. (1973) Pattern of regional metamorphism suggests a sinistral displacement of 160 km along the Great Glen Fault. *Nature, Phys. Sci. (London)* **246**, 81–84.

Winchester, J. A. (1976) Differing Moinian amphibolite suites in northern Ross-shire. *Scott. J. Geol.* **12**, 187–204.

Winchester, J. A. (1984) The geochemistry of the Strathconon amphibolites. *Scott. J. Geol.* **20**, 37–51.

Winchester, J. A. (1985) Major low-angle fault displacement measured by matching amphibolite chemistry—an example from Scotland. *Geology* **13**, 604–606.

Winchester, J. A. and Floyd, P. A. (1984) The geochemistry of the Ben Hope Sill suite, northern Scotland, U.K. *Chem. Geol.* **43**, 49–75.

Winchester, J. A. and Max, M. D. (1984) Geochemistry and origins of the Annagh Division of the Precambrian Erris Complex, NW Co. Mayo, Ireland. *Precambrian Res.* **25**, 397–414.

Winchester, J. A. and Max, M. D. (1987*a*) A displaced and metamorphosed peralkaline granite related to the late Proterozoic Labrador and Gardar suites: the Doolough Granite of Co. Mayo, NW Ireland. *Can. J. Earth Sci.* **24**, 631–642.

Winchester, J. A. and Max, M. D. (1987*b*) The Pre-Caledonian Inishkea Division of NW Co. Mayo, Ireland: its geochemistry and probable stratigraphic position. *Geol. J.* **22** (in press).

Winchester, J. A., Max, M. D. and Long, C. B. (1987) Trace element geochemical correlation in the reworked Proterozoic Dalradian metavolcanic suites of the western Ox Mountains and NW Mayo inliers, Ireland. In Pharaoh, T. C., Beckinsale, R. D. and Rickards, D. (eds.), Geochemistry and mineralization of Proterozoic volcanic suites. *Spec. Publ. geol. Soc. London* **33**, 489–502.

Ziegler, P. A. (1986) Geodynamic model for the Palaeozoic crustal consolidation of western and central Europe. *Tectonophysics* **126**, 303–328.

Index